D0397983

HOME

ALSO BY JOHN S. ALLEN

The Omnivorous Mind:
Our Evolving Relationship with Food

The Lives of the Brain:
Human Evolution and the Organ of Mind

HOME

HOW HABITAT MADE US HUMAN

John S. Allen

BASIC BOOKS
A Member of the Perseus Books Group
New York

Published by Basic Books,
A Member of the Perseus Books Group

Library of Congress Cataloging-in-Publication Data
Allen, John S.
Home : how habitat made us human / John S. Allen.
pages cm
Includes bibliographical references and index.
ISBN 978-0-465-03899-2 (hardback) — ISBN 978-0-465-07389-4 (e-book)
1. Home (Psychological aspects) 2. Place attachment. I. Title.
GT2420.A55 2015
392.3'6019—dc23

2015027697

10 9 8 7 6 5 4 3 2 1

To everyone who has had the dubious privilege of sharing a home with me, especially my wife, Stephanie, and our sons, Reid and Perry

CONTENTS

INTRODUCTION

We humans are a species of *homebodies.* Home is universal, and around the world, people create places for themselves where they feel at home. From the outside, these places can look very different: houses, apartments, tents, shacks, thatched huts, and so on. From culture to culture, and even within a culture, housing and living situations vary tremendously. From Downton Abbey to a downtown flophouse, an igloo to an adobe pueblo, the human penchant for building shelters can be expressed in seemingly infinite ways. But no matter what they look like on the outside, they all fulfill some basic human needs.

Describing humans as homebodies, however, is somewhat misleading: we are more fundamentally *homeminded.* Home is not simply a location on the landscape where a person lives; it has a privileged place in our cognition. Home brings on feelings of comfort, security, and control. Home does not provide a jolt to the mind like the button-pushing pleasures of sex, drugs, or eating something that tastes really good, nor is it likely to elicit one of the "big six" basic emotions (happiness, fear, anger, sadness, disgust, or surprise). However, home has a powerful influence on mood and mind. Home is where we go to recover from our exertions in the outside world. It is a place that is different from all others, and when we say we "feel at home," it means something important. But how did feeling at home come to be so important? That's the question I want to explore in this book.

Building and living in houses has helped us humans and some of our ancestors survive and reproduce successfully. In a sense then, houses, or the ability to build and use them, are an *adaptation,* as we say in evolutionary science. When it comes to the human world, adaptations can be biological or cultural. Clearly, cultural traditions dictate the structure or building in which a person makes a home. People are not genetically programmed to make houses in the way that wasps are to make nests. Instead, the psychological motives and general intellectual ability to make a house and create a home emerge from our brain biology.

So houses and other kinds of homely shelters are undoubtedly cultural products that help humans adapt to the world around us. But the human cultural propensity to shelter ourselves is also reinforced biologically; thus, home is a biological adaptation as well. While houses around the world look different viewed from the outside, the people living in them share some basic patterns in thought and behavior that form the cognitive basis of home life. We share these patterns because over the course of human evolution, our ancestors who developed these feelings about the home environment have likely been more reproductively successful than those who did not. Building a concept or feeling of home in our minds has been critical. We may understand houses from the outside looking in, but home is made from the inside out.

Feeling at home is a biological phenomenon because feelings themselves are the product of the human mind and body. Feelings and emotions evolved over the course of millions of years of evolution to help us regulate and monitor our internal state in relation to the external environment. Anger, fear, sadness, and the other emotions and feelings prime the body for action and help the mind make decisions about what we should or should not do. The feelings we associate with home are built on a cognitive foundation that we inherited from our ancestors. A feeling for home coupled with the inclination and ability to build a shelter for our bodies makes for a profound biocultural adaptation. This adaptation helps people survive in all manner of environments, far from the African woodlands and savannas where the human evolutionary journey first started some six million years ago.

Homesickness and the Homely Mind

I can drive home the importance of our feeling for home by considering our reaction to its loss. Not too long ago, my older son went off to college. He very much looked forward to going to university, and the school he was going to was a top choice for him. Naturally, when he got there, he had a period of adjustment. He was anxious about classes and majors, making new friends, even finding his way around.

All this was to be expected, and after a few weeks, he had things pretty much under control.

He was also homesick; again, not really a surprise. After all, and as I will develop more fully throughout the book, home has a special place in our minds, feelings, and emotions. Bouts of homesickness provide a window into these implicit sensations and perceptions that often lie just below the cognitive surface. What surprised me, however, was that when I talked to other parents, friends, and acquaintances about his adjustment to school, I found that there were some who avoided the "homesick" label like the plague. If I said he was homesick, then they might reply that he was lonely or unsettled or nervous. Homesickness was just not something they wanted to acknowledge. I thought about why this might be the case.

One possible reason is that people see homesickness as a sign of weakness, and therefore it was something they did not think it was polite to project on to my son. Historian Susan Matt has chronicled the interesting history of homesickness in the United States.[1] In the eighteenth and nineteenth centuries, America, a nation of immigrants and movers during a time when relocating meant severing nearly all ties with one's homeland, was a country awash in homesickness. This was widely acknowledged and even expected; medical practice of the time recognized homesickness as a potentially dangerous and primary form of mental illness.

Over time, however, homesickness generally lost its standing as a medical illness, and homesickness itself came to be seen as a sign of weakness. As Matt writes: "The modern attitude towards homesickness [is] an attitude predicated on the belief that movement is natural and unproblematic and a central and uncontested part of American identity."[2] Sentimental attachment to past people and places is acceptable, but under the guise of nostalgia (which incidentally is a word that was co-opted from the nineteenth-century medical science term for severe homesickness). Homesickness as weakness makes it very unattractive, and in American culture, something that deserves pity more than sympathy.

Perhaps a more sympathetic, or even empathic, way to look at homesickness is to see it not as weakness but as a form of bereavement. Psychologist Margaret Stroebe and her colleagues consider the possibility that homesickness may actually represent "mini-grief."[3] In their view, homesickness is an emotional reaction to the temporary loss of significant relationships, with parallels to the reactions that are felt when mourning for someone who has died. An important aspect of grief and bereavement is that people vary in how they express and experience them. Culture and personality both influence how a person reacts to a significant loss. Dealing with loss requires some adjustments to the new situation, and some people adapt more easily to changed circumstances.

Despite this variation, there is a core emotion at the heart of grieving. Evolutionary psychologists have long argued that the transient depression associated with grief reflects and reinforces important relationships for humans and other primates. Homesickness is both about relationships left behind and the new social environment in which new relationships will be formed. As social primates, human beings have a long evolutionary history of forming intensive relationships with others in their groups. A result of this is that people want to feel that they belong among the people with whom they interact on a regular basis. Susan Watt and Alison Badger look at homesickness from the perspective of the human "need to belong."[4] They argue that the need to belong underlies the feelings of (mini-)grief that people feel when relationships are disrupted. In a study of Australian college students, Watt and Badger found that homesickness was greater among students who had a higher need to belong (as measured by a psychological scale), and that those who felt more accepted in their new environment had less homesickness.

Homesickness is much more than a sign of weakness. It is a powerful indicator of the importance of place in the emotional lives of human beings. I suspect that the real reason that some Americans might not want to address or acknowledge homesickness may be related to the fact that they are not all that comfortable with mourning or grief in any form. Hearing about someone else's homesickness evokes memories of homesick experiences, just as hearing about the death of someone else's loved one may bring back one's own memories of mourning. Homesickness, provided it does not go on for too long, is just as natural a reaction to the loss of home as mourning is for the loss of a loved one. In both cases, grief signifies the loss of a significant and, in evolutionary terms, adaptive relationship. Luckily, since we carry home with us in a cognitive sense, we can reestablish a feeling for home in a new environment. But the transition can be difficult, since we are temporarily without one of the main biocultural adaptations that we use to live our lives.

A Natural History of Home

Until we lose it, it is easy enough to take home for granted. In fact, words such as *home* and *homely* can signify human life at its plainest and most mundane. Like the air we breathe or the food we typically eat, home is central to our existence but not too exciting. When home is working for us, it fades easily into the cognitive and emotional background. We take and find comfort in home both literally and figuratively. And part of what makes home comfortable is that it is not something that needs to be thought about and nurtured on a day-to-day basis.

Scholars and scientists of various persuasions have done some thinking about home, but it is not an overcrowded area of intellectual exploration. Home has attracted some students of culture and cultural history, because home gives shape to, and is shaped by, prevailing cultural forces. Decades ago, Witold Rybczynski used home as a vehicle to track social and cultural changes in the private lives of Western Europeans.[5] In their summary of the growing literature on home studies, Alison Blunt and Robyn Dowling show that in the contemporary world home has a role in a variety of cultural dynamics, providing a spatial rendering of power and gender relationships.[6] Home is the basic unit of human societies, and it reflects the greater forces that underlie cultural variety both within and between societies.

Home is not just a building or a shelter, but a vehicle for carrying emotions and status.

Like spoken language, home is a human universal that shows great cultural variation. But even scholars interested more in the cultural facets of home acknowledge that home emerges from something more basic than culture. As Rybczynski writes: "Domestic well-being is a fundamental human need that is deeply rooted in us, and that must be satisfied."[7] We all know that when we "feel at home" we are in some way satisfying this need, but where does the need originate? Before home became a cultural phenomenon it was a personal one, rooted in the biology and cognition of individual humans.

The transformative emotional and cognitive power of home is reflected in this quote from the architects Charles Moore, Gerald Allen, and Donlyn Lyndon: "Its site is only a tiny piece of the real world, yet this place is made to seem like an entire world."[8] Part of feeling at home is to give in to a perceptual trick, one that divides the world in two: a domestic domain and everything else. This simplification of the complexity of the "real world" has clearly been adaptive for the human species. From a cognitive standpoint, how do we embody this simpler world that gives us a feeling of home?

This book is to some extent a natural history of home. Given the ubiquity and importance of home, there are many possible stories to tell about it, both natural and unnatural. My focus here will be in trying to answer the two main questions I have raised: How did home evolve to become a ubiquitous feature of human life, and what does it mean when we say we "feel at home"?

The Evolution of Homely Feelings

The human need to feel at home has its roots in our evolutionary biology and is reinforced today by our cognitive psychology. As it does for many animals, home provides us with shelter and an arena for forging and maintaining social relationships. The feeling for home is built from sensory inputs (smell and vision), memory, and emotions that create in the mind a holistic sense of the self in a specific place.

I begin this exploration of home by first addressing the "how" question (Chapter 1): How is "feeling at home" manifested in our minds, brains, and bodies? This gets us into the realm of cognitive neuroscience. Recent research has given scientists an unprecedented window into the workings of the brain, including those once-elusive (from a scientific perspective) things known as feelings. Feeling at home may be a matter of feeling settled and comfortable and secure, but perhaps we should think about it in the opposite way: that feeling settled and comfortable in any setting is simply making oneself at home. Feelings are often sensed below conscious detection, in our guts as much as our minds, and yet they have a profound impact on decision making and other aspects of "higher" thinking. Given the central place of home in most lives, a feeling of home may be one of the cornerstones of human cognition, less obvious than language, but critical nonetheless.

In Chapters 2 through 4, I investigate the evolutionary origins of home. Why should a feeling or desire for home have ever evolved? How are the basic needs for shelter, a place to rest, protection from predators, and access to mates and resources met by feeling at home? Animals such as prairie dogs give us some insights into why building and living in a shelter in general helps animals meet some basic needs. Our closest relatives, the other primates, are actually not known for making use of shelters, but aspects of the behavior of the great apes may hint at our transition to home. The fossil and archaeological record of human evolution provides evidence of several key evolutionary transitions. What is the evidence for home being one of them? I will spend some time looking at the Neandertals, our very close, very human relatives, to see if their remains offer any clues to the origins of home.

In Chapters 5 through 7, I address the important question: So what? How does understanding the evolutionary and cognitive bases of feeling at home shed light on contemporary issues? Can we use these insights to make our home lives better for ourselves? I believe that the answer to both questions is "yes," and that there are several areas in which understanding the feeling of home, both its evolutionary and its cognitive roots, can give us insights into our lives today. In developed countries, we have recently gone through a period of unprecedented turmoil in the housing markets. How do people's feelings about home influence their economic decisions concerning houses as investments? Developed countries also have a problem with homelessness. Some people are homeless as a direct result of economic factors, while others find themselves living on streets, or in parks or shelters, as a result of the sometimes malignant mixture of social and medical factors surrounding the care and treatment of people with mental illness and/or substance addiction. How do feelings about home influence the homeless themselves and the attitudes of more fortunate people toward the homeless?

Housing instability is also on the rise in many developed countries. Is this necessarily a bad thing? In complex societies, economic, social, or professional mobility all go along with residential mobility. Getting ahead often means getting on the road. A different kind of housing instability affects the youngest in society: children of divorce or who are in foster care often have disrupted home lives. We are not born walking or talking or able to navigate through a difficult social situation. These are all sophisticated abilities that are honed in appropriate environments, during critical periods, as we are growing up. The complex feelings that provide the cognitive background for all the thinking and deciding we do every day are also trained and developed during childhood. How do disruptions in the home environment during this critical period affect our ability to feel at home? What are the costs in adulthood if the ability to feel at home is never properly developed?

In the end, I plan to show how a feeling of home extends into other areas of life, and how understanding it illuminates a variety of modern problems and issues. However, there are lessons here that extend beyond problems and issues: when we understand how things go wrong, we can also learn how they can go right. Like sex, eating a satisfying meal, or listening to someone tell a good story about people we know, we are predisposed to enjoy the feelings associated with home. Relationships with close family and friends, security, comfort (however a culture wants to define it)—these are all things that can readily give us pleasure, even if it is the kind of pleasure that we are more likely to be aware of when it is absent. My hope is that by being made aware of the feeling of home, we can use that awareness to enhance its associated pleasures, and maybe even enjoy them while they are present, and not retroactively, when they are gone.

Chapter One

FEELING AT HOME

The vast and harsh landscapes of the deserts of north central Australia seem to be anything but homey. The dry heat and dry cold make the environment inhospitable for most plants, save for a few hardy species, such as the dominant spinifex grass. Animal life is present but sparse, and water sources are widely separated. People are scarce on the ground, living in stations and small towns separated by hundreds of kilometers of rough roads and tracks. For the Warlpiri people, however, whose ancestors have lived there for countless generations, this desert is home. It is the place of *wiri jarrimi*—growing up—where children are born, initiated into adulthood, and learn how to be a member of a culture that has mastered life in an unforgiving environment.[1]

The Warlpiri maintain a striking attachment to life on their land, despite efforts to "improve" their lot with more modern (i.e., Western) conveniences. A certain reluctance to embrace these efforts is understandable: after all, nearly two centuries of official government policy at various times encouraged their extermination, subjugation, and persecution.

Times and policies have changed, but many Warlpiri do not seem comfortable adjusting to mainstream Australian culture. Anthropologist Michael D. Jackson spent years among the Warlpiri trying to understand their view of domestic life. In the end, they taught him how "to be at home in the world . . . to experience a complete consonance between one's own body and the body of the earth."[2] This consonance can be achieved in whatever form the home takes. For the Warlpiri still connected to their traditional cultural values, that home is indeed made in the land they occupy, more so than in any structure. As one of Jackson's Warlpiri friends evocatively put it: "A house is just like a big jail."[3]

The Warlpiri feel at home *there,* in the Australian desert; I feel at home *here,* in my house in the American middle South. What we share, and what the vast majority of humans share, is an ability to form an emotional relationship with the places in which we live, the spaces we occupy. Over the course of our evolution as a species, home became the building block for human cultures. From the outside looking in, these building blocks are different in size and composition: the actual structures (i.e., houses) that provide the physical setting of home vary from culture to culture, and the designs of households are determined by cultural traditions.

From the inside looking out, however, a more universal picture emerges. People form relationships with their home spaces. These relationships are typically not expressed or understood in the same emotional terms as, say, a hot and heavy love affair or the devotional love between a parent and child. Instead, it all comes down to a feeling—the feeling that this particular place, whatever its shortcomings, is something special compared to all other places.

To understand how we feel at home, it is necessary to look at feelings in more detail. Obviously, artists and writers have explored feelings for millennia, but scientists, including psychologists and their fellow travelers, were less than enthusiastic to take up this topic (Darwin was a notable exception). In fact, for much of the twentieth century, psychology as a discipline, dominated by behaviorism and mechanistic theories of behavior (picture rats running in mazes), totally rejected the idea that scientists had the ability to get inside the heads of people and study the "internal brain-mind states in behavioral control."[4] This viewpoint basically took emotion and feeling off the table for scientific study. However, in the latter part of the twentieth century, a greater appreciation for the importance of evolved, species-specific behavioral patterns helped to knock behaviorism off its pedestal. The door for the scientific study of emotions was reopened.

Revolutionary advancements in brain imaging came in the 1990s, giving scientists the ability to observe the workings and structures of living, healthy brains. The science of emotions was invigorated by these new techniques. The conscious feelings that result from the combined effects of changes in mind and body are now accessible to direct study. Emotions and feelings that cannot be described in words by those experiencing them can be studied directly via the brain processes that produce them.

In this chapter, I will explore what emotions and feelings are, why they may have evolved in the first place, and how they are generated in our brains. With this emotional foundation laid, I will look in detail at how feelings associated with home could emerge from our own brains. These feelings of home tend to be quieter than the big emotions, such as fear and joy, that we are readily aware of. They reside more in the background, linking body and mind on a moment-to-moment basis.

Emotional Evolution

The strange, new worlds of the various *Star Trek* franchises contain a host of worldly and otherworldly characters. Interestingly, amid all the fundamental ways of conveying alienness, emotions, or lack thereof, play a major role in defining two of the main, nonhuman characters. In the earliest incarnation of *Star Trek,* Spock, the half-human, half-Vulcan science officer, follows the Vulcan practice of suppressing emotions, allowing for a more dispassionate and "logical" approach to life in general and decision making in particular. Spock's apparent coldness is (constantly) juxtaposed against the hotness of his emotional human colleagues, especially the captain, James Kirk, and the medical officer, Dr. McCoy. In *The Next Generation* reboot of the *Star Trek* franchise, the emotionless central character is embodied in the android Data, who, unlike Spock, wishes he could experience emotions in order to be more fully human. Like Pinocchio, Data strives to be a real boy, or man, and it is the absence of emotions that keeps him from achieving this dream.

One of the messages conveyed by these emotionless characters is that emotions are good for something, essential not only for being human but for reacting appropriately to crisis and chaos. Sheer intellectual ability (Spock) or computational processing power (Data) can compensate only up to a point for the absence of emotional inputs. Yes, emotions can get in the way sometimes, and yes, they can mislead as much as lead. But in the *Star Trek* universe, emotions drive action, and action is what gets things done. Indeed, that seems to be the case in our universe, as well: the capacity of emotions to stimulate animals to action may underlie why they evolved in the first place.

What, then, is an emotion? Scores of scientific definitions have been offered up over the years.[5] Psychologist Robert Plutchik provides a definition that not only accounts for how emotion takes place within an individual mind and body, but that is also evolution-friendly in the sense that it is easy to generalize to animals all along the zoological spectrum (humans included). He writes:

> An emotion is not simply a feeling state. Emotion is a complex chain of loosely connected events that begins with a stimulus and includes feelings, psychological changes, impulses to action and specific, goal-directed behavior. That is to say, feelings do not happen in isolation. They are responses to significant situations in an individual's life, and often they motivate action.[6]

The various definitions of emotion generally agree that an emotional episode is a sequence of events: stimulus → mind/body response → action. One thing to keep in mind is that "action" can take many forms, including, in some cases, doing nothing at all.

For most animals, emotions stimulate actions to fulfill certain basic needs, such as finding food or mates, avoiding predators or dangerous situations, and exploring their surrounding environments. In parenting animals, emotions help to create the relationship between parent and offspring. In social animals, emotions, both observed and experienced, serve to coordinate and direct the behavior of individuals in order to reap the benefits of group living. In humans, with our special form of social living dominated by cultural traditions, beliefs, and practices, emotions are shaped by and shapers of the cultures in which people live.

Do all animals have emotions? Most researchers would likely agree that having a brain is probably a prerequisite for having emotions, so that would rule out the simplest animals (such as single-celled organisms). Neurologist Antonio Damasio suggests that the simplest animals are best thought of as having reflexes rather than emotions, but that animals with only a small and rudimentary brain, such as fruit flies or sea slugs, show evidence of having emotions. What these very simple animals might not have are feelings: an ability to internally experience emotions.[7]

For more complex animals, emotions and feelings evolved to serve the essential role of motivating appropriate actions in response to environmental stimuli. Not all actions are the result of emotional prompts, of course—we are all a bit like Spock or Data when undertaking routine or habitual actions. But not everything can be routine. Environments are not fully predictable, and emotions help us respond in different ways to a variety of stimuli. The *fear* we feel when confronted with danger, the *anger* at encountering an obstacle, the *joy* of gaining a valued object, the *sadness* at losing one, the *disgust* after consuming or viewing something unpalatable, and the *surprise* that stops us in our tracks when something unexpected happens—these are just some of the major emotions that make behavioral flexibility possible. For us conscious humans, these emotions are not simply acted upon, but experienced and remembered and contextualized as feelings. To not express or experience emotions is a significant handicap.

And the Emotions Are . . .

Several decades of work in cognitive psychology has generated a taxonomy of basic emotions, including happiness, anger, fear, sadness, disgust, surprise, and more, which has a certain face-validity. These basic emotions were recognized by folk psychology long before psychological scientists became interested in understanding them, of course—indeed, until the 1960s, many academics thought it was possible that the emotions were unique to each culture and did not reflect an underlying common neurological or cognitive foundation. Today, however, there is no doubt in modern cognitive science that emotions are "real" and universal psychological, neurobiological, and psychophysiological phenomena, and that, while culture is seen as providing the context in which emotions are expressed, culture is not their ultimate source.

The universal view of human emotion was bolstered in the 1960s by the landmark work of psychologists Paul

Ekman and Wallace Friesen. They showed that people across very diverse cultures (Western cultures versus an isolated group from Papua New Guinea) were similar in how they read emotions in photos of human faces.[8] Although universality in and of itself is not a demonstration of a strong biological foundation, it suggests that the search for such a foundation may be warranted. Subsequent research has strengthened the biology-based view of emotions.

Beyond the importance of biology, however, there remains considerable debate over how the biology of emotions plays out. Ekman himself has remained a strong advocate for the idea that discrete, basic emotions exist. Based on a variety of criteria, Ekman identifies the following as being basic emotions: amusement, anger, awe, contempt, contentment, disgust, embarrassment, excitement, fear, guilt, interest, pride in achievement, relief, sadness, satisfaction, sensory pleasure, and shame. Each of these basic emotions has innate and particular qualities that have been shaped by evolution.[9] Ekman sees each of these as representing the biological, evolved core of a "family" of related emotions. Variations on a theme emerge as a result of learning, cultural and otherwise, and individual differences.

Other schemes to classify emotions also try to reconcile the idea of basic emotions with the obvious variation that exists in emotional expression. Robert Plutchik organizes the emotions into something like a color wheel, in which related emotions are linked together as though they are different hues of a basic color.[10] Emotions and their opposites sit on opposite sides of the emotion wheel: admiration versus loathing, sadness versus joy, anticipation versus surprise, and so on. Implicit in the emotion wheel is that there are emotions that can be mixed from those representing the basic spokes of the emotion wheel.

The color-wheel-like representation of the emotions is a kind of hybrid between the basic emotion model and a quite different model, one in which the emotions are represented as products of the interaction between two psychological processes—valence (whether it is positive or negative) and arousal (high or low). In this dimensional model, the emotions emerge as combinations of these two more basic and fluid psychological dimensions, which are combined with "cognitive processes that interpret and refine emotional experience according to salient situational and emotional contexts."[11] In this view, the basic emotions we perceive as experiencing are not really all that basic or special when we consider the vast range of complex and interrelated emotional experiences we typically have. As Jonathan Posner and his colleagues note, people do not "recognize emotions as isolated, discrete entities, but they rather recognize emotions as ambiguous and overlapping experiences."[12] Identifying basic emotions may therefore be a somewhat arbitrary way of dividing up a continuum.

Some cognitive scientists have tried to clarify this situation by looking systematically at the vast quantity of data available from functional imaging studies of the emotions. Since the mid-1990s, thousands of people have participated in studies during which the activity patterns of their brains were measured while they did some sort of emotional task. One way to combine and align these disparate studies is to conduct meta-analyses, whereby the results from large numbers of experiments are combined in an explicit and statistically rigorous way. The hope is that a signal can emerge from the noise, giving an objective basis to what would otherwise be subjective impressions.

The meta-analyses of emotion in the brain definitely clarify some things. For one, it is quite clear that there are no specific areas in the brain reserved for specific emotions. Decades of research has shown that the amygdala—a small, almond-shaped collection of neurons located within the temporal lobes of the brain—is critical for experiencing fear, but the amygdala is also activated in networks involved with anger, disgust, happiness, and sadness.[13] The classic emotional areas of the limbic system and parts of the cerebral cortex also participate in these networks but not in an exclusive way. On the other hand, as Stephan Hamann has emphasized, neuroimaging studies have shown that the basic emotions are consistently correlated with specific activation networks.[14] This is consistent with the idea that basic emotions evolved across species, employing common networks in the brain. However, as Kristen Lindquist and Lisa Feldman Barrett have pointed out, the context in which emotions are expressed, as well as the behavioral responses to those emotions, varies tremendously across species. This suggests that the neural networks for even the basic emotions must also be highly variable.[15]

Somewhere between the hidden brain mechanics and the overt expressions of emotions that others can observe, there is a person who actually experiences those mental events. To experience and be aware of emotional events is to *feel* them. Everyone knows that feelings are subjective and private, but cognitive scientists are beginning to get a better feeling for feelings.

Nothing More or Less Than Feelings

Most of the time, for most people, it is not very hard to figure out what emotion another person is experiencing (assuming that the other person is making no effort to hide it). This is especially true of the "big" emotions, such as anger or joy. It can be a little harder to detect the more "social" emotions, such as guilt, pride, or envy, but really, with some understanding of the context in which the emotion is being expressed—of the relationship between the person expressing the emotion and those with whom he or she is interacting—we should be pretty good at detecting these, as well. In humans, both the primary and the social emotions are part and parcel of social interaction. We have evolved to be good at reading them and reacting to them, and cultures impose rules for their appropriate expression.

Neurologist and cognitive scientist Antonio Damasio identifies another group of emotions that are a bit harder for others to read. He calls these "background" emotions. These emotions include "enthusiasm . . . subtle malaise or excitement, edginess or tranquility." As a rule, these are much more obvious to the person experiencing them than to others observing them, reflecting an ongoing expression of self-regulatory processes in the body. Damasio writes that "these include metabolic adjustments associated with whatever internal need is arising or has just been satisfied; and with whatever external situation is now being appraised and handled by other emotions, appetites, or intellectual calculation."[16] Background emotions are distinguished from mood in that they do not reflect a persistent, ongoing affective state, but rather, like other emotions, signal the current state of the mind and body.

Reading background emotions is more difficult than figuring out, for example, that someone who just discovered he was holding a winning lottery ticket is full of joy. Nonetheless, by picking up on subtle nonverbal cues, including facial expression and how the whole body is positioned or moves, the background emotions of another person need not always be opaque. Of course, given that these background emotions are subtle, it helps to be familiar with the person whose emotions are being read. Compared to basic emotions, background emotions do not stand out as prominently from the baseline personality. They are therefore probably prone to being more easily misread or misinterpreted.

Here's an example. I was once at a swim meet watching one of my sons swim a race. After the race, another parent, who did not know me well, said, "Oh, you're so calm during the race. I could never be that way." I mumbled something back to her about having seen a lot of races, which was true, but I did not say that I was anything but calm. Someone who knew me better would have known that my stillness and body rigidity and lack of yelling were signs of tension not serenity. I knew, of course, because these background emotions are fully available to me, if not to others. They are what we all use to make the vital assessment of "how we feel," but they are not in and of themselves feelings.

So what are feelings? Damasio's concept of background emotions takes us closer to the level of feelings, in that both are intimately concerned with the status of the body. He has defined feelings as "the perception of a certain state of the body along with the perception of a certain mode of thinking and of thoughts with certain themes."[17] More succinctly, they are "mental experiences that accompany a change in body state."[18] Like most emotion researchers, Damasio sees an emotion as a response to a stimulus that initiates an action by an animal. He sees a feeling as the perception of that emotion. So for example, fear, the emotion, is that set of physiological responses we have when we confront something that is a threat to our well-being, while fear, the feeling, is the conscious perception of that emotion. They can arise in response to both external stimuli and wholly internal changes in body state.

Like the emotions, feelings are influenced by multiple networks in the brain.[19] At the core of feelings, however, is the perception of the body's current state—the assessment of the range of homeostatic processes that govern whether or not things are going well for us. A structure known as the brain stem, located as the name suggests at the base of the brain as it extends from the top of the spinal cord, is critical for maintaining homeostasis in the body. Information about the body is conveyed up the spinal cord to the brain stem, through which it passes on the way to the thalamus, and then on to the cerebral cortex. Several discrete regions within the brain stem are implicated in the generation of emotions and feelings. These regions link body and behavior via basic life-support functions, such as breathing, ingestion, elimination, and so on. The internal monitoring of the body's state, known as *interoception,* within the brain stem provides the basis for the "emergence of feeling states."[20]

Feelings are not simply perceptions of our emotions, however; Damasio has referred to feelings as "interactive perceptions."[21] Feelings arise during the perception of external objects and events in combination with assessment of the current state of the body. In turn, the state of the body may be influenced by many factors, including memories and attitudes toward the object or events being perceived. The external object being perceived is constant, while the internal object, the body, is dynamic. Damasio emphasizes that this combination means that feelings themselves are dynamic phenomena.

Consider a person looking at Picasso's powerful painting *Guernica* for the first time. Anyone can see the technical mastery and highly emotional content in the painting. But say we told the first-time viewer of *Guernica* that it is named after a girl who dumped Picasso when he was eighteen years old. That viewer's feelings about the painting might range toward puzzlement or confusion—given the scale and content of the work, it would seem to be a bit of an overreaction. Then we tell the viewer that in reality it was painted as a memorial to the small Basque town that was heavily air-bombed in April 1937 by the combined Fascist forces of Germany and Italy, at the request of the Spanish Nationalists under the direction of Franco. Presumably, the feelings of the viewer would change and more reflect those that Picasso intended any viewer of the painting to have.

Human emotions and feelings are strongly influenced by our environments: developmental, historical, cultural, and psychological. Certainly, there are some basic emotions that are brought out by very basic stimuli—fear in response to an animal noise that surprises us at night, for example. Some foods and flavors to which we seem to have an innate attraction, such as sweet or salty or crispy/crunchy, bring on joyous emotions when we consume them. But consider that with a little cultural and biomedical conditioning, a person can be turned against chocolate cake or a fatty steak, despite their overload of button-pushing flavors, to see these foods not as attractive but as "evil" or "disgusting."[22] In this case, the neurobiology of disgust remains constant while culture dictates when that disgust comes into play.

I bring up the importance of context and variability in feelings because the notion that "feeling at home" is widely shared by all people across cultures implies a certain constancy between an external object and a feeling. It could be that our brains are biased toward forming specific feelings toward specific things, perhaps in the same way that our brains are biased to form certain networks that lead to the production of spoken language. Alternatively, we could have a range of emotional capabilities that combine under certain conditions to yield a feeling that we associate with home. The former would be consistent with a more hardwired evolutionary viewpoint, while the latter would be in line with a more interactive view between neurobiology and the environment in general. Certainly, such an interactive perspective could still be consistent with an evolutionary one, but at a cultural level. And of course, given the complexity of human cognition and behavior, it does not have to be an either/or proposition. Indeed, the example of the Warlpiri shows that the notion of feeling at home can be quite expansive—that home is a place defined not by a structure but by activities and relationship, by memories and information.

Having discussed the foundations of emotions and feelings, we are now prepared to explore feeling at home from a cognitive perspective. Like the human moral sense, feeling at home is not an emotion per se, but draws on a range of emotions and cognitive processes interacting in certain environments.[23] We will see that feeling at home is not one thing, but several, and isn't simply about the feelings of individuals about themselves, but about everyone else, too. Home provides a place for human beings to prepare to face the outside world. Part of this preparation involves forging critical relationships with others who will help us succeed outside the home. Being prepared also means being rested in body and mind, and home is, of course, the preeminent place where we recover from our worldly pursuits. Feeling at home therefore emerges from the intertwined feelings we experience around rest, restoration, and relationships (with people and places).

Feeling at Homeostasis

Imagine that for the first time you are invited to your intimidating boss's house for an important work-related dinner. When you arrive, you toss your coat over a chair, kick off your shoes, pull off your socks, and flop onto the nearest couch. Scratching yourself delicately, you say, "Wow, I really feel at home here!" Your boss would probably be somewhat taken aback by your behavior, which could be construed as overly familiar, disinhibited, and ultimately disrespectful. But the topper would be in saying that you feel at home. Inappropriate behavior is one thing, but an invasion is another. To claim that you feel at home in her house is to violate the relationship your boss has with her own personal space. Such a claim might even suggest an unbalanced or malfunctioning psyche. After all, everyone knows that a normal person cannot instantaneously feel at home. Feeling at home means something, and in reality, it is a feeling that must be earned and not simply taken.

The essayist Verlyn Klinkenborg writes that home is "a place we can never see with a stranger's eyes for more than a moment."[24] A stranger looks at someone else's home and sees a structure and objects, a space defined by material goods. When we look at our own homes, we see those same things, of course, but our interactive perception of those objects generates a feeling that is shaped by memories and relationships and by the sense of relief that comes with the body's rest and recovery. Home provides protection and control, more illusory than real in many cases, but enough to provide a respite from all the stresses found in the surrounding environment.

The sense of home does not need to be explicitly taught. It is for the most part shaped by largely intuitive feelings. We are not born with these feelings fully formed, not any more than we are born with the ability to speak a specific language or to sense right from wrong. Rather, the feelings of home, and the ability to experience and make use of those feelings, are nurtured and absorbed in a proper developmental environment. The difference between the home sense and language or morality is that home is really about what goes on inside our heads, about our feelings we have about a place. Humans have a feeling for home, a sense of home, but we are not biologically determined builders. Our genes have not programmed us to build a certain structure like an ant hill or beehive, in a certain place, to engage in a very stereotyped and specific activity. We do not have a house-building instinct. We are makers and users of tools. A house or a hut or an apartment building—even a cave partitioned into activity areas—all represent our technological intelligence, nurtured within cultural traditions. Yet we have arrived at a place in our evolution with the ability to think of a place as a home. I think it is very likely that the home sense—the ability to feel at home—is like the moral sense: the product of several different emotional and homeostatic networks in the brain. There is no single "home center" in the brain, just as there is no single morality center. Instead, a certain central location in our lives becomes cognitively special, as feelings and emotions coalesce around the place we call home.

For many of us, home is the place where our minds and bodies recover from the challenges we face in the outside world. This is as true in today's megacities as it would have been when all people lived as hunter-gatherers of one kind or another. The trials and tribulations we face in the outside world, which can wear us down and leave us feeling out of kilter, are redressed at home. Home is necessary to maintain equilibrium in our lives. It is essential for our whole-being homeostasis.

What do I mean by "whole-being homeostasis"? In classical physiological terms, homeostasis refers to an assortment of body-regulating processes that maintain equilibrium, involving things like feeding and drinking behavior, digestive activities, sexual behavior, temperature regulation, and so on, that are controlled by that critical structure in the brain, the hypothalamus.[25] The hypothalamus receives connections from the autonomic nervous system of the body (the nerves that connect to gut and cardiac muscles and glands), and via the limbic system extends these connections to the cerebral cortex. The sense of stability and balance we get from being at home is not exactly the same as physiological homeostasis, but those processes certainly contribute to the overall whole-being homeostasis that we achieve when we have a home to retreat to. It is part of feeling at home.

Antonio Damasio organizes the body's homeostatic systems into a hierarchy of processes.[26] At the lowest level are basic reflexes, systems of metabolism, and the immune system. At the middle level are behaviors associated with pain and pleasure, and then above them, the basic drives and motivations, processes closely associated with the notion of *appetites.* At the highest level are the emotions and feelings. These motivate actions and reactions of people, often at an automatic level rather than as a result of conscious decision making. Of course, conscious decision making can combine with emotions and feelings to result in a person taking an action.

Home is a place where we can balance our lives with the imbalance imposed by the outside world. This whole-being homeostasis connects to the homeostasis of the body at the level of feelings. Feeling at home contributes to maintaining the body's homeostasis by encouraging us to place our bodies in a (relatively) protected or controlled environment where physiological homeostasis can be achieved more efficiently.

When we think of home helping us maintain balance in our lives, two things that probably first come to mind are eating and sleeping. Eating and appetite are regulated by a complex web of connections between the brain and gut that researchers are just beginning to unravel.[27] Feelings of hunger and satiety are not necessarily linked directly to home, but one of the innovations of human evolution is that food is often acquired in one place but prepared and consumed in another (the home).

Sleep may be more strongly tied to home. We feel the homeostatic benefits of sleep at the whole-being level as sleep brings about the transition from fatigued to rested. Sleep is clearly important to brain function. Studies have shown that disrupted sleep interferes with learning and other cognitive functions. In addition, many neurological diseases, such as Alzheimer's disease, are associated with sleep disturbances.

How does the brain's homeostasis benefit from sleep? Several ideas have been offered.[28] One is that while we are awake and doing or learning something, we inevitably increase the activity of synapses (connections between neurons) to a level that is unsustainable. During sleep, these synaptic connections are downregulated, as the physiologists say, to a level that is energetically sustainable. Another hypothesis is that sleep is necessary to consolidate connections and pathways that have formed during wakefulness.

This makes sleep a necessary part of learning and memory formation. On a different track, recent research has shown that during sleep, toxic waste products produced at the cellular level during wakefulness are washed away. While we sleep, there is an increase in the exchange between cerebral spinal fluid and the extracellular fluid within the brain that facilitates this washing. In all these models, sleep provides the homeostatic balance to the neurological activity of wakefulness.

Hunger and fatigue are powerful motivators, and they signify that fundamental needs of the body must be addressed. One aspect of feeling at home is the expectation that this is the setting in which these needs are met. Expectations, hopes, and beliefs can be powerful cognitive forces. The well-known placebo effect shows just how powerful expectations can be, and how they can modify not only behavior but also the biology of the brain and body.

Like emotions and feelings, the placebo effect was once thought to be beyond scientific study. It was a variable that had to be controlled for in drug studies, but not the object of study. For example, it is generally held that a third of the people treated for pain will show a placebo effect after treatment; this figure is so large that it establishes a very high bar for determining whether or not an actual treatment is effective. The magnitude of placebo effects for the same drug can vary from culture to culture, demonstrating the complex interactions among expectations, patient–doctor relationships, and attitudes toward drugs.[29] With new brain-imaging techniques and other advances in tracking the molecular physiology of living bodies, placebo effects can now not only be observed at the clinical level but the mechanisms underlying their effectiveness can also be established. So now we have a better idea of how placebo influences the treatment of pain, anxiety, depression, addiction, Alzheimer's disease, and Parkinson's disease, and how it can affect the performance of the respiratory, cardiac, endocrine, and immune systems of the body.[30]

A fascinating neuroimaging study by Pedrag Petrovic and his colleagues demonstrates the interaction between reward expectation and emotion in placebo.[31] The experiment basically consisted of showing the participants a series of photos, some neutral but others disturbing or unpleasant (e.g., mutilated bodies), and then asking them to rate the overall unpleasantness of the experience. To build up the expectation of reward, on the first day of the experiment, the participants looked at a group of pictures, first without any medication, then after taking an antianxiety drug, and then again after taking the antidote to the drug. As expected, they rated looking at the pictures as the most unpleasant during the no-drug trial, with greatly reduced reported unpleasantness after taking the antianxiety drug. The subjects were fully informed about the sessions before they started and thus were able to learn to expect that the drug would modify their perception of the images. On the second day of the experiment, the participants again looked at a series of photos, although this time while undergoing neuroimaging. They were led to believe that they would be receiving a drug (either the antianxiety drug or its antidote) as they had the previous day, but in reality they received a placebo (saline instead of the drugs in their intravenous feed).

As a group, the participants showed a substantial reduction in rating the images as unpleasant when they received the placebo—that is, they experienced a reduction in anxiety or emotion despite only receiving saline. There was also an expectation effect: participants who showed a greater effect of the drug during the first day subsequently had a greater placebo effect, as well. The neuroimaging results identified a network that encompassed parts of the anterior cingulate cortex and the orbitofrontal cortex, areas often active in the higher-level processing of emotion, which are also active during pain modulation with placebo. Petrovic and his colleagues suggest that there is a general placebo network that is shared by both anxiety and pain modulation. In addition, the link between expectation of reward and the strength of the placebo effect indicates the importance of reward circuits in the brain, especially those mediated by the neurotransmitter dopamine.

More than anything, placebo demonstrates the power of expectation. The ability of placebo to affect different systems of the body illustrates how expectation can shape many aspects of our lives. Home is not a homeostatic mechanism. It cannot provide us food when we are hungry nor rejuvenate our bodies when we are tired. However, it is the place where, over eons of evolutionary time and over the course of our relatively short lives, we have come to expect these fundamental needs to be met. That expectation, linked to two of the most fundamental homeostatic systems of our bodies, contributes to the visceral power of feeling at home.

Home is one of the places where people usually achieve homeostasis in their physiological lives, so it comes as no surprise that words like *comfortable* or *stable* or *secure* are often used to describe our feelings for it. They are background emotions brought to the fore by conscious introspection, encouraged by questions such as, "How do you feel?" For home life, an equally important question is "How do you expect to feel?" Entering your boss's house and making yourself feel at home too quickly and too enthusiastically isn't just a social faux pas, it betrays a fundamental misunderstanding about what a person's intuitive expectations should be about a place. In a sense, a place—let's say a house—"earns" the right to be a home by demonstrating that it meets our homeostatic needs over time.

The Brain's Rest Home

For many people living busy lives in the modern world, the least home can offer is a place to relax, chill out, and have a few moments free to let the mind wander. Maybe it has been this way for a very long time. For both the hunter-gatherer and the young urban person-about-town, home is defined at least in part by the absence of stimulation and novelty that can be found in the outside world.

When I sit in my home office, staring out the window at the birds at a feeder, watching them but not really seeing them, my mind skipping from one thing to another, what exactly is my brain doing? Nothing or something? Psychologists have long recognized something called brain states.[32] These are different from mood, although mood can certainly interact with brain states. Alertness is a notable brain state—you can be happy and alert or sad and alert, but the continued heightened sense of awareness characteristic of the alert brain is maintained by a certain set of neural networks. Though transitioning between brain states is usually done unconsciously, meditation teaches people how to consciously transition from one brain state to another. Brain-imaging studies show that the more expert a meditator becomes, the less effortful is the transition between brain states, with decreases in activation, for example, in parts of the brain that are active in monitoring attention (i.e., to keep the mind focused).

The brain state that characterizes a wandering mind or daydreaming is known as the *resting state.* When we are in the resting state, it is nothing like sleeping, in that we are fully awake and aware. Our eyes can be closed or open but not really visually attending to anything. We are usually quiet, relaxed, and at repose. During the resting state our conscious awareness is filled with stimulus-independent thoughts—in other words, they originate within our heads rather than from what surrounds us. When people are daydreaming, they can appear "far away," and they might as well be, because their focus of attention is not on their surroundings.

The resting brain is not physiologically at rest. One of the most misleading things about those functional brain-imaging pictures you see in media reports, the ones that show colorful blobs on the parts of the brain that are active during some task, is that they are created by subtracting out all of the baseline activation in parts of the brain that are not positively activated during a task. Surprisingly, researchers have found that the increase in energy consumption in brain networks during a task may only be about 5 percent greater than the activity of the brain while it is in a resting state.

Neuroimaging pioneer Marcus Raichle and colleagues were among the first to look in depth at the activity of the resting brain.[33] One very basic discovery they and others looking at this issue have made is that the "intrinsic activity" of the resting brain is not simply random noise. There exists something that Raichle named the brain's "default mode network" (DMN). This is a network of brain regions that tends to reliably reduce its activation when the brain starts doing a goal-directed task, no matter what brain areas the task positively activates. This sign of the activity of the resting brain encompasses parts of the cortex associated with high-level visual processing and also with the mediation of emotion with cognition, among other regions.

In addition to the DMN, the resting brain's activity displays quite extensive fluctuations that at first glance might appear random, but which more in-depth analysis revealed to be something quite striking and nonrandom. Emerging from the noise of the resting brain are signals that correspond to functional brain networks. The resting brain very briefly activates these networks, one after another, even in the absence of an external task or stimulus to bring them on. One study, comparing resting brain activation in a group of thirty-six subjects to a database of tens of thousands of functional brain studies, found that the resting activations could be classified into ten separate functional domains (those associated with doing some cognitive task or sensory processing, such as those involving aspects of language, memory, or visual perception), each of which could be divided into multiple functional subnetworks.[34] Resting brain activation studies can even provide evolutionary insights. One study compared the resting brain networks that emerge in monkeys to those in humans.[35] Although there is overlap, human-specific networks were identified in the frontal and parietal lobes, the two regions of the brain that have undergone the most significant enlargement over our evolution.

What do we do with all this resting brain activity? One thing is certain: people spend quite a large percentage of their waking hours in such a state. Psychologists Matthew Killingsworth and Daniel Gilbert used an iPhone app to survey thousands of people (mostly Americans) at random times of the day, to ask them what they were doing or feeling. They found that people spend 46.9 percent of their time in a state of mind-wandering. In other words, even when we are awake, our brains are in a resting state about half of the time.

One of the things that we do as our minds wander is mentally time travel. Psychologist Michael Corballis has emphasized that the ability to look backward and forward in time, to make plans and rehash plans that have gone right or wrong, is likely a unique human attribute. Other animals have some planning abilities, especially when it comes to negotiating their spatial environments, but as Corballis writes, our mental time travels draw on memories of "individual people, actions, objects, emotions, and so forth . . . present in different combinations in different episodes."[36] The resting brain is by definition a brain freed, at least momentarily, from the constraints of external stimuli and the internal requirements of processing motor actions. When our minds wander, we engage multiple worlds, past and present.

Just as home is the place where we sleep, it is also likely the place where we can most safely and reliably engage the resting brain state. Given how much of our lives are apparently spent in this state, can we say that it is something that we *need* to do? Undoubtedly, the intrinsic activity of a resting brain is a sign of a healthy, properly functioning brain. In the relatively short time that investigators have been studying the topic, disruptions of resting brain activity have been observed in Alzheimer's disease, depression, schizophrenia, autism, ADHD, and other conditions. One hope is that for progressive diseases, such as Alzheimer's disease, deficits in the default mode network may precede more obvious clinical manifestations of the disease and therefore serve as an early biomarker for the condition, which would in turn allow for clinical treatment before too much permanent damage is done. Home is also a critical environment for brain development, and the default mode network matures as the brain matures.[37] At this point, we do not know if a disrupted developmental environment can alter the resting brain state or what the possible cognitive consequences of such a disruption would be later on in life.

Being at home is to be part in the world and part out of it, just as a resting, wandering mind is engaged more with the internal rather than external stimuli. The most salient connection between home and the resting brain state is that to be at repose and not attending to external stimuli is not just safer but also more acceptable in the home environment. Besides all the other distractions and stimuli of the external world, we are a social species, and to disengage socially can have consequences. Subjectively, with too much stimulation in the outside world, most people seem to appreciate the downtime that home offers. Perhaps they are seeking a homeostatic balance. In the long run, the somewhat increased energy requirements of a brain performing a task or attending to something in the environment may eventually take a toll, which can be repaid by some relative downtime at home.

The Relaxing Brain

A resting brain is not necessarily a relaxing brain. Lying in an MRI machine, eyes closed, mind wandering, waiting for the "real" experiment to start, may produce a brain at rest, but the situation is not inherently relaxing. People often see their home environments as a place not just to rest and recharge but to relax, or at least that would be an ideal. Of course, in reality, home life is not always relaxing, and those who can find relaxation away from the home may choose to do so. But harkening back to more traditional times, there may not have been an option to relax elsewhere, no corner bars or fitness centers or public baths. Just as home provides a place conducive to sleep and rest, separate from the intrusions of the outside world, it is also linked to the possibility, at least, of relaxation.

Studying the truly relaxing brain in an experimental setting is difficult. However, one proxy available to us may be found in the cognitive studies of meditation. But, as neuroscientist Antoine Lutz and his colleagues have pointed out, meditation is not simply relaxation.[38] There are, in fact, many different types of meditation, which all have distinct goals as to the kind of mental transformations they are hoping to achieve. One form, known as open monitoring (OM) meditation, may be particularly evocative of home-style relaxation. The goal of OM meditation is not to devote attention to an external object, but to gain an enhanced sense of the "usually implicit features of one's mental life." Open monitoring training allows practitioners to become aware of background emotions, bodily sensations, and wandering thoughts, while at the same time not having any of them become the target of attention. Awareness is sustained without any attentional focus. Perhaps paradoxically, awareness of these potential external and internal distractors allows them to be controlled and ignored, freeing the mind to engage in quite complex activities. As one OM practitioner describes writing a blog post: "As I type, I allow myself to be aware of my breathing, of my fingers touching each key, and the feeling of warmth of my back against the chair. I am also aware of random thoughts . . . I observe them and let them pass. All of this activity keeps me grounded in the present . . . the words flow effortlessly."[39]

The functional neuroimaging literature on meditation is quite complex and not altogether consistent.[40] A meditative state (as opposed to a nonmeditative state) typically shows changes in activation and deactivation of multiple brain regions; more experienced meditators can be distinguished from less experienced ones; and even short-term training can change brain activations in novice meditators. There may be some overlap in brain activations/deactivations observed during meditation with those seen in the default mode network, but given that these networks are quite extensive, the significance of the overlap may be limited. Interestingly, there have been actual changes in the brain's anatomy found in long-term meditators, with increases in gray matter in parts of the frontal lobe and the hippocampus observed.[41] Meditation can shape the plastic brain not just in a functional way but structurally, as well.

Putting all this neuroimaging data together (and also including support from brain wave [electroencephalograms, or EEG] studies on meditation), Antoine Lutz and colleagues suggest that the meditating brain should be looked at as another brain state, like the resting brain or the alert brain.[42] With practice, they suggest that meditation can make an enduring alteration in the brain's baseline function. Such a change could subsequently influence a range of brain processes.

The brain state induced by OM meditation gives us some insight into the state of relaxation that we might experience in our homes. When we feel relaxed, we are monitoring our minds and bodies but not closely attending to them, nor are we overly distracted by what is going on around us. The relaxed state of mind is such that we can still engage in goal-directed behavior, maybe even more efficiently since we are less distracted than usual. The big difference between being relaxed and doing OM meditation is that the meditators are trained to deal much better with negative or highly distracting or anxiety-provoking sensations or phenomena. They maintain a relaxation-like brain state under conditions where most people could not.

Feeling at home is not just one thing. At a minimum, it involves three separate brain states—expectation of physiological homeostasis, rest, and relaxation—all of which are more likely to be achieved at home than in the outside world. When we say that we can make ourselves at home at a place that is not home, we are likely feeling or predicting that any or all of these brain states can be experienced. Of course, it can be inappropriate to quickly feel at home in a place where one really is not at home. It might even be dangerous, especially in the distant past. But feeling at home is about more than the internal brain state of an individual: it often involves relationships among people sharing a home.

Home as a Sphere of Empathy

Former vice president Dick Cheney and Ohio senator Rob Portman are both Republicans whose status in American politics as right-wing, conservative politicians would be hard to refute. But during the 2000s, they both parted ways with the large majority of their party members on the issue of gay marriage, which they both publicly supported. Underlying their support for gay marriage is the fact that they each have a gay child: Vice President Cheney's daughter Mary and Senator Portman's son Will. It is difficult to escape the conclusion that their enhanced empathy for the plight of their children has influenced their political position on gay marriage. In fact, there is something known as the "Rob Portman effect." Pollsters have found that the percentage of Americans supporting gay marriage has increased almost directly with the increase in the percentage of those who say they have a family member or close friend who is gay.[43] This could just as easily be referred to as the "empathy effect."

Of course, empathy only goes so far. Mary Cheney's own sister Liz took an anti-gay-marriage position during her brief Senate run in Wyoming. And in Indiana, state representative Milo Smith took the lead in advancing a bill that would change that state's constitution to ban gay marriage, prompting a public remonstration from his son Chris, a gay man living in a domestic partnership in California. Nonetheless, the power of empathy to modify even deeply held political and religious beliefs is apparent in the striking gay marriage conversions of Vice President Cheney and Senator Portman.

Empathy is an emotional response, arising from understanding another's emotional state or condition, that is similar to what the other person would feel or be expected to feel. It is a stronger emotion than sympathy in that while both require an accurate perception of the emotional plight of others, empathy is much more about feeling some aspect of that emotion rather than simply showing concern for the needs of the other. Psychologist Jean Decety is one of the foremost investigators of the empathic brain.[44] He argues that the basic neuromechanisms for empathy, along with those for other emotions, arose long ago over the course of mammalian evolution. Evidence of "emotional contagion" can be seen in rodents; for example, mice display freezing behavior more readily after observing pain in a close relative rather than in a stranger. Our closest primate relatives show a range of appeasement and social behaviors in response to the emotional distress of others.

In some mammals, such as ourselves and the other primates, empathy appears most fundamentally in the bond between mothers and their offspring. As Decety writes, "Attachment cannot survive without empathy in the sense that the caregiver must necessarily be empathic with the infant."[45] In humans, as most of us are fully aware, strong and ongoing emotional attachment bonds are not just limited to a mother and her (currently youngest) offspring, but can extend to her older offspring and to her reproductive partner. Strong empathic bonds can form between fathers and their children and among siblings.

Like other complex cognitive phenomena, empathy develops as a child grows into adulthood. Decety's functional neuroimaging study of children and adults has charted how the brain's empathy networks shift with increasing maturity. For both adults and children, looking at people in pain empathically activated the regions that are part of the "pain matrix" of the brain. When looking at pictures of people suffering accidental pain, activations were seen in regions associated with emotions and feelings, such as the insula and the anterior cingulate cortex, and parts of the motor and sensory cortices, among other areas. When viewing pictures of people who were harmed by another, activations in several regions of the prefrontal cortex were also observed, in addition to increased activation of the amygdala, suggesting the even greater emotional content of these images. The younger the participant, the stronger the activation of the amygdala. This is consistent with the idea that children's emotional lives are dominated more by early-maturing limbic structures, rather than later-maturing, higher cortical regions.

Empathy is influenced by group membership or identification as well as by individual relationships. We might expect this to be true in, for example, an overtly racist individual, but there may even be subconscious effects, as well. In one neuroimaging study by psychologist Xiaojing Xu and colleagues, Chinese and Caucasian subjects were shown images of people in painful situations who were from either their in-group or out-group.[46] When viewing the painful images from members of their in-group, both groups showed brain activations consistent with an empathic response, involving the anterior cingulate cortex, insula, and parts of the frontal cortex. However, when viewing images of out-group members in pain, there was a remarkable reduction in the activity of the anterior cingulate cortex. This was true for both Chinese and Caucasian subjects. When subjectively reporting on the level of pain intensity or unpleasantness of the images, the subjects showed no group bias at all. So at a conscious level, the Chinese and Caucasian subjects looked on all subjects in pain in the same way, but at a subconscious level one could argue that their empathy for members of their in-group was more emotional, more deeply felt.

Race and ethnicity are not the only factors that influence empathic response.[47] Other neuroimaging studies have shown differences in empathy response depending on whether or not a subject imagines a loved one or a stranger in a painful situation, or if a fan of a soccer team is deciding to help a fellow fan or the fan of a rival team out of a pain-inducing situation. These results are not surprising: human sociality is not random. We live and interact in enduring groups.[48]

The most fundamental in-group in human society consists of those with whom we share our home. Although we most often share our homes with family members, this is not always the case. In fact, since our ability to know who our relatives are is based on knowledge and context, not a gene detector, home itself is important in shaping our impressions of who is and isn't family. Some modern workplaces (something for which there is no traditional hunter-gatherer equivalent) start to "feel like home" for some workers, who "feel like family." This is not just a metaphor—time, proximity, and familiarity can make a workplace feel the same way a home feels.

Empathy may be the most powerful emotion that emerges from home, and thus empathic feelings are a critical part of feeling at home. The sphere of empathy that comprises a home encompasses its residents in emotionally vested, mutually beneficial relationships. It is not surprising perhaps that conservative American politicians like Vice President Cheney and Senator Portman would empathize with their children, leading them to disavow the policies of the other most significant in-group in their lives. Maybe it is more surprising when politicians choose social policy over family. At the very least, without looking too deeply into these home lives, most of us know from experience that such an empathic transgression cannot be a good thing for relationships within a family.

Synchronized Lives

Like many people, when I was younger I lived in a variety of housing situations with somewhat randomly assembled roommates. Sometimes, we were friends before moving in together, or friends of friends, or total strangers (but from the same general socioeconomic stratum of society). Over time, if things went well, we would come to develop a homey camaraderie, not quite like that of a family who shared a home for years, but with some elements of it.

One thing I noticed that prevented the development of a nicely integrated household—besides intractable personality disorders or questionable hygiene—was if some roommates were simply not on the same schedule as the others. They were not synchronized with the goings-on of the majority of the household. This could be due to a work or study schedule, or just to a propensity for living an extremely nocturnal lifestyle. Whatever the underlying reason, and no matter how well they might have otherwise meshed with the group, not being on the same schedule prevented them from becoming a fully integrated member of the home.

When you hear about people in synchrony who share a household, the first thing that might come to mind is menstrual synchrony. This phenomenon was first reported in 1971 by Martha McClintock (then an undergraduate, she later became a prominent academic psychologist), who found that women sharing a dormitory at an eastern US women's college appeared over time to start to synchronize their menstrual cycles, or at least have them come increasingly into alignment. This raised the possibility that pheromones could be influencing human menstrual cycles just as they were known to influence those of mice, or perhaps that some other aspect of domestic proximity was aligning the cycles. Unfortunately, subsequent research on a range of populations and greater consideration of statistical issues in analyzing this type of data have pretty much shot down the idea that humans (or captive nonhuman primates) have menstrual synchrony.[49]

The idea of menstrual synchrony is intriguing for many reasons, one of which must be based on an intuitive sense that sharing living quarters brings people into alignment, whether it is a women's dormitory or a military barracks. People who live together do become synchronized in psychologically and even physiologically important ways. For example, one study has shown that there is "coregulation" of mood and cortisol (stress hormone) levels in cohabitating married couples.[50] This synchrony was especially pronounced in couples who were less satisfied in their marriages. Negative mood was "contagious" and it was accompanied in both spouses with relative increases in cortisol levels, linking negative mood to increased stress.

Cognitive scientists have long known that subconscious behavioral mimicry and emotional synchrony play a vitally important role in human social interaction. Psychologist Ruth Feldman has emphasized that this pattern begins in the earliest interactions between parents and infants.[51] Her functional neuroimaging studies have shown that the same brain regions, a widely dispersed network encompassing areas associated with reward, stimulation, and social understanding, are used by mothers to detect synchrony in others and the expression of emotional synchrony in their infants. Parent–infant synchrony may play a major role during the critical period when young children are learning how to be social beings. At the same time, this gives social synchrony a central place in human social interaction.

Social synchrony may begin with behavioral mimicry, which is a powerful, albeit somewhat covert, force in human relationships. Observational psychological studies of people spontaneously interacting have shown a wide range of behaviors that people copy from one another, including mannerisms, gestures, postures, and other motor movements. Behavioral mimicry is more common within groups of people who already know each other well, and among those wishing to become affiliated with a group or otherwise establish rapport. Individual variation also plays a role, with "high-empathy" individuals more likely to engage in behavioral mimicry. Mimicry facilitates positive social interactions, and as Tanya Chartrand and Jessica Lakin write, it "creates liking, empathy, and affiliation between interactants . . . the 'social glue' that brings people together and bonds them."[52]

Mimicry implies a leader and a follower, while social (or interactional) synchrony suggests a mutual arrival at a similar behavior. Like behavioral mimicry, social synchrony is a clear facilitator of positive social interaction. Studies have shown that social synchrony increases liking; enhances perceptions of similarity, closeness, and rapport; promotes conformity, helping behavior, and cooperation; enhances the memory for information about an interacting partner; and even increases pain thresholds.

People like doing things with other people. Social synchrony is psychologically rewarding and socially beneficial. We see this in cultures throughout the world, expressed in a multitude of ways. Over evolutionary time, social synchrony in its various guises has served to help organize and stabilize the increasingly complex societies that our ancestors created for themselves. Cognitive archaeologist Steven Mithen argues that making music is one of the most striking avenues of social synchrony that has ever evolved. He writes:

> Those who make music together will mold their own minds and bodies into a shared emotional state, and with that will come a loss of self-identity and a concomitant increase in the ability to cooperate with others. In fact, "cooperate" is not quite correct, because as identities are merged there is no "other" with whom to cooperate, just one group making decisions about how to behave.[53]

For Mithen, music-making is one of the prerequisites for the evolution of language. For example, pitch and rhythm are important components of language, as well as music. He argues (for a variety of reasons) that only modern humans possess true language. Our ancestors and cousins the Neandertals were simply "singers," but their singing filled a vital role in emotional communication and in maintaining social harmony.

Reread Mithen's quote but replace the word "music" in the first sentence with "a home." The benefits of social synchrony are congruent with the benefits of home life. When people live together in a home, they typically share more than just a space carved out within the larger cultural and ecological environments. Their schedules and activities also become entwined and overlapping. This kind of synchrony appears to have been evolutionarily advantageous in the past and is psychologically rewarding in the present.

The feeling of home emerges from the feelings we have at home. These feelings encourage a sense of control and stability at three levels. For our bodies, home is the place where we seek and expect to be able to recover from the labors and challenges of the outside world. For our brains, home is a place where we feel able to rest and relax, to experience downtime that is probably essential (in some as-yet-to-be-determined way) for optimal brain function. Finally, for everyone, home begins, or should begin, in the powerful, empathy-driven social relationship between mother and child, as well as other significant relationships involving members of a shared household. Sharing a household means sharing a life, synchronized not only in terms of activity patterns but ultimately in shared goals and common worldly activities.

A house cannot become a home instantaneously. However, the ability to feel at home, learned in childhood, makes it possible for us to be at home in different locations and situations throughout our lives. Our ability to experience the benefits of home is transferable. We know we have been successful in creating a new home when the sum total of the feelings relating to body homeostasis, expectations of recovery from physical and mental fatigue, and the establishment of social synchrony make a dwelling feel like the one place in the world that belongs more to us than to anyone else. We feel at home there. But how did home come to be? In the following chapters, I will trace our journey as a species from homelessness to homely.

Chapter Two

NESTING AT HOME

Animals other than humans also seem to like to make themselves at home. I know this from personal experience: we've shared our own home with animal guests both invited and not invited. Dogs, cats, hamsters, lizards, and fish belong to the former category; the latter is populated by the likes of amphibians, rodents, spiders, and lots of insects. One reason humans build houses and other shelters is to protect themselves from the outside environment, a goal we share with countless other species. For many animals, the world is a dangerous place, full of predators and other creatures that would do them harm. Animals build and opportunistically make use of a variety of shelters to protect themselves from these dangers. A shelter is also a place where food can be stored, reproductive partners can be housed, and offspring can be trained. It is a place for rest or even hibernation.

How did home become a human universal? How did we make the evolutionary journey to home? As we will see, it was in many ways not a straightforward journey. In fact, our closest primate relatives do very little to physically shelter themselves from the surrounding environment. Although a shelter is not essential to feeling at home, there is a strong relationship between home (a concept in the mind) and house (a physical structure) in the vast majority of human cultures. It may be useful to consider the sheltering habits of some other species that are not particularly closely related to us, to see what benefits a home, or at least a sheltered life, has to offer.

What Is Home Good For? A Rodent's View

When I was about five years old, I loved going to the zoo located in the main city park of Iowa City, Iowa, where I was born. These little zoos hardly exist anymore—they really can't be maintained at the level now considered appropriate for housing wild animals. I don't remember too many of the animals. Maybe there was a bear in a cage and a buffalo in a pen. But the one species I really remember well might be a surprising one: the black-tailed prairie dog (*Cynomys ludovicianus*).[1]

The zoo had a great prairie dog display. There was a colony of them, and they occupied a plot of ground where they had burrows, just like in the wild. They would pop up and out of the various holes in the ground, look around, and go back down again. It was clear that they were a cohesive family, with some of them busy looking out for predators (there weren't too many to worry about in the zoo), while others went about their business. The whole display was much more "real" than a cage. More importantly, in this child's eyes, it was like looking at a self-contained, miniature household (or a village of several households), a rodent reflection of human home life.

Prairie dogs are reddish-brown rodents of the squirrel variety, who weigh between one and two pounds as adults.[2] Their natural range is across the central plains of the United States, leaking over into southern Canada and northern Mexico. There used to be lots of prairie dogs: it was estimated in 1900 that there were five billion of them, and a single colony in Texas may have contained four hundred million. Since they were regarded as pests by farmers and ranchers, they were enthusiastically exterminated, reaching a point in 1974 where they were placed on the endangered species list. They have made something of a comeback since then.

Although the prairie dog colony at the zoo that I saw looked like a household, in the wild, colonies are obviously much larger. They are subdivided into smaller, more household-like units (coteries), however, which are vigorously defended by their occupants. Coteries are made up of a core of two or three related females, a male, and their offspring. The females spend their entire lives in these coteries, while the males are more nomadic, which has the positive effect of reducing inbreeding, a universal concern in social animals. Prairie dog burrows have multiple entrances, and are divided into specialized rooms, such as those for sleeping or raising and nursing young.

As it is for human houses and dwellings, the main purpose of a prairie dog burrow is protection, both from the elements and from predators. Prairie dogs are diurnal and retreat to the burrow to sleep at night. They do not hibernate, and thus burrows provide protection from severe weather year-round. Many different animals, ranging from coyotes to snakes to birds of prey, like to eat prairie dogs. Therefore they spend as much as a third of their time looking out for animals that might want to make a meal of them. When a potential predator is spotted, prairie dogs let out a loud, repetitious call, which sends both close and distant kin scurrying for the nearest burrow opening. The colony burrow system has clearly worked for prairie dogs (and other burrowing rodents) as a predator-avoidance strategy. The prairie dog's coordinated social activity, centered on an artificially modified environment, is a wonderful example of how natural selection shapes and creates adaptive behavior over millions of years of evolution.

What Is Home Good For? An Insect's View

Every spring and summer, the organ-pipe wasps make their presence felt around my house. Or perhaps I should say they make their presence heard, because I usually first notice them by their loud buzzing. They buzz around the outside of the house, under eaves and along deck railings, and in the garage. The females are looking for a somewhat, but not entirely, secluded surface, free from direct sunlight, on which to build a nest. They build their nests out of very fine-grained mud, "weaving" it to form a half-tube that is four to five inches long and half an inch wide, affixed to a flat surface. This surface can be a brick or wood wall, the back of a lawn chair, the leg of a workbench, the fold of a sun umbrella—almost anything, including many things to which one might prefer not to find a mud nest attached. They can build these tubes remarkably quickly, in very little time laying down several inches of soft mud that dries to form a surprisingly hard shell. They seem to buzz particularly loudly while they are laying down the mud. Although each tube is a solo project for the female building it, several females may choose to make nests in the same location. When this happens, the group of tubes can indeed resemble the collected tubes of a pipe organ.[3]

Each tube serves as the nursery for a single organ-pipe wasp offspring. The mother lays an egg within, it hatches, and the offspring passes through the larval and pupa stages within the tube. For food, the mother helpfully provides a selection of paralyzed spiders stuffed en masse into the tube, on which the larva can feed. After reaching a certain size, the larva pupates and eventually emerges into full wasp-hood.

There is an elegance in the one-to-one correspondence between species and structure in organ-pipe wasps. It's not simply a matter of the nests being the most distinguishing feature of the species—that's what they are named for, after all—instead it's how biology and behavior and building all come together in an efficient and immutable package. I find myself marveling at the industriousness of the wasps and the beauty of their constructions, even as I scrape their nests off of a brick wall with the blunt end of a broom handle, watching the dried mud, dust, larvae, and paralyzed spiders rain on the ground below.

The organ-pipe wasps, along with other wasps, bees, and ants, belong to a group of insects (the order Hymenoptera) that are known for the nests that they build. They are among the most well-known "animal architects," as James and Carol Gould have called them,[4] which also include in their ranks a great variety of bird species, other insects and spiders, and some mammals. Humans could be included among these animal architects, but really we are a special case. Humans may be builders and architects, but the motley collection of buildings and structures we as a species construct lacks stylistic cohesion. Although we humans can look on how we build as a virtue—what better evidence is there of our supreme tool-making and tool-using ability than the varied objects we design to transform the landscape?—from the perspective of a wasp or a bee, the human-constructed world might appear to be chaotic, instead of a pure, clean, "natural" extension of our species-hood.

It is not surprising that humans are not natural builders; we come from a long line of nonbuilders. Humans are members of the order Primate, and on the whole, the lemurs, tarsiers, monkeys, and apes who are our closest cousins do not build. It is unfortunate that other primates do not build houses: it would be great to see a range of primate houses—starting with simple and small ones for the lemurs, somewhat more elaborate ones for monkeys, larger ones for the great apes, and finally McMansions for well-off, highly evolved, suburban humans. If that were the case, then we could link the evolution of house and home to broader patterns of primate evolution and maybe learn something about the deep origins of home. Even though there is little shelter building to speak of among primates other than humans, the origin of home as a feeling or mental space likely emerges from some aspects of our primate ancestry. Thus, exploring our primate identity may help us find the beginning of the path toward the evolution of home. As we will see, our closest relatives, the great apes, provide the most intriguing evolutionary clue of all.

What Is Home Good For? A Primate's View

To understand where the human feeling for home comes from, it is important to consider the zoological neighborhood in which we have evolved. We humans are just one among about three hundred primate species around the world.[5] Primates have long been divided by scientists into two major groups, although the composition of those groups has changed over the years (this reflects changes in the scientists, not the primates themselves). In the more classical scheme, primates are divided into the prosimians and the anthropoids. The prosimians are made up of the lemurs, lorises, galagoes, pottoes, and tarsiers, all small-bodied, more "primitive" versions of primates, while the anthropoids consist of the monkeys and apes. Over the last couple of decades, anthropologists have come to agree that there are several anatomical features (such as a dry as opposed to wet, doglike nose) that favor placing the tarsiers with the monkeys and apes. Recent molecular genetic studies based on comparing whole genomes definitively place the tarsiers with the monkeys and apes.[6]

Genetic studies also show that all primates share a common ancestor separate from all other mammals. In this sense, we can be confident that they all belong together in an evolutionary sense. Primates possess several anatomical features that help define them as a group. Our bodies exhibit several of these features. Look at your hands—their ability to grasp using an opposable thumb, as well as the fact that you have fingernails instead of claws, marks you as a primate. The very good depth perception you have when looking at your hands is also a primate feature. Stereoscopic vision is possible because the visual fields of our two forward-facing eyes overlap. Compared to other mammals, neither primate bodies nor teeth are particularly specialized, retaining a generalized form. However, there are subtle features of the skull that unite primates as a group separate from other animals (these not-very-obvious features can be important in trying to figure out if a very early, rodent-like fossil from sixty million years ago was actually a primate ancestor).

Another important feature all primates share is that females typically have single births. Compared to other animals, mammals invest heavily in the raising of their young; this begins with mothers providing milk to their offspring. With single births and slow maturation times, primates take this trend to an extreme. Primates take time growing up because life as an adult primate can be relatively complicated; anthropoid primates typically live in highly interactive social groups that require a certain level of intelligence and training to navigate through. In keeping with this, they are quite literally brainier than typical mammals. This trend is much more characteristic of the apes and monkeys rather than the prosimians, and becomes more pronounced in the great apes and then ourselves. A larger brain takes time to train and also requires a relatively large amount of energy to grow and maintain.

All of these factors contribute to why bearing single offspring who are intensively reared has become a primate feature. The distant, evolutionary foundations of home, going back to when all primates looked like modern prosimians, were likely built on the relationship between the primate mother and her child. One of the main functions of a home is to serve as a nursery. Human intelligence, both social and technological, requires years of nurturing to fully develop. Raising of young depends not only on the mother but also on other individuals, and a home becomes the venue for a more collective (compared to other primates) form of child rearing.

But I am getting ahead of myself here. Before leaving the prosimians behind entirely, I want to discuss one other aspect of their behavior that is also relevant to home. With the exception of some lemur species on Madagascar (who evolved in isolation from other primates), prosimians are not particularly social and they are most active at night. In those ways they are not like humans at all. Like us, however, they make use of the broader spaces in which they live, and they do so from a specific place in the environment.

A Prosimian Fix-Point

The everyday—or more typically, the everynight—lives of prosimians are not all that well understood. There's a good reason for this: most of them are small and active at night. So while hundreds of PhDs have been earned over the last fifty years by graduate students watching monkeys and apes do what they do in broad daylight, the lives of nocturnal prosimians are relatively unstudied.[7] We know that like most mammals, they have an area over which they forage or hunt to get their food, find others of their kind to mate or socialize, and so on. This is known as a "home range," a simple but fundamental concept that has long been at the heart of ecological studies of animal behavior. Within a home range, animals have certain favored places to which they return on a more than chance basis. These favored places are referred to as "fix-points" by some researchers.[8]

Animals can have all sorts of fix-points within their home ranges, places where they return repeatedly for food or water, to socialize, or to rest. For some animals, a specific fix-point in the home range comes to take on more significance than any other. The ground squirrel's burrow and the robin's nest are fix-points that serve multiple purposes: places for rest, for protection, to give birth, and to raise young. These are fix-points that come closer to our notions of home.

Though primates are not builders, and setting humans aside for a moment, do we find any of these special sorts of fix-points in any primate's home range? Many years ago, pioneering animal psychologist Heini Hediger considered the topic of nests and homes in primates, and he expressed some astonishment at the fact that nesting in primates could be found only at the "two extremes of the phylogenetic [evolutionary] tree."[9] On one hand, the great apes habitually build and use sleeping nests (more on that below), but such structures are not known at all among the gibbons (i.e., lesser apes) and the monkeys. In the more "primitive" primates, the prosimians, however, there are many examples of nesting behavior.

Several prosimian species have been shown to make use of leaf nests or to occupy tree hollows for extended periods of time. Primatologists sometimes use these nests to locate and track prosimians in the wild, since they are difficult to observe at night when they are active. It is important to keep in mind how small these prosimians are. Except for some of the lemurs, who have been able to evolve larger body sizes in isolation in Madagascar, prosimian species top out at between three and four pounds, with many of them averaging around half a pound.[10] They do not require very extensive housing to keep themselves covered. Given that these species are nocturnal, one of the reasons that they need some sort of shelter is simply to keep out of sight during the day, where a sleeping little prosimian would make a nice meal for all kinds of predators.

The nesting habits of prosimians may also be influenced by how mother prosimians transport their offspring. Most anthropoid primates carry their newborn offspring in their arms, and from a very young age, the newborns themselves can assist this process by grasping the mother's fur. Across all primates, there is a general trend toward extending gestation length and thus advancing the physical maturity of the offspring at birth (with long gestation time and a relatively helpless infant, humans are an exception to this trend). As for other primate trends, the prosimians present a mixed picture in this regard. Some prosimians, such as the lorises, have relatively long gestation lengths (compared to other mammals their size), and the newborn infants are mature enough to be carried in the mother's arms. Gestation lengths are shorter in the bushbabies and galagoes, the offspring are born at a less advanced stage, and the mothers carry their young in their mouths. The various tarsier species exhibit one pattern or the other.[11]

As it turns out, how prosimian mothers carry their infants is directly related to whether or not the infants are deposited in nests. Prosimian mothers who carry their babies in their arms do not leave the child behind in a nest while they go out hunting or foraging; those that carry them in their mouths do leave them unattended in the nest. One could imagine that carrying a baby in arms could pose quite a burden on a mother looking for food, but from a relatively young age, the baby is actually clinging to the mother's fur, leaving her arms relatively free.

Prosimians who carry their babies in their mouths have nests that serve at least three critical functions: nursery, sleep, and protection. In this sense, their nests are more complex fix-points than the sleeping nests of the prosimians who carry their babes in arms. But it is generally acknowledged that, in evolutionary terms, mouth-carrying is a more primitive behavior than arm-carrying, which is a novel or derived feature. Primate mothers carrying their babies in arms is associated not only with longer gestation times, but also with intensifying the relationship between the mother and her single infant. Increasing the link between mother and child may have been one of the most critical factors in setting the stage for the evolution of larger brain size and greater intelligence in primates. It takes time and energy to grow and train a big brain, and primate mothers carrying their babies in arms made sure that they were in a position to make these investments in their young.

The anthropoid primates almost undoubtedly evolved from an infant-in-arms prosimian, and along the way, they got rid of their nest-building and nest-sleeping habits. For tens of millions of years, it is likely that no species that scientists consider to be a monkey or ape made regular use of a nest or nestlike structure to rest or do anything else important in their daily lives. However, sometime between fifteen and twenty million years ago, along the evolutionary lineage that would eventually give rise to the great apes and ourselves, the habit of sleeping in a nest evolved in primates once again. And so today we have the astonishing situation, as Heini Hediger described it, where among the primates you can find a bushbaby weighing less than 200 grams (less than a half pound) and a male gorilla weighing more than 200 kilograms (more than 440 pounds) both bedding down for a day or night of sleep, in something that is reasonably called a nest. But the branches between them on the primate evolutionary tree are for the most part free of nests of any kind.

The Nesting Apes

In the 1960s and early 1970s, three young women separately entered the tropical forests of East Africa and Indonesia to begin long-term studies of our closest relatives, the great apes. In Tanzania, Jane Goodall began her legendary studies of the common chimpanzee (*Pan troglodytes*), Dian Fossey embarked on her ultimately ill-fated scientific journey with the mountain gorilla (*Gorilla beringei*) in Rwanda, and Birutė Galdikas started her lifelong commitment to the study and preservation of the orangutans (*Pongo pygmaeus*) of Borneo. Around the same time, dozens of other researchers also took to the field to gain a greater understanding of primates of all kinds, leading to a revolution in how scientists look at primate lives in the wild. But it was Goodall, Fossey, and Galdikas who captured the imagination of the public, with their work and images featured in popular magazines, television documentaries, and even movies.

All of the women in this famous triumvirate were also mentored and to some extent directed by the famed Anglo-Kenyan paleoanthropologist Louis Leakey.[12] They are even known today as "Leakey's Angels." Leakey knew that to better understand human evolution, and to be in a better position to interpret the various fossil apes and humans that he and other paleontologists were uncovering in East and Southern Africa, anthropologists needed more information about how contemporary primates lived in the wild. Thus Leakey was very supportive of initiatives to get researchers out in the field, and he used his influence to fund and place them.

When Goodall, Fossey, and Galdikas went into the field, one thing they all saw was that before going to sleep every night, their ape subjects took a few moments to prepare a nest or bed of sorts (Western researchers prefer the term *nest* while Japanese primatologists tend to use the term *bed*, which may actually be more accurate[13]). All three of them described ape nest-building in their scientific memoirs. I think it is worth quoting from each of them in some detail to get an idea of what is involved in making a nest. Let's begin with Jane Goodall and her chimpanzees:

> I found that every individual, except for infants who slept with their mothers, made his own nest each night. Generally this took about three minutes: the chimp chose a firm foundation such as an upright fork or crotch, or two horizontal branches. Then he reached out and bent over smaller branches onto this foundation, keeping each one in place with his feet. Finally he tucked in the small leafy twigs growing around the rim of his nest and lay down. Quite often a chimp sat up after a few minutes and picked a handful of leafy twigs, which he put under his head or some other part of his body before settling down again for the night.[14]

One stray observation: it is probably easier to follow this technique if one has four hands instead of two.

Gorillas also make nests, although as Dian Fossey notes, often on the ground:

> Adult nests are sturdy, compact structures, sometimes resembling oval, leafy bathtubs. . . . Construction is concentrated on the rim of the nest, which is composed of multiple bent stalks, the leafy ends of which are tucked around and under the animal's body for a more "cushiony" central bottom. Nests can be built in trees as well as on the ground, but because of the adult gorillas' great weight nests are more commonly found on the ground. Favored nesting locations during the rainy season are in the sheltered hollows

> of tree trunks and nests may be made only of moss or loose soil. . . . Nests built by immatures are often only flimsy clusters of leaves until practice enables the construction of a solid, serviceable nest.[15]

This last statement shows that nest-building in apes is a skill that must be developed as a child grows up. Obviously, it is a skill that young apes learn while observing their mothers, one they may have a predisposition to acquire.

Finally, Biruté Galdikas shows that orangutans are as adept at making nests as their cousins, although young orangutans also take some time to master the skill:

> I was beginning to appreciate the orangutan's skills as an engineer. The ease with which orangutans build nests belies the complexity of the structure. First, the orangutans must select a sturdy support—a strong limb or, less frequently, the tops of two trees brought together or the crotch of a large tree. Next they bend and twist the smaller branches of the limb at ninety-degree angles, folding them into a springy, circular platform, like a box spring. This rarely takes more than three or four minutes. But they spend up to a half-hour picking loose, leafy branches, and piling them atop one another to form a mattress almost a foot thick. The end result is a bouncy arrangement, quite comfortable. . . . Young orangutans exhibit a natural tendency to bend branches over and under their bodies, but building nests is not instinctive (as in birds). Juveniles practice for years starting with play nests, and only later graduate to the real thing.[16]

From Galdikas's account, it seems as if the orangutans may make the most elaborate nests of all, though she may have just been the researcher most impressed with her primates' ability to do so.

The great apes (and you can throw in the pygmy chimpanzee or bonobo and ourselves into the mix) are all pretty obviously cousins, both anatomically and genetically. But in one way or another each lineage has gone its own evolutionary way. Orangutans split from the African apes around fifteen million years ago, and chimpanzees and gorillas separated around seven million years ago. These three great apes have not only evolved distinctive appearances, including marked changes in body size, but each of them has evolved a distinct social structure. Chimpanzees live in multi-male/multi-female communities; gorillas are characterized by a harem-like social structure with a single adult male associated with multiple females; and orangutans live largely solitary lives, with the exception of females and their young. And yet through all these behavioral and anatomical changes, nest-building remains a common feature of the great apes' lives. Maybe this should not be so surprising: everyone needs a good night's sleep.

Natural Histories of Sleep

Sleep is critical to all of us. Daily nest-building in apes shows that, for them, sleep is important enough that a good night's sleep is worth a modest but consistent investment in time and energy. So to ask what home is, it would seem, also means asking what sleep is.

Sleep is controlled by the brain, specifically the hypothalamus, which is the site of the master circadian clock that tells the body when it should be asleep and awake. Within the hypothalamus, a tiny collection of neurons (only ten thousand on each side—a mere drop in the brain's ocean of eighty to ninety billion neurons[17]) known as the suprachiasmatic nucleus (or nuclei—there is one for each hemisphere) is actually responsible for initiating daily sleep cycles. Regulator proteins produced by these neurons initiate an "upward" cascade of activity through various brain regions, ultimately including parts of the cerebral cortex, that determine sleep or wakefulness. The clock of the suprachiasmatic nucleus does not run in isolation. Pathways that run directly from the retinas of the eyes to the nuclei mean that ambient light and darkness directly influence sleep cycles.

As most of us are aware, whether or not we are sleepy depends not only on circadian rhythms but also on the physical state of our bodies. Thus sleep scientists consider sleep to be governed by two separate processes: a circadian cycle via the suprachiasmatic nuclei and a homeostatic cycle. The homeostatic cycle cannot be traced to a single brain locality. As befits a process that takes into account the ongoing state of the mind *and* body, a variety of hormones, circulating substances, and neurotransmitters play a role in the homeostatic regulation of sleep. For example, the role of adenosine, a molecule that is a by-product of energy metabolism, in sleep regulation has been appreciated since the early 1900s.[18] Caffeine keeps people awake because it blocks the activity of adenosine in the brain (the average daily caffeine intake of Americans exceeds the dose of caffeine known to increase wakefulness). Adenosine levels in the brain drop during sleep, and sleep deprivation leads to increases in adenosine levels, which in turn promote sleepiness. One way adenosine encourages sleepiness is to inhibit neurons that use the neurotransmitter acetylcholine, which in turn inhibits the brain's arousal pathways that depend on this neurotransmitter. The adenosine-acetylcholine pathway is just one part of the homeostatic control of sleep.

Sleep itself is divided into two main types: non-rapid eye movement sleep (NREM) and rapid eye movement sleep (REM).[19] Normal sleep begins with NREM sleep and then alternates with REM sleep, cycling back and forth every ninety to one hundred minutes. NREM sleep is divided into four stages. Stages 3 and 4 of NREM sleep, which are also known as slow-wave sleep (based on the characteristic brain waves it produces on electroencephalograms [EEGs]), form the deepest portion of sleep, and are the stages when sleep problems such as night terrors, bedwetting, or sleepwalking, are most likely to occur. The more sleep-deprived a person is, the longer stages 3 and 4 will be, suggesting their importance in compensating for lack of sleep. REM sleep differs from NREM sleep in that some of the EEG output is indistinguishable from an awake state. And although the eye muscles are activated during REM sleep, the skeletal muscles are inhibited, leaving the body in a temporarily paralyzed state called atonia. Dreaming occurs in both NREM and REM sleep, but it is more vivid and realistic in REM sleep.

We all know that some people need more sleep than others. I can get by on a little less than eight hours a night, while my wife requires a little more. My younger son, an early teenager, needs nine hours or more to function well. If our household had an infant and an elderly person in it, we might see them sleeping sixteen and six hours per day respectively. Within the human species, there is variation in sleep duration at the individual level and by age. Nonetheless, for normal adults, a figure around eight hours characterizes the average amount of sleep humans as a species need. Of course, whether we get it is another matter.

Other species need different amounts of sleep. Basically all animals need sleep in some form, but only mammals and birds engage in cyclical NREM and REM sleep. This pattern of sleeping evolved independently in the two groups (convergent evolution), since it was not present in their common reptilian ancestor. The convergent sleep patterns of birds and mammals go along with convergences in brain structure and function.[20] Sleep researchers in general agree that sleep is necessary for its restorative effects on the brain and nervous system, and the multiphase form of sleep seen in birds and mammals undoubtedly somehow reflects the particular needs of larger, more complexly wired brains.

All animals may need sleep, but they clearly do not need the same amount of sleep.[21] Most anthropoid primates (and many other mammals that are generally active in daylight) need somewhere in the neighborhood of nine to eleven hours of sleep. Nocturnal owl monkeys, sleeping up to seventeen hours a day, are very sleepy mammals, a characteristic they share with other nocturnal mammals, such as brown bats, giant armadillos, and the North American opossum, who all sleep more than eighteen hours per day. At the other extreme, large, grazing, herbivorous mammals appear to sleep the least, with horses, donkeys, and elephants needing three to four hours of sleep per day. Giraffes get by on less than two hours per day. What these varying figures show is that while sleep is universal, both the amount of and presumably the behavior surrounding sleep are shaped by how an animal goes about living its life. Sleep patterns evolve in response to an animal's ecology and environment.

The group of mammals known as cetaceans—whales, dolphins, and their kind—illustrate this in a striking way. Cetacean sleeping is unique among mammals.[22] The amount of time they sleep is not unusual: studies of different species of captive dolphins and the beluga whale indicate that they sleep anywhere from eight to eleven hours over a twenty-four-hour period. However, when they do sleep, it is only with one hemisphere of the brain at a time, a phenomenon known as unihemispheric slow-wave sleep (USWS).

Why have they evolved this way of sleeping? It could be to keep an eye out for predators or to allow mothers to keep an eye on their young while they both rest. Sleeping with half their brains at a time also allows cetaceans to keep moving. Some researchers argue that they do this in order to continuously be able to come to the surface to breathe oxygen. Oleg Lyamin and his colleagues accept that breathing may have been an important factor in USWS evolution, but they also point out that there are other solutions to the problem of getting oxygen.[23] Instead they favor the idea that USWS evolved in cetaceans to keep them moving, which helps keep them warm. Water is a much more efficient medium than air for heat transfer away from the body, and warm-blooded marine mammals have to deal with this issue on an ongoing basis. Blubber is one way cetaceans keep warm, protecting the internal organs from heat loss. During USWS, cetaceans keep moving, contracting their muscles and thereby generating heat. Cetaceans also do not appear to have REM sleep; thus their bodies are never subject to atonia, further bolstering the idea that cetacean sleep physiology has been shaped by the need to remain warm (via movement).

Ecology and sleep habits are linked by evolution. Cetacean sleep reflects how cetaceans have adapted to life in water. From a physiological standpoint, great ape sleep is not an unusual form of mammalian sleep. But the nests apes build are unusual, and they reflect the apes' adaptation of their sleep to their environments.

The Risks of Sleep

It is a fact of nature that some animals eat other animals. One way to make life as easy as possible for predators would be to have their prey immobilized and rendered insensitive to the surrounding environment. In other words, a sleeping animal makes for an easy meal. The restorative powers of sleep must really be substantial in order to make up for the risks of being eaten (or killed) while sleeping.

So with sleep both a necessity and a risk, there is a tendency for animals to evolve ways to reduce their risk of being eaten while sleeping. Sleeping less is obviously one solution, but given that there are undoubtedly real constraints on the minimum amount of sleep an animal needs, it could never be a universal solution. Hiding away while sleeping, in a burrow or a hollow in a tree, works well for small animals but less well for larger ones. One of the benefits of being a social animal is the extra pairs of eyes or ears that serve as predator detectors for the whole group. Social dynamics must sometimes play a role in shaping the sleep habits of social animals.

Biologists Niels Rattenborg and Steven Lima have looked in depth at the problem of sleep and predation.[24] One possible solution to the problem of being eaten while sleeping they have explored is to sleep with only part of the brain at a time. As I mentioned above, one of the explanations of unihemispheric sleep in cetaceans is that it serves to help maintain vigilance against predators. Although cetaceans are unique among mammals in having this ability, it turns out that some birds sometimes also sleep with only half their brain at a time and with one eye open. Rattenborg and Lima did an experiment with four sleeping mallard ducks. They arranged them in a row and monitored their sleep. They found that the ducks on ends of the row engaged in more unihemispheric sleep than those in the middle, and they also tended to sleep with the outward-facing eye open. Since the outside ducks would have perceived themselves to be more at risk of predation (although this would have only been the case if the experimenters had gotten hungry), this suggests that their unihemispheric sleep was an antipredation adaptation.

Unihemispheric sleeping is a form of partial sleep, in that only half of the brain is engaged in sleep at one time. Given that sleep has multiple components, other forms of partial sleep are possible by engaging different brain regions in different ways. Some people become uncomfortably aware of this during sleep paralysis, when conscious awareness overlaps with the atonia of REM sleep—they wake up out of REM sleep but (temporarily) can't move. Rattenborg and Lima asked, why don't animals make more use of partial sleep as a means of predation avoidance? Their theoretical modeling suggests that while partial sleep may allow for increased vigilance during rest, it can also compromise the restorative powers of sleep. Global sleep means that sleep can be shorter, which can itself be a defense against predation, while partial sleep provides at best a compromised form of vigilance.

In mallard ducks, sleeping with part of the brain is a trade-off between sleep quality and predation awareness. Ape nest-building may reflect another kind of sleepy trade-off to prevent predation. Primatologists have recently studied chimpanzees in different sites to see if nest-building patterns are affected by predation pressure.[25] Chimpanzees do sometimes get eaten by predators, although it is apparently not a common occurrence, and direct observations of great apes being taken by predators are exceedingly rare. Great ape remains have been observed in the feces of leopards and lions, and indirect evidence at some sites suggests some predation pressure. Chimpanzees are certainly aware of the dangers posed by big cats: there are many accounts of chimpanzees throwing rocks and sticks at leopards and lions, and some groups may also have specific alarm calls that they use when they see big cats.[26]

With the exception of very large gorillas, almost all primates sleep in trees. Sleeping at the ends of branches may be one way that non-nest-building primates avoid predators.

The nests of the great apes may improve upon the more common primate model by providing a physical barrier or by shielding view of the apes from below. Looking at chimpanzees specifically, primatologists have attempted to show that nest-building protects against predation by comparing nest distribution patterns in field sites that have greater and lesser amounts of predation pressure. Chimpanzee living sites have varying amounts of predation pressure, for a variety of reasons. In some parts of Africa, humans have basically eliminated big cats, providing a low-predation environment for chimpanzees; forested sites located on the edges of savanna grasslands tend to have more predation pressure than densely forested areas.

According to Fiona Stewart and Jill Pruetz, there are several ways that chimpanzee nesting habits could reflect relatively increased predation pressure.[27] Chimpanzees could choose to nest in more dense forest patches; they could nest closer together or tend to do so in the same tree; they could choose nesting sites that have more "escape routes" (i.e., to another tree rather than to the ground); and they could nest higher up in trees (after controlling for overall tree height). In their study, comparing chimpanzee sites in Tanzania and Senegal, Stewart and Pruetz found that in the more predated-upon site, chimpanzees built nests higher in trees and on more peripheral branches. Other studies have produced somewhat mixed results when looking at the effects of predation on chimpanzee nesting.[28]

Many factors influence great ape sleeping patterns and nest-building practices.[29] It is important to keep in mind that the large-bodied, social great apes have clearly evolved a range of strategies for avoiding predation—so much so that today even a great ape group that has a relatively high amount of predation pressure will not appear to be all that different from one that has none. This makes it difficult for primatologists to show that variability in nesting alone corresponds to variability in predation pressure. Nonetheless, Stewart and Pruetz assert that nest site selection in chimpanzees is influenced by the dangers posed by terrestrial predators; this seems reasonable based on the available data and the fact that ape nests fulfill the general goal most animals have of providing a predator-safe place to sleep. No one wants to be eaten while sleeping, and the nests of the great apes help them avoid becoming a substantial meal for some substantial predator.

A Better Night's Rest

In the modern world of the early twenty-first century, if you speak to a physician or a mattress salesperson about having trouble sleeping, the term *sleep hygiene* is likely to come up. Are you doing the things you need to do to prepare your body and sleep environment to increase your chances of having a good sleep? Advice about sleep hygiene ranges from the commonsensical to the clinical. Get a firm mattress and a supportive pillow. Avoid stimulants such as caffeine close to bedtime, and also do not eat too much too late at night. Exercise, but in the morning or afternoon, not before going to bed. Avoid napping, and adhere to a regular bedtime and waking time. Expose yourself to natural light during the day to help set those circadian clocks in your brain. Make sure that your sleeping space is not too bright or loud or full of distractions, especially electronic ones, and that the temperature is comfortable.

The great apes in the wild live in a considerably more natural state than do contemporary human city-dwellers. The apes' sleep hygiene is in some ways much better than ours. They get plenty of exposure to natural light, and their sleep schedule is dictated by the rise and fall of the tropical sun. They are not prone to obesity, live free of electronic distractions, and rarely ingest concentrated forms of brain stimulants. But tree nest-building may serve to enhance their sleep hygiene and comfort in other ways. For example, I think not having elephants around in one's sleeping space is in general a good idea, and gorillas seem to agree. Caroline Tutin and her colleagues studied nesting in a group of gorillas in Gabon.[30] They found that while the gorillas generally preferred to sleep in nests on the ground that were made from soft and flexible herbaceous plants, when elephants were around, feeding on the same plants as the gorillas, the gorillas were more likely to make nests in the trees. Heavy rainfall also encouraged them to nest in the trees. So sleep hygiene recommendations for these gorillas would include finding a dry, elephant-free place to make a bed.

Chimpanzees are also sensitive to their environmental conditions when they choose a place to sleep. Humidity in particular seems to be something that chimpanzees look to avoid when sleeping. At a mountainous study site in Guinea, West Africa, Kathelijne Koops and her colleagues observed that the more humid the conditions, the more likely it was that chimpanzees would seek trees at higher altitude in which to make their nests.[31] They also tended to make their nests higher up in trees when it was more humid. There was no evidence in this study that choice of chimpanzee nest sites was influenced by the avoidance of parasites or other small pests. Most researchers agree, however, that the apes' habit of changing nests every night is probably an effective way of avoiding certain parasites.

Trying to find a comfortable mix of temperature, humidity, parasites, and elephants would influence the sleep hygiene of the great apes whether or not they made nests. But comfort is important for achieving a truly restful sleep, and that is where the nest-building habit of the great apes may have started. Support for this notion comes from a recent study comparing the sleep of baboons and orangutans in the Indianapolis Zoo.[32] David Samson and Robert Shumaker analyzed hundreds of hours of video of the sleeping primates, and found that by every measure (duration, sleep motor activity, total waking time, sleep fragmentation, and so on), the orangutans had deeper and more efficient sleep than the baboons. As is the habit of most monkeys, the baboons slept upright, sitting on their ischial callosities (or butt pads, as they are otherwise known). In contrast, the orangutans generally slept on their backs, in a relaxed, "insouciant" position, looking very comfortable. Although they did not have materials to make proper nests, the orangutans invariably made use of materials in their environments (bedding, straw) to create a bedlike sleeping platform; this kind of behavior was never seen in baboons. Samson and Shumaker suggest that the "next-day" cognitive benefits of deep sleeping in our ancestors could have played a role in making possible the enhanced cognitive abilities that emerged in human evolution.

Modern great apes evolved from a smaller, tree-sleeping ancestor. For whatever reason, larger body size evolved in this lineage, such that lying on the tops of branches or sitting upright against tree trunks were no longer effective means of resting. The ape's bed likely evolved in part to enhance comfort in this arboreal environment, a habit that may have been essential to later human cognitive evolution. The evolution of nest-building also points to the deep roots for controlling and creating our human sleep environments. As much as comfort, the sense of security that comes with creating a sleeping place that is in some ways removed from the dangers of the natural world may provide the evolutionary link between the ape's nest and the human home.

Falling from the Nest

I would not want to pay $70,000 for a faux bird's nest large enough to sleep in, but evidently a "corporate raider" from the wealthy southern California suburb of Bel Air was willing to shell out that much money for one of artist Roderick Wolgamott Romero's signature human-sized nests. In fact, currently, artists and ecologically oriented furniture makers making nestlike structures for sleeping or communing seems to have become a "thing."[33] In Big Sur, not far from Esalen, the famous 1960s countercultural retreat, there is a resort where a customer can pay $110 per night to sleep in a large nest, feeling the Pacific winds blow through the gaps in the eucalyptus twigs from which the nest is woven. Couples have honeymooned in this nest; babies have been conceived in it.

Emulating animal architects to build structures in which people can regain access to a more natural world is a somewhat charming notion. However, while birds and their nests are evolutionarily closer to us than organ-pipe wasps, our natural past did not really take us through the woven nests built by passerine birds. As we've seen, the nests in our past were simpler, but perhaps cushier, and meant for short-term sleep and rest. The bird's nest is a structure primarily concerned with reproduction and the rearing of young. Complex bird nests converge on constructions we humans use to adapt to the natural world. The novelty of sleeping in one, the exposure to the sounds and smells of the night air, can probably be obtained without resorting to sleeping in wicker suspended in a tree.

But would we get a more realistically natural experience sleeping in an ape-type nest in the wilds of suburbia? Certainly, there would be more exposure, but a nice, mattress-like bed of soft branches and leaves might be more comfortable for recumbent sleeping than a giant bird's nest. I would be tempted to try it, but since I am between a chimpanzee and a gorilla in weight, I would have to make a ground nest. Unfortunately, we have chiggers in our area, and so a ground nest is not appealing.

As a thought experiment, however, I am tempted to conclude that sleeping in an ape's nest would have no more special appeal than sleeping in a bird's nest. Yes, there would be the general novelty of sleeping in the night environment and in a structure meant for someone built more like us and less like a bird. But we humans have put ape nests behind us. What we have likely retained, however, is the sense of place associated with making a nest every night. Along with a sense of place come feelings of security, comfort, and control. I believe that in the ape's nest we find the foundation for the evolution of a feeling for home in human beings. As we will see in the next chapter, the home itself, as a place where people live over an extended period of time, was still to come.

Chapter Three

THE STONE AGE TRANSITION TO HOME

Evolution is about transitions, and when anthropologists reconstruct the last six million years of human evolution, they often focus on a few key transitions.[1] These transitions encompass the anatomical and behavioral changes that separate us from our nearest ape relatives. Some of these transitions are pretty obvious: the great apes spend most of their time walking on four legs, while humans walk on two; humans have large, bulbous heads with flat faces, while apes have smaller heads and more protruding faces with large teeth; we are the naked apes, they are the regular old hairy apes; and so on.

But "obvious" depends on the available data. For more than a century, scientists strongly believed that another important transition in human evolution involved the development of tool use. However, after just a few years of studying chimpanzees in the wild, Jane Goodall observed that they were also capable of making and using tools. Other researchers have since shown that chimpanzees have a fairly rich and varied material culture, making humans a little less unique in this regard.[2] The tool use transition, while still important, was a little less essential to becoming human than once thought.

My focus here is on the transition from the ape pattern of living in temporary sleeping nests within a home range to the more human pattern of living a life based around a single home. Like any complex transition that happens over time—in this case, over a very long time—it is not always clear when the transition has been completed. It's obvious to all when it is over, but it can be very hard to identify a threshold that signals completion. Let me offer a homely analogy.

Over the past twenty-plus years, I have lived with my family in three different houses. Each move to a new house was a pretty big one, both in distance and in the change in culture and climate. With each move, we left behind a *home* and found ourselves instead in a new place that at first was not a familiar, emotionally comforting oasis from the world around us. Over time, fortunately, each of these new houses became a home for us. We made this transition without being fully aware that it was happening. You have probably had this experience. One day the relatively new place you live in still feels strange and unfamiliar, but sometime later, you realize that the newness has worn off, the strangeness is gone, and that for better or worse, you are now at home. Although relocating to a new place can happen overnight, the more complex cognitive and emotional transition to home typically takes longer, developing in small increments over time.

Here is an evolutionary example of a tricky transition. A topic I have thought and written about quite a bit over the past several years is the evolution of the human brain.[3] A large brain is something that unequivocally distinguishes us from our closest ape relatives. The size of the brain has tripled over the course of human evolution, resulting in an average human brain having a volume of 1,300–1,400 cubic centimeters (cm^3), compared to 400–500 cm^3 for a chimpanzee. Although brains typically do not fossilize, there is, unsurprisingly, a strong correlation between brain size and the size of the cranium in which the brain sits (measured as cranial capacity). So if nothing else, the fossil record provides us with information about the evolution of brain size. It tells us that for the first several million years of the evolution of hominins (members of the human family after splitting from the apes), the brain remained more or less ape-sized. However, starting about two million years ago, hominins with brains somewhat larger (especially for their body size) than would comfortably fit in the ape range started to appear. Within a few hundred thousand years, brain volumes around 1,000 cm^3 began to appear. Over the last million years or so, brain volumes have increased until reaching modern dimensions.

The transition from ape-sized to human-sized brain therefore took place over a couple of million years. But when did the brain become more human-like than ape-like? This is a somewhat arbitrary question that students of human evolution have wrestled with for more than a century. On one hand, everyone recognizes that to draw a line in the neurological sand and say, "Once a brain reaches this size, it is more human than ape," is probably unrealistic. Yet there has been a strong tendency to place any fossil hominin with a brain that clearly transcends that of the great apes (i.e., 600 cm^3 or more) into our own genus, *Homo.* In other words, a line has been drawn, suggesting that a threshold, a "cerebral Rubicon" as it has been called, has been crossed. This line may be helpful for naming new species or helping to identify if a species is a true human ancestor, but from the brain-evolution perspective, it is likely an arbitrary division in a continuous process.

Identifying important transitions in human evolution is one thing; figuring out how and why they happened is another. Anthropologists make hypotheses, sometimes very strong ones, about human evolution based on the wealth of data from the contemporary world of living humans, great apes, and monkeys. Ultimately, however, data and hypotheses need to be placed in a real timeframe, based on evidence that emerges from the past—we want to know the single and unique story of how people came to be. This is not always easy; what is preserved of the past is just a pale representation of what once was. Nonetheless, paleoanthropologists and archaeologists, with a little help from geologists—who are continually improving the methods of dating ancient materials—are making real progress toward providing a framework for retelling the biological history of the human species. Home emerged from the evolving lives of our hominin ancestors. Fire and family, sex and meat, the relatively slow road to adulthood that human children take—these all had their roles in the evolution of home. Looking backward, it may be difficult to pinpoint a time—an evolutionary line that we crossed—when we transitioned into being a home-living and -loving species. Eventually and undeniably, however, we as a species arrived at home. Before looking at that journey in more detail, however, I'll provide some of the evolutionary background for this transition to home.

From Four-Legged Ape to Two-Legged Hominin

Any chimpanzee and any human alive today can trace their shared common ancestor to an ape who lived somewhere in Africa about six million years ago.[4] This common ancestor would look to us more or less like a great ape. It would not be a chimpanzee, of course, since they have been evolving just as we have. But humans have certainly diverged more from that common ancestor than the chimpanzee has. In fact, our ability to reconstruct the behavior and biology of the common ancestor, or of any early hominins, is enhanced by our knowledge of contemporary great apes in conjunction with the fossils and the geological settings in which they are found.[5]

Early anthropologists were certain that the triad of large brain size, bipedal locomotion (walking on two legs), and making stone tools all evolved together in some interactive way to yield modern people. These factors were all critical, of course, but the information coming out of the quarries, canyons, valleys, and deserts of East and Southern Africa has shown that they did not all evolve simultaneously. Human evolution has also not been a straight shot from the last common ancestor with the chimpanzee to people today: that transition can be better represented as a bush than as a tree with a straight and narrow trunk.

Although some older fossil remains have been somewhat controversially assigned to the subfamily Homininae, the best widely (but still not universally) accepted candidates for earliest hominin belong to members of the East African genus *Ardipithecus.*[6] If you were to see an *Ardipithecus* adult today, it would look quite ape-like, which is not too surprising, since that's basically what it was. It would weigh around 110 pounds and have a cranial capacity of about 300–350 cm^3, which would be on the smaller side for a great ape today. Like great apes today, and unlike modern humans, *Ardipithecus* had large canine teeth, but the wear pattern on their canines may link them more to later hominins than to the great apes. Most critically, certain aspects of *Ardipithecus* anatomy, such as the pelvis and the position of the skull atop the vertebral column, indicate a clear step toward bipedality. The arms of *Ardipithecus* were still long and adapted for life in the trees, but that is to be expected in a form that is still basically an unusual ape.

Ardipithecus gave way to a diverse group of hominins known as the australopithecines (after the genus name *Australopithecus,* which was first proposed by Raymond Dart in 1924). Evolving from their presumed ancestor *Ardipithecus,* the australopithecines lived in East and Southern Africa between about four and one million years ago, likely occupying woodland and woodland/grassland boundary areas rather than the open savanna grasslands that were once thought to be the incubation environment of early humans. As many as ten different species can be included among the australopithecines; recent discoveries, along with differing perspectives among anthropologists, have complicated what was once a simpler picture.[7] As a group, the australopithecines were bipedal, relatively small-bodied compared to modern humans (66–110 pounds), and had brain volumes that generally fit within the range for modern great apes (400–500 cm^3). Given their small body size compared to the contemporary great apes, the australopithecines had brain sizes that were relatively large; a larger brain-to-body-size ratio is generally considered to be a possible sign of the increasing cognitive development of australopithecines.[8]

The diversity among australopithecine species teaches us an important lesson: human evolution went down some unexpected paths. One whole group of australopithecines, known as the "robust" australopithecines (sometimes placed in their own genus, *Paranthropus*), is characterized by the presence of massive molar and premolar teeth. Robust australopithecines used these large back teeth to eat tough grasses and hard objects such as seeds. Studies indicate that they ate other foods, as well, but may have initially evolved large molars as a feeding adaptation for when other foods were scarce.[9] They generated tremendous chewing forces with their large jaw muscles, which required a reworking of their skulls with the addition of crests and ridges for muscle attachments. They were robust in their heads, not their bodies, which were about the same size as those of other australopithecines.

The robust australopithecines show that our family tree (or bush) was surprisingly diverse. In addition, they provide a valuable insight into how human evolution unfolded:

bipedality was not destiny. In other words, we modern humans, with our large brains, spoken language, complex cultural traditions, and sophisticated material culture, were not an inevitable outcome of a process that started when an ape became two-legged. The robust australopithecines showed that there was another way, one that never would have been predicted based on what scientists know about modern humans and apes. The robust australopithecine way worked for a while (their kind lasted from about 2.7 to 1.2 million years ago), but ultimately, they could not compete with the two-legged apes who went another direction.

The Human Transition

About 2.6 million years ago, a hominin living in what is now Ethiopia knocked some flakes off a stone core and used them as cutting or scraping tools.[10] We do not know what species the tool-maker belonged to: he or she could have been a member of an australopithecine species or perhaps the very earliest member of genus *Homo.* No matter who made the tools, their appearance marks the beginning of an extraordinary transitional era in human evolution. At 2.5 million years ago, the hominin fossil record contains a variety of australopithecines, whose direct relationships to us today are either unclear or definitely more at the cousin (rather than great-great-great-etc.-grandparent) level. By about 1.8 million years ago, however, the fossil record unequivocally reveals that a more human-like lineage had been established.

Probably not coincidentally, also around 2.5 million years ago, glacial cycles became more pronounced, leading to changes not only in climate but also in sea level. Archaeologist Richard Potts argues that our increasingly cognitively sophisticated ancestors, more so than other hominins, were able to thrive and expand their range due to their behavioral flexibility during these ecologically variable times.[11] Our own genus *Homo* emerged in Africa by around 2 million years ago out of this climatic instability. The first members of the genus, which were given the name *Homo habilis* (or "handyman") by Louis Leakey and his colleagues, were not all that different from the (nonrobust) australopithecines.[12] To us, they would probably have looked like shortish apes walking on two legs, with arms that might have still seemed a bit too long to be truly human, but their crania had an average volume of about 650 cm^3—substantially larger than the average for any great ape or australopithecine. *Homo habilis's* relatively larger brain is the main reason most anthropologists see them as the most likely fabricators of the stone tools found from this time period.

Several possible early *Homo* species may have been living around two million years ago, giving paleoanthropologists plenty to argue about today. But as we move forward in time, a species called *Homo erectus* indisputably emerges as the primary descendant of early *Homo* species. In its "classic" form, *Homo erectus* is one of the most well-represented human ancestors. First discovered more than one hundred years ago in Asia (Java and China), *Homo erectus* is now known to have lived throughout the Old World, from about 1.8 million to as recently as possibly fifty thousand years ago. It clearly overlapped with australopithecines and early *Homo* when it first appeared, and with early and modern *Homo sapiens* at the end of its run. *Homo erectus*'s cranial capacity ranged from under 900 cm^3 to about 1,100 cm^3.[13] The skull of *Homo erectus* has several features that distinguish it from both early *Homo* and australopithecines as well as modern humans.[14] Although its capacity is relatively large, the shape of the cranium is long and low, and the forehead is sloping. The cranium is thick-boned compared to earlier and later species, and the brow ridges can be quite massive. *Homo erectus* individuals were not as large as modern humans, but they were substantially larger than australopithecines or early *Homo.*

The *where* of *Homo erectus* is almost as important as the *what.* They evolved initially in Africa, but by 1.7 million years ago, they were beginning their migration out of Africa. The earliest non-African evidence for *erectus*—for any hominin—comes from a site called Dmanisi in the Republic of Georgia. Eventually, *Homo erectus* would become the first hominin species to demonstrate the ability to traverse large distances and survive in nontropical environments. As I will discuss later, fire may have been part of the technological arsenal that allowed them to do this. They are also associated with a specific tool form, the hand axe—a teardrop-shaped, multipurpose implement chipped with symmetrical precision from stone—which clearly represents a cognitive advance on the simpler tools of early *Homo.* In addition, just as we expect for earlier hominins, *Homo erectus* likely made use of a whole range of nonstone implements that would not be preserved in the archaeological record.

I have written elsewhere that *Homo erectus* is in some ways the most "distant" hominin species, at least when it comes to trying to understand how they lived.[15] For the earlier hominins, we can use the great apes to model aspects of their lives; this is not to say that they were exactly like great apes, but in evolutionary terms, early hominins certainly were more similar to them than *erectus* was. At the other end of the time scale, we can use ourselves as a model to gain insights into how Neandertals and our other, more immediate, ancestors might have lived. *Homo erectus* sits somewhat isolated and distant from ourselves and our ape cousins. Nonetheless, the foundations for what we think of as humanity were surely laid during *erectus* times. Language, a cultural existence, a reliance on technology as much as on our own physicality—these human features likely shaped, to a greater or lesser extent, the daily lives of *erectus* individuals. The transition to fully modern humanity was likely well under way during *Homo erectus* times.

I think that a feeling for home also began to take form during the critical early *Homo–Homo erectus* stage of human evolution. Besides providing a protected setting for sleep, home is a place where people prepare, eat, and share food. It is also the physical and emotional center of reproductive life. Anthropologists have used different kinds of information, from the archaeological and paleontological records, as well as from observations of the few remaining contemporary hunter-gatherers, to try to reconstruct how human patterns of eating and reproducing may have evolved. Since direct evidence of the evolution of home can be hard to come by, we can rely on some of these other insights into the human ways of food and sex to try to learn more about how home, in a cognitive sense, may have been built.

The Home Base Insight

In the modern world, home can be defined by the things we don't do in it. More often than not, people work, find food, become educated, are healed, and meet reproductive partners outside the confines of the home. Aided by technology for venturing ever further into both the physical and virtual worlds, people have expanded the size and scope of a home range beyond its classic zoological limits. But home itself still remains a privileged place in this potentially vast home range. It is a base from which people engage with the wider world around them.

In the 1970s, archaeologist Glynn Isaac identified the development of the "home base" by early *Homo* around two million years ago as a critical innovation and signpost in human evolution.[16] Isaac argued that the home base reflected important changes in social behavior, which signaled a switch from an ape way of living to a more human way. Isaac's model for the evolution of the home base reflected his own view of a multifaceted Paleolithic archaeology as a truly anthropological discipline that would draw on knowledge of other primates and modern humans, plus experimental studies, to create evolutionary hypotheses about the past. Isaac's approach has been extremely influential, and he remains a relevant figure in archaeology nearly thirty years after his untimely death.

Isaac developed his home base model by emphasizing the differences between the great apes and traditional hunter-gatherer humans in how they go about obtaining and dealing with food. For humans, food is a "corporate responsibility," while in the great apes it is a much more individualistic matter. Chimpanzees hunt cooperatively and can share the meat from those hunts, but at most only 10 percent of their diet comes from animal foods (and much of that from insects rather than cooperative hunting).[17] In contrast, Isaac emphasized the greater importance for modern humans of hunting and animal protein, both as a percentage of the overall diet and as a central aspect of social life. This social life is facilitated by language, a communication system not found in the great apes. Isaac also pointed out the human propensity for carrying objects, including food and tools, something that quadrupedal apes are much less adept at doing.

All of these various aspects of humanity came together for Isaac, both literally and figuratively, in the home base. Isaac points out that all human societies have places where people meet up after they have either individually or collectively done whatever they needed to do in the surrounding environment. People carry raw materials to the home base to make them into tools; food is carried to the home base for processing and distribution; the home base is where people talk to one another to make plans about what they are going to do in the outside world. The home base is thus the precursor to home.

That our hominin ancestors went through a home base stage of living before developing something more like a home is quite plausible and uncontroversial.[18] In fact, given what we know about how modern apes and humans live in their environments, such a stage was probably unavoidable, and it is interesting to think about how it might have taken shape.

But Isaac's home base model was more explicit than this: he identified the origins of home bases in the two-million-year-old early *Homo* archaeological sites of East Africa. Isaac drew three main conclusions from his work on these sites, two of which are now generally accepted. First, he asserted that the early stone tool-makers carried and transported tools or raw materials; second, that these tools were used to butcher large animals (something that is demonstrated by cutmarks made by stone tools on animal bones).

Isaac's third assertion was much more controversial. He argued that the distribution of stone tools and animal bones on the ground showed that these hominins carried the animal carcasses to a central place for butchering and sharing. In Isaac's view, the concentration of these materials in the archaeological site was a direct representation of a home base from two million years before. Alas, this interpretation of how these materials came to be deposited in this way was not well received, and was subjected to criticism both withering and judicious.[19] The formation of these archaeological sites resulted from the interplay of past geological, ecological, and zoological factors, along with hominin activity. Simply put, other archaeologists had a hard time seeing home bases (or "central places," as Isaac sometimes referred to them) in the deposits of Olduvai Gorge.

Isaac's harshest critic may have been Lewis Binford, an archaeologist who never shied away from an academic debate. When I was an undergraduate at Berkeley in the early 1980s, I saw a lecture by Binford, with Isaac (then a professor at Berkeley) in attendance, in which Binford addressed the home base model. As I recall, he pulled no punches, and although Isaac had some rebuttals on particular points, in the end, the home base model did not hold up well. Despite being in a large lecture hall, it was a very uncomfortable situation. Not long before, Binford had written that Isaac saw early *Homo* with their home bases "as a kind of middle-class genteel protohuman," a characterization that was as unfair as it was revealing.[20] I say "revealing" because I think that it shows just how powerful the concept of home is. Beyond the archaeological merits of the argument, Binford reacted to the notion of home in Isaac's model in terms of his own life as a middle-class academic (albeit one who was not so genteel at times).

Despite the fact that Isaac may have overreached in his interpretation of these very early archaeological sites, I think that his home base model is very useful in highlighting the critical transition in early *Homo* from a great ape way of using the environment to one that would ultimately be based around the home. Isaac's model has also spurred interesting research getting at the issue of how homeish activity may be preserved in ancient East African archaeological deposits. For example, Ellen Kroll's careful examination of how artifacts and animal remains are distributed in these sites has led her to hypothesize that shade trees were a potentially important feature on the landscape.[21] Early *Homo* may have sought to escape the heat of the sun by working and resting in the shade of large trees, leaving behind detritus (stone tools, flakes, animal bones) over the course of multiple visits. Humans working in hot, open climates seek whatever shade is available, and it is likely our ancestors did the same. It's not home or a home base, but it may have been a beginning.

A completely different approach to exploring the transition to home comes from archaeologist Jeanne Sept.[22] In order to better understand how traditional camps and homes leave their mark on the ground, some researchers have studied the refuse and imprints left by contemporary hunter-gatherers. Sept noted, however, that many of these habitations would not leave any lasting remains in the archaeological record, and perhaps more critically, she argued that using the human pattern as a standard would bias interpretation of past (possible) home site remains. In order to provide a more balanced perspective, in the late 1980s she embarked on one of the earliest studies of chimpanzee nest "archaeology," mapping their locations and use patterns and the residue associated with them. This research has led to numerous studies of nesting behavior in the great apes, some of which I discussed in the previous chapter. One outcome of this research is that we now know that reuse of nesting sites in the great apes is not random, but can be linked to specific environmental or social conditions. Sept also suggests that given that chimpanzees hunt small animals, greater attention by archaeologists to small-animal hunting in early *Homo* could provide another means of evaluating their behavior with reference to great ape behavior rather than modern human behavior.

Nesting behavior takes us back to the trees. It is important to remember that the bodies of many earlier hominins, including early *Homo,* while fully adapted for bipedality, also appear to retain adaptations for tree climbing. These included features such as long, curved fingers and long arms relative to leg length. Earlier hominins may or may not have spent significant time in the trees—we can't know that. But we do know that one thing that would be hard to do while in a tree is to make and maintain a fire. So if early *Homo* or its immediate predecessor initially spent a significant amount of time in trees, the adoption of fire might have brought them to the ground.

Which Came First, the Hearth or the Home?

> *I secretly understood: the primitive appeal of the hearth. Television is—its irresistible charm—a fire.*
>
> —John Updike in *Roger's Version* (1986)

In the early 2000s, a group of anthropologists and archaeologists from UCLA's Center on Everyday Lives of Families (CELF) intensively studied the material culture (i.e., stuff) in the homes of thirty-two families in the Los Angeles area.[23] They looked at how the residents of those homes related to the things they assembled around themselves. Unsurprisingly, they found that televisions loomed large in the material culture of American homes (and in homes throughout the developed world, of course). The CELF researchers observed that in the modern home, the location of the television often defines the main arena for social interaction. These areas are in turn shaped and defined by the position of the television. Second and third televisions—over half of American homes have more than one television—are usually found in bedrooms. People thus put their televisions where they gather and where they rest.

Like Updike, the CELF researchers reached for an inescapable metaphor: "For all of its influence on the design and organization of space, the TV may as well be a hearth, which until quite recently in human history exerted the most influence on the spatial distribution of social interactions and activities inside homes."[24] Actually, maybe I am mischaracterizing these statements as metaphorical. As a centralized source of light, sound, contained movement, and warmth, the television is literally firelike, which is what Updike is getting at. For the modern house-dweller, understanding the importance of the hearth for less technologically privileged peoples may be achieved by contemplating life without television.

Hearth and home. The connection is ancient and universal. Archaeologists looking at the earliest urban sites—dating back ten thousand years—identify houses, among the relatively undifferentiated structures of mud and stone, by the presence of a hearth or evidence of controlled fire.[25] Archaeologists studying the array of artifacts and refuse spread over the living floors of the nonurban sites of hunter-gatherers or traditional agriculturalists also look for evidence of fire. The presence and position of fire helps them identify how the living space may have been oriented. In the same way, when we walk into a room in a modern house, devoid of its occupants, the position of the television lets us imagine the typical positions of the residents as they use and interact within that space. So if hearth in some way corresponds to dwelling, then identifying the origins of the controlled use of fire is very helpful toward understanding the evolution of the feeling for home. Unfortunately, identifying fire in the archaeological record is, like so many things, "obvious" in the best and most recent cases, but gets a bit trickier when we go back in time. Eventually, we reach a point where one researcher's controlled fire is another's accidental burning.

Many archaeological sites older than several hundred thousand years of age have evidence of fire in them. However, fire has many natural sources, including lightning strikes and volcanic activity, which makes it difficult to discern whether or not hominins were responsible for the evidence of fire left among their tools and the bony remains of their meals. Also, in Africa, the drier woodland savanna environment in which early *Homo* lived would have been quite susceptible to natural fire, which of course complicates the matter of detecting hominin-controlled fire. Archaeologists have a long history of claiming evidence of controlled fire in sites of great antiquity; there is an equally long history of other archaeologists arguing that the evidence of controlled fire is insufficient, even if fire was present at the site.[26]

Working backward from the present, the evidence is clear that both modern humans and Neandertals, living at least 100,000 years ago (in Africa and Europe respectively), used fire not only for cooking but also to shape raw materials (stone, bark pitch) as an aid in tool-making.[27] The common ancestor of modern humans and Neandertals (more about this time period in the next chapter), who lived several hundred thousand years ago, was undoubtedly a fire user, although not necessarily at this sophisticated level. Evidence for the use of fire in Europe goes back at least 400,000 years, in China at Zhoukoudian (where "Peking Man" was discovered), to perhaps 500,000 years ago, and at a site called Gesher Benot Ya'aqov in Israel, good evidence of the controlled use of fire dates back to 790,000 years ago.[28] The collective evidence for fire at these sites, combined with the actual stones and bones found at them, strongly suggests that *Homo erectus* was a fire user, at least sometimes, in some places. But when did use of fire start?

Recent research at a site called Wonderwerk Cave in South Africa suggests a beginning that dates back at least a million years.[29] Twenty-five meters deep into the cave, Francesco Berna and his colleagues have identified the ashed remains of plant material and bone, along with flaked stone tools that show evidence of fire exposure, in an archaeological layer dating to 1.07–0.99 million years ago. This layer sits in an archaeological sequence that dates back as far as 1.8 million years, which is likely associated with early *Homo erectus.* There is no evidence that there was a fire in the cave caused by ignited bat guano or that the ashed material was carried in by some other disruption, such as flooding; the distance of the site from the mouth of the cave makes an incursion of a wildfire very unlikely. This recent, apparently robust, discovery of controlled fire at 1.0 million years ago may lend credence to earlier studies that suggest that fire at even older archaeological sites may have been controlled. Several notable sites in East Africa, dating to 1.4 million years of age and older, have definitive evidence of fire.[30]

The idea that fire was critical in the emergence of the genus *Homo* (around 1.8 million years ago) has recently been championed by primatologist Richard Wrangham.[31] In Wrangham's view, fire meant cooking, and cooking made available an increased range and quantity of foods to our ancestors. These foods included not only the abundant protein found in the muscle of large animals, which would otherwise be very tough and difficult to ingest, but the stores of carbohydrates and calories present in tubers. The brain is a very energy-hungry organ, and Wrangham argues that the increasing brain size we see in *Homo* species at this time required a higher-quality diet to support its growth and maintenance. He is almost certainly correct that fire would have made an enhanced diet much more achievable. Wrangham thus gives fire a critical role in the transition from a two-legged ape to something more comprehensively human. Equipped with fire, this near-human was able to leave tropical Africa and enter more temperate climates in Eurasia starting 1.7 million years ago. At least this is one possibility—uncontestable evidence of controlled fire outside of Africa earlier than 790,000 years ago would help prove this scenario.

Making or using fire not only would have encouraged early *Homo* to abandon life in the trees, but also would have committed the species to an investment of time and energy in a specific place on the ground. A fire, at least for its duration, could have become the focus of life in a home base, at least as much as butchering an animal or crafting tools. And even if fire began as a technology to process food, the critical importance of food and sharing, within the confines of a shared space, eventually transformed it into something more, leading ultimately to the identification of fire with home that we see in cultures around the world today. In a review generally supporting the hypothesis that controlled fire was important for early *Homo,* the eminent archaeologists Desmond Clark and Jack Harris wrote: "Fire—every bit as much as food-sharing or meat-eating or new forms of sexual behavior . . . helped to weld early hominid groups into the coherent family units that are the characteristic of human society."[32] More than a million and a half years ago, home may have emerged from the union of fire and family.

Family Households

Families leave little evidence in the archaeological record. Certainly their presence can be inferred, maybe even felt, when looking at the remains of the past, but the set of relationships and social dynamics that make a family is not preserved in stones or even bones. To gain insights into the origins of family requires interpolating between the behavior of modern humans and great apes, and adding some associations between behavior and anatomy that can be derived from the analysis of mammals in general. The study of family dynamics across species makes clear that the origins of home are likely tied to family.

One thing to remember is that the search for the origins of family is not a search for the cozy little nuclear family of post-WWII America, the kind of family that so many people still think of as "natural." Anthropologists have shown that while kin relations are central to human social structure, there is a good deal of cultural fluidity in how the kin are arranged and organized. At the heart of family, however, are some basic relationships, which evolutionary anthropologists have focused on in trying to reconstruct how human families may have come about.

For the great apes, the relationship between a mother and her child is fundamental, but males have relatively little to do with raising or providing for their offspring. At some point in hominin evolution, males became responsible for feeding more than just themselves, and started to have a significant role in providing food for their reproductive partners and children. As I mentioned earlier, the increasingly large hominin brain required a higher-quality diet; in addition, a larger brain capable of more complex cognition takes a longer time to train and grow. If hominin mothers had remained totally on their own, they would have been in quite a bind. Compared to a great ape, they would have had a more (nutritionally) demanding offspring for a longer period of time. By bringing fathers into the picture, our ancestral mothers were able to provide not only for a single, large-brained offspring, but likely for more than one (at different ages) at the same time. This would have placed them at a reproductive advantage over hominins who may have retained an ape-like pattern of child rearing—one at a time, birthed several years apart.

The evolution of the household must have been critical in how hominins solved the dilemma of raising slowly maturing children while having more of them dependent at the same time. Karen Kramer and Peter Ellison suggest that, at the most basic level, the human household began, and can still be understood, as a venue for pooling energy resources.[33] One can focus on how "Man the Hunter" brought back large game animals for butchering and sharing at a home base, or how "Woman the Gatherer" maintained a steady supply of food to subsidize the more erratic productivity of males, or how children pay back the energy invested in them during their years of growth by maintaining long-term associations with family when they are adults, sharing their knowledge, experience, and effort. But in the end, humans increased their access to resources by combining different pathways of energetic exchange and transfer depending on ecological conditions, including culturally influenced subsistence habits. Despite abundant ecological and cultural variation, these energetic pathways ultimately came to converge in a place that is defined both physically and cognitively: the home.

Of course, households are defined not simply by how energy and resources are exchanged within them, but by the relationships shared by the exchangers. Mother/child is the primary relationship, of course, but that predates the household. Some researchers, most notably the paleoanthropologist Owen Lovejoy, have hypothesized that the formation of a pair-bond between a male and a female was at the center of the evolution of family life and critical for the separation of our lineage from the great apes.[34] Pair-bonding, as seen in various bird and some primate species, theoretically benefits females because it commits the male partner to provision her and her offspring; in turn, the male increases his certainty that the offspring he is helping to raise are indeed his own. However, the idea that pair-bonding was a significant development in human evolution (especially in early hominin evolution) has not been well received by most anthropologists. Mating systems in modern humans are more fluid than dedicated monogamous pair-bonding would suggest, and in looking at the fossil record, our ancestors' bodies did not conform to what would be expected of a monogamous species (e.g., differences in body size and other anatomical features persist between the sexes). Nonetheless, there can be little doubt that male and female humans often form mutual emotional attachments in the context of having babies together and then raising them. This is indeed uncommon among our closest ape relatives and may indicate an evolved feature of hominin home life.

Around the world, even in male-dominated societies, the home is considered to be the domain of women. Anthropologists Donna Leonetti and Benjamin Chabot-Hanowell theorize that women have been at the center of creating and defining home since home began.[35] Looking at the home as a place of pooled resources, Leonetti and Chabot-Hanowell emphasize the importance of intensive food processing (cooking, grinding, chopping, mixing, etc.) that would necessarily be centered at a home-type base. In hunter-gatherers today, these activities are dominated by women, including not only current mothers but also younger girls and older women. In the evolutionary past, such processing activities, which would have been vital for expanding the diet of early *Homo* and *Homo erectus,* served to stabilize the home environment. Leonetti and Chabot-Hanowell suggest that males became part of the dual-sex household not only because they were physically attracted to females, but also because of the stable resource base they provided via food-processing activities. This stability complemented high-risk/high-reward food-collecting activities such as large-animal hunting.

The mother/child relationship within the human household is the most basic one, as it is for almost all other primates. But at some point in human evolution, mother/child became parents/children. The development of the household helped early *Homo* parents obtain the necessary nutrition, over an extended period of time, to raise their multiple, overlapping offspring. It is very likely, however, that parents were not the only ones involved in this enterprise.

Anthropologist Sarah Blaffer Hrdy characterizes humans as *cooperative breeders,* which simply means that individuals other than the parents provide a significant amount of the care and provisioning of the young.[36] Many other species, ranging from insects to elephants, are cooperative breeders, as are some other primate species. The colobine (leaf-eating) monkeys, found throughout Asia and Africa, are noted for the amount of care and interest infants receive from females who are not their mothers. Colobine infants often have strikingly different coat colors from the adult forms. Blaffer Hrdy notes that as the infant coloring fades, so too does the interest of others in helping to care for the infant. This indicates that the stimulus response to the coat color, encouraging attention toward infants, has been shaped by natural selection.

According to Blaffer Hrdy, modern humans, as evidenced by the abundant care that mothers in traditional cultures receive from other women, especially their female kin, can easily be classified as cooperative breeders. Although human babies, with their large heads and plump bodies, elicit attention from mothers and others, cultural practices (dressing the baby in distinct clothes, or even decorating them with makeup) enhance the mothers' chances of receiving care for their infants from others. Cooperative breeding of this kind is pretty much absent from our closest ape relatives. Blaffer Hrdy suggests that during the critical period of early *Homo* and *Homo erectus* evolution, when our ancestors were evolving both larger brains and bodies, it was cooperative care as much as any other cognitive or behavioral development that differentiated hominins from other apes.

Cooperative breeding may have been *the* factor in the cognitive and emotional evolution of humans or one among several (I would tend toward the latter view), but the existence of a home base or place or camp, or whatever we want to call it, would have gone hand in hand with it. Caregivers are only useful if they can be found when needed, and mothers need to be able to find their children when they return from whatever they were doing. So the first "room" in the human home may have been a nursery, rather than a kitchen, a dining room, or a bedroom.

The Homely Imagination

I want to end this chapter with a thought experiment. Picture in your mind's eye a group of *Homo erectus,* say a dozen or so, living 1.4 million years ago in East Africa, milling about and resting in the shade of a stand of a few trees, surrounded by moderately tall savanna grassland. Some are standing upright on two legs, while others sit, and maybe one sleeps in the crook of a large branch in a tree. They are shorter than most modern people but taller than most great apes. They are more or less hairy (scientific reconstructions typically give them a pretty good coat of fur, but we don't really know how furry they would have been). Their faces are relatively ape-like, with large brow ridges and a sloping forehead, but with a substantially flatter profile. From a distance, in this wild environment, they might look ape-like as a group.

Now imagine that this group returns habitually to this spot, and they are not using it simply as a place to rest during the day as they move around a large home range. The trees are still there, but the grass surrounding them has been trampled down. On this trampled floor, signs of stone tool manufacture can be seen—areas where flakes and cores have accumulated over time. A couple of adult males and an adult female are working on stone tools now. Some children emerge from behind the trees, and you notice a couple of babes in their mothers' arms.

Add two small campfires to the scene. Each has one or two females tending to it. At one, you see a mother pass her small infant to an older female, freeing her hands so that she can poke the fire with a stick. Two large tubers sit in the fire, slowly roasting, while a small pile of other tubers sits under a tree in the cleared area. A young adult male arrives carrying a small gazelle on his back. He drops it next to the fire tended by the woman with the baby. The older children come over to look at the gazelle, while the young man goes over to the stone knappers and selects a fresh and sharp flake from their pile. He returns to the gazelle, sits down, and starts skinning and butchering it. He nibbles on prize bits of organ meat (liver, brain, kidney) as they emerge, sharing them with the two women at the fire. Eventually, he gives small pieces to the children standing around him. The adults working on the stone tools make some noises that suggest they are not altogether happy with the situation, and the young male takes a piece of liver over to them. Eventually, the older female hands the young infant back to her mother. She now starts to work with the young hunter to separate the legs from the gazelle carcass, which will make the animal easier to cook on the fire.

This slice of life has taken a decidedly human and homely transition from where we started. The long-term occupation of a spot of ground, the stone tool-making station, the campfires and cooking, the special relationships between some of the adults, the accumulation of children at different stages of development, the trading and pooling of resources—these are all of the hallmarks of humanity that likely evolved during *Homo erectus* times. We can create a very evocative and intellectually seductive picture of a day in the life of *Homo erectus,* one that on the surface, in its focus on a home place for food preparation, resource sharing, and child rearing, is not all that different, at a fundamental level, from life as we know it today.

Of course, we know that we don't quite know what *Homo erectus* was up to on a daily basis. As I said at the beginning of the chapter, we really can't know when all the elements of home coalesced to form this very human institution. But

I think we can be sure in saying that the pieces were evolving before the emergence of fully modern humans or near-humans. *Homo erectus* was a significant and revolutionary departure from not only earlier hominins and the great apes, but from all forms of life that had ever lived on Earth. Home may well have been part of this revolution.

Chapter Four

AT HOME IN A NEANDERTAL'S GRAVE

A burial is the most permanent home a person can have. It is so permanent, in fact, that the buried dead tend to accumulate around us. Within about ten miles of my home in the Bluegrass region of rural central Kentucky there are several cemeteries or group burials. Closest to my house is a Native American (Shawnee) burial ground that is mentioned in a 1928 survey of archaeological sites in the state.[1] Longtime residents of the county have told me that there are also ad hoc burials from the American Civil War in the area. These are supposedly the work of Confederate guerilla groups burying their comrades after small-scale skirmishes. Scattered about in the county are tiny cemeteries built as private burial grounds by some of the first European American families to come into the area in the late eighteenth and early nineteenth centuries. These typically have a half dozen or so gravestones, surrounded by a low, decorative wrought-iron fence. They can be incongruously placed in a time of rapid development—I have noticed one next to a department store parking lot, another in a traffic island of a busy intersection.

In the hamlet (consisting of a couple of hundred residents) closest to my house, there is an old community cemetery that dates from this same time period. It is a motley collection of gravestones, varied by time and fashion and means, arrayed into family groups. The cemetery sits on a pleasant rise on the edge of the small village.

The two towns in our county have larger cemeteries, befitting their relative status as urban centers. These cemeteries are like any one might see in countless American towns and cities. Although they were once perched on the edge of town, they have long ago been surrounded by urban development (of a modest kind). Finally, the grandest cemetery in our area is a US national cemetery for armed forces veterans. Different from the municipal cemeteries, the gravestones in the military cemetery, arrayed in an orderly way, are a uniform white and undecorated. In death, as in their service in life, the soldiers sacrifice some of their individuality for a greater, collective cause.

Cemeteries house the dead. Across the world, cultures are invested in death, both emotionally and symbolically. Thus, burial of the dead, wherever it occurs, represents a particular kind of housing investment. The symbolic content of the veterans' cemetery is obvious to all, while community cemeteries provide a means to express images of wealth, spiritual devotion, or familial togetherness, whether or not they reflect life as it was actually lived. Although burial of the dead can sometimes be primarily a matter of body disposal, more typically it is an act filled with symbolic significance. When we look to the past, at burials from tens of thousands of years ago or more, we cannot help but see them as symbolic acts (at least in part), even if they were not.

The ability to think and communicate using symbols is usually considered a hallmark of modern humanity. What do I mean by "modern" humanity? Simply put, I mean people like us, who all belong to a single, potentially interbreeding species known as *Homo sapiens.* We all know what people are like and what people can do: among other things, we have languages, cultures, religions, use all sorts of tools, and even if we move around quite a bit, we define the environments we live in as domestic spaces, creating our own habitats in the wider and wilder environment. In terms of anatomy, even though we vary biologically according to sex, age, geographic origins, and individual genetic makeup, we are all clearly part of a single hominin species. We are the final product (so far) of a particular evolutionary journey.

The home stretch of the human evolutionary journey involved the transition from a *Homo erectus*–type of hominin to one that looks like us. The fossil and ancient DNA record from this time period is quite complex, leading to much debate about how modern humanity finally came to be. Complicating matters, there were the Neandertals, our cousins by virtue of the fact that they were also descended from *Homo erectus.* Neandertals were anatomically distinct from modern humans, and it makes sense that the transitional forms from *Homo erectus* to Neandertals were different from those that were transitioning to us. Those differences were quite subtle, however, especially during the earliest stages of the transition.

Along with changes in anatomy, there were also changes in behavior. We expect that the general *Homo* trends toward more elaborate social behavior, increased cooperative food gathering and sharing, and more shared responsibility (including the father and other relatives) for raising children all continued to play out during the transition to modern *Homo sapiens.* Unfortunately, the archaeological record, until relatively late in the process, does not give us all that much to work with to track these behavioral changes. Eventually, however, we do see the hints of modernism (in a Stone Age sense) that indicate that something had changed.

One of these hints is burial of the dead. Burial of the dead is not a universal practice today—cultures have all sorts of ways of dealing with the mortal remains of family, friends, and foes alike.[2] It was certainly not universal in the past. However, when anthropologists have found evidence of intentional burial from the distant past, they have quite reasonably focused on it as a significant remnant of someone's behavior. My interest here, of course, is in the evolution of home. One thing I can say for certain is that while the buried dead are housed, they cannot feel at home, since they obviously cannot feel anything. But when the living choose to deal with a body by burying it—a more or less novel behavior in a zoological sense—they obviously do have feelings. I suggest that one of those feelings, a feeling for home, may have been expressed in the earliest hominin burials. Before getting to that in more detail, it is probably best to go over just what we know about how *Homo erectus* transitioned into modern people.

Toward the Evolution of People Like Us and Not Quite Like Us

Home erectus had a good evolutionary run, beginning 1.8 million years ago in East Africa and ending as recently as fifty thousand years ago on the island of Java. Anthropologists have found *Homo erectus* remains across Asia and Africa from during this time period. But even as some *Homo erectus* populations carried on, maintaining the "classic" anatomy and presumably living life the *erectus* way, others were evolving. These transitional forms are sometimes referred to as "late phase" or "advanced" *Homo erectus.* They were presumably progressing toward *Homo sapiens*, so the imprecise term "archaic *Homo sapiens*"—which reflects some scientific uncertainty—has informally been used to classify them. Over the past two decades, more fossils and better dates for fossils have helped to improve our view of hominin evolution during this dynamic time period.[3]

These transitional forms date from 600,000 to 150,000 years ago. They are known from sites throughout the Old World. Many are isolated finds, but at the extraordinary cave pit site at Sierra de Atapuerca in Spain, the remains of more than twenty-eight individuals, from children to adults, have been found, dating to more than half a million years old. Most paleoanthropologists agree that these specimens are distinct enough from what came before and after them to be given a proper species name (rather than something like "archaic *Homo sapiens*"). The name most commonly applied to these specimens dates back to 1907, when a very robust hominin lower jawbone (mandible) was found in the German village of Mauer, near Heidelberg. The anatomist who analyzed the jawbone correctly recognized that it was distinct from both *Homo erectus* and modern *Homo sapiens* and gave it a new species name: *Homo heidelbergensis.*[4]

Compared to the Mauer mandible, a modern human lower jawbone looks positively delicate, with relatively tiny teeth and a distinct chin (which Mauer lacks). In general, the crania of *Homo heidelbergensis* are characterized by thick bones, including formidable brow ridges that are unmatched in the hominin fossil record. The overall shape of the cranium is long and low, similar to the shape in *Homo erectus.* However, and this is one of the critical factors in separating out these specimens, the cranial capacity of *heidelbergensis* ranges between 1,000 and 1,400 cm^3—larger than *erectus* proper and approaching what we see in modern humans. Along with their heads, their bodies were also robust: the remains of the disarticulated and damaged skeletons from Sierra de Atapuerca show that these hominins were strongly built and well muscled.

Unfortunately, the archaeological record for *Homo heidelbergensis* does not tell us much about what they were doing with their larger brains. Some *heidelbergensis* groups certainly engaged in cooperative large-animal hunting, as sites in Spain and England attest.[5] A remarkable find in Germany of wooden spears in association with the remains of horses is indicative both of hunting and of the material culture beyond stone tools that we generally do not have access to today.[6] Presumably, the various trends toward increased cooperative behavior, enhanced parental care, and growing sophistication in food-gathering and -processing technologies, which we expect emerged during *Homo erectus* times or earlier, all continued to develop for *Homo heidelbergensis.* Whatever they were doing, however, it worked, at least in an evolutionary way. Two new cousin species emerged from *heidelbergensis* populations, each of which displayed a significant cognitive advance over all hominins that had come before: *Homo sapiens* and *Homo neanderthalensis* (the Neandertals).

Neandertals were first discovered in 1856 in the Neander Valley of Germany (actually a specimen was discovered in Gibraltar before this but no one knew what it was).[7] The earliest "classic" Neandertals lived about 140,000 years ago and the last of their kind may have survived until 27,000 years ago—not too long ago at all. Anthropologists have discovered specimens in Europe, the Middle East, and into western Asia. Superficially, the strongly built Neandertals (or at least their bones) bore a general resemblance to *Homo heidelbergensis*.[8] However, there were some important changes that set Neandertals apart from *heidelbergensis*. Certain dental features and structural aspects of the cranium and jaw were unique to Neandertals, differentiating them from both *heidelbergensis* and modern humans. Like *heidelbergensis,* Neandertals had large brow ridges, but they were generally not as large and were distinctly double-arched with a more pronounced *m*-shape. Most critically perhaps, the Neandertal cranial capacity was in the range of 1,300–1,600 cm^3—larger than what would be seen in most modern human populations. In general, the Neandertal head and body looked to be adapted to living a hard, active life in colder climates.

The modern human head and body represent another variation on the late hominin theme. Our bones and bodies are less strongly built than those of *heidelbergensis* or *neanderthalensis,* and may reflect our origins in a warmer rather than colder climate. The human face (including the nose) and jaws are reduced in size, and a pronounced chin is evident, something not seen in other *Homo* species. Most people do not have brow ridges, and when these are present, they generally do not compare in size or development to those seen in Neandertals. Our cranial capacity, in the 1,400 cm^3 range, is consistently larger than those seen in *heidelbergensis,* but a bit smaller than the Neandertal measure. However, the entire cranium of modern people is reformed—gone is the long and low cranial vault, replaced by a bulbous skull with a high forehead. Modern humans first appeared in East Africa about two hundred thousand years ago, and remains have been found in the Middle East dating to around one hundred thousand years ago (this region was alternately occupied by Neandertals and modern humans, depending on the climate conditions). They are present in Australia by fifty thousand years ago, suggesting a gradual move into and through Asia. Modern humans are not present in Europe much earlier than forty thousand years ago, perhaps unable until then to supplant the Neandertals on their home turf.

Over the years, based on the fossil evidence, there has been much to-ing and fro-ing about whether or not the anatomical differences between Neandertals and modern humans warrant putting them into separate species.[9] The prevailing (although not unanimous) opinion in anthropological circles today is that it is OK to put Neandertals in their own species. In the late 1990s, the first ancient DNA studies of Neandertals began to appear. The genetic evidence suggests a primary split between Neandertals and modern humans dating back several hundred thousand years. The estimates vary depending on what genetic system is studied, but a date of about 400,000 plus or minus 150,000 years would probably be generally acceptable today. These dates indicate that it is very likely that some *Homo heidelbergensis* specimens, especially those in Europe, are pre-Neandertals, while some of those in Africa are pre-modern humans.

For several years, genetic studies seemed to provide a tidy solution to how Neandertals and modern humans were related to one another—at a cousin level, but not closer than that. This view of things changed in 2010, following the preliminary sequencing of the Neandertal genome.[10] Studies now show that the gene pools of modern human European, East Asian, and Melanesian populations have a 1–4 percent contribution of Neandertal genes, and that some sub-Saharan African populations also have a small representation of Neandertal genes. According to the geneticists' models of how these genes were distributed, the intermixing and gene flowing that occurred between Neandertals and modern humans were not ongoing over hundreds of thousands of years, but occurred over a period eighty thousand to fifty thousand years ago. This admixing likely occurred in the Middle East, where we know that Neandertal and modern human populations overlapped.

The genetic studies generally support the idea that modern humans and Neandertals were two separate, very closely related species, capable of interbreeding, but that they were not particularly profligate about it. While it is fair to say that a fairly large number of modern humans have some Neandertal ancestry, at a species level, the Neandertals were not the ancestors of modern humans. Largely separate from *Homo sapiens,* they charted a different, parallel path toward humanity. But what did Neandertal humanity look like? Did they express this humanity in an environment that was homelike or was that something that only we surviving moderns are capable of?

How Modern Was Neandertal Thinking?

An old thought experiment in anthropology classes was to imagine a Neandertal on a subway train, all cleaned up and nicely dressed. Would he or she look so out of place? I tend to think that most people would find a Neandertal at the very least to be a notable occupant of a subway car. His or her heavy, muscular body, topped by a large, sloping cranium, combined with a chinless face carrying an extremely large nose and brow ridges—these features viewed in combination would be fairly striking. I am not saying that the Neandertal would necessarily be the most notable occupant of any subway car—that would depend on the particular city, line, and time of day or night. However, it would be hard not to notice a Neandertal, standing, hand on overhead bar, looming over you while you tried to read your text messages.

Physical appearance is only one thing Neandertals would have to overcome to ride a subway unnoticed: their behavior would also have to be unnoticeable. That is, they would have to master the social and cognitive skills necessary to ride in a contemporary public transport system, blending in with the modern human general public. I do not think they would have much trouble controlling their behavior in this particular social setting (with proper training or experience, of course). They lived in highly interactive, cooperative social groups that required at times some measure of social discipline and restraint. I also do not think they would have had trouble managing the folk physics necessary to roll with the subway car at high speeds or during braking. After all, they were very strong and agile enough to hunt large game.

Could they learn to say "excuse me" after bumping into a fellow rider? Again, I think they would have had the social sophistication to recognize a minor faux pas and make an equally minor gesture to address it. Speech is another matter. Most researchers interested in language evolution would probably agree that Neandertals used some form of spoken language (given large brain size and some of the behaviors we know they were capable of). But whether or not it was language in the modern human sense is perhaps unknowable. Over several decades, there have been attempts to reconstruct Neandertal speaking ability from their bony remains, by using them to reconstruct the soft structures in the throat that produce language.[11] Some researchers have concluded that Neandertals were only capable of a limited range of sounds, much fewer than humans can produce. Others have argued either that such reconstructions are inaccurate or that a limit on sound production would be no barrier to producing quite sophisticated spoken language.

Language as a marker of humanness is important because it reflects a mind that is capable of reconfiguring the world in terms of symbols and what they refer to. The extent to which Neandertals were capable of symbolic cognition is a vital issue in assessing their relative humanity. They may have had spoken language, but we do not know if their languages were fully symbolic and grammatical in the same way that human languages are.

Beyond language, we do not know if the Neandertal mind was capable of making and using the kinds of symbolic associations that modern humans use every day. Let's return to our subway-riding Neandertal, and say that he is on a New York City subway. To gain entry to the subway trains, a patron must swipe a Metrocard with sufficient credit through an electronic reader at a turnstile, which will then open to let him gain access to the platforms. I am sure that a Neandertal could master this task—there would be no issue with manual dexterity or in understanding the cause-and-effect relationship between swiping the card and the turnstile allowing him through. However, the task of using a credit card at an automated kiosk to add funds to the Metrocard might be a different matter. Consider all the symbolic understandings necessary to conduct such an exchange, and the implied reciprocity with a totally abstract entity, for services to be rendered at some undefined point in the future.[12] The great diversity of modern humans who have used the New York City subway successfully—some of whom may not have shared a common ancestor for more than one hundred thousand years—demonstrates the universal capacity of modern humans to deal with very abstract and symbolic concepts, including those involving a projection of future exchanges.

Did Neandertals have this same capacity for symbolic or abstract thought? Most anthropologists and archaeologists would probably be reluctant to claim that they did, for the basic reason that there is very little evidence of symbolic thinking of any kind in the Neandertal archaeological record.[13] When we look at the archaeological record of the earliest modern humans, beginning two hundred thousand years ago in Africa, it is not hard to draw connections between their cognitive abilities and our own. There is evidence of personal adornment, symbolic use of pigments, even musical instruments. Beyond this, the earliest humans demonstrated a marked increase over their hominin ancestors in technological sophistication, with stone tools becoming more varied, more finely made, and obviously designed to be part of compound tools (e.g., hafted with wood). They also made use of novel tool-making materials, such as bone, antler, and ivory. Humans exhibited a mastery of their landscapes, both domestic and wild, as they ranged further in pursuit of food, sometimes following the seasons, and establishing trade networks in order to obtain precious raw materials and goods.[14]

The so-called cognitive revolution in Europe thirty thousand to forty thousand years ago was in part a function of Neandertals and their material culture being replaced by modern humans and their material culture over a relatively short period of time. Compared to what came before them, the Neandertals were technological innovators, but it was on a much more limited scale than what modern humans would eventually prove to be capable of. Neandertals are usually associated with a tool culture known as the *Mousterian,* which first appeared about three hundred thousand years ago. It was characterized by hand axes, like those used by older *Homo erectus* populations, but also included a variety of more finely crafted stone tools made from flakes, including small points and scrapers. A method to systematically produce uniform flakes was the most impressive *Mousterian* innovation. Neandertals apparently did not make tools of bone or antler, although presumably they made ample use of wood and skins. There is very little evidence (other than a few contested artifacts) that they made anything decorative or what could be construed as artistic or symbolic.[15]

Neandertals also exhibited, as archaeologist Steven Mithen describes it, an "immense cultural stability."[16] From approximately 250,000 to 35,000 years ago, the Neandertal tool kit remained more or less the same. This kind of stability is much more in keeping with what we see in *Homo erectus/heidelbergensis* than in modern humans. In less than 200,000 years, modern humans have gone from stone tools to smartphones, while at the same time maintaining an extraordinary level of cultural variability. The only substantial change in Neandertal material culture archaeologists have identified comes near the end of their time in Europe, around 30,000 years ago, when some groups in France seem to have adopted or copied a tool culture, known as the *Chatelperronian* (including stone implements and decorative beads), more consistent with anatomically modern humans.

Neandertals were not entirely "the other" in relation to modern humans. The genetic data show that we were closely related indeed, and Neandertals were responsible for producing those *Chatelperronian* tools and objects, demonstrating the cognitive wherewithal to transcend their "traditional" cultural bounds. Over the long history of Neandertal research and study, no matter how far outside the bounds of proper humanity someone might want to place them, there was one thing that made it hard to push them too far away: they buried their dead.

Burying Neanderthals at Home

In the United States, there are basically three acceptable ways to deal with a dead body: burial, cremation, and in a very small percentage of cases, donation to an institution for scientific research or training (i.e., dissection). As Peter Metcalf and Richard Huntington have noted, in the grand anthropological scheme of things, American ways of death are remarkably uniform given the nation's cultural diversity. Other cultures show much more variation according to class, caste, gender, age, and other pertinent variables. Taken collectively, the human species subjects bodies to a remarkably broad range of postmortem treatments. As Metcalf and Huntington write: "Corpses are burned or buried, with or without animal or human sacrifice; they are preserved by smoking, embalming, or pickling; they are eaten—raw, cooked, or rotten; they are ritually exposed as carrion or simply abandoned; or they are dismembered and treated in a variety of ways."[17]

We can be fairly certain that we know of at least two ways that Neandertals, or at least some of them, dealt with their dead. The first is that they ate them. Evidence for this comes from Neandertal sites that have yielded abundant but fragmentary bony remains, sometimes showing evidence of cutmarks and burning. At these sites, all long bones were broken open (for marrow), and cranial bones were smashed, presumably to get access to brains. The Neandertal remains were often mixed in with those of other animals, suggesting a butchery dump rather than a mortuary processing center.[18] Select instances of Neandertal cannibalism does not mean that all Neandertals were cannibals, any more than the fact that some human cultures engaging in cannibalism means that the entire species should be characterized as cannibalistic. We do not know if these Neandertals ate their friends or foes; modern humans have been known to make a meal of either.

Besides sometimes eating them, we know (or think we know) that Neandertals also buried their dead. Or as is the case for humans, that some Neandertals sometimes buried their dead. In a comprehensive overview and analysis of the evolution of hominin mortuary practices, archaeologist Paul Pettitt identifies more than twenty instances of Neandertal burials (dating from about seventy thousand to thirty-four thousand years ago) that are generally accepted by scholars.[19] Some of these are single burials, others are in groups (that is, multiple single burials at one location); they have been found in multiple sites in Europe and in the Middle East. These are not necessarily the oldest burials we know about: some early *Homo sapiens* burials predate these. However, given the largely separate evolutionary trajectories of modern humans and Neandertals, evidence of burial in both groups could suggest that the cognitive basis of burial, whatever that might be, predates the separation of these two species.

Pettitt identifies several patterns in these Neandertal burials, which also generally apply to the earliest modern human burials as well. They were found in caves or rock shelters, and were typically associated with other archaeological signs of living or occupation; bodies were buried in a wide range of positions; they could be covered by dirt or stone slabs; some bodies were deposited in shallow excavations while others were "cached" into preexisting, natural hollows or depressions; adult males and females, as well as juveniles, were buried; although in rare cases, buried bodies were found in association with artifacts or the remains of other animals within the grave, there is no convincing evidence for grave goods or "offerings"; burials could be of whole bodies or parts of bodies.[20]

There is some academic debate about whether or not Neandertal burials were really burials rather than natural depositions, and beyond that, what intentional burials tell us about their behavior.[21] Many anthropologists look at the large number of complete or nearly complete Neandertal skeletons that have been discovered, including those of very small children or infants with delicate bones, as indication that they were intentionally interred. Going back over the millions of years of hominin evolution, the recovery of even partial skeletons is rare indeed; the fossil known as "Lucy" (*Australopithecus afarensis*) is famous for a reason. All of a sudden, with Neandertals (and early modern humans), whole bodies start to be found, and at numerous sites spread over time and space.

Many Neandertal burials were excavated before modern techniques, especially the careful recording of the archaeological and geological contexts of the burials, were developed. Critics of the idea that these were intentional burials argue that many of these records are so sketchy and incomplete that we cannot really trust the picture of intentional burial they present.[22] This is certainly a fair enough criticism in some cases, and going back over the records of decades- or even century-old excavations can be a trip down a bumpy memory lane. For example, a recent reappraisal of a Neandertal child from Roc de Marsal, a cave in southern France, calls into question its status as a deliberate burial, despite the fact that it has been identified as such in the scientific literature since not long after its discovery by an amateur archaeologist in 1961.[23] Dennis Sandgathe and his colleagues found that the primary descriptions of the Roc de Marsal site in the initial publications about the child were limited, and the original field notes and photos concerning the discovery were schematic at best. New excavations and analyses of the site indicate to Sandgathe and colleagues that the child was not buried but was instead positioned in a natural fissure within the cave. The child could have been placed there intentionally or not, but they provide strong evidence that this was not a burial.

The Roc de Marsal child is one Neandertal burial that may not be one, but on the whole, there still remain several other burial sites that have been well excavated and are likely to stand up to scrutiny. Archaeologists have in general not decided, when confronted with an "obvious" burial, if the burden of proof lies with those who question the burial or those who support it. At any rate, as Harold Dibble, one of the leaders of the team reexcavating Roc de Marsal, says, the big question is really, are we looking at "a burial or a funeral?"[24] The former would reflect nothing more than the "disposal" of the body, while the latter is necessarily laden with symbolic significance.

Many anthropologists, while conceding that Neandertals buried their dead, have been content to look upon the practice as something "more akin simply to corpse disposal."[25] Fueling this viewpoint is the fact that not long after the last of the Neandertal burials, some burials associated with modern humans (especially those of the European Upper Paleolithic between twenty thousand and thirty thousand years ago) start to take on a very recognizable and modern cast: deep burials in open ground not caves, complete with grave goods and evidence of decorated clothing or body ornamentation, sometimes with more than one body interred in a common grave. In some regions, a symbolic culture of burial seems evident in the common patterns observed. In contrast—and this contrast may or may not be fair—the Neandertal burials can be defined by the absence of such features.

I think that the case for intentional Neandertal burial is quite compelling. On the other hand, I also agree that there is no reason to ascribe to this practice some deeper symbolic significance—there is simply no good evidence for that. This is not to say that there was not something symbolic going on; we just can't say that there was. Neandertal burial stands out, however, as a "modern" behavior in a behavioral repertoire that, based on the bulk of available archaeological evidence, is not quite modern.

The one conclusion about Neandertal burial that I find quite hard to accept is that it was "simply" corpse disposal. First, there were lots of other ways available to Neandertals to dispose of a corpse. Eating was one way; burning was another. Or taking the body away from the living area and leaving it exposed to scavengers would also have worked. Considering that many burials were of very small children, carrying them some distance would not have been a problem. Neandertal populations were not particularly high-density, so it is safe to assume that there was ample room to simply dispose of a body somewhere on the landscape. Burial of a corpse is one way to get rid of it, of course, but it is a method that requires the expenditure of a considerable amount of time and energy. Far from being simple, it is a complex activity that takes a certain amount of planning, and in the case of burying a fully adult individual, would benefit from cooperative behavior.

In other words, burial—even Neandertal burial—is not so simple an activity that it should be taken for granted. But perhaps this is easy to do when considering Neandertal burial from a fully human perspective. In the context of the elaborate mortuary and funerary activities that modern people engage in, plain old burial can look quite simple indeed. However, viewed in the context of human evolution overall and primate behavior in general, simple burial emerges as a novel, complex, and cognitively advanced activity. Paul Pettitt makes the case that burial did not arise de novo, but followed from various activities, such as postmortem body processing (represented by cutmarks made by stone tools on bony remains) and body caching in natural structures, that can be traced over the course of millions of years of evolution.[26] I think that burial itself can be seen as the culmination of these practices that emerges only very late in human evolution.

Viewed from the bottom up, it is easier to worry less about whether or not Neandertal burial is a "marker" of symbolic thinking. Instead, we can worry about what burial signifies for other aspects of Neandertal cognition. Consider that the deaths of loved ones, friends, or colleagues typically bring out powerful feelings and emotions. Even social anthropologists such as Peter Metcalf and Richard Huntington, who are fundamentally interested in the cultural diversity of responses to death, acknowledge that there is a "universal impact of death" leading to reactions that are "always . . . meaningful and expressive."[27] This emotional response to death goes way back in our history as primates. Primatologists have observed many instances in which nonhuman primates, especially the great apes, were clearly negatively affected by the deaths of loved ones.[28] Their displays of subdued or distressed behavior—singly and in groups—are difficult not to interpret as akin to mourning in some cases; in others, normal behavioral patterns are clearly disturbed. Most evocative to us human observers may be the instances in which a mother refuses to part with the corpse of a dead infant, carrying it with her until it can be carried no longer.

I think it is safe to assume that in at least some instances, Neandertals mourned, or at least had emotional responses to, the deaths of their loved ones. But did mourning extend to burial? I think the fact that young children were buried increases the likelihood that this was a practice reserved more for family and friend rather than foe or stranger. We know that burial accomplishes corpse disposal. But what else does it accomplish, especially if those doing the burying are filled with emotion, confronting the loss and separation of someone close to them?

For one thing, burial can be just as much about retaining a corpse as disposing of it. Burial keeps the body—the person—in a kind of contact and under control, safe from further degradation at the hands of other, less benign forces. Neandertal burials were in caves, which often show evidence of having been living spaces. A body stored in this setting is still very much a part of the group that survives it and continues to occupy that space. Burial also quite literally binds an individual to a place. Neandertals likely occupied multiple locations within a home range as they pursued different kinds of foods from season to season. It is also likely that such occupation sites were reused over time. A burial in one of the sites may suggest that the deceased somehow belongs there: if he rested there in life, he can continue to do so in death.

My view of Neandertal burial is quite literally a very homely one. I see in it an expression of the emotions and feelings that surround and emerge from a sense of home: the corpse is linked to a group of intimates who share a significant place. Burial potentially extends these significant relationships beyond death. Obviously, Neandertals, like people today, had many ways to deal with the deceased, including eating them. I suggest that a prerequisite for dealing with them in this one particular way, burial, was a feeling for home. Rather than being simply a time- and energy-consuming way to dispose of a corpse, burial may have functioned to soften the emotional blow of separation by keeping the dead in their place(s).

Death and burial may have been emotional events for Neandertals, but that does not mean that they were rife with symbolic content. In fact, my view of burial as an extension of Neandertal home life implies that burial could have been a powerful cultural practice with no symbolic content intended or included whatsoever. Instead it may have been a very concrete extension and continuation of life as it was lived. This response sheds light on the Neandertals' relationships to people and place, and their ability to feel at home.

Home but No House

Over millions of years of primate and hominin evolution, the cognitive components of home life came together slowly. If a feeling for home was indeed a driving force in the practice of the burial of the dead, then more rather than fewer of these components may have come together even before the primary separation between Neandertals and modern humans more than three hundred thousand years ago. The one thing that is missing in this tidy scenario, however, is archaeological evidence of houses, or even spaces that look organized in a domestic sort of way (i.e., hearth here, waste there, detritus from stone tools in a nice little area in the corner, etc.).

As I have said, for my purposes here, I will try to keep house and home separate. As we will see, even among people today, a feeling of home does not need to be tied to a specific structure. On the other hand, it would be helpful to have the remains of some houses to establish the antiquity of home. Unfortunately, the archaeological record dating earlier than tens of thousands of years ago has not yet yielded anything even close to definitive on this matter. Two sites, Terra Amata in France and Bilzingsleben in Germany, have been seen by some archaeologists as providing evidence of structures or living spaces as early as four hundred thousand years ago, well within *Homo erectus/heidelbergensis* times; many others view these claims with skepticism.[29]

It is obvious that no one would expect the earliest structures to be substantial ones. Rather they would tend to be small and temporary shelters, not unlike some of the short-term living structures built by hunter-gatherers today (or at least until recently). Identifying these simple structures poses some exceptional challenges for archaeologists. As Sally McBrearty and Allison Brooks write, there are many potential biases in interpreting the early archaeological record of housing: "Ephemeral structures may have little formal structure imposed by their inhabitants, but sites occupied for long periods may have any existing patterns 'smeared' by later inhabitants. Caves may preserve occupation features better than open sites, but they also circumscribe the occupation area and impose space limits."[30]

Archaeologists excavating modern human sites have definitively identified thousands of living spaces, but any comparison to earlier patterns is biased by just how recently these were formed compared to, for example, the sites of the earliest modern humans in Africa, fifty thousand to two hundred thousand years ago. Unfortunately, as archaeologist John Yellen has pointed out, "Paleolithic Pompeiis are unknown."[31] His own analysis of a ninety-thousand-year-old site (Katanda 9 in Zaire) from the Great Rift Valley in East Africa, which may provide evidence of living floors of two family groups, shows just how hard it is to advance beyond probability statements when trying to draw conclusions about these types of ancient home sites. At first glance, it is easy to "see" a family living and working there, creating the spread of materials that have been preserved. But it is not much more difficult for someone else to show that such an accumulation could have resulted from natural forces or reflected multiple occupations over an extended period.

There are certainly no Neandertal Pompeiis, and some would argue that this is not simply a matter of an absence of fortuitous preservation. As Richard Klein writes, "When Neandertals occupied sites outside caves, they left no persuasive evidence for substantial 'houses' even though the people faced often extraordinarily cool conditions."[32] Preservation aside, can we say with any certainty whether or not Neandertals had the cognitive ability to construct houses or shelters or dwellings of any kind?

Let me visit Neandertal burials one last time. Some archaeologists have looked at how stone tools are made and have reconstructed the number of steps, amount of planning, and other aspects of their construction in order to gain insights into the cognitive capacity of the stone tool-makers (e.g., *Homo habilis* or *Homo erectus*).[33] Similarly, we can consider the steps necessary to produce a burial: a site must be selected taking into account not only its position but also its ability to accommodate the body being buried; the excavation needs to be conducted with some measure of planning as to its final future dimensions; this excavation may be accomplished by using a tool or tools; a decision must be made as to when the floor or foundation of the grave has been reached; a visuospatial assessment of the fit between the body and the excavated hole must be made (although trial and error could work, this would be difficult with a large corpse); the body must be placed in the excavated hole; loose fill is used to cover the body; and stones or slabs are selected and placed to seal the grave. In sum, constructing a burial has an *architectural* aspect, potentially requiring a good deal of planning in its execution.

If Neandertals buried their dead, then I would argue that they also had the cognitive capacity to build simple shelters. Does that mean that I think they did build them? It would not surprise me, but we have no strong evidence that they did. I do not think that the fact that Neandertals buried their dead means that they had a symbolic concept of housing the dead or providing a home for them. I suggest that Neandertal burial occurred in the context of home but does not extend beyond that. Conceptions of burial as providing a home for the dead, where they can undertake eternal rest, surrounded by their loved ones as in life, in cemeteries constituting cities of the dead—these deeply symbolic renderings of postmortem existence belong to modern humans alone. They speak not only to the emotive power of death but also to that of the home life that has been left behind.

The emotional power and presence of home are still with us in the modern world. Now that we have established the cognitive and evolutionary foundations of home, let's look at some of the ways that these very human but still "animal spirits" remain important and relevant today.

Chapter Five

CAN *HOMO ECONOMICUS* EVER FEEL AT HOME?

Feelings for home are connected to economics via the sometimes harsh realities of the housing market. I probably learned this as a kid, watching television shows such as *Good Times* and *The Jeffersons*. Both were family comedies about the 1970s "black experience," with *Good Times* set in a Chicago housing project and *The Jeffersons* in a luxury Manhattan apartment. Part of the message of the shows was that issues involving home and family were a constant across economic (and racial) lines. The comparative framework these shows provided also made it clear that economic circumstances shaped home life, for both the rich and the poor.

In the modern world, there are few alternatives to buying or renting the home you live in. Housing is not just part of the economies of developed countries—in many cases, the buying, selling, and construction of housing are major engines of economic growth and development. Housing is big business, so politics naturally follows. Government actions to suppress or stimulate the housing market, or to encourage or discourage private home ownership, or to allow building in some areas but not others—all of these make housing not just about shelter but about the accumulation and distribution of power.

Housing is a commodity, and while humans are adaptable enough to feel at home in houses big or small, fancy or plain, the impact of the economy on our feelings of home, and vice versa, is inescapable. In the old days, before market economies or socially stratified societies or even agriculture, people lived in dwellings that reflected environmental constraints and immediately available resources. Although there was tremendous variation in housing forms between cultures, there was much less variation within them. This does not mean that the romantic ideal that traditional hunter-gatherers were living in some kind of egalitarian collective is true—certainly there existed profound differences in power and resources (not necessarily material) within these societies. However, housing was not a typical way to mark power or success.

The vast majority of people in developed countries are neither wealthy nor powerful, so they are not typically "players" in the larger housing economy game. Instead, their relationships with their dwellings are influenced not just by economic factors but by the more basic feelings for home that have evolved over millennia. It is a two-way street, with basic needs and desires about housing interacting with economic forces, but that does not mean that everything is in a nice homeostatic balance. As an analogy, consider food and eating.[1] The dietary needs and preferences of human beings have been shaped over quite literally tens of millions of years of evolution. One thing that makes humans a unique primate is the extent to which we are omnivores, extremely flexible in what we can and will eat. But that omnivory is shaped by some basic preferences, and our physiology and psychology are attuned toward maintaining body weight and health in environments where food in general, and certain dietary components in particular (e.g., fat, salt, and simple sugars), are not overly abundant.

In developed countries, industrial production gives people what they want: cheap and abundant quantities of those foods for which we have strong innate preferences (again, e.g., fat, salt, and simple sugars). Beyond this, the food industry, through marketing and research, shapes the dietary preferences of its customers. And via the government-food industrial complex, certain foods gain a privileged position with regard to hygienic regulations, transport, subsidies, and sale. The result is that individuals are no longer in dietary homeostasis in this new food world, resulting in a massive and unsustainable public health problem—obesity—with severe economic and health costs.

Modern housing interacts with our evolved sense of home in a similar way, and the twentieth and twenty-first centuries have given us two important examples of the mismatch between our evolved tendencies and the world as it is today: housing bubbles and public housing projects. There is no doubt that many complex sociological, political, and economic factors have a role in the business of home and housing. But at the core of all these powerful forces shaping the housing industry is an individual who wants or needs to feel at home. Let's see how those feelings contribute to the shape of housing booms and busts.

Animal Spirits

In 1922, a young Columbia University economist named Rexford Tugwell, twelve years away from appearing on the cover of *Time* magazine as a member of Franklin Roosevelt's New Deal "Brain Trust," considered then-current economic theory in light of what psychologists, anthropologists, and sociologists were beginning to learn about human nature. He wrote:

> Man is equipped with the psychical and physical make-up of his first human ancestors; he is the sort of being who functions best in the exhilarations and fatigues of the hunt, of primitive warfare, and in the precarious life of nomadism. He rose superbly to the crises of these existences. Strangely and suddenly he now finds himself transported into a different milieu, keeping, however, as he must, the equipment for the old life, largely useless now. . . . Life on the instinctive level is unthinkable in a culture having as its most important features urbanization, depersonalization of human contacts, diversity of tasks, restrictions on freedom and solitude, monotonies, fatigues, and incomplete expressions.[2]

Tugwell suggested that reason and rationality helped humans overcome some of their shortcomings to adjust to the new world in which their minds and bodies found themselves, but the message was that the basic "equipment" remained the same. He argued that economic theory was focused on "price" with a poorly developed concept of what constituted human "welfare."

Tugwell anticipated by some seventy years what evolutionary psychologists would call the "environment of evolutionary adaptedness" (EEA) or what comedian Rob Becker refers to as "caveman days."[3] The EEA is a useful concept because it is undeniably true that most of human evolution, both in terms of our bodies and our behavior, occurred in an environment that was very different from the environment of complex, urbanized societies that most people live in today. As I have discussed in previous chapters, home evolved in a "traditional" environment, before the advent of agriculture, civilizations, or the industrial revolution. So there is likely a "mismatch" between the EEA of home and the world today, or to put it less strongly, some aspects of our evolved feelings for home may not always work to our advantage in some current environments. Conversely, some current environments may not accommodate or encourage the development of feelings of home.

The EEA concept is useful, but it can become an explanatory crutch. Every time you see in the media that some study or another shows that women evolved to do *X* and men evolved to do *Y*, therefore things today are all messed up, you are seeing the EEA mismatch idea being used as an explanatory shortcut. This is especially true if these explanations, based on population-level investigations, are carried down to the individual level. Detecting evolved patterns generally involves identifying statistical patterns that emerge after cultural or individual variability is somehow accounted for. Of course, for any given individual, cultural or personal histories cannot be eliminated, so while evolution may be a factor that helps us understand a given individual's behavior, the behavior cannot be "explained" by evolution. In fact, to use evolution as an explanation for an individual's behavior, say for an American man who serially cheats on his significant others, is to focus on the story of a species instead of an individual.

That said, economic phenomena emerge from the behavior of groups of people, behavior that has been shaped by millions of years of evolution in environments far different from those of the economic world today. While government policies, financial institutions, and all manner of markets give shape to economic systems, at their core is individual psychology in all its messy and evolved complexity. Economists George Akerlof and Robert Shiller have championed the view that economic theorists have for too long ignored the *animal spirits.*[4] Economists know that there are human factors that influence financial systems beyond those that can be directly measured by economic data, but they do not usually know how to incorporate those factors into their econometric models and theories. So, as Akerlof and Shiller point out, they have ignored them, at their, and ultimately our, peril.

What are Akerlof and Shiller's animal spirits? They take the phrase from John Maynard Keynes, who used it in the 1930s when he recognized that businessmen and other actual participants in the economy did not, as economic theory would tend to have it, always base their decisions on rational, quantitative analysis. In fact, they tended to be more spontaneous rather than deliberative in their decision making, driven by these animal spirits. Over the years, animal spirits in economic parlance came to represent the ambiguous or uncertain elements in the economy, which remained difficult to observe or account for in economic data. Rather than setting these animal spirits aside, Akerlof and Shiller have identified them as critical drivers in the global economy. These drivers are rooted in human psychology. Akerlof and Shiller focus on five as being especially useful for explaining economic phenomena.

Ideas about *fairness* are especially important in how people judge appropriate wages and compensation. Concepts of fairness in general are derived from norms of social behavior. Reciprocity and exchange are at the center of human social life, and we are highly attuned to violations of social expectations about fairness. Another reality of evolved social life is that some members of society will try to get ahead by cheating, colluding, or favoring some members of the group over others. Akerlof and Shiller thus identify *corruption and antisocial behavior* as being major psychological factors influencing economic systems. Corrupt behavior can put stress on economic systems, and economic crises often have corrupt cores. One reason that corrupt individuals can exploit economic systems is that people are not very good with money, especially with regard to understanding the effects of inflation and deflation over time. Akerlof and Shiller argue that *money illusion* undermines the ability of individuals to make rational economic decisions, which in turn undermines economic theory based on rational decision making. Akerlof and Shiller suggest that money illusion is one reason that homeowners tend to overestimate the value of real estate as an investment compared to other long-term strategies. Homeowners often remember the buying and selling prices of a home more so than other investments, and do not keep track of the sources of equity in their homes (payments, inflationary growth, actual growth) or the interest costs a mortgage incurs over a long period of time.

While money illusion is important for deciding among investments, I think that two other factors identified by Akerlof and Shiller are especially relevant at the intersection of the economics of housing and feelings of home. First, there is *confidence,* which Akerlof and Shiller place as the "cornerstone" of their theory. At its heart, confidence is about trust and belief, two things that suggest immediately that quantitative theories or arguments are going to have a hard time accounting for their effects. Confidence is subject to both multiplying and dampening factors, which can have extraordinary effects, for example, on financial markets.

Finally, there are *stories.* People construct narratives to make sense of the many interacting components of their environments, to communicate with others, and to understand their place in the world. Stories begin with individuals, but they are contagious, spreading like viruses from person to person, until they start to take on a life that transcends any single storyteller-and-listener pair. Akerlof and Shiller argue that economic stories can have a profound influence on confidence, both positively and negatively. The cumulative effects of stories can lead to epidemics of confidence.

It will become clear later why I think that confidence and stories have critical roles in the relationship between home and the economics of housing. But before getting to that, I want to briefly consider another, complementary approach to the psychological origins of how people make their economic decisions.

Neuroeconomics: The Brain on Money

Since the turn of the twenty-first century, a small but growing group of cognitive scientists has turned its attention toward understanding the neural basis of economic decision making.[5] While animal spirits encompass the psychology of individuals as they are expressed and magnified in groups, the neuroeconomists want to understand the individual, cognitive basis of economic actions. Neuroeconomics builds on a rich body of psychological research on decision making, some of which has already had an influence on economic thinking. However, armed with the latest neuroimaging methods, neuroeconomics looks to provide a cognitive foundation for economic theory.

Akerlof and Shiller's animal spirits perspective makes it clear that people do not always, or even often, base their economic decisions on rational thinking. This is a top-down approach, looking to understand how economic systems are influenced by individual psychology. The neuroeconomists work in the opposite, or bottom-up, direction, but start with the same basic assumption: that people's economic decisions are not always based on rational, deliberative decision making. The neurocognitive networks of decision making in the human brain are influenced by many factors. So many, in fact, that the rational weighing of the pros and cons of any particular economic action may take a relatively minor role in economic decision making.

What do the neuroeconomists tell us about decision making? One basic insight is that decision making relies on dual systems in the brain (not always defined in exactly the same way by psychologists), one more emotional or automatic (or "hot") and the other more deliberative or reflexive (or "cold").[6] Psychologist and pioneering behavioral economist Daniel Kahneman divides them into System 1 and System 2.[7] System 1 is the automatic system, which kicks in effortlessly without voluntary control. Some of what drives this system is emotional, such as reacting to a hostile act, but some is based on innate or highly ingrained skills. System 2 requires "effortful mental activities" that subjectively involve active concentration or thought. At the very least, the cognitive activities associated with System 2 require some manner of focused attention. Systems 1 and 2 do not operate independently. As Kahneman says, System 1 is always feeding "suggestions" to System 2. For example, System 1 feeds intuitions and impressions to System 2, which may or may not lead to voluntary actions or be incorporated into systems of belief.

George Loewenstein, Scott Rick, and Jonathon Cohen suggest that modern neuroimaging methods can potentially clarify the complementary roles different cognitive systems (dual or multiple) play in economic decision making, and indeed, many experimental studies provide support for a dual-system model.[8] Neuroimaging data, as well as studies of patients with brain lesions, support the dual-system model in decision making for the assessment of risk and ambiguity, in assessing rewards or losses that are distributed over time, and in social decision making. Decision making involves emotional brain systems (incorporating the amygdala and other limbic and paralimbic structures), along with parts of the frontal cortex associated with these structures and other frontal executive regions. Cortical regions outside the frontal lobes may also be involved in specific risk-assessment situations.

The reward pathways of the brain involving the neurotransmitter dopamine are also critical in some aspects of decision making. For example, decisions concerning short-term, immediate rewards are more likely to activate dopamine-rich areas of the limbic system, while decisions about delayed rewards activate higher-level frontal and parietal regions.[9] The dopamine system is important in the pathology associated with substance abuse, mood disorders, and schizophrenia. Carla Sharp, John Monterosso, and Read Montague have suggested that the kinds of research tasks that neuroeconomists have designed to look at decision making may also be useful for uncovering cognitive markers for behavioral disease.[10] This gets at the basic idea that while the decision-making networks of the brain may be functioning properly, problems in the reward system may give the appearance of a primary decision-making deficit where one does not exist. I will come back to this in the next chapter when I discuss homelessness and the mentally ill.

Both animal spirits and neuroeconomics suggest that the psychology of home and housing has deep roots. The term *animal spirits* is not meant to literally invoke animal behavior (it was originally used in the sense of the Latin term *animus,* suggesting a life force). However, the psychological factors highlighted by Akerlof and Shiller are clearly those of an animal with a highly evolved and complex social life, and a less evolved grasp of even moderately difficult math. Neuroeconomics traces the roots of economic behavior to the hardware in our heads, which has been programmed both by evolution and experience. Animal spirits and neuroeconomics help us understand a whole range of economic phenomena. They should be especially useful, however, for examining the economics of those phenomena—such as home—that are themselves a product of a deep evolutionary past.

Home Bubbles

Whether it was with tulips in early seventeenth-century Holland or baseball cards in the 1980s and 1990s in the United States, economic bubbles have been an inescapable feature of relatively unregulated capitalism. Each boom and bust has its own story, of course, depending on the historical and sociological contexts. Causes and effects vary. But economists, sociologists, and psychologists have sought common causes rooted in human behavior for why these bubbles occur. One thing that is universally true of economic bubbles is that they make more sense when we look at them with twenty-twenty hindsight. Looking backward, everyone can see how "obvious" it was that a given bubble was going to burst and can express some wonderment at the credulity of those who got caught in them.

The housing bubble of the first decade of the twenty-first century was notable not only for its magnitude, but also for the role it played—mostly via "sophisticated" financial instruments—in precipitating the worldwide recession of 2008–2009.[11] Beginning in 2000, there were significant housing price increases in Australia, Canada, China, France, Hong Kong, Ireland, Italy, New Zealand, Norway, Russia, South Africa, Spain, the United Kingdom, and the United States. Between 1997 and 2007, housing prices in the United Kingdom tripled. In the United States, the average sales price of a new home was $181,900 in 1998 and reached a peak of $313,600 in 2007, after which it started to decline. The US home ownership rate peaked in 2004 at 69.2 percent. Beginning in 2005, there was a market slowdown in sales of existing and new housing, which accelerated through the bursting of the bubble over 2007–2009.

The turn-of-the-century housing bubble was not of the same magnitude in all these countries. Even within countries, the size of the housing bubble could vary significantly among cities and locations. In the United States, the cities and suburbs of coastal California led the way. In 2005, the top ten areas for housing price increases in the United States were all in California, and the median sales price for a house in San Francisco reached $765,000, triple the national average. Significant housing bubbles were also experienced in major cities such as Boston, New York, and Washington, as well as in Sun Belt cities, such as Phoenix, Las Vegas, and some Florida communities. Although there was a national housing boom, some areas (such as in the Midwest and Texas) saw more modest increases and subsequently more modest decreases when the bubble burst.

During the boom, housing price rises were truly exceptional in the "big glamour cities," as Robert Shiller calls them, of the world. London, Sydney, Paris, New York, Boston, Moscow, and so on participate in a global housing market in which prices are driven up not only by local demand but by the demand for excitement or a safe financial haven for the world's transnational rich. As economic journalist Faisal Islam writes, "London in particular has become the preferred residence of the world's wealthiest people, from Russian oligarchs to Arab oil sheikhs."[12] These international pressures eventually affect local markets, as price increases in major centers reverberate through their national economies.

The potential influence of international markets on local housing markets is something relatively new. Robert Shiller has shown in his groundbreaking analyses of historical house price data that housing prices over long periods of time have tended to be relatively flat, after controlling for general inflation and the fact that houses themselves have an increasing list of standard amenities that must be included in their cost (e.g., running water, electricity, garages, etc.). There were plenty of housing bubbles in the past, but they tended to be local and tied to regional developments, such as the introduction of a new railroad line. The international pressures on housing prices reflect the new world of the twenty-first century, a world that is much more connected and interconnected, both in terms of information transfer and the transfer of wealth. Housing has long been considered a great investment for the middle class to save money. Now, individual buyers often look at housing to meet all the basic needs of home, along with being a speculative investment. Shiller writes, "The increasingly large role of speculative markets for homes, as well as other markets, has fundamentally changed our lives."[13]

The housing bubble of the 2000s was not unlike other financial bubbles, and it probably shares some of its potential root causes with them. For Robert Shiller, the animal spirits played their roles in this situation as they have done in countless others. Stories about successful speculation in the housing market spread, encouraging an atmosphere of urgency around house buying. Optimism engendered by advances in the stock market spread to the housing market. The basic money illusion, which has always made housing seem like a better investment than it is, created an impression of housing as a safe speculative vehicle. For conservative economists like Thomas Sowell, the artificial manipulation of housing supply and demand by local governments was a culprit. In some communities, restrictive zoning laws meant to deter growth or encourage preservation of natural environments limited the expansion of housing stock, fueling increased prices. Cities with a more laissez-faire approach to housing development generally (but not always) had a more muted boom and bust. Sowell also points out that in several markets, investors rather than individual homeowners were the main drivers of the bubble.

For the housing bubble of the 2000s, almost all economists, journalists, and other observers agree that the relaxation of lending standards and the bundling of prime and subprime mortgages into obtuse investment instruments played a critical role in the collapse of home prices. In the United States, government policy instituted in 1999 directed the government's semiautonomous mortgage corporations to increase the percentage of mortgages purchased that were made to lower-income families.[14] On the ground and in the neighborhoods, policies like this, along with optimism about the health of housing and other markets, encouraged lenders to put a premium on selling new mortgages. At the same time, lenders became much less concerned about the long-term ability of borrowers to make their payments. During this time (in the United States and other countries), there was a vast increase in available credit with a corresponding decrease in the standards for borrowing, leading to a dangerous expansion in subprime lending. Again, for lenders the emphasis on selling new mortgages outweighed concerns about the ability of borrowers to pay. This led to the spread of "predatory lending" in some areas, in which subprime borrowers fell victim to mortgages they could not really afford (and which would prove financially disastrous when the bubble burst). One question is, why were the prey of predatory home lenders such easy marks? Was it just their greed, their desperation, or something else?

The Home Ownership Meme Can Be Very Powerful

Most of us are familiar with the concept of the "internet meme." Someone does something, say "planking" on a fast food counter, and uploads a video of him- or herself doing this to a public or social media site. Other people see the video, emulate the activity, and post themselves doing this to even more sites. Before you know it, the activity has spread like an epidemic throughout the world. Another internet meme has been born.

Evolutionary biologist Richard Dawkins coined the term *meme* decades ago, but it really did not become a widespread meme itself until the internet age.[15] Dawkins's concept of meme was that it is an analog to gene—a replicator that is subject to natural selection and other evolutionary forces just as genes are, but in the cultural rather than the biological environment. Although memes spread through the cultural environment, competing with other memes and undergoing changes or mutations that make them more or less appealing, the ultimate selective environment for them is the human brain. This means that cultural memes that tap into the fundamental cognitive drives and preferences of the mind may be quite powerful and pervasive and important. For example, there has been much investigation of the evolutionary "memeplex" of religious belief systems, which are obviously of critical importance to our species as a whole.[16]

I suggest that another very important meme, at least in the United States, the United Kingdom, and some other countries, is "home ownership is good." This cannot be a particularly ancient meme, because home ownership is a concept that emerges from relatively modern capitalist economies. On the other hand, the notion that people somehow control and possess the structures (or living spaces) they inhabit is probably much more ancient. Indeed, the security and stability that are part of feeling at home likely derive from a sense of possession of a structure, if not ownership in a legal or economic sense.

From its founding, what with the emphasis on individual self-determination, it seems only natural that the United States would foster an ideology of home ownership. By owning their own property, farmers on the expanding frontier could be free of the feudal oversight of landlords and control their own destinies. Of course, farms are both home and business, which further encourages ownership on the part of their occupants. As historian Lawrence Vale has pointed out, by the second half of the nineteenth century, more and more Americans were living in towns and cities, and fewer were living where they worked. Nonetheless, the frontier mentality still held, and the idea of home ownership as a good, American thing remained common. Although only 7 percent of Bostonians were homeowners in 1880, that figure rose to 25 percent by 1900 and 35 percent by 1910.[17] This increase was made possible by the development of new neighborhoods and suburbs away from the original city center.

At the beginning of the twentieth century, nonfarm home ownership became a driving force in housing policy in the United States, at several different levels of government. Zoning laws and tax codes increasingly encouraged home ownership, and realtor and builder trade organizations began to promote home ownership as the apotheosis of American values. As Vale writes, "the home could be lauded as the superiority of individualism to anarchosyndicalism or other socialist or communist movements."[18] This all came together in an extraordinary promotional campaign launched in the early 1920s by the National Association of Real Estate Boards (NAREB), working with the US Department of Labor and other groups. The "Own Your Own Home" campaign made home ownership practically a patriotic duty. Not only that, home ownership was equated with manliness and power, the rugged frontiersmen of the nineteenth century reborn in the suburban bungalows of the twentieth. This campaign, along with other policy initiatives, consistently denigrated renting or tenancy, and offered little to address the plight of the slum-dwelling poor (who were in fact vilified). The government's role in housing was seen to be only as a facilitator, not as a primary provider.

Undoubtedly, the "Own Your Own Home" campaign both shaped and reflected public opinion. We see that the ideology of home ownership present in the US housing bubble of the 2000s had deep roots, both in the culture and in decades-old government policy. In the United Kingdom, home ownership in the twentieth century has been seen as a foundation of democracy. Margaret Thatcher in fact hailed the "property-owning democracy," and under her direction, two million government-owned residences were sold to private buyers. Since her time in the 1980s, successive British governments strongly encouraged home ownership. These policies had the dual effect of temporarily expanding home ownership while supporting an extraordinary rise in prices. For many years, this was not seen as bad news. As Faisal Islam says, "Housing is the only basic human need for which rapid price rises are met with celebration rather than protest."[19]

The power of the home ownership meme helped make some people exceptionally vulnerable to predatory lenders during the housing boom. There is nothing new about lenders who prey on the desperation of their clients: loan sharking has long been a cornerstone of organized crime, payday lenders and check-cashing services make their money on people who need their money sooner rather than later, and pawnbrokers have been around forever. But lenders for buying houses are different, or at least they used to be. To paraphrase one British banking executive, during the subprime lending boom, it was as if mortgage lenders changed from being like doctors, with the best interests of the client (at least in terms of the ability to actually pay for a loan) at the forefront, to being like bartenders, plying customers with more and more whether or not they could handle it.[20] Statistics during the housing boom in the United States showed that subprime loans were most common in the most overheated housing markets of California, Florida, Nevada, and Arizona, where they accounted for between five and ten new mortgages per one hundred housing units in 2005. Subprime loans were also disproportionately concentrated in zip codes with larger African American and Hispanic populations.[21]

To be in a position to obtain a mortgage, even a subprime one, buyers usually have some income and are almost certainly not homeless. They are far from hitting economic rock bottom; although they may be poorer or have worse credit than average, they are not conventionally financially desperate. During the boom, the vulnerability of subprime borrowers to their lenders was not based on an immediate need of financial rescue or resolution. Instead, they were seduced by cheap credit, or at least money that looked cheap (remember money illusion), and the possibility of substantial financial gains in a booming housing market.

Prey become vulnerable to predators for all sorts of reasons. It would be easy to blame greed for why borrowers took on more than they could handle, but I suspect that fear was also an important factor. For poorer first-time buyers and those whose bad credit had kept them out of their own house, the booming real estate market would make the prospect of joining or rejoining the ranks of homeowners seem ever more distant. We have seen that the home ownership meme is very powerful and pervasive—to not be a homeowner is to not be a full participant in democratic society. The fear of not ever being able to own a home, to be permanently priced out of a market, is a great motivator to buy *now.*

Some regions of the United States, such as the large urban centers of the Northeast, have long accepted renting as an acceptable and realistic alternative to home ownership. It is perhaps not surprising that the subprime crisis was not much of a crisis in these regions. Instead, it hit hardest in the Sun Belt states, with geographically expansive real estate markets and an abundance of new home construction. These areas also have more migration into them—they make up some of the new frontiers of American life. The home ownership meme may be more powerful in these areas, and the failure to live up to it more acutely felt. The idea of home ownership is culturally constructed, but the power of feelings related to home runs much deeper. They affect our emotions, which in turn affect our ability to make decisions, financial or otherwise.

Confidence and the Security Illusion

Even economists recognize that there is more to a house than its monetary value as a physical shelter. As Thomas Sowell writes, "'Home' is, after all, a word with emotional overtones, quite aside from its physical or economic significance."[22] Sowell brings this up to reinforce the point that decisions about housing policy made at the government level are often based on emotional pressures and ideological visions. At the level of the individual home buyer, emotion certainly plays a role. As I discussed above, the feeling of anxiety about being priced out of a housing market in a boom can influence a decision about whether or not to buy using a loan whose terms, even if they are fully understood, are quite risky.

Acting on fear or anxiety during a boom may be bad, but an even more dangerous feeling for a house buyer during a boom may be confidence. George Akerlof and Robert Shiller argue that confidence, or more specifically, overconfidence, is one of the animal spirits that play an important role in a booming market. One thing that happens during a boom is that circumstances and context increase baseline levels of confidence. These amplifiers or multipliers turn confidence into overconfidence, fueling increases in price and reducing the level of risk aversion in investors.

What are some of these confidence amplifiers? In a series of surveys taken from the 1980s until the 2000s, Shiller found that over this period investor confidence in both stocks and housing generally increased, although it eroded somewhat during periods of decline.[23] One thing these data show is that people's viewpoints are a reflection of their experiences. That is, their optimism or pessimism is not strongly shaped by an understanding of broader historic trends. In fact, Shiller identified an ongoing optimistic bias: investors are much more likely to believe that markets that have gone down will recover within a few years but not that markets that have shot up will just as likely retreat. Booming markets also generate lots of success stories, which in turn enhance investor confidence. Shiller suggests that economic bubbles are like "naturally occurring Ponzi schemes." Rather than being misled by a fraudster manipulating profits and promoting his "success," investors are emboldened by real short-term profits and the publicity about them.

Shiller's view of investor confidence fits in well with what Daniel Kahneman has called the "what you see is all there is," or WYSIATI, rule.[24] At the heart of this rule is that people are prone to making judgments based on the evidence available to them, whether or not that evidence is particularly good or comprehensive. In fact, Kahneman says that most people do not even allow for the possibility that critical evidence for making a judgment might be missing. This is not to say necessarily that ignorance is at the root of confidence, but rather that our intuitive thinking depends on the suppression of doubt and ambiguity. As Kahneman says, the subjective confidence people have in their beliefs "depends mostly on the quality of the story they can tell about what they see, even if they see little."[25]

Human cognitive evolution occurred for the most part in a WYSIATI world. By this, I mean that for most of our evolution, people's understanding of the world was based on their own experiences and on the experiences of their closest relatives. At the zoological level, human language allows for an unprecedented level of information transfer and accumulation. But for most of our evolution, the information available could only be obtained firsthand or shared among a group of closely associated individuals. Although in our current Age of Information the volume of WYSIATI-based information any traditional hunter-gatherer group could possess seems quite meager, it is important to keep in mind that that amount of information far outstrips the amount available to any other animal. Our ability to both store and process information placed us in our own cognitive niche.

Our feelings of home, both in an evolutionary and in a personal sense, are very much the product of the WYSIATI world. The feelings of stability, safety, and security that we associate with the place we live in can really only be obtained through firsthand experience. The evolutionary environment that shaped our propensity to have these feelings was wholly local and immediate, as were the security and safety offered by home. In the traditional world, home did not provide a real defense against the most dangerous of threats: severe weather, disease, attacks by other groups of humans, and so on. This is not to say that the feelings of control, safety, and stability offered by home are illusory, but they are limited.

In the past, home ownership was encouraged because it fostered stability and security, values that reflect the feelings we associate with home. But in the latter part of the twentieth century, owner-occupied real estate in the United States, the United Kingdom, and other countries became as much about speculation as about saving. In this environment, feelings of home likely joined other confidence amplifiers in fueling buyer participation in these overheated markets. Real estate investors may have played a critical role in driving up prices in these booms, but owner-occupants were then forced to participate in them, often at great financial risk. As critics of predatory lending claim, many borrowers did not understand the risks they were dealing with or were deliberately misled about them. As Shiller and Kahneman make clear, however, overconfidence has many possible sources, reflecting how our intuitive decision making is shaped. I suggest that at the cognitive overlap between house and home, the feelings of control and security we associate with home can serve to lessen our perception of financial risk when signing a house mortgage (especially one, for example, that has a balloon payment several years down the road). In the WYSIATI world, home provides us with protection, but this can be a security illusion in the face of high-risk financing.

American Public Housing: A Mostly Dismal Story

Among Akerlof and Shiller's five animal spirits at the psychological root of ebbs and flows in the economy are stories. In a boom, positive stories about the successes of others fuel increased participation in rising markets. No one wants to miss out on a potentially big return, or to be left behind as prices rise out of reach. Conversely, negative stories can have a dampening effect on the economy, accelerating downward trends or preventing recoveries. In a down economy, such as that following the recession of 2008–2009, underconfidence can be even more damaging than overconfidence. People hear stories about laid-off friends and family, declining real estate values, and hard-to-obtain credit; alarmist political debates spread stories about what terrible things *could* happen if one side or the other gets its way.[26] When negative stories of these kinds become dominant in people's conversations, spending and investment inevitably suffer.

Why are stories so important? Anthropologist Robin Dunbar has hypothesized that the evolution of language can be traced back to the importance of telling stories about people—in other words, gossip.[27] In most primates, mutual grooming serves to enhance group cohesion and stability. Dunbar suggests that language took the place of grooming in our ancestors, allowing for increased group size (since language does not require one-on-one contact). Although language can be used to convey information about all sorts of things and concepts, Dunbar's studies show that people spend the highest percentage of their time talking about other people.

Even if gossip is not the primary factor behind the evolution of language (there are plenty of other ideas about how that happened),[28] there can be no doubt that telling stories about other people is one of the most important ways people use language. According to literary scholar Brian Boyd, storytelling—or the use of narrative in communication—has evolved to become perhaps the most essential way that people convey information to one another. We can share information and facts without embedding them in narrative, of course, and many kinds of information can even be transmitted nonverbally. But as Boyd points out, narrative is what allows us to take full advantage of the brain's complex cognitive abilities. He writes, "With narrative we could, for the first time, share experience with others who could then pass on to still others what they found most helpful for their own reasoning about future actions . . . with narrative we can be partially freed from the limits of the present and the self."[29] And as Boyd also points out, narrative also allows us to be freed from real things and events, opening the door to a world in which fiction has an important and useful place.

Stories with emotional content resonate most strongly with us. Conversely, it may also be true that we match our most important emotional relationships to preexisting narratives. Psychologist Robert Sternberg suggests that stories frame the relationships that couples have with one another.[30] No one comes to a new relationship as a blank slate: expectations, preferences, previous experiences, cultural traditions, all contribute to an individual's implicit story of how to be in a relationship with someone else. Sternberg argues that these stories can change over the course of a relationship, becoming more elaborate, but that they remain the frame through which one partner views the other. Couples in successful relationships can have quite different stories, but they need to be compatible ones, and relationship partners must fill their expected roles. Sternberg says that there are a potentially infinite variety of relationship stories. The key thing is that everyone has a relationship story of some kind.

Stories are critical in how people communicate within and think about their relationships of all kinds, so it is no surprise that Akerlof and Shiller identified stories as one of the animal spirits that underlie economic booms and busts. The spread of information and, perhaps more importantly, disinformation provides the fuel for these cycles. At an individual level, stories are clearly the most effective, even the most natural, way to get information. It is one thing to read about housing prices going through the roof in California, but to hear a friend tell a story about how she has borrowed $100,000 against the equity in the house she has owned for only a year, that can get your attention.

Getting to public housing—alas, increasing equity is not part of its story. The stories here are bleaker, and get titles like *Blueprint for Disaster: The Unraveling of Chicago Public Housing* by D. Bradford Hunt; *Purging the Poorest: Public Housing and the Design Politics of Twice-Cleared Communities* and *From the Puritans to the Projects,* both by Lawrence J. Vale; and *American Project: The Rise and Fall of a Modern Ghetto* by Sudhir Alladi Venkatesh. The story these books collectively tell about public housing in the United States, especially during the second half of the twentieth century, is mostly one of failure, waste, and loss. Although there are certainly inspirational individuals who figure in this story, they emerge mostly as paragons of futility, unable to convert good intentions into workable solutions. These books are not escapist fare. They are crucial, however, to understanding the mismatched role of home in modern society.

Lawrence Vale, in his masterful study of public housing in Boston, which mirrors the experiences of other urban centers in the United States, details the complex relationship between communities and their governments and the poorest, most ill-housed members of those communities.[31] Even before the development of a modern, capitalist economy, America was a country of individuals, thus dependence was always viewed with suspicion. In earlier times, care of those too poor or infirm to house themselves was a strictly local issue. Christian charity, or sympathy for familiar members of the community who had fallen on hard times, would see to it that the poorest had a roof of some kind over their heads. By the 1820s, governments took a more systematic approach to reform and housing both of the "worthy" poor and those deemed less worthy. Large almshouses, poorhouses, and orphanages were built in or outside many cities, along with penitentiaries and insane asylums.

In a country where rugged individualism and home ownership are both celebrated, the notion that there were "worthy poor" among those needing housing assistance was somewhat paradoxical. Even so, governments recognized that affordable housing could be an issue even for responsible, gainfully employed members of the community. Lawrence Vale divides the history of public housing in the United States in the twentieth century into three "experiments."[32] During the first experiment, beginning in the Great Depression in the 1930s, public housing programs targeted a "submerged" middle class—lower-class people with middle-class aspirations. An income threshold was placed on who could live in public housing, and with it, thresholds for race or ethnicity, citizenship, and family structure (single mothers need not apply) were also enforced.

Until the 1950s, the occupants of public housing were typically white, two-parent families and were employed in lower-middle-class occupations (such as fireman or policeman).

In the 1950s, the white middle class began their exodus from city centers to outlying suburbia, and city housing administrations had more and more difficulty filling their units with "acceptable" candidates. At the same time, the civil rights movement put cities under pressure to provide decent housing for all of their residents. Thus began what Vale sees as the second experiment in public housing, in which it was less a step up and more a welfare mission.

Clearing of slums or ghettos had long been part of many public housing schemes. Although this was sometimes more about making land available for lucrative development, there was also a concern for providing poor people with better, more modern dwellings. During the 1950s and 1960s, the epitome of more modern housing was the high-rise and high-concept housing project. These built-from-scratch communities were supposed to be both efficient and, with large open spaces between structures, salubrious. Unfortunately, many of the grand designs were not funded adequately, and they turned out to be a nightmare to maintain, manage, and police. In addition, the concentration of poverty in these projects was staggering: by 1992, three-quarters of public housing residents received no income from employment.[33] In Chicago's notorious Robert Taylor Homes project, 95 percent claimed government social welfare as their sole source of income.[34] D. Bradford Hunt, in his history of Chicago public housing, writes that there was no intent "to 'warehouse' the low-income poor in unattractive 'vertical ghettos,' though this was certainly the end result."[35]

Due to inadequate funding for maintenance, the physical decline of these high-rise projects began almost immediately after they were built in the 1960s. The quality of life in the projects did not necessarily decline so quickly, however. Indeed, as sociologist Sudhir Alladi Venkatesh makes clear in his historical ethnography of the Robert Taylor Homes, many residents entered the 1970s with a sense of optimism.[36] Shortcomings in services and communication between tenants and city and federal housing authorities had been addressed by the creation of Local Advisory Councils (LACs). These LACs served to empower the residents and gave them a voice in the management of their own buildings and their own lives. However, over the course of the 1970s and 1980s, economic stagnation and the fact that the projects became a catchment for the poorest of the poor were both hard to overcome. Eventually the projects became dominated by organized gangs and criminal activity.

The widely perceived "failure" of public housing in the 1980s ushered in what Vale calls the third experiment, which in effect was a retrenchment. The number of public housing units in the United States peaked at 1.4 million in 1991, and has been in decline ever since.[37] Hastening the decline was the systematic demolition in many American cities of the high-rise projects themselves.[38] The reduction in available housing units meant that many of their original inhabitants were displaced. As Vale points out, this allowed governments to again focus their attention on housing the "worthy poor." The replacement projects, reflecting a "new urbanism" emphasis on smaller-scale, more intimate housing, were designed "to appeal to and reassure" the middle-class community. Many of the poorest remain dependent on a patchwork of government and charitable programs in their search for affordable housing.

So the story arc of public housing in the United States in the twentieth century is a rather sad one. But I think there are actually several stories that are relevant to this larger one. First, there are happy stories to be found among the projects. Obviously, high-rise housing is no impediment to successful home life, and the acceptance of public housing in countries such as New Zealand show that, despite some political and economic tensions, it can be successful and socially acceptable.[39] I lived in New Zealand during the 1990s, coming from the United States at a time when housing projects were regarded by middle-class observers with both fear and loathing. I was surprised that many middle-class New Zealanders nostalgically extolled the virtues of "state housing" (under political pressure at the time due to a short-lived experiment at charging more market rates for them), and even recommended buying old state houses due to their sturdy construction. In his interviews with people who had grown up and lived in the Robert Taylor Homes during the 1960s and early 1970s, Sudhir Alladi Venkatesh found many who truly felt at home in the projects and looked back fondly on those times. Things changed as the environment became more menacing, but for a while, despite many shortcomings, it was not obvious that the project was doomed to failure.

The meme of home ownership imposes a certain kind of story on the economic relationship between people and their dwellings. In the United States, home ownership goes hand in hand with individualism, independence, and aspirations of social and economic advancement. Public housing can have a role in this story if it is seen as helping someone achieve any of these goals. More often than not, however, public housing is now seen as the landing spot in a downward spiral, when aspirations are gone and dependence a permanent state of being. These two stories are difficult to reconcile, and the destruction of many of the large housing projects from the 1960s suggests that federal and local governments have given up trying to fit large-scale public housing into the story of the American Dream.

Robert Sternberg suggests that stories give shape to our romantic relationships. I think that it is also very likely that stories give shape to emotional relationships of all kinds, whether they are with people, objects, or other entities. Home is about the emotional relationship of a person with a place. That place was once primarily a shelter and a location in a small-scale social world. Today, our living places are part of a much larger world, important cogs in economic systems, and the object of government promotion and regulation.

Our personal stories about our homes will be influenced by government policies toward housing. The poorer you are, the more true this will be. In a country like New Zealand in the second half of the twentieth century, where there was support for the notion that the government should provide a reasonably large safety net for the basic needs of its citizens, there was widespread acceptance of public housing, and living in it had a limited stigma. On the other hand, in the United States, where the government subsidizes home ownership at a much higher level than it does public housing (mortgage interest tax deductions are more than double the budget of the Department of Housing and Urban Development),[40] to be so overtly dependent on the government for housing makes for a story in a damaged setting.

Looking at the recent history of public housing in the United States, it is easy to get the sense that the conflict between home comfort and cultural expectations has taken a perhaps permanent toll on this institution. The demolition of the 1960s high-rise projects might have appeared to signal, in a literal and figurative way, the demise of large-scale government investment in American public housing. But unlike written stories, the stories that shape relationships are ongoing and can be changed and modified as circumstances change. The problem of housing the poorest in a capitalist liberal economy is an ongoing one. Can it change for the better? Sociologists, psychologists, city planners, architects, and others learned much from the failed experiments in public housing of the second half of the twentieth century. This knowledge can help make for a better, more people-friendly, version in the new century. As always, the question is if there is sufficient political will and basic empathy to see that it does.

Can *Homo economicus* Ever Feel at Home?

The short answer is, no. This is not to say that having more money rather than less cannot make a materially better home life possible. More money always helps to pay for a mortgage or rent, buy a bit more space, buy better furniture or appliances, pay for taxes or for a nicer shared lobby or private garden, and maintain a property and fix those broken windows that are the bane of declining inner-city neighborhoods. And in a free-market economy, poverty can always be a barrier to obtaining adequate and safe shelter. So with money in his or her pocket, *Homo economicus,* the economists' beloved rational, decision-making participant in the economy, can be in a better position to make decisions about how to spend money on his or her dwelling.

But as Daniel Kahneman and George Akerlof and Robert Shiller make clear, *Homo economicus* often falls short as a model for how people make decisions and behave in the real world. Although it represents a somewhat limited explanatory framework for human behavior, we can say that we are all, at least sometimes, *Homo economicus.*[41] Our behavior is also a product of the norms of society in which we live: we are also *Homo sociologicus,* as it is sometimes called. And our evolutionary past also shapes what we do—let's call that part of us *Homo darwini.*

So when I say that *Homo economicus* cannot feel at home, I mean that these feelings do not emerge as a result of rational cognition, but from a different part of our evolved psychology. As I have discussed in this chapter, there are ways in which our feelings for home can influence our economic decision making. They can be confidence amplifiers or suppressors, or be the basis of our emotional stories surrounding home. Conversely, the economic world can amplify or suppress our various feelings about home. It would be foolhardy to say that our homely well-being is not influenced at least in part by the quality of our material surroundings and all of the aspects of economic status and power that go along with them. Ultimately, however, our feelings of home emerge from the inside out. The old cliché is that you can't buy happiness. Likewise, despite what a realtor might say, you can buy a house—a nicer house, a bigger house, a house with granite countertops and automated zone heating—but you can't buy a home. That you have to build yourself, according to the blueprints drawn from your evolutionary history, cultural traditions, and personal experiences.

Chapter Six

WITHOUT HOME

Being without a home, with all its cognitive and emotional trappings, is to be on the margins of society. This is true whether we are talking about a small-scale, traditional village or a massive urban center. In capitalist economies, short-term homelessness represents economic failure, and chronic homelessness is a sign of total capitulation to whatever ailment, condition, or circumstances brought it about. The homeless attract sympathy and support, but rarely empathy.

The Universal Declaration of Human Rights was adopted by the United Nations General Assembly in 1948. Its Article 25(1) reads in full: "Everyone has a right to a standard of living adequate for the health and well-being of himself and of his family, including food, clothing, housing and medical care and necessary social services, and the right to security in the event of unemployment, sickness, disability, widowhood, old age or other lack of livelihood in circumstances beyond his control."[1] Each country decides whether or not to accept United Nations declarations and recommendations. Like many others, this one has been generally ignored. Certainly, a right to housing is not widely recognized. Scotland provides a notable exception, with legislation there guaranteeing that from the end of 2012, "all those assessed as unintentionally homeless . . . are entitled to settled accommodation as a legal right."[2] On the Scottish government website, the word *historic* is frequently used in describing this legislation, reflecting just how novel this initiative is (and a certain amount of deserved national pride).

Homelessness is often considered to be akin to a disease. For an individual, chronic homelessness is something for which a cure must be found. For a community, homelessness is an ailment of the social body as a whole. But like many nonterminal bodily ailments, it can be ignored if things are going well enough otherwise. An international opera company, great restaurants, or year-round fine weather can go a long way toward taking people's minds off the homeless in their community. In addition, the very marginal status of the homeless makes them easy to be ignored. They are often physically removed from heavily trafficked areas, and when they are not, they are easy enough to not see. Perception and attention are easily biased by a multitude of factors, and the biases against the homeless are legion.

How big a problem is homelessness? Obviously, it varies from country to country and city to city. Counting the homeless poses more than a few difficulties. In many ways, it has all the problems epidemiologists run into when they try to figure out how common a disease is. Homeless counts are biased toward those in shelters, jails, hospitals, or other facilities; those living on the streets, under overpasses, or in cars are harder to locate. Similarly, disease counts are biased toward cases that are seen in hospitals and clinics, which is fine for counting very serious diseases, but means it is easy to miss chronic conditions, such as diabetes. How the counts are reported can also make a difference. Cases can be measured over a shorter (a day) or longer (a year) period of time. It is easier to do a count over a short period of time, but that does not necessarily give an accurate picture of just how big a problem a disease, or homelessness, is.

The US Department of Housing and Urban Development (HUD) produces an annual report on homelessness nationwide.[3] Their survey showed that on a single night in January 2013 there were 610,042 homeless people, of whom about two-thirds were in some kind of shelter and one-third were living outside shelters, in cars, on the street, and so on. About one-third of the total were families rather than single individuals. Again, and as advocates for the homeless will point out, nonshelter counts of the homeless are inevitably low, missing not only those living totally unsheltered but also those staying with friends or in motels. In the late 1990s, the National Law Center on Homelessness and Poverty estimated that there were 840,000 people homeless on any given day.[4] The HUD figures suggest that there has been a steady decline in the number of homeless in the years approaching 2013, reflecting some success in dealing with the issue.

Advocates for the homeless argue that figures from a single point in time may paint a misleading picture.[5] Those at risk for homelessness are often intermittently housed, so a single day's survey only captures a proportion of the total number of people who may be homeless over the course of a year. Based on various studies, an annual homeless figure can be derived from the single date number, by multiplying it by a factor of between four and five. Six hundred thousand homeless on a single day indicates about 2.7 million homeless for the year. The US Census Bureau found that 46 million Americans were living below the poverty line in 2012, which means about 6 percent of people in poverty in the United States experience homelessness in a given year. Nearly 40 percent of these are children.

Without worrying too much about exact numbers, it is clear that homelessness, or the threat of homelessness, is an ongoing, daily concern for millions of Americans. Similarly, statistics released in 2013 by the European Union suggest that 4.1 million people in their member nations are exposed to homelessness each year.[6] Around the world, the United Nations estimated in 2005 that there were at least 100 million homeless people.[7] There are many causes and effects of homelessness.

In developed countries, the immediate cause of economic homelessness is fairly obvious: housing costs money and when people do not have enough money to pay for housing and lack a personal family or social safety net, they are at risk for homelessness. Although one can explore the deeper societal and economic causes of homelessness, for many individuals, not having enough money to pay for shelter is what puts them on the street for the night. Other people wind up homeless not because of primary economic factors, but because they are mentally ill or are substance abusers. In the United States, surveys indicate that 26 percent of the homeless have a mental illness, 38 percent are dependent on alcohol, and 26 percent abuse other drugs (there can be overlap among these categories).[8]

Homelessness is generally defined as not having a permanent shelter, but one can be without a home when moving between places, and people suffering from the mini-grief of homesickness can feel psychologically homeless. Displaced people, such as refugees, are also often sheltered but without a home. From a cognitive standpoint, for most people, being without a home—or even living with the threat of losing a home—is to be in an emotionally vulnerable and distressed state, even if free from the physical dangers of true homelessness. What may also be possible is that if feeling at home is at least in some sense a basic part of our cognitive repertoire, there are people who have a diminished capacity to feel at home. Clearly, the generalized effects of mental illness and substance abuse undermine some people's ability to obtain a home, but they may diminish the motivation to do so, as well. An interesting question is, are there individuals who cannot feel at home not because of some general mental illness but because of some more specific deficit—they are physically housed, but cognitively homeless?

Home Insecurity

"Home insecurity" sounds like a phrase a house alarm system company might use to sell their products. It describes, however, the status of most of the people at risk for economic homelessness in a country like the United States. The majority of these individuals and families are not actually homeless, at least not on any given day, but the specter of homelessness looms over them on a persistent basis. It is an ongoing stress in their lives, which takes its toll on their emotional and psychological well-being. Read or listen to any media account of someone about to lose his or her home, and the anxiety and distress are often palpable.

Home represents one of our most basic needs. Another basic need is food. For decades, researchers have used the concept of "food insecurity" to study the chronic effects of undernutrition or malnutrition (short of starvation).[9] In a wide range of studies conducted in many different settings, food insecurity has been shown to be associated not just with physical ailments but with a wide range of psychological issues. Most prominent among these are anxiety and depression. The ongoing stress of an uncertain food supply, or of a food supply consisting of the cheapest calorie-rich but nutrition-poor foods available, takes a long-term toll. Home insecurity has been less studied than food insecurity, but there are studies that show that these two forms of chronic insecurity share some debilitating effects.

In a study conducted by Yong Liu and colleagues, both housing and food insecurity were examined in a survey of more than sixty-eight thousand people in twelve of the United States.[10] They found that housing insecurity and food insecurity each independently increased the reported levels of "mental distress" (stress, depression, problems with emotions) by about threefold. For those with food insecurity, the rate of mental distress jumped from 7.7 percent to 23.5 percent, and for housing insecurity, it went from 6.8 percent to 20.1 percent. Both housing and food insecurity were also associated with a doubling in the rate of sleep problems. Other studies have shown that lower income in general is associated with sleep problems, as is depression.[11] It is very likely that poverty, food insecurity, and housing insecurity all interact to undermine both physical and mental health. To top it off, in the United States, food and housing insecurity are associated with poorer access to health care.[12]

Foreclosure is almost always linked to home insecurity. With between two and three million residences foreclosed upon each year in the United States from 2008 to 2010, the Great Recession has provided an unfortunate but timely opportunity for behavioral scientists and public health specialists to study the psychological effects of foreclosure.[13] Of course, not every foreclosed-upon household winds up homeless. Given that obtaining a mortgage requires some income (not much in the case of some predatory mortgages) and economic stability, it is probably reasonable to assume that foreclosure is not a direct route to homelessness, although it may be the start of a longer downward spiral. The events surrounding foreclosure no doubt foster home insecurity, however, even if homelessness is not an imminent danger.

Studies have shown that foreclosure is definitely depressing, or at least that people undergoing foreclosure have higher rates of depressive symptoms. In an ongoing longitudinal health study conducted in the Detroit, Michigan, area, residents were surveyed in 2008 and again in 2010. Epidemiologist Kate McLaughlin and her colleagues found that those who had undergone a foreclosure between the two surveys were significantly more likely to suffer from symptoms of major depression and generalized anxiety disorder than others in the survey who had not. McLaughlin and colleagues point out that among life stressors, foreclosure may be a particularly bad one, in that it takes quite a while to play out, and when it is over, it leaves its "victims" potentially socially isolated, with unstable housing, and in the worst cases homeless.[14]

A different longitudinal study of health trends, this one focused on Americans fifty-seven years of age and older, also shows that foreclosure is depressing, even when it is happening to someone else.[15] Kathleen Cagney and her colleagues looked at health survey data collected in 2005–2006 and 2010–2011, bookending the worst of the Great Recession and US housing crisis. Focusing on the Los Angeles, New York, and Chicago urban areas, they looked to see if there was a correlation (after controlling for other variables) between increased foreclosure-related activity in a neighborhood and depressive symptoms in older adults. Breaking their data down by zip codes, Cagney and her colleagues indeed found that there was an association between increased foreclosures and depression, which was more pronounced in the oldest group of individuals: depression rates increased in zip codes/neighborhoods where foreclosure was more common. Cagney and colleagues did not find an association between depression and increases in poverty or visible disorder (i.e., signs of structural decay in a neighborhood), suggesting that foreclosure itself is a critical and depressing sign of disorder and decline on its own. Feelings of home insecurity can be fostered not only by an at-risk personal residence but by an entire neighborhood in distress.

Foreclosure, being surrounded by foreclosure, and even the threat of foreclosure are just a few of the underlying causes of home insecurity. Chronic uncertainty about home leads to anxiety and depression. Given the importance of home for our species, it should come as no surprise that anxiety should result when home is challenged. I think it is possible, however, that the depressive aspects of home insecurity might be particularly pronounced in capitalist societies in which houses and living structures are also displays of status. Evolutionary psychologists have long argued that depression is tied to aspects of rank in dominance hierarchies; that in its more "normal," milder form, depression helps in the adjustment to loss of status.[16] The loss of home, or even the threat of home loss, can be a profound blow to social status, and homelessness itself works to undermine social identity entirely.

The Vulnerability of Children Without Home

A child without a home almost always represents a tragic story. While statistics show that children make up a quarter of the homeless population, there are many more children who are sheltered but lack a proper home.[17] I don't mean this in a subjective or judgmental way: much more common than children who are literally homeless are those in foster or some other form of transitional care. They do not have a permanent home, at least not one that fulfills the emotional and psychological needs of a growing child. How many children fall into this troubling category?

In a country like the United States, where foster care and adoption are regulated by a myriad of governmental and private agencies, it has been difficult to obtain a good idea of how many children are at risk for foster care placement. Using demographic methods, Christopher Wildeman and Natalia Emanuel have provided an estimate that indicates that it is much more common than typically perceived.[18] Wildeman and Emanuel's mathematical model is based on the number and ages of children who first entered foster care, for each year over the period 2000–2011. Combining these numbers with overall census data allowed them to estimate the lifetime risk for children of being placed in foster care at some point in their lives. Wildeman and Emanuel found that 5.91 percent of all children in the United States were placed in foster care between birth and age eighteen. Disturbingly, the rates for African American children (11.53 percent) and for Native American children (15.44 percent) were much higher than the average. Wildeman and Emanuel argue that these figures indicate that "researchers and policymakers must give far greater attention to this vulnerable group of children."[19]

What exactly makes children without home so vulnerable? As we all know, the environments in which we spend our childhoods can have a profound influence on the adults we become. Developmental psychologists and neuroscientists have long recognized that there are "critical periods" during which we acquire important cognitive and perceptual abilities.[20] Although these abilities emerge under the strong influence of our genes and physiology, they need to be nurtured in a proper developmental environment. Basic senses, such as seeing and hearing, require exposure to stimuli to develop in an appropriate way. We acquire our first language effortlessly as a child during the critical period of language development; it is much more difficult to master language if exposure during childhood was limited. Our "natural" skills for social interaction are nurtured during childhood. The list goes on and on. Children raised in environments that do not meet the adequate needs of cognitive development are vulnerable to behavioral and psychological issues later in life.

Child welfare advocates and researchers highlight "family stability" as a key to healthy child development.[21] Family stability can be defined in terms of family structures (e.g., one or two parents), the mental health of parents or other caregivers, the stability of relationships within the family, and family cohesion. The emotional environment of the household is also considered, along with more general aspects of the home environment, such as the amount of positive stimulation and activity available to a child. Family stability also means not moving from home to home, which is something that children in foster care can have trouble avoiding.

It is almost self-evident that being literally homeless has devastating effects on children, even when they remain with family members. Studies done on homeless children (mean age of becoming homeless, 7.6 years) in Los Angeles suggest that over three-quarters of them have psychiatric, behavioral, or academic problems.[22] Another study showed that a similar proportion of three- and four-year-old homeless children in a New York daycare showed delayed speech and language development, hyperactivity, and difficulties with attention. In their 2013 review of this literature, Roy Grant and his colleagues point out, however, that homeless children do not seem to be much worse off when compared to other poor children. Most studies do show that they have more problems, but homelessness is seen as another risk factor in a life filled with risks. One heartening trend identified by Grant and colleagues is that studies done after the year 2000, compared to those done in the 1980s and 1990s, tend to show a reduced difference between homeless and nonhomeless poor children, which likely reflects an improvement in shelters and resources available for homeless children.

On a less positive note, one reason that there might not be that big a difference between homeless and nonhomeless poor children is that any sort of housing insecurity can be quite damaging. A large-scale study led by Diana Becker Cutts, based on a survey of over twenty thousand low-income households in seven major US metropolitan areas, showed that the health of young children (three years and under) is challenged by housing insecurity.[23] Housing insecurity is typically defined as multiple moves within a set period of time or household overcrowding (e.g., living with another family). Cutts and colleagues found that compared to those in the survey with secure housing, children in insecure households were more likely to have fair or poor (as opposed to good or excellent) health, were more likely to be classified as "at risk" in an evaluation of standard developmental markers, and had lower weight for their age. Household food insecurity is also more common in the housing insecure. Other research has shown that housing insecurity is a barrier to interventions to improve childhood health.[24] This suggests that until a family can address housing and other material needs, children's health will languish.

Foster Care and Psychic Homelessness

So we see that if a child is in a poor housing situation but remains with his or her family (however that might be constituted), there are impacts on health and development. Children in foster care experience something like the opposite situation: they are separated from kin caregivers, but once in foster care their material housing situation improves, at least in the short term. Decades of research has shown that children enter foster care with far worse health than that of their socioeconomic peers, many of whom presumably also live in a state of housing insecurity. As the American Academy of Pediatrics summarized the data in 2002: "Compared with children from the same socioeconomic background, [children entering foster care] have much higher rates of serious emotional and behavioral problems, chronic physical disabilities, birth defects, developmental delays, and poor school achievement."[25]

It is clear that children who enter foster care often have a range of problems, but how do they do when they leave? In general, children who experienced foster care are far more likely than other children to later be incarcerated, have substance abuse problems, or become homeless.[26] However, as I just noted, these children were already at risk when they entered the child welfare system. Theoretically, foster care could provide them with a less abusive or more supportive home environment than what they had come from. An analysis by Joseph Doyle based on a unique longitudinal data set from Illinois shows that foster care itself may lead to negative outcomes.[27] The data set links children with a range of governmental programs and the criminal justice system, and allows the identification of children in similar circumstances (at the "margin of placement"), some of whom were placed in foster care and others who were not. Doyle compared outcomes for these two groups. He found that children who had been in foster care were two to three times more likely to enter the criminal justice system as adults, and that teen pregnancy and juvenile delinquency in general were also more common. Although even Doyle cautions that these results based on this one data set may not apply everywhere, they are consistent with other studies reflecting poor outcomes for many children in foster care.

From start to finish, while a very necessary institution, or group of institutions and programs, foster care falls far short of providing an ideal home situation for growing children. Almost by definition, it cannot provide the child with an environment that leads to the formation of critical attachments to people and place, which provide the foundation for normal development. One of the basic structural problems with foster care systems is the lack of stability. Children come into foster care from different backgrounds and with different needs, but it is generally thought that all will do better if their living situation is more stable and less volatile. Studying the effects of greater or fewer number of moves on a child within foster care has been difficult. Different systems have different practices, and there are many reasons why a child might be moved from one caregiver to another.

There is also a chicken-and-egg issue: children who have more behavioral problems may be the cause of instability rather than the result of it.

A research project conducted by Rae Newton, Alan Litrownik, and John Landsverk in San Diego looked at the effect of number of moves on the expression of behavioral problems in foster children by following a cohort of them over time.[28] This study was designed to eliminate some of the confounding variables that had clouded results gained from other studies. One critical innovation was that Newton and colleagues were able to conduct baseline behavioral assessments of the children at the time of first entry into the foster care system, and then use the same instrument at follow-up eighteen months later. Over this time, the children had been in an average of about four placements each (foster homes or group homes), with a range from one to fifteen(!). As expected, children who had behavioral problems at intake had more placement changes than those who did not; these are children who would be predicted to have attachment issues with caregivers. More strikingly perhaps, Newton and colleagues also found that for the children who did not have any behavioral problems at intake, whether or not they developed behavioral problems later was strongly correlated with the number of placements they experienced. They concluded that behavioral problems are both a cause and an effect of a child in foster care having to make a large number of moves.

No matter how many placements a child has in foster care, the goal is always to move that child into a permanent, or forever, home. In the United States, about half of foster children typically wind up back with their original families and around 30 percent go to live with other relatives or are adopted.[29] Of the remaining children, some enter legal guardianships, run away, or are transferred to other government agencies. But nearly 10 percent of the total "age out" of foster care, reaching the age of eighteen without becoming a permanent part of a family. These children are in a tremendously vulnerable position. After already experiencing a volatile childhood (by definition, or they would not be in foster care), they then enter the adult world without the safety net of family or, in most cases, a government program designed to meet their particular needs. Over half of the kids who age out of foster care experience homelessness or unstable housing later in life, and about 30 percent are incarcerated. They experience high levels of unemployment or underemployment, with only 2 percent graduating from college (about one-tenth the population average). Although some proportion of these children may have been destined to have problems in adulthood, it is also very likely that a stable housing situation would have gone a long way toward preventing problems for many of them. One way policy makers have addressed this issue is to encourage federal and state governments to provide more incentives for the adoption of older children. Given the high costs of adult dependency, this straightforward intervention would be both humane and cost-effective.

For foster children, emotional and behavioral problems and attachment issues are to be expected given the uncertain nature of their home lives.[30] Children in foster care who do not gain a permanent home cannot, of course, feel at home.

One question I have is, do disruptions during a critical period of development undermine their ability to feel at home as an adult? As I discussed earlier, home conveys a whole range of benefits, both material and psychological. One of the psychological benefits is that home provides a place to escape the stress of "outside" life. It is a place for rest and recovery.

The brain and body's response to stress emerges from a set of structures that communicate with each other using a variety of hormones. The hypothalamus (the part of the brain dedicated to regulating many important body processes), the pituitary gland, and the adrenal glands make up what is called the hypothalamic-pituitary-adrenal (HPA) axis. With its connections to parts of the brain stem, the amygdala, and the cortex, the HPA regulates the stress response. There are several feedback and feedforward mechanisms at play in the HPA system, but at the heart of it is cortisol, a steroid hormone released from the adrenal glands. This hormone is released in response to stress, triggering processes (such as glucose metabolism and the breakdown of lipids and proteins) that make energy available to respond to the threat. At the same time, increased cortisol levels suppress growth and the immune system.

In the absence of a stressor that causes a spike in cortisol levels, our bodies maintain a very regular pattern of cortisol production over the course of a day. Cortisol levels, which can be measured quite easily from a saliva sample, are highest in the morning, with a peak about thirty minutes after waking. There is then a rapid decline for a couple of hours, followed by a slower decline until bedtime, when it reaches its lowest level. Disruptions to the HPA system are indicated by higher- or lower-than-average levels of cortisol at various points during the day, by unusual patterns of cortisol rhythms over the course of a day, and by enhanced or suppressed cortisol response to stressful events.

Maintaining high cortisol levels is fundamentally damaging to the body's tissues and overall health; the stress response is about activating the body in the short term, not keeping it going over the long term. Altered levels of cortisol are also associated with some psychiatric conditions.[31] People with posttraumatic stress disorder have low waking and evening cortisol levels, but enhanced response to stressful events. High levels of cortisol are associated with depression. Other conditions, such as substance abuse and conduct disorder, may also be associated with lower daily cortisol levels, although the data are not entirely conclusive.[32] Unusual daily cortisol patterns are generally taken as evidence of some problem with the HPA stress response system, although at this point, the physiological link between the altered stress response and pathology is generally not clear.

Since 2006, when a team led by psychologist Mary Dozier first looked at the issue, there has been much research on cortisol levels in foster children.[33] Separation from parents is a stressful occurrence for almost all children, prompting short-term changes in cortisol levels. Dozier and colleagues looked to see if there were more long-term changes associated with foster care, which represents a profound kind of parent-child separation. Looking at children in foster care between the ages of twenty months and five years, Dozier and colleagues found that there was indeed evidence of changes to the HPA system. The children in foster care had significantly lower levels of cortisol at wake-up, reaching roughly equivalent levels with the comparison children at bedtime. Dozier and colleagues noted that lower levels of cortisol have been associated with conduct disorder in children (in general) and emerging antisocial behavior in adolescents. As they concluded: "Our results suggest the possibility that conditions associated with foster care may foster a neurobiology that predisposes to conduct disorder and psychopathy."[34]

Subsequent research has generally supported these findings, but there have been some twists. A later study by Dozier's group compared the cortisol profile of children in foster care not only with that of low-risk children, but with that of children who had been in contact with child protective services while remaining with their parents.[35] They found that while the children from both groups in contact with protective services had significantly lower morning cortisol levels than the low-risk children, the children who had remained at home had even significantly lower levels than the children in foster care. In other words, foster care does seem to be an intervention that improves (without normalizing) neuroendocrine function, by removing children from abusive, neglectful, or absent parents. Work by Philip Fisher and colleagues makes clear that for children in foster care, a placement change (obtained by measuring cortisol levels both before and after the change) leads to a blunted daily cortisol pattern.[36] However, children who had been involved in an intensive, multiple-component, therapeutic support program did not show such a decline with a placement change.

For a child in dire straits at home, foster care is often the only home alternative available, at least in the short term. It changes the stress response system, but therapy may help return that system to a more normal level, and in fact, monitoring cortisol levels of children in protective care may someday (but not yet) provide a means to evaluate their health and well-being.[37] This may be quite important for adult, as well as child, health. Long-term, follow-up studies on children adopted from Romanian orphanages, and on adults who had been separated from their parents when they were children, have shown that parental or maternal separation can alter cortisol levels and the stress response in later life—even as many as sixty years later.[38] Although the full significance of these findings is yet to be determined, they show that a child's developing stress response system is sensitive to disruptions in home life at a young age, and that these disruptions can persist.

Many of us have experienced, at least for short or transitional times in our lives, the sensation of "being" at home without really "feeling" at home. This is a milder version of something adoption researcher René Hoksbergen has called "psychic homelessness." He writes that when someone feels at home, he or she "feels secure under a certain roof, feels safe, and exhibits an emotional bond with that home and the people who live there."[39] Psychic homelessness results when, on an ongoing basis, this important feeling is somehow not achieved or is less than complete. In his own work, Hoksbergen has studied foreign adoptees in the Netherlands, examining how many of these children suffer from psychic homelessness as a result of reactive attachment disorder

(RAD). RAD is a psychiatric condition in which children have disrupted social interactions, leading to being either actively antisocial (e.g., being hypervigilant, ambivalent, or needlessly contradictory) or prone to forming indiscriminate attachments. Although many foreign adoptees do not have problems feeling at home, many of them do develop problems that ultimately stem from issues involving RAD and their inability to form appropriate relationships within their households. Although RAD can be an issue for any adopted child, especially those adopted at a later age who came from a situation with severe deprivations, the problem of psychic homelessness may be more acute for foreign adoptees with RAD. Their social "microreference" (as Hoksbergen calls it) of the family echoes the broader disconnect they feel with society as a whole.

Feeling at home is only partly a function of the social relationships we have with our home-mates. But if those relationships are disrupted, for whatever reason, psychic homelessness may result. For most people, if there is no fundamental problem with forming relationships, bad relationships can be healed or new ones formed, so relationship issues need not be a permanent barrier to feeling at home. For people suffering from the long-term effects of something like RAD, it may be much harder later to overcome this barrier, and home may become something that is elusive. Similarly, if home is a place where we escape the stresses of the outside world, relax, recover, and rest, then alterations to our stress response may prevent home from truly functioning as home. Disruptions to the developing stress response system, as seen in foster children and others who have suffered various deprivations, may undermine the ability to benefit from home, and the feelings of home, that most people are capable of enjoying.

Mentally Ill People Without Home

In a landmark paper published in 1976, anthropologist Jane Murphy, drawing on her direct experience studying mental illness in several cultures, most notably among the Eskimos of the Bering Sea and the Egba Yoruba of Nigeria, made the case that the major mental illness schizophrenia was seen in virtually all cultures.[40] Rather than simply being a deviant form of behavior defined and labeled differently by different cultures, Murphy argued that it was a biological "affliction shared by virtually all mankind," albeit at a relatively low rate within any group. Murphy's work provided an on-the-ground perspective to complement the genetic, pharmacological, and neurobiological studies that were then providing the foundation for the developing field of biological psychiatry.

Murphy's paper contains a striking photograph of a thin young man from Gambia reclining on what looks to be a large rock. She describes the story behind the picture:

> The case is of a man, identified as insane, who lived some 500 yards outside a village. The villagers lived in thatched mud houses. The madman lived on an abandoned anthill. It was about 2.5 meters long and 1.5 meters high and the top had been worn away to match the contours of his body. Except for occasional visits to the village, he remained on

> this platform through day and night and changing weather. His behavior was said to have become odd when he was a young man, and when I saw him, he had not spoken for years, although he sometimes made grunting sounds. In one sense, he was as secluded and alienated from his society as patients in back wards are in ours. On the other hand, the villagers always put food out for him and gave him cigarettes.[41]

This village "madman" was not in the back wards of a psychiatric hospital, but he was homeless, by the standards of his community. There was no sense that he was being forced to live on an anthill, but that it was a choice dictated by his illness. He may have had alternatives, but it seems that none of them were as appealing.

The issue of choice and homelessness was raised in an interview by Ronald Reagan in 1988, near the end of his presidency.[42] For those of you too young to remember it, the "homeless problem" was one of the biggest ongoing political and cultural issues in the United States during the 1980s. There was a sense that the "trickle-down," laissez-faire economic policies that Reagan advocated were creating an increasingly large class of economic refugees, who were not receiving adequate support for housing from the government. At the same time, large urban public housing projects initiated in the 1960s were being perceived as near-total failures and not particularly good policy for housing the poor. And underlying the expansion of the homeless population, especially among those who were chronically homeless, was the deinstitutionalization of those suffering from major mental illnesses (bipolar disorder and especially schizophrenia). The combination of these three factors no doubt helped fuel a real rise in homelessness, as well as a rise in the perception of homelessness as a problem.

In his interview, Reagan absolved his administration of any responsibility in creating the homeless problem. He put more of the blame on groups such as the American Civil Liberties Union for fighting for the rights of the mentally ill, facilitating their release from institutions and hospitals. He said people with mental illness "make it their own choice for staying out there . . . there are shelters in virtually every city . . . and those people still prefer out there on the grates or the lawn to going into one of the shelters." Reagan claimed that a "large proportion" of the homeless were mentally ill. Whether or not this claim was true, Reagan never showed much sympathy for "those people" in the streets.

A "large proportion" can mean different things to different people. But it is clear that people with mental illness make up a substantial minority of the homeless population. A review of twenty-nine surveys conducted in Western countries showed that, among the homeless, the average prevalence of psychotic illness is 12.7 percent and of major depression is 11.4 percent. The prevalence of alcohol dependence is 37.9 percent and of drug dependence, 24.4 percent. Of course these categories can overlap, and it is known that there is substantial overlap for mental illness and substance dependence in homeless individuals. Rates of psychopathology are higher among the young adult homeless. A more specific survey of data from several cultures put the average prevalence of schizophrenia among homeless people at about 11 percent.[43] This means that people with schizophrenia are about ten times more common among the homeless population compared to the general population. But given the absolute numbers, it is also safe to say that the large majority of people with schizophrenia or substance abuse problems are not homeless. It is also reasonable to infer that a substantial majority of people who experience homelessness at some point do not have a mental illness and are not substance abusers.

But schizophrenia and substance abuse are definitely risk factors for becoming homeless. Why should this be the case? Obviously, economic competency is one reason. Housing costs money, and people without money for whatever reason are at risk of homelessness if there is not some sort of safety net provided to catch them. At a more fundamental level, schizophrenia and substance abuse lead to difficulties maintaining relationships, and home is about the relationship of a person with a place. With any form of substance abuse or addiction, the dopamine reward system of the brain becomes focused on the cognitive and emotional payoffs that can be derived from a specific chemical or behavior.[44] As I have alluded to earlier, home provides psychological rewards—derived from subtle feelings of security and comfort—that cannot compete with those that are derived from more cognitively powerful stimuli, such as drugs or even food. A home without relationships is just a shelter—important still, of course, but lacking an important cognitive connection.

The delusions, auditory hallucinations, disorganized thinking, and motivational deficits associated with schizophrenia all combine to make it difficult for people with the disease to form appropriate relationships. Paradoxically, the antipsychotic medications introduced in the 1950s and 1960s, the first that reduced the symptoms of schizophrenia (albeit in many cases with some serious physical side effects), precipitated the homeless problem for people with schizophrenia. The medications allowed vast numbers of patients to be released from the large residential mental hospitals where they were once "warehoused." Ideally, the meds meant that many of those with schizophrenia could be treated on an outpatient basis, requiring hospitalization only during acute psychotic relapses. Unfortunately, despite government policies that were in many cases well intentioned, the medical support in the community for these patients was and remains inadequate.

As psychiatrist E. Fuller Torrey, the most noted critic of public health policies relating to the community-based treatment of major mental illness, wrote in the 1980s: "The policy of deinstitutionalization has been a disaster whose dimensions are apparent everywhere."[45] At the heart of his ongoing critique is that federally funded Community Mental Health Centers have not focused attention on those with major mental illnesses but instead on the "worried well." Unfortunately, the decline in the homeless population during the 2000s does not necessarily reflect improvements in the community treatment of the mentally ill, since most homeless people do not have mental illness. In fact, changes in policy toward housing tend to focus on economic issues, rather than the more complex problems associated with how to balance the care and treatment of the severely mentally ill in the community while at the same time respecting their civil liberties.

In 1982, journalist Susan Sheehan published a best-selling biography of a woman with schizophrenia living in the New York City area called *Is There No Place on Earth for Me?*[46] The title was a question once asked by the subject of the book, Sylvia Frumkin (a pseudonym), of her mother as she was being transported in an ambulance from one hospital to another. For Frumkin and many others with schizophrenia, this question is far from rhetorical. Frumkin herself was usually housed, either with family or in a variety of community or assisted-living homes or in hospitals when she was having an acute psychotic episode. Unfortunately, her schizophrenia was generally resistant to treatment with any of the conventional medications, even those introduced in the early 1990s that were helpful to many whose disease was previously drug resistant (as Sheehan revealed in a later article). As told by Sheehan, Frumkin's life was an exhausting series of episodes and events, involving understaffed hospitals, uncaring bureaucrats, and her emotionally battered family. One thing is certain: Sylvia Frumkin never really could feel at home in a way that most people can. Her inner world and the world immediately surrounding her were simply too unsettled, chaotic, and uncomfortable.

That Sylvia Frumkin never became one of the chronically homeless mentally ill is due to the support, constant if not uncomplaining, of her parents, and after their deaths, a dedicated older sister. But just as the symptoms of schizophrenia leave many of its sufferers unable to distinguish between reality and delusions, the disease also often leaves them unable to control or assess their own living conditions. Again, this probably comes as no surprise in the context of an illness that can so deeply alter perceptions of the surrounding world. But especially for the large number of people with schizophrenia who do not have adequate financial support, making home a priority may not make sense. Anthropologist Sue Estroff, who spent an extended period of time doing fieldwork among the clients (most of whom had schizophrenia) of a community mental health center in a medium-sized US city, described it this way:

> The only spaces over which clients had even occasional control were living space and mental space. Living space was under the ultimate control of landlords and others who often paid the rent. Mental space was often intruded on by medications and psychotherapy. On the whole, however, the clients' inner, mental space, being relatively infinite, was more under their control and was perhaps more attractive than any form of external space. Some clients felt that they had no place to go and that they did not belong anywhere, though such people were in the minority.[47]

Yes, it is only a minority of people with schizophrenia who become homeless, but that number is still substantial and reflects the much higher risk of becoming homeless for someone with the disease compared to someone who does not have it. Without adequate treatment and support, for someone with schizophrenia, life on a street grate or under a highway overpass or on an anthill may become the most satisfying choice among a collection of bad ones.

The Strength and Fragility of Home

When we think of people being homeless, we usually visualize them without a physical structure to house them. That's a perfectly accurate concept of homelessness. But people can be thoroughly well housed but still homeless in a cognitive sense. As I discussed earlier, if you are homesick, you could certainly feel housed but homeless. Homesickness can result when the feelings of home are not in sync with the place you live. The mismatch between feelings and place causes the mini-grief of homesickness, whether for college students or cultural refugees. But to be homesick, no matter how bad it feels, is not to be homeless in either a cognitive or literal sense.

Extreme hoarders are a group of people whom I think can be both well housed and cognitively homeless. American television audiences of the early twenty-first century cannot seem to get enough of hoarders, as at least three different programs have been devoted to them. But this fascination with hoarders goes back much further. In New York City in the 1930s and 1940s, the Collyer brothers, who filled their Harlem brownstone with tons of hoarded items, were the subject of media and public scrutiny both before and after their linked deaths in 1947. Their story has been retold many times in books and films.[48] I think of the Collyer brothers every time I see on my local television news a house being emptied of tons of stuff and garbage by face-masked hazmat workers, its former occupant(s) now deceased or taken somewhere safe to be looked after. These are sad stories, not so much because they represent people separated from their home, but because they signal the inevitable degradation of home life when it becomes overwhelmed by hoarding.

In the fifth edition of its *Diagnostic and Statistical Manual of Mental Disorders* (DSM-5), the American Psychiatric Association separated hoarding disorder from the more general obsessive-compulsive disorder, giving it independent status as a psychiatric condition. Hoarding is characterized by "difficulty discarding or parting with possessions regardless of the value others may attribute to these possessions. . . . [Individuals who hoard] accumulate a large number of possessions that often fill up or clutter active living areas of the home or workplace to the extent that their intended use is no longer possible."[49] Hoarding behaviors begin on average at around thirteen years of age, but given the progressive nature of the condition, it becomes a greater problem with increasing age.[50] People who hoard tend to be socially isolated and anxious, have few friends, and are often unmarried. It is also the case that the maintenance of social relationships within a household can act as a curb on hoarding behavior. For example, in an elderly individual, the loss of a spouse can lead to an exacerbation of hoarding behavior that may have been kept in check while the spouse was alive. Hoarding is also associated with low socioeconomic status, and as hoarding progresses, there may be health risks associated with unsanitary or unsafe living conditions.

The American Psychiatric Association says that hoarding disorder interferes with the "intended use" of home, and it would be hard to disagree with that.[51] Throughout this book, I have been exploring what some of the intended uses of home might be, defined from the inside out, and storing great piles of useless stuff is not one of them. Certainly, a house or dwelling is a place to store things, but when storage overtakes home's other roles—especially the storage of items defined as useless by the prevailing culture—then something more profound than simple living space has been lost. Psychologists Randy Frost and Gail Steketee have written that their research on hoarding has allowed them to see that "we may own the things in our homes, but they own us as well."[52] In taking the storage of objects to such a degree, extreme hoarders do not have a home by any conventional—evolutionary or cultural—definition of home. For them, the place where they live is defined by the mass of objects and materials they possess. They may say that they feel at home, but it cannot be in the same way that most people feel at home.

The Shame of Homelessness

The word *shame* comes up often in the writings and speeches of advocates for the homeless. They often make reference to the shame of the powers that be for not addressing the issue with more dedication. Such rhetoric turns the undeserved shame of homelessness back on those who may be in a better position to do something about it. Stigma goes hand in hand with shame, and there is a stigma associated with homelessness, with living life at the margins of society. I think that the shame and stigma associated with homelessness are evidence of the implicit power and importance of home. Home is not just where we live, but it is a signifier of our very humanity.

The feelings of shame and the reactions to those who possess stigma are visceral, but what is shameful and stigmatized is largely determined by culture. These cultural preferences influence the brain pathways that generate shame and perceive stigma. As one of the "moral emotions" (along with embarrassment and guilt), shame functions to inhibit social transgressions.

Two interesting neuroimaging studies, one done in Japan and the other in Germany, but using the same methods, show that guilt and shame incorporate overlapping but also distinct networks in the brain.[53] The experiment(s) involved having subjects read sentences that had a neutral content (e.g., "I washed my clothes"), guilty content (e.g., "I sent a computer virus by e-mail"), or shameful or embarrassing content (e.g., "I soiled my underwear"), while their brains were being scanned with functional MRI (fMRI). Brain activations were extensive for either the guilt or shame conditions, incorporating parts of the cortex associated with emotion processing, other higher-level cognitive regions (in frontal and temporal lobes), as well as the visual cortex. Comparing their results to those of the older Japanese study, the German researchers, led by Petra Michl, concluded, "we interpret the findings as indicating that shame manifests itself similarly across cultures, whereas guilt is based more on specific social standards."[54]

Both the Japanese and German studies found that shame prompted overall greater activation, incorporating a wider network, in the brain than did guilt. One reason for this may be that shame is a feeling that depends on social interaction. In a sense, feeling shame requires an audience, real or imagined, whose perceptions must be accounted for or predicted. This may be a more complex process than simply recognizing or feeling guilt, which does not require social interaction of any kind. The German researchers found several differences between their subjects and the Japanese subjects in activation for guilt, which led them to conclude that guilt processing was more culturally variable than shame processing.

Like shame, stigma also arises fundamentally out of social interaction. As the pioneering sociologist Erving Goffman wrote in the mid-twentieth century, stigma refers "to an attribute that is deeply discrediting."[55] At the neurobiological level, regulation of the negative emotions associated with perceiving stigma is distinct from general emotional regulation, requiring more activation in the association areas of the prefrontal cortex.[56] Anne Krendl and her colleagues suggest that one reason for this could be that the emotional response to stigma may be relatively complex, since viewing individuals with stigma (e.g., homeless people or substance abusers) may elicit fear, pity, or disgust, or some combination of the three. Conversely, attempting to control the negative "gut" responses to such images may also require more emotional processing.

Two studies demonstrate the complex interaction among stigma, social interaction, and empathy, and how this interaction plays out in our neural processing. It is generally recognized, and supported by psychological research, that people with AIDS and obese people carry a strong stigma in modern Western cultures. To explore the cognitive underpinnings of the AIDS stigma more fully, Jean Decety and his colleagues charted the neural networks of subjects while they watched short videos of hospital patients suffering pain from tinnitus (painful ringing in the ear).[57] Subjects were told that some of the patients developed the tinnitus while healthy after chance exposure to an unnamed virus, but that others developed it as a complication of AIDS. The AIDS patients were further divided into those who had contracted it from intravenous drug use or from a blood transfusion. Contrary to the experimenters' hypothesis, there was not a clear separation in response to the AIDS versus non-AIDS patients. Rather, subjects responded most strongly, both in terms of their reported emotional response and the strength of activation of the brain networks associated with processing pain, to the AIDS transfusion patients. In effect, their empathy for this potentially stigmatized group exceeded that of the nonstigmatized subjects or the AIDS drug patients.

In a study of the neurological response to obesity stigma, R. T. Azevedo and colleagues conducted fMRI on study participants (all normal-weight) while they watched videos of obese and normal-weight individuals either suffering pain (by needle stick) or being touched with a Q-tip.[58] The participants were told that the obese subjects either had a hormonal imbalance that caused their obesity or that they became obese for unknown reasons. Azevedo and colleagues correctly predicted that there would be an obesity stigma effect evident when these normal-weight participants watched the videos, which was seen in reduced activity of pain-processing areas of the brain while watching the obese subjects suffering pain compared to those of normal weight.

Azevedo and colleagues also predicted, following the Decety results with the AIDS patients, that there would be a greater empathic response to the hormonally obese rather than the obese of unknown origins. In fact, Azevedo and colleagues found just the opposite: a reduced response for the hormonally obese compared to the other obese subjects. Subjectively, the participants professed to have more "pity" for those they thought had a hormonal condition, but this did not translate into empathy. There are big differences between AIDS and obesity in terms of prognosis and treatment, and so the two conditions are far from directly comparable, which no doubt influences how people respond to them. However, taken together, the two studies show that stigma influences empathy processing at a basic level, but that this response can be modified by a range of factors.

The stigma associated with clinical or medical conditions may be quite different from more "everyday" forms of stigma. Stigma is a form of stereotyping. The stereotype content model (SCM) is a scheme that psychologists use to categorize stereotypes of people along two dimensions: warmth (a friend–foe judgment) and competence (basic capability). A high–low matrix of these two qualities produces four quadrants—in the high–high quadrant are (for example) groups such as middle-class people and Olympic athletes; the low competence–high warmth quadrant is represented by the likes of elderly and disabled people; in the high competence–low warmth quadrant are rich people and business professionals; and finally, the low–low quadrant is where homeless people (naturally) and drug addicts are slotted. Lasana Harris and Susan Fiske used fMRI to map brain activity of participants while they looked at photographs of people representing each quadrant in the SCM matrix.[59] They found that when photos of those in the low–low quadrant, such as the homeless, were viewed, they prompted relatively low activation in the medial prefrontal cortex, a region associated with social cognition; in addition, these same images caused increased activation in the amygdala and insula, consistent with a high emotional/disgust response. Harris and Fiske argued that this pattern of brain activation is consistent with the "dehumanization" of these doubly stigmatized groups.

We don't need neuroimaging studies to tell us that stigma and shame are real, or how we should deal with them. They are fundamentally part of any human society. I think it is safe to say that if you needed a stimulus in a neuroimaging experiment on the general topic of stigma, no one would argue if you chose to use images (still or moving) of homeless people. Homelessness is not just another stigma in modern societies; it is the archetypal stigma. The neuroimaging studies show us that the varied responses to any stigma can be modified, in sometimes complex ways, by mitigating factors. For diseases or physical conditions that carry stigma, their underlying causes change how the stigma is perceived. Home is so intertwined with the fabric of our lives that the loss of home, or the inability to maintain a home, can be seen as the result of "failure" in any of several areas. Overcoming the stigma of the homeless—one step in their "rehumanization"—may be made more difficult due to the range of its underlying causes.

Most people master the ability to feel at home without effort. It does not require great intellect or creativity, an acute social intelligence, or a particularly high level of technological skill. As with spoken language or the ability to walk on two legs, we achieve the ability to feel at home because over the course of our development, our brains are trained to put the cognitive pieces together to make this possible. Home encompasses a range of emotions and capabilities, which we generally take for granted when they are all working well together.

The fact that the Universal Declaration of Human Rights mentions housing as something to which humans are entitled as a right should be seen as more than just a matter of shelter and protection—although it is that. Housing and home are very much part of what we need to fully exist and live as humans. When external forces, economic or political or natural, work to rob people of their homes, that is a tragedy. It is also tragic when people have difficulty finding home due to their own cognitive makeup. Home becomes a harder place to find, in a cognitive sense, when it is no longer a place to escape the stressors of the outside world, when no psychological reward can be derived from its quiet pleasures, or when the very existence of home in the context of the real world is no longer obvious. In a broad sense, home is a very robust psychological phenomenon, but we see that it can be derailed at several different junctions. In the worst cases, this results in people who are both literally and cognitively homeless. For most people, there may be moments when home seems less than truly homely, and for them (us) it may be worth remembering in those times that home is more complex than it seems—a psychological house of cards that has been shaped by evolution, development, and culture to be built and rebuilt.

Chapter Seven

HOME IMPROVEMENT

Most people seek home improvement by visiting stores such as Ikea, Home Depot, and Komeri. These chains, with their massive buildings filled with products and tools designed to maintain and improve the home, have become more and more popular. Maybe their popularity is a function of houses becoming increasingly important as economic investments rather than simply as places to live. After all, when a house represents your main financial asset, it makes sense to work to maintain it and to convert elbow grease into profit. The house one lives in can be a great signifier of rank and success, so perhaps a desire to improve it is tied to a desire to improve rank. Conversely, giving up on home maintenance can be a way to signal an unwillingness to "play the game," at least in this context. Finally, there will always be those who improve their houses with the more fundamental goal of making them more attractive, secure, comfortable, and healthy environments in which to live.

Some people more than others are probably just drawn toward improving their home. After all, people vary in how they orient themselves to their surroundings. There are those who are focused on the near and immediately accessible, while others take in a broader perspective of their environments. Those who take a more expansive view of their place in the world may be more focused on the external structure they occupy, rather than on the organization of the space within. Perhaps the former are more likely to seek home improvements than the latter. More fundamentally, however, people vary tremendously in how they manage their personal relationships with others. Some are more easily satisfied, while others are constantly looking for improvement. Some people derive pleasure from stability, while others seek novelty, even within the context of an established and satisfying relationship.

As I have emphasized throughout this book, home is about our relationships with the places we live. Looking at those relationships from evolutionary, cognitive, and psychological perspectives, it is easy to see that they are complex and multifaceted. Our relationship with home can seem simple and straightforward because for most of us, it comes quite naturally. But that "naturalness" belies the cognitive machinery that must be kept in balance to make home possible. At times, when we are not feeling at home, we consciously sense that something is not right. But when home is right, it lives up to its reputation as stable, secure, and in the best way possible, a little bit dull.

There is room for improvement in most things, and home is no exception. In the interest of facilitating home improvement, in this chapter I am going to distill some of the points about home that I have covered in this book as I have explored its cognitive natural history over the course of human evolution. Some of these points may be insightful, and some may be fairly implicit, but if home is truly the universal experience/feeling that I—and many others—claim it is, then they should be as relevant to your homely experience as to my own. The ancient Greek saying "know thyself" can be interpreted as meaning "know your place" (originally with reference to the gods of the Greek pantheon). I offer up a modification of this ancient aphorism: "know your home." Home is, after all, where self and place combine to form something that is unique to each person.

Shelters, Wasps, and Prairie Dogs

At the most basic level, most of us would probably agree that home provides shelter. To be homeless is to be out in the open, exposed, without protection of any kind from the environment. We share a propensity to seek and depend upon shelter with all manner of animals. They are not even animals that are particularly like us—wasps and rodents, for example—and yet it is a little easier to identify with animals when they occupy a space of their own making.

For the organ-pipe wasp, the shelter they build for themselves is a sign of their unity as a species. In contrast, the great range of houses, huts, tents, and apartments that humans call home can be taken as evidence of great diversity within our species. This diversity should not blind us to this universal technological attribute of humanity: we build things to live in. Or more properly, since most of us in the developed world are not literally house builders, we live in structures that are human-made. These structures do not all look the same, because our house building is guided less by our genes than by our cultures. But in general we seek shelter and have the wherewithal to produce it ourselves.

In the modern world, our houses—made with materials far stronger than straw, wattle, and mud, and kept temperate with central heating and cooling—should not make us forget the level of protection that even the simplest structures can provide against the elements. The sun can be less draining, the wind less drying, and the rain less soaking with even the flimsiest of shelters around us. At night, heat from the body or a fire can be focused and husbanded as protection from the cold. And predators can be kept at something greater than arm's length. The extraordinary value of even simple shelters may be one of the reasons that humans seem to have an almost evolved propensity to build them. There is no common blueprint in the human mind for a "house," however. This is where we split from most other animals that shelter themselves. Our houses and dwellings emerge from a sense of home and our general technological ability.

Home, Family, and the Prairie Dog's Education

Prairie dogs have to learn how to be prairie dogs. Their education is centered on the burrows in which they are born and raised. They learn how to use the space around their burrows and how to be vigilant for predators. The calls they use to warn each other are not something they know out of the womb; instead, they master them over time and with experience. Each learns how to live as an appropriately socialized prairie dog. A prairie dog is always home- or burrow-schooled.

They are also always schooled by family or their close relatives. This is not schooling with explicit teaching, of course, but rather learning by being in proximity to others. The odors they smell as young pups teach them who is and is not family. The linking of burrow with kin is very important in an evolutionary sense: the shared investment in time and space could only have evolved among animals related to one another and who therefore shared the same genetic goals. Ultimately, family can be too much of a good thing, due to the genetic costs of inbreeding. But the male prairie dogs generally move on, mating with less closely related females in other burrows.

A human home is really nothing much like a prairie dog burrow, but they share some fundamental features. First, home is a place where children will learn things, deeply and without effort. Obviously, some things are easier to learn than others, and our biology biases us in certain directions, but home defines the space of the key educational relationships of early childhood. Second, home is a place that families share, and humans, like prairie dogs, avoid oversharing in the form of inbreeding. The psychological mechanisms that encourage outbreeding in humans demonstrate a particular form of social learning that goes on within homelike settings. This is learning without teaching occurring at an implicit level.

Home is clearly a special learning environment. Although this learning evolved in the context of related individuals imparting knowledge (parents with children, siblings with each other, etc.), the mechanisms governing this learning do not include a gene detector. Homelike learning can clearly occur among unrelated individuals. Home constitutes an educational environment distinct from more formal spheres of education, reflecting a more "natural" place for learning. That does not mean that it is an ideal place for learning all things, but there are certain things that we may be psychologically biased to learn in a homelike setting.

Natural Selection Has Shaped Home Behavior

When we look at other animals that build structures or dig holes to live in, it is easy to see these largely instinctive behaviors as adaptations shaped by natural selection. For the organ-pipe wasp, the structures they build in which to rear their young are a direct, functional extension of their "species-hood." Burrowing rodents are not quite the automatons that wasps are, but their burrow making nonetheless represents adaptive behavior that enhances their survival and ability to successfully raise offspring to continue their genetic lineages. "Home" for many animal species is the direct result of natural selection.

I think it is safe to say that the houses and other dwellings that humans build to live in are not adaptations in a biological sense but are instead cultural adaptations shaped by tradition. They are most certainly adaptive, providing protection against the elements, predators, and other threats in the environment. Houses enhance comfort and survival.

Dwellings also play a critical role in the context of human societies. They can indicate significant relationships within larger groups and can carry symbolic meaning embedded into their structure and design. These aspects of housing contribute to the overall workings of society as a whole, reflecting an adaptive role in a strictly cultural context. Taken collectively, all the structures that humans build to live in are a product of our shared technological ingenuity and the ability to protect ourselves from the environment, without reflecting an instinctive, hardwired urge to build.

If housing itself is not a biological adaptation, then what about home? Throughout this book, I have made the argument that there is a collection of emotions and cognitive processes that contribute to our feeling at home. There is also ample evidence that home—a special place in the environment in which important tasks can be undertaken, relationships can be maintained, and physiological recovery is enhanced—contributes strongly to human survival, and in fact, aspects of it predate the emergence of modern humanity. In this sense, home is an adaptation, shaped by natural selection, even if it emerges mostly wholly formed in our own personal feelings for the place where we live. By definition, this feeling is not directly accessible to others. But I think we can be certain that our feeling for home is a distillation of several cognitive processes that together form an adaptation encouraging our relationship with place. Other animals are wedded to the places they live, but as it arises out of the context of the unique cognitive and intellectual environment of the human mind, the human home is a zoologically unique place.

The Value of Homesickness

Nothing of value is given up lightly, and we mourn the loss of those things we hold most dear and to which we feel most strongly attached. The value of home can be seen in the miserable depths of homesickness. The mini-grief many of us experience when we are separated for an extended period of time or permanently from our homes is not a trivial feeling, although in the highly mobile modern world there may be a tendency to trivialize it as a way to lessen its impact. When people are homesick, they are generally not pining for a dwelling, specific things, or even particular people, but for the totality of the ongoing experience of home.

Homesickness shows that home is valuable and important to us, but what is the value of homesickness? Like other forms of normal depression or grief, the simple sadness associated with the loss of anything important has long been recognized as potentially adaptive. Negative emotions discourage actions that lead to the loss of valuable resources. Although we tend to immediately think of depression as a clinical entity with ongoing debilitative effects, normal, short-term depression is an important aid for animals learning to deal with a variable and complex environment. In a species such as ourselves, this environment includes the complex social world we all inhabit. Sadness helps us learn to avoid future mistakes.

To trivialize homesickness is, I think, to trivialize home itself. Homesickness is a perfectly normal reaction to the loss of a significant relationship in our lives. But like other forms of normal sadness and negative feelings, it can have debilitating effects if it goes on too long. Being without a home is to be in a maladaptive place. Homesickness can encourage us to rectify the situation by moving us forward, even as we experience loss for what was left behind.

Home as a Fix-Point in Our Home Range

We all have a pretty intuitive feel for what the term *home range* means. In zoological studies, it refers to that part of the wider environment that an animal uses on a more or less habitual basis, over which it travels, obtains resources, and rests. A home range is not quite the same thing as a territory, an area that is more actively defended or at least monitored on a regular basis. I would say that the grocery store is an important part of my home range, but it is not part of my territory.

A fix-point is a location within the home range that an animal visits frequently and uses more intensely than other places. For most people, home represents the most significant fix-point in the environment. The same features that attract animals to a fix-point, such as access to water or food, protection from predators, or providing a good spot to rest, would have also appealed to our human ancestors. Over time, however, our ancestors began to focus on and enhance, both in a technological and social way, a particular fix-point in the environment. This was the beginnings of home.

One of the ongoing sources of tension for people living in the modern world emerges from trying to maintain a balance between home and work. Although this can be about relationships and responsibilities, it is also about place. For countless millennia, in our hunter-gatherer and agrarian pasts, home was largely unchallenged as the primary fix-point in our home ranges. Whether home is a more psychological or literal place, it has long been the center of human everyday life. Today, most people have multiple fix-points—nothing new about that. But the primacy of home is often challenged by the place of work. Can people have a primary relationship with two different places at the same time? Some people probably can, but for many, I suspect that our evolved psychology is more in tune with a single primary fix-point that becomes home, wherever its physical location.

The Ape's Nest, Home, and Sleep Hygiene

One of the most striking and little-appreciated adaptations seen in the great apes is the making of their nightly nest. This behavior, which is not seen in any other apes or monkeys, requires some skill and foresight, and it takes a certain amount of time before young apes are able to make a truly successful nest. It is almost impossible to say whether or not the ape's nest is a direct precursor to home. The fact that the apes rarely make a nest in the same spot twice, and that the nest is not really a location for intensive social interaction or food processing, might speak against that. However, like home, it is most definitely a place to sleep.

Although researchers are still trying to figure out why exactly animals need to sleep, we see that sleep patterns can be shaped by natural selection to fit the particular needs of animals with different ways of living. One thing that the great ape's nest suggests is that something changed in that lineage as it split from other ape lineages millions of years ago. One possibility is that increasing body size simply required a more secure platform for sleeping in trees, and that this platform was retained even as gorillas eventually evolved a body size so large that they were sleeping mostly in nests on the ground. Another possibility is that something changed in ape cognition that meant that a better night's sleep was more essential than ever for the brain maintenance that happens during sleep.

Humans are a greatly modified great ape, and nest-building is probably part of our evolutionary heritage. At its core, nest-building is about modifying the environment to obtain a secure or comfortable place to sleep. Even without nests, humans typically sleep in an environment they have modified to improve the quality of sleep. So if the ape's nest is a precursor to home, it is at this very fundamental level.

A great irony of modern life is that we can now have homes that meet every possible human need from a technological perspective, yet that technology can undermine the quality of sleep. From artificial lighting to the internet, the sleep environment within the home is in many ways under technological assault. Scientists and doctors increasingly warn of the dangers of too little sleep, in effect advocating that people take back the night for the sake of sleep. From an individual perspective, the most basic role a home fills in daily life is in providing a safe place to sleep. For the great apes, preparing a place to sleep at night is a focused activity. Perhaps for the sake of their own health, more people need to focus on sleep as a critical "activity" of the day.

Hearth and Home

Although technology today may interfere with sleep, going back in time, one of the most basic ways of identifying home was that it was a place where technology was used and created. We do not necessarily think of fire as a form of technology, but its creation and preservation certainly depend on an understanding of how the materials of the natural world can be modified for human use. The equation of home with hearth has been an enduring one. In cultures throughout the world, both extant and extinct, home is identified as the place where fire is used and maintained.

We do not know why fire was first used. It could have been to keep members of the early genus *Homo* warm or perhaps to keep predators away. But for whatever reason it started being used, we know that its enduring value in human evolution comes from the fact that cooking with fire opens up a whole new world of nutritional possibilities. Without fire, the protein-rich but tough meat of large animals would not be nearly as edible nor would calorie-rich tubers be quite as palatable. Fire increased the value of tools, stone or otherwise, that were used to obtain these foods.

Cooking is a process, and fire is at the center of a home-centered process of dining that may be a defining feature of humanity. Food is brought home, it is cooked, it is shared and eaten, and then fire remains, a focus for social interaction and companionship. Across cultures, this is a deeply satisfying event, one of the rewards, both physiological and psychological, for creating and maintaining a home.

Home Bases, and Should We Demand More of Home?

The word *base* connotes a sense of preparatory activity. Armies and air forces don't typically fight battles on their bases, but that is where they get ready for them. A base camp is where climbers get ready to make their assault on some imposing mountain peak. Most of their time may be spent at the base camp getting ready rather than actually climbing. Bases are also places that are safe and where we can recover. In baseball parlance, being on base is to be "safe," and to be at home is to be safest of all.

In Chapter 3, I discussed the "home base" hypothesis of the evolution of early *Homo.* Although it may be nearly impossible to identify such a home base in the very ancient archaeological record, it makes sense, based on what we know about how traditional people live(d), that activity areas such as home bases may have been established before humans built houses or structures of any kind. Following from the more generalized notions of what a base is, a home base can be seen as a safe area of generalized activity, where early humans did all the things they needed to do to make their lives possible. This could include sleeping, eating, making tools, raising children, telling stories, beginning and ending relationships, making plans, and so on. Before there were larger societies, the home base was society—it was where everything happened.

As cultures became more complex, the home, however it was constituted, became the basic unit in a larger social structure. And as societies have increased in size and complexity, the basic unit of home has become simpler and simpler. Home is a place for sleep, meals, hygiene, and passive entertainment. It is reserved for the closest familial relationships. It is not typically a place for making tools, significant food production and processing, major ritual activity, formal education, and so on. Obviously, this varies somewhat, but for many people in modern society, home by definition is where nothing much is happening. It is the farthest thing possible from a home base.

Although there may be a need today for home to be a place of escape from the chaos of the urban world, centering activities more on home should pay dividends. The cognitive rewards of home are to some extent vested in the feelings of control and security. We depend on home for maintenance of our minds and bodies. Returning home to its more traditional place as a center of productive activity—in whatever way, shape, or form—should enhance our positive feelings for it.

Attachments and the Shared Energy of Home

Some people live alone, and quite successfully and happily. That's all well and good, but most people live all or most of their lives with other people. This social habit is likely a holdover from our ancestors. Our closest relatives among the great apes, the chimpanzees and gorillas, also live mostly in social groups, albeit with somewhat different compositions. One thing that we think happened during human evolution is that overall group size increased compared to groups of apes, so much so that eventually we started to come together in important groups-within-a-group, namely, households.

As a unit, the household typically brings together individuals who essentially work to pool their resources. In other words, there is a sharing and exchange of energy. An important innovation during human evolution was that fathers became involved in raising their offspring. This involvement occurred and still occurs in different ways, either through direct interaction with his child, or as some theorists have suggested, initially by providing food or other support to the mother of the child. Humans also routinely bring in other caregivers to assist the parents in raising children, so much so that we may be classified zoologically as "cooperative breeders." Even in adulthood, children remain associated with their families, "paying back" the energy invested in them as they grew up. The household is thus not simply a unit in larger society: it is a setting for energetic exchange among mutually supportive individuals.

Of course, all this sharing of energy does not happen in an emotional vacuum. Attachments form between the members of a household, which help change it from being just a forum for energy exchange to something more like a home. Obviously, the most important relationship within the home, as it is with all primates, is between a mother and her offspring. Close relationships among siblings, echoing those seen among chimpanzees who share the same mother (father unknown), are also important. The novel (among the great apes), ongoing, emotion-based relationship between a reproducing human male and female is for many people the primary one that defines a home. This is especially true in the atomized households of the developed world, each typically consisting of one nuclear family. But we have to keep in mind that, for much of human history, households were more variable, consisting of kin of varying degrees of relatedness spread over generations. Such complex households reduce the importance of any single coupled relationship within them.

A feeling for home evolved based on the relationships among a group of people with common genetic goals. Compared to many traditional households, modern households reflect a simplified roster of relationships. In the same way that we actually do less in our homes today, we do what we do in the context of fewer basic relationships. This is not necessarily a bad situation, but I think it suggests that home can sustain and nurture a greater number of rewarding relationships than we might expect based on our experience in the nuclear-family world. And even though the origins of home relationships were among family and other kin, such relationships in the current world need not be limited by genetics.

Home World

We live on Earth, but we live in our homes. And the homes we live in can be found throughout the world, in climates hot and cold, wet and dry. While our extraordinary adaptability as a species relies on many things—technological intelligence, an omnivorous diet, the ability to share information and store it collectively—our ability to make the place we live in another significant other in our lives has surely helped this cause. By investing home with emotional significance, we enhance its cognitive status.

As I have said before, home is less about a dwelling or a house and more about relationships and associations we have with a certain place. Many of these relationships no doubt predate the development of any sort of living structure, and probably predate any particular association with a specific place. But at some point, such an association did become important. As I suggested in Chapter 4, the strongest ancient evidence we might have for an emotional association between an extinct hominin species and a place may be the existence of Neandertal burials. Obviously, there are other interpretations of these burials, but since we do not have evidence of Neandertal living structures, their burials might be a proxy for them.

A home represents a melding of our enriched cognitive lives with our technological prowess. As such, it may not be too surprising that this combination emerges relatively late in human evolution—in fact, at this point, definitive living structures in the archaeological record are only associated with modern humans. But for many of the benefits of home, a roof is not a necessity. Certainly, a structure enhances security and comfort and makes home life better, and from a societal standpoint, it defines the place where we live. Yet I think it is important to remember that the biological, cultural, and cognitive evolution of home proceeded on different fronts, and that our feelings of home emerge from the interplay of these histories.

The Long, Relatively Painless Road to Feeling at Home

Like spoken language, a moral sense, or the ability to read other people's minds (i.e., to be able to predict how other people will act), the ability to feel at home is something learned as a child while growing up. All of these cognitive skills have multiple components and engage multiple brain regions. They are to some extent hardwired, but the wiring only works properly if it is trained up in a reasonably supportive and "normal" developmental environment. There is learning involved in mastering these skills, but often we think of "acquiring" them since the process is so gradual and typically free of formal teaching.

Although spoken language is a robust skill that most people more or less master, it is well-known that people vary in their language abilities, and that some people, many of whom are still able to communicate adequately, have discrete language deficits. Similarly, while most people develop a moral sense or an ability to understand the motives and emotions of others, cognitive scientists are increasingly detailing the ways deficits in these areas interact with more overt behavioral problems.

Very little work has been done on variation in the ability to feel at home, but I suspect that there is real variation present. Sources of this variation could be biological, reflecting truly hardwired differences in cognition, or could be due to the environment. An absence of models for home life during childhood would be a handicap in developing an ability to feel at home as an adult. As I discussed in Chapter 6, among the mentally ill and substance abusers, particular cognitive deficits may interfere in different ways with maintaining a cognitively normal home life. The main thing a developmental perspective teaches us is that it is probably very difficult to try to explicitly teach a child to engage in home; instead it is a skill that needs to be nurtured by example over time in a sufficiently homely environment.

Home and the Life in Balance

Many people view a healthy life as one that, over the long run, is balanced. Work and play, sleep and wakefulness, food consumed and energy expended, and so on—too much one way or the other can upset the body's overall well-being. Such a viewpoint, a standard part of traditional medical cultures in Europe, the Middle East, and parts of Asia, probably goes back to ancient Greek ideas about how the four humors (black bile, yellow bile, phlegm, and blood) need to be in balance to maintain health.

Home is a great source of balance in our lives. After all, to take an extreme example, if a person does not have a home at all, then he or she is totally subjected to forces in the environment that cannot be controlled or limited. For most people, this would be a difficult situation over the long term, in part because we expect home to be a place where balance can be regained. As I discussed in Chapter

1, the scientific term for physiological balance is *homeostasis*. Home and homeostasis are intrinsically linked. Home is the place where our bodies are able to achieve homeostasis with at least some protection from the distracting or energy-sapping stimuli of the outside world.

Earlier, I argued that the expectation for home to be a place of homeostatic balance is so strong that it may function almost in the same way as a placebo. This expectation, with the actual restorative powers of home, makes for a powerful combination. At the core of feeling at home is a sense that it is a place of recovery. Food and rest, of body and mind, prepare us to face the world again. If a home does not regularly make homeostasis possible, then something is wrong. For many people today, home may primarily be a place of homeostasis. Many of us work, socially interact, become educated, and even eat, largely away from the home. For those who live lives that are focused less on home and more on the outside world, if home does not provide rest and balance, then it is not fulfilling its most basic role.

Home Is the Best Place to Travel Across Time and Space

Home is the place where our mind wanders best. When the functional neuroimaging revolution allowed scientists to finally begin looking at the workings of the normal and healthy human brain, most of the attention—with good reason—was paid to the "active" brain. This is the brain at work, doing something. It took some time for researchers to start studying the brain at rest. It is quite clear that we spend much of our time doing nothing in particular—certainly not doing anything that resembles the kinds of focused tasks that are often used in functional neuroimaging experiments. This is not really "nothing," but rather our brain activity during these periods is not a direct result of external stimuli.

A brain at rest is a brain free to wander, to test and retest ideas, to review past experiences, and to make plans for new ones. Maybe other animals can do this, but what is key in human evolution is that these private "experiences" do not need to remain private. With language, these mind wanderings can become the basis of plans, stories, and even technologies that can be shared among the members of a social group.

To some extent then, home provides a place critical not only for resting but also for thinking (in the broadest sense of the word). By removing us from the distractions and stimuli of the outside world and providing a wholly predictable environment, home gives us an opportunity to use our intellectual powers to better deal with that world. There is a trade-off, however, in that home allows some people to completely escape from the outside world, its pressures, and its problems. Such a withdrawal is unlikely to be beneficial for either physical or mental health. But for those who do not seek total escape, the short, daily escape that home can provide is likely essential for living life outside the home.

Team Home

Many of us know the phrase "Familiarity breeds contempt," and we know about families and households that have ripped themselves apart at the seams, from the inside out. But in the vast majority of cases, a home embodies a reasonably stable group of people, usually but not always related in some way, who share a living space and assist one another in achieving shared or individual goals. From the outside looking in, the shared residents of a home can look very much like a team. We expect that it is only natural and normal that they should be supportive of one another, favoring the residents of their own home over those of other homes.

The familiarity of those sharing a home is more likely to breed empathy than contempt. If nothing else, the relatively synchronized lives of those sharing a household encourages attachment and empathy. Numerous studies have shown that we humans, a highly social species, are very much influenced positively toward each other by doing things together. Home is one of the prime venues for synchronized behavior. Organizations or groups that want to encourage solidarity and conformity often make their members live, sleep, and eat on the same schedule, mimicking the "natural" situation we see in homes. The cognitive mechanisms of synchronized home life probably evolved among related individuals, but these mechanisms operate among those living together, whatever their genetic relatedness, and thus can be "fooled."

In many of the developed countries of the world, families and households are getting smaller, and it is not just that there are fewer children, but that there are fewer extended relatives present in households. Divorce and remarriage also play a part in breaking up the home team. As household numbers get smaller, the opportunities for synchronized behavior become more limited. Not only are there fewer opportunities for synchrony in smaller households, but there is less unity within these families in terms of their goals in the outside world. After all, in the developed world, children are less and less expected to follow in the occupational footsteps of their parents. It is well-known that simply sharing a family meal every day has many positive benefits, especially for children and youth.[1] A planned, shared daily family meal may be the best way to maintain some degree of synchrony, and thereby derive both the psychological and the physical benefits of being on a home team.

Animal Spirits in the House

For many people in the developed world who are influenced by the home ownership meme, home is as much about economics as anything else. Thoughts of cost, affordability, equity, depreciation, and appreciation dominate their minds when they consider a house purchase. For renters, too, economics is a critical factor in deciding where to live. From the perspective of feeling at home, I don't think this is necessarily a bad thing. The one thing we see in the natural history of home is that this feeling is quite robust and resilient. People can make themselves at home in all manner of structures, in all kinds of environments, and with all types of housemates. Even if a house is bought for strictly economic reasons, that should be no barrier to feeling at home.

It is the flip side that may be the bigger concern. The feelings we have for and about home may undermine our economic decision making when it comes to housing. Our expectations about the nonmaterial benefits of home may make us overconfident when investing in housing. The feelings of safety and comfort we associate with home are not meant to be extended to the rational, economic realm. Feeling at home belongs among the animal spirits that keep people from acting like the idealized rational consumers of economic theory.

At a personal level, we can rationalize excessive housing cost both as an economic investment and as an opportunity to improve our home lives. Part of the allure of spending more on housing is in moving up—to a nicer house, in a better neighborhood, with good schools and less crime. These goals are all fueled less by the intrinsic feelings of home and more by the external economic and social status that is associated with housing. As too many people learn after the fact, these extrinsic improvements do not necessarily improve life within the home. They are certainly not a barrier, but they do not guarantee an improved feeling of home. Most people probably sense this is the case. What is less obvious is that the combination of home and housing economics can mix certainty and uncertainty in quite different ways, throwing off our ability to assess risk of both the monetary and emotional kinds.

You do not need a hammer and paintbrush to make home improvements. No one doubts that the physical setting of home is important, even meaningful, in ways beyond comfort and protection. Home life and home feelings can be improved by structural enhancements. But these kinds of enhancements are fundamentally limited. At its core, home is not about a building or a structure of any kind. As Dorothy said in *The Wonderful Wizard of Oz:* "No matter how dreary and gray our homes are, we people of flesh and blood would rather live there than in any other country, be it ever so beautiful. There is no place like home." Our physiological, emotional, and cognitive well-being depends to a great extent on our evolved "flesh and blood" feelings for home. By understanding what these feelings are and where they come from, I believe that we can achieve a profoundly personal kind of home improvement.

EPILOGUE: THE STORY OF HOME

I am getting near the end of telling my version of the story of home. I know that such a story could be told from other perspectives—sociological, cultural, geographical, archaeological, architectural, historical, and so on. I have read many of those stories, and they all have something to offer. In looking at home from an evolutionary and cognitive perspective, however, I have tried to get at something more universal than the home that appears in most other treatments. I know there is a consensus that there is something universally human about home—everyone says that before they get on with whatever story of home they want to tell. But that is a given, part of the background to the story, like the subtle emotions and feelings we associate with home itself. I hope that I have brought that background to foreground, and showed how home became the critical foundation of our lives today.

Home is the focus of our most important relationships and most private moments. Our own stories of home may not be ones that we want to share with the outside world (at least not every detail), but there should be value in reflecting on them to ourselves. After all, many forms of therapy involve exploring the past, focusing on, and in some cases, psychologically excavating, relationships that center on the home. In the developed world today, these stories typically involve parents and siblings, but in other times, they could have involved a wider range of characters.

The emotions surrounding home provide the background for the most formative experiences of our lives and also for the most mundane ones. Antonio Damasio has likened consciousness to a "movie-in-the-brain" that we both create and observe simultaneously.[1] One component of consciousness is the "autobiographical self" that is generated from memories of an individual's past experiences and thoughts of the anticipated future. The autobiographical self provides the movie-in-the-brain with a story that is pieced together from the events of a life. Many underlying neurological processes work together to maintain continuity of consciousness, placing events in an ongoing narrative. Nonverbal feelings, such as those associated with home, are essential to this process.

Over the course of our lives, feeling at home provides cognitive continuity in the face of a surrounding environment that is highly variable and that we cannot control. Counting them up, I have lived in nineteen different dwellings during my life. Through all these living situations, with family and strangers (including family members who were strange and strangers who became like family), in houses or apartments small and large, warm and cold, a feeling of home permeates my memories of them. I lived in very different places over the years, and I was in many ways not the same person at different points in this timeline, yet home provides emotional and cognitive continuity to the movie-in-my-brain. Home is one of the fix-points around which my autobiographical self is organized. It is one of the pillars of my conscious being.

Loss of self can go along with loss of home. Earlier, I discussed the tragedy of homelessness and how being without a home puts a person at the edges of human society. Homelessness often results from economic hardship or public health inadequacies, but sometimes it comes from within, when an individual can no longer cognitively maintain a home life. This occurs in Alzheimer's disease and other forms of dementia. Even well before the end stages, when cognitive impairment imposes living dependence, many Alzheimer's patients exhibit "wandering" behavior.[2] Such wandering can be very distressful to family and caregivers and puts the patients themselves at considerable risk. Underlying wandering is the loss of the ability to recognize familiar places and objects. In many people with Alzheimer's, there is an awareness that something is wrong, of a loss of home. Some of them become anxious or paranoid, while others wander, seeking a home that is no longer available to them. When a person becomes unmoored in this way, he is no longer who he once was.

This is a somewhat depressing image, but it leads us to an important insight: home is important. Again, this is something that everybody knows, but I hope that over the course of this book you have come to know that home is important in a new way. Home is part of our evolutionary heritage; it helped us become the species we are today. It is part of our shared humanity. Home serves our most basic physiological and psychological needs, and provides the core of our lives as social beings. By both evolutionary hardwiring and environmental nurturing, home has a privileged place in our cognition. In sum, home is about maintaining and enhancing our well-being. Improving home—feeling better at home—can improve lives.

I have a friend who once consulted a feng shui specialist to see if she should rent a certain apartment. Was it oriented in the right way to its surroundings? How was the flow? Another friend, an architect, designed a home for wealthy clients who were dedicated followers of the teachings of Maharishi Mahesh Yogi and the precepts of transcendental meditation. The clients wanted a house that was fully modern but built along ancient Vedic design principles and were quite willing to pay whatever it took to get it right. The people who built and designed the house I live in now seemed to be dedicated to the proposition that no television was too large, and that no one should ever be out of the light cast by one. We all seek enlightenment in different ways, but the trip always begins at home.

For people in all cultures, home represents not just the center of the physical universe but the center of the cognitive one, as well. Humans have been such a successful species in part because we live in environments of our own making. From these relatively safe havens, we plan explorations into the less friendly, less secure, less homely world. Paradoxically, our success as a species in conquering this greater world, with all its dangers and rewards, may be due to the fact that we have a powerful, implicit desire to feel at home.

ACKNOWLEDGMENTS

I would like to thank Hanna and Antonio Damasio, the keepers of my long-term intellectual homes, first in the Department of Neurology at the University of Iowa and more recently at the Dornsife Cognitive Neuroscience Imaging Center and the Brain and Creativity Institute at the University of Southern California. This work would not have been possible without their support. Thanks also to Andrea Wiley and Tom Schoenemann, who have provided me with supportive scholarship and friendship in the Department of Anthropology at Indiana University. Thanks to Craig Stanford of the University of Southern California and Peter Sheppard of the University of Auckland, who never failed to provide answers to queries that fell within their considerably broad areas of expertise. Elie Canetti of the International Monetary Fund provided me with very helpful guidance on the subject of home economics, as did Sarah Gerstenzang of the New York State Citizens' Coalition for Children on foster care and adoption.

My agent, Jill Marsal of the Marsal Lyon Literary Agency, nurtured this project through the brainstorming and proposal stages with admirable wisdom and patience. My editor, Thomas Kelleher at Basic Books, took over from there and made this into something much more cohesive and readable than what it was before his multiple, careful readings of the text. I thank them both for their essential contributions to this book. Thanks also to project editor Collin Tracy for coordinating the production of the book and to copyeditor Deborah Heimann for her many excellent suggestions to improve the clarity of the text.

Finally, I thank the humans who have helped make my house a (usually) happy home: my sons, Reid and Perry, and my wife and housemate of nearly thirty years, Stephanie Sheffield.

NOTES

Notes to Introduction

1. S. J. Matt, *Homesickness: An American History* (New York: Oxford University Press, 2011).

2. Ibid., 4.

3. M. Stroebe, T. van Vliet, M. Hewstone, and H. Willis, "Homesickness Among Students in Two Cultures: Antecedents and Consequences," *British Journal of Psychology* 93 (2002): 147–168.

4. S. E. Watt and A. J. Bader, "Effects of Social Belonging on Homesickness: An Application of the Belongingness Hypothesis," *Personality and Social Psychology Bulletin* 35 (2009): 516–530.

5. W. Rybczynski, *Home: A Short History of an Idea* (New York: Penguin, 1986).

6. A. Blunt and R. Dowling, *Home* (London: Routledge, 2006).

7. Rybczynski, *Home,* 217.

8. C. Moore, G. Allen, and D. Lyndon, *The Place of Houses* (Berkeley: University of California Press, 1974), 49.

Notes to Chapter 1

1. M. Jackson, *At Home in the World* (Durham, NC: Duke University Press, 1995).

2. Ibid., 110.

3. Ibid., 84.

4. J. Panksepp, "Toward a Cross-Species Neuroscientific Understanding of the Affective Mind: Do Animals Have Emotional Feelings?" *American Journal of Primatology* 73 (2011): 545–561. Quote from p. 559. Panksepp answers the question he posed in the title of this article very much in the affirmative. Appendix A in this paper provides a very succinct account of the rise of behaviorism in psychology during the nineteenth and twentieth

centuries. See also J. Panksepp, *Affective Neuroscience: The Foundations of Human and Animal Emotions* (New York: Oxford University Press, 1998), Chapter 1.

5. P. R. Kleinginna and A. M. Kleinginna, "A Categorized List of Emotion Definitions, with Suggestions for a Consensual Definition," *Motivation and Emotion* 5 (1981): 345–379. The Kleinginnas provide their own definition of emotion on page 355.

6. R. Plutchik, "The Nature of Emotions: Human Emotions Have Deep Evolutionary Roots," *American Scientist* 89 (2001), 345.

7. A. Damasio, *Looking for Spinoza: Joy, Sorrow, and the Feeling Brain* (Orlando: Harcourt, 2003), 40–42.

8. P. Ekman and W. V. Friesen, "Constants Across Cultures in the Face and Emotion," *Journal of Personality and Social Psychology* 17 (2003): 124–129.

9. P. Ekman, "All Emotions Are Basic," in P. Ekman and R. J. Davidson (eds.), *The Nature of Emotion: Fundamental Questions*, 18 (New York: Oxford University Press, 1994).

10. R. Plutchik, "The Nature of Emotions."

11. J. Posner, J. A. Russell, and B. S. Peterson, "The Circumplex Model of Affect: An Integrative Approach to Affective Neuroscience, Cognitive Development, and Psychopathology," *Development and Psychopathology* 17 (2005), 730.

12. Ibid., 719.

13. K. A. Lindquist, T. D. Wager, H. Kober, E. Bliss-Moreau, and L. F. Barrett, "The Brain Basis of Emotion: A Meta-analytic Review," *Behavioral and Brain Sciences* 35 (2012): 121–143; K. A. Lindquist and L. F. Barrett, "A Functional Architecture of the Human Brain: Emerging Insights from the Science of Emotion," *Trends in Cognitive Sciences* 16 (2012): 533–540.

14. S. Hamann, "Mapping Discrete and Dimensional Emotions onto the Brain: Controversies and Consensus," *Trends in Cognitive Science* 16 (2012): 458–466; K. Vytal and S. Hamann, "Neuroimaging Support for Discrete Neural Correlates of Basic Emotions: A Voxel-Based Meta-analysis," *Journal of Cognitive Neuroscience* 22 (2010): 2864–2885.

15. Lindquist and Barrett, "A Functional Architecture of the Human Brain."

16. A. Damasio, *Looking for Spinoza*, 44.

17. Ibid., 86.

18. A. Damasio and G. B. Carvalho, "The Nature of Feelings: Evolutionary and Neurobiological Origins," *Nature Reviews Neuroscience* 14 (2013), 144.

19. Ibid.; A. D. Craig, "Significance of the Insula for the Evolution of Human Awareness of Feelings from the Body," *Annals of the New York Academy of Sciences* 1225 (2011): 72–82.

20. Damasio and Carvalho, "The Nature of Feelings," 147.

21. Damasio, *Looking for Spinoza,* 91.

22. J. S. Allen, *The Omnivorous Mind: Our Evolving Relationship with Food* (Cambridge: Harvard University Press, 2012).

23. C. Boehm, *Moral Origins: The Evolution of Virtue, Altruism, and Shame* (New York: Basic Books, 2012); J. Haidt, "The New Synthesis in Moral Psychology," *Science* 316 (2007): 998–1002.

24. V. Klinkenborg, "The Definition of Home," *Smithsonian Magazine*, May 2012, www.smithsonianmag.com/science-nature/the-definition-of-home-60692392/.

25. J. Nolte, *The Human Brain: An Introduction to Its Functional Anatomy* (St. Louis: Mosby, 2002), 568–569.

26. Damasio, *Looking for Spinoza.*

27. H.-R. Berthoud and C. Morrison, "The Brain, Appetite, and Obesity," *Annual Review of Psychology* 59 (2008): 55–92.

28. R. Szymusiak, "Hypothalamic Versus Neocortical Control of Sleep," *Current Opinion in Pulmonary Medicine* 16 (2010): 530–535; M. Gorgoni, A. D'Atria, G. Lauri, P. M. Rossini, F. Ferlazzo, and L. De Gennaro, "Is Sleep Essential for Neural Plasticity in Humans, and How Does It Affect Motor and Cognitive Recovery?" *Neural Plasticity* (2013): 103949; L. Xie, H. Kang, Q. Xu, M. J. Chen, Y. Liao, M. Thiyagarajan, J. O'Donnell, D. J. Christensen, C. Nicholson, J. J. Iliff, T. Takano, R. Deane, and M. Nedergaard, "Sleep Drives Metabolite Clearance from the Adult Brain," *Science* 342 (2013): 373–377.

29. D. E. Moerman, "General Medical Effectiveness and Human Biology: Placebo Effects in the Treatment of Ulcer," *Medical Anthropology Quarterly* 14 (1983): 13–16.

30. D. G. Finniss, T. J. Kaptchuk, F. Miller, and F. Benedetti, "Biological, Clinical and Ethical Advances of Placebo Effects," *The Lancet* 375 (2010): 686–695; M. Beauregard, "Mind Does Really Matter: Evidence from Neuroimaging Studies of Emotional Self-Regulation, Psychotherapy, and Placebo Effect," *Progress in Neurobiology* 81 (2007): 218–236.

31. P. Petrovic, T. Dietrich, P. Fransson, J. Andersson, K. Carlsson, and M. Ingvar, "Placebo in Emotional Processing—Induced Expectations of Anxiety Relief Activate a Generalized Modulatory Network," *Neuron* 46 (2005): 957–969.

32. Y.-Y. Tang, M. K. Rothbart, and M. I. Posner, "Neural Correlates of Establishing, Maintaining, and Switching Brain States," *Trends in Cognitive Sciences* 16 (2012): 330–337.

33. M. E. Raichle, A. M. Macleod, A. Z. Snyder, W. J. Powers, D. A. Gusnard, and G. L. Shulman, "A Default Mode of Brain Function," *Proceedings of the National Academy of Sciences* 98 (2001): 676–682;

M. E. Raichle, "Two Views of Brain Function," *Trends in Cognitive Sciences* 14 (2010): 180–190.

34. S. M. Smith, P. T. Fox, K. L. Miller, D. C. Glahn, P. M. Fox, C. E. Mackay, N. Filippini, K. E. Watkins, R. Toro, A. R. Laird, and C. F. Beckmann, "Correspondence of the Brain's Functional Architecture During Activation and Rest," *Proceedings of the National Academy of Sciences* 106 (2009): 13040–13045.

35. D. Mantini, M. Corbetta, G. L. Romani, G. A. Orban, and W. Vanduffel, "Evolutionarily Novel Functional Networks in the Human Brain?" *Journal of Neuroscience* 33 (2013): 3259–3275.

36. M. C. Corballis, "Wandering Tales: Evolutionary Origins of Mental Time Travel and Language," *Frontiers in Psychology* 4 (2013): article 485, 3; see also T. Suddendorf and M. C. Corballis, "The Evolution of Foresight: What Is Mental Time Travel and Is It Unique to Humans?" *Behavioral Brain Sciences* 30 (2007): 299–351.

37. D. A. Fair, A. L. Cohen, N. U. F. Dosenbach, J. A. Church, F. M. Miezin, D. M. Barch, M. E. Raichle, S. E. Petersen, and B. L. Schlaggar, "The Maturing Architecture of the Brain's Default Network," *Proceedings of the National Academy of Sciences* 105 (2008): 4028–4032.

38. A. Lutz, H. A. Slagter, J. D. Dunne, and R. J. Davidson, "Attention Regulation and Monitoring in Meditation," *Trends in Cognitive Sciences* 12 (2008): 163–169.

39. K. Crosland, "Open Your Mind with Open Monitoring Meditation," January 7, 2011, http://psychologyofwellbeing.com/201101/open-your-mind-with-open-monitoring-meditation.html.

40. A. Lutz et al., "Attention Regulation and Monitoring"; K. B. Baerentsen, H. Stødkilde-Jørgensen, B. Sommerlund, T. Hartmann, J. Damsgaard-Madsen, M. Fosnaes, and A. C. Green, "An Investigation of Brain Processes Supporting Meditation," *Cognitive Processes* 11 (2010): 57–84; S. W. Lazar, G. Bush, R. L. Gollub, G. L. Fricchione, G. Khalsa, and H. Benson, "Functional Brain Mapping of the Relaxation Response and Meditation," *NeuroReport* 11 (2000): 1581–1585.

41. E. Luders, A. W. Toga, N. Lepore, and C. Gaser, "The Underlying Anatomical Correlates of Long-Term Meditation: Larger Hippocampal and Frontal Volumes of Gray Matter," *NeuroImage* 45 (2009): 672–678; S. W. Lazar, C. E. Kerr, R. H. Wasserman, J. R. Gray, D. N. Greve, M. T. Treadway, M. McGarvey, B. T. Quinn, J. A. Dusek, H. Benson, S. L. Rauch, C. I. Moore, and B. Fischl, "Meditation Experience Is Associated with Increased Cortical Thickness," *NeuroReport* 16 (2005), 1893–1897.

42. A. Lutz et al., "Attention Regulation and Monitoring," 166–167.

43. CNN Political Unit, CNN poll: "'Rob Portman effect' fuels support for same-sex marriage," March 25, 2013, http://politicalticker.blogs.cnn.com/2013/03/25/cnn-poll-rob-portman-effect-fuels-support-for-same-sex-marriage/.

44. J. Decety and M. Svetlova, "Putting Together Phylogenetic and Ontogenetic Perspectives on Empathy," *Developmental Cognitive Neuroscience* 2 (2011): 1–24.

45. Ibid., 6.

46. X. Xu, X. Zuo, X. Wang, and S. Han, "Do You Feel My Pain? Racial Group Membership Modulates Empathic Neural Responses," *Journal of Neuroscience* 29 (2009): 8525–8529.

47. Y. Cheng, C. Chen, C.-P. Lin, K.-H., Chou, and J. Decety, "Love Hurts: An fMRI Study," *NeuroImage* 51 (2010): 923–929; G. Hein, G. Silani, K. Preuschoff, C. D. Batson, and T. Singer, "Neural Responses to Ingroup and Outgroup Members' Suffering Predict Individual Differences in Costly Helping," *Neuron* 68 (2010): 149–160.

48. R. Dunbar, *How Many Friends Does One Really Need? Dunbar's Number and Other Evolutionary Quirks* (Cambridge: Harvard University Press, 2010).

49. M. K. McClintock, "Menstrual Synchrony and Suppression," *Nature* 229 (1971): 244–245; H. C. Wilson, "A Critical Review of Menstrual Synchrony Research," *Psychoneuroendocrinology* 17 (1992): 565–591; J. M. Setchell, J. Kendal, and P. Tyniec, "Do Non-human Primates Synchronize Their Menstrual Cycles? A Test in Mandrills," *Psychoneuroendocrinology* 36 (2011): 51–59.

50. D. Saxbe and R. L. Repetti, "For Better or Worse? Coregulation of Couples' Cortisol Levels and Mood States," *Journal of Personality and Social Psychology* 98 (2010): 92–103.

51. R. Feldman, "On the Origins of Background Emotions: From Affect Synchrony to Symbolic Expression," *Emotion* 7 (2007): 601–611; S. Atzil, T. Hendler, and R. Feldman, "The Brain Basis of Social Synchrony," *Social Cognitive and Affective Neuroscience* 9 (2014): 1193–1202.

52. T. L. Chartrand and J. L. Lakin, "The Antecedents and Consequences of Human Behavioral Mimicry," *Annual Review of Psychology* 64 (2013), 293.

53. S. Mithen, *The Singing Neanderthals: The Origins of Music, Language, Mind, and Body* (Cambridge: Harvard University Press, 2006), 215.

Notes to Chapter 2

1. Evidently I was not the only one who liked the prairie dogs at this zoo: see Sarah's Notebook, June 20, 2011, http://trundlebedtales.wordpress.com/2011/06/20/iowa-city-park-zoo/.

2. J. L. Hoogland, "*Cynomys ludovicianus*," *Mammalian Species* 535 (1996): 1–10.

3. J. L. Gould and C. G. Gould, *Animal Architects: Building and the Evolution of Intelligence* (New York: Basic Books, 2007).

4. Ibid.

5. This overview of the primates is based on C. Stanford, J. S. Allen, and S. Antón, *Biological Anthropology: The Natural History of Humankind* (Upper Saddle River, NJ: Pearson, 2013); B. B. Smuts, D. L. Cheney, R. M. Seyfarth, R. W. Wrangham, and T. T. Struhsaker (eds.), *Primate Societies* (Chicago: University of Chicago Press, 1986); Karen B. Strier, *Primate Behavioral Ecology*, 3rd ed (Boston: Allyn & Bacon, 2006).

6. N. M. Jameson, Z.-C. Hou, K. N. Sterner, A. Weckle, M. Goodman, M. E. Steiper, and D. E. Wildman, "Genomic Data Reject the Hypothesis of a Prosimian Clade," *Journal of Human Evolution* 61 (2011): 295–305. This study also estimates that the last common ancestor of all primates can be dated to between 64.9 and 72.6 million years ago.

7. E. J. Sterling, N. Nguyen, and P. J. Fashing, "Spatial Patterning in Nocturnal Prosimians: A Review of Methods and Relevance to Studies of Sociality," *American Journal of Primatology* 51 (2000): 3–19.

8. H. Hediger, "Nest and Home," *Folia Primatologica* 28 (1977): 170–187; C. P. Groves and J. Sabater Pi, "From Ape's Nest to Human Fix-Point," *Man* 20 (1985): 22–47.

9. Hediger, "Nest and Home," 172.

10. S. K. Bearder, "Lorises, Bushbabies, and Tarsiers: Diverse Societies of Solitary Foragers," in B. B. Smuts, D. L. Cheney, R. M. Seyfarth, R. W. Wrangham, and T. T. Struhsaker (eds.), *Primate Societies*, 11–24 (Chicago: University of Chicago Press, 1987).

11. Ibid.

12. V. Morell, *Ancestral Passions: The Leakey Family and the Quest for Humankind's Beginnings* (New York: Touchstone, 1995).

13. See, for example, Y. Iwata and C. Ando, "Bed and Bed-Site Reuse by Western Lowland Gorillas (*Gorilla g. gorilla*) in Moukalaba-Doudou National Park, Gabon," *Primates* 48 (2007): 77–80.

14. J. Goodall, *In the Shadow of Man* (New York: Mariner Books, 2000 [1971]), 29.

15. D. Fossey, *Gorillas in the Mist* (New York: Mariner Books, 2000 [1983]), 47.

16. B. M. F. Galdikas, *Reflections of Eden* (Boston: Little, Brown and Company, 1995), 111.

17. F. A. C. Azevedo et al., "Equal Numbers of Neuronal and Nonneuronal Cells Make the Human Brain an Isometrically Scaled-Up Primate Brain, *Journal of Comparative Neurology* 513 (2009): 532–541. Estimates of one hundred billion neurons in the human brain, and many more times that of nonneuronal cells, seem to have been an overestimate.

18. T. E. Bjorness and R. W. Greene, "Adenosine and Sleep," *Current Neuropharmacology* 7 (2009): 238–245.

19. S. W. Lockley and R. G. Foster, *Sleep: A Very Short Introduction* (Oxford: Oxford University Press, 2012), 7–11.

20. N. C. Rattenborg, D. Martinez-Gonzalez, and J. A. Lesku, "Avian Sleep Homeostasis: Convergent Evolution of Complex Brains, Cognition and Sleep Functions in Mammals and Birds," *Neuroscience and Biobehavioral Reviews* 33 (2009): 253–270.

21. Lockley and Foster, *Sleep: A Very Short Introduction*, 48–49.

22. O. I. Lyamin, P. R. Manger, S. H. Ridgway, L. M. Mukhametov, and J. M. Siegel, "Cetacean Sleep: An Unusual Form of Mammalian Sleep," *Neuroscience and Biobehavioral Reviews* 32 (2008): 1451–1484. This is an outstanding review article that contains a wealth of information on this fascinating topic.

23. Ibid.

24. N. C. Rattenborg, S. L. Lima, and C. J. Amlaner, "Facultative Control of Avian Unihemispheric Sleep Under Risk of Predation," *Behavioral Brain Research* 105 (1999): 163–172; S. L. Lima, N. C. Rattenborg, J. A. Lesku, and C. J. Amlaner, "Sleeping Under Risk of Predation," *Animal Behaviour* 70 (2005): 723–736; S. L. Lima and N. C. Rattenborg, "A Behavioural Shutdown Can Make Sleeping Safer: A Strategic Perspective on the Function of Sleep," *Animal Behaviour* 74 (2007): 189–197; N. C. Rattenborg, S. L. Lima, and J. A. Lesku, "Sleep Locally, Act Globally," *The Neuroscientist* 18 (2012): 533–546.

25. J. D. Pruetz, S. J. Fulton, L. F. Marchant, W. C. McGrew, M. Schiel, and M. Waller, "Arboreal Nesting as Anti-predator Adaptation by Savanna Chimpanzees (*Pan troglydytes verus*) in Southeastern Senegal," *American Journal of Primatology* 70 (2008): 393–401; K. Koops, W. C. McGrew, H. de Vries, and T. Matsuzawa, "Nest-Building by Chimpanzees (*Pan troglodytes verus*) at Seringbara, Nimba Mountains: Antipredation, Thermoregulation, and Antivector Hypotheses," *International Journal of Primatology* 33 (2012): 356–380; F. A. Stewart and J. D. Pruetz, "Do Chimpanzee Nests Serve an Anti-predatory Function?" *American Journal of Primatology* 75 (2013): 593–604.

26. See review in Stewart and Pruetz, "Do Chimpanzee Nests Serve an Anti-predatory Function?" 594–595.

27. Ibid., 596.

28. Pruetz et al., "Arboreal Nesting as Anti-predator Adaptation by Savanna Chimpanzees"; Koops et al., "Nest-Building by Chimpanzees."

29. J. R. Anderson, "Sleep, Sleeping Sites, and Sleep-Related Activities: Awakening to Their Significance," *American Journal of Primatology* 46 (1998): 63–75; C. B. Stanford and R. C. O'Malley, "Sleeping Tree Choice by Bwindi Chimpanzees," *American Journal of Primatology* 70 (2008): 642–649.

30. C. E. G. Tutin, R. J. Parnell, L. J. T. White, and M. Fernandez, "Nest Building by Lowland Gorillas in the Lopé Reserve, Gabon: Environmental Influences and Implications for Censusing," *International Journal of Primatology* 16 (1995): 53–76.

31. Koops et al., "Nest-Building by Chimpanzees."

32. D. R. Samson and R. W. Shumaker, "Orangutans (*Pongo pp.*) Have Deeper, More Efficient Sleep Than Baboons (*Papio papio*) in Captivity," *American Journal of Physical Anthropology* 157 (2015): 421–427.

33. P. Green, "Twigitecture: Building Human Nests," *New York Times,* June 19, 2013.

Notes to Chapter 3

1. For a general discussion of the concept of transitional forms in evolutionary biology, see R. J. Raikow and R. B. Raikow, "Transitional Forms," *Reports of the Center for Science Education* 33, no. 3 (2013), http://reports.ncse.com/index.php/rncse/article/view/226/314.

2. W. C. McGrew, *Chimpanzee Material Culture: Implications for Human Evolution* (Cambridge: Cambridge University Press, 1992).

3. J. S. Allen, *The Lives of the Brain: Human Evolution and the Organ of Mind* (Cambridge, MA: Belknap Press of Harvard University Press, 2009).

4. This is the generally accepted date based on decades of work in comparative molecular genetics. Recently, a somewhat older divergence date of seven to eight million years has been suggested in K. E. Langergraber, K. Prüfer, C. Rowney et al., "Generation Times in Wild Chimpanzees and Gorillas Suggest Earlier Divergence Times in Great Ape and Human Evolution," *Proceedings of the National Academy of Sciences* 109 (2012): 15716–15721. I use the six-million-year date (see Langergraber et al. for references and discussion) since that is still the most accepted estimate as of this writing. Whether it is six or seven million years may not matter much in the grand scheme of things; however, fossils older than six million years that have been controversially assigned to the family Homininae cannot simply be dismissed as "too old" if they fall within the range of the molecular divergence date.

5. C. B. Stanford, "Chimpanzees and the Behavior of *Ardipithecus ramidus,*" *Annual Review of Anthropology* 41 (2012): 139–149.

6. T. D. White, B. Asfaw, Y. Beyene, Y. Haile-Selassie, C. O. Lovejoy, G. Suwa, and G. WoldeGabriel, "*Ardipithecus ramidus* and the Paleobiology of Early Hominids," *Science* 326 (2009): 75–86. The entire contents of this October 2, 2009, issue of *Science* was devoted to papers analyzing the *Ardipithecus ramidus* remains.

7. See C. Stanford, J. S. Allen, and S. Antón, *Biological Anthropology: The Natural History of Humankind* (Upper Saddle River, NJ: Pearson, 2013), 315–335, for a discussion and review of the australopithecine material.

8. What that significance is, is hard to assess. See Allen, *The Lives of the Brain.*

9. T. E. Cerling, E. Mbua, F. M. Kirera, F. K. Manthi, F. E. Grine, M. G. Leakey, M. Sponheimer, and K. T. Uno, "Diet of *Paranthropus boisei* in the Early Pleistocene of East Africa," *Proceedings of the National Academy of Sciences* 108 (2011): 9337–9341.

10. S. Semaw, M. J. Rogers, J. Quade, P. R. Renne, R. F. Butler, M. Dominguez-Rodrigo, D. Stout, W. S. Hart, T. Pickering, and S. W. Simpson, "2.6-Million-Year-Old Tools and Associated Bones from OGS-6 and OGS-7, Gona, Afar, Ethiopia," *Journal of Human Evolution* 45 (2003): 169–177. The discovery of even older stone tools from Kenya has recently been announced. See S. Harmond, J. E. Lewis, C. S. Feibel et al., "3.3-Million-Year-Old Stone Tools from Lomekwi 3, West Turkana, Kenya," *Nature* 521 (2015): 310–315.

11. R. Potts, *Humanity's Descent: The Consequence of Ecological Instability* (New York: Avon, 1996).

12. L. Leakey, P. Tobias, and J. Napier, "A New Species of Genus *Homo* from Olduvai Gorge," *Nature* 202 (1964): 7–9. Discussion based on G. C. Conroy, *Reconstructing Human Origins: A Modern Synthesis* (New York: Norton, 1997), 254–281; Stanford, Allen, and Antón, *Biological Anthropology,* 342–348.

13. For a discussion see Allen, *The Lives of the Brain,* 62–63.

14. Discussion based on Conroy, *Reconstructing Human Origins,* 283–343; Stanford, Allen, and Antón, *Biological Anthropology,* 348–375.

15. Allen, *The Lives of the Brain,* 80–81.

16. G. Isaac, "The Food-Sharing Behavior of Protohuman Hominids," *Scientific American* 238(4) (1978): 90–108. See also J. Sept and D. Pilbeam (eds.), *Casting the Net Wide: Papers in Honor of Glynn Isaac and His Approach to Human Origins Research* (Oxford: Oxbow Books, 2011).

17. K. Milton, "The Critical Role Played by Animal Source Foods in Human (*Homo*) Evolution," *Journal of Nutrition* 133 (2003): 3886S–3892S; C. B. Stanford, *The Hunting Apes* (Princeton: Princeton University Press, 1999).

18. See discussion in J. D. Moore, *The Prehistory of Home* (Berkeley: University of California Press, 2012).

19. L. R. Binford, *Bones: Ancient Men and Modern Myths* (New York: Academic Press, 1981); R. Potts, *Early Hominid Activities at Olduvai* (New York: Aldine de Gruyter, 1988); a very nice concise discussion of some of these issues, including the problem of tracking evolutionary change across

sites separated by time and space, can be found in R. Potts, "Variables Versus Models of Early Pleistocene Hominid Land Use," *Journal of Human Evolution* 27 (1994): 7–24.

20. Binford, *Bones,* 295.

21. E. M. Kroll, "Behavioral Implications of Plio-Pleistocene Archaeological Site Structure," *Journal of Human Evolution* 27 (1994): 107–138.

22. J. M. Sept, "Was There No Place Like Home? A New Perspective on Early Hominid Archaeological Sites from the Mapping of Chimpanzee Nests," *Current Anthropology* 33 (1992): 187–207; J. Sept, "A Worm's Eye View of Primate Behavior," in J. Sept and D. Pilbeam (eds.), *Casting the Net Wide: Papers in Honor of Glynn Isaac and His Approach to Human Origins Research*, 169–192 (Oxford: Oxbow Books, 2011).

23. J. E. Arnold, A. P. Graesch, E. Ragazzini, and E. Ochs, *Life at Home in the Twenty-First Century:* 32 *Families Open Their Doors* (Los Angeles: UCLA Cotsen Institute of Archaeology Press, 2012).

24. Ibid., 120.

25. S. Gardiner, *The House: Its Origins and Evolution* (Chicago: Ivan R. Dee, 2002).

26. See, for example, J. D. Clark and J. W. K. Harris, "Fire and Its Roles in Early Hominid Lifeways," *African Archaeological Review* 3 (1985): 3–27, versus S. R. James, "Hominid Use of Fire in the Lower and Middle Pleistocene," *Current Anthropology* 30 (1989): 1–26.

27. K. S. Brown, C. W. Marean, A. I. R. Herries, Z. Jacobs, C. Tribolo, D. Braun, D. L. Roberts, M. C. Meyer, and J. Bernatchez, "Fire as an Engineering Tool of Early Modern Humans," *Science* 325 (2009): 859–861; W. Roebroeks and P. Villa, "On the Earliest Evidence for Habitual Use of Fire in Europe," *Proceedings of the National Academy of Sciences* 108 (2011): 5209–5214.

28. Roebroeks and Villa, "On the Earliest Evidence for Habitual Use of Fire in Europe"; J. A. J. Gowlett, "The Early Settlement of Europe: Fire History in the Context of Climate Change and the Social Brain," *Comptes Rendus Palevol* 5 (2006): 299–310; S. Weiner, Q. Xu, P. Goldberg, J. Liu, and O. Bar-Yosef, "Evidence for the Use of Fire at Zhoukoudian, China," *Science* 281 (1998): 251–253; N. Goren-Inbar, N. Alperson, M. E. Kislev, O. Simchoni, Y. Melamed, A. Ben-Nun, and E. Werker, "Evidence of Hominin Control of Fire at Gesher Benot Ya'aqov, Israel," *Science* 304 (2004): 725–727.

29. F. Berna, P. Goldberg, L. K. Horwtiz, J. Brink, S. Holt, M. Bamford, and M. Chazan, "Microstratigraphic Evidence for in Situ Fire in the Acheulean Strata of Wonderwerk Cave, Northern Cape Province, South Africa," *Proceedings of the National Academy of Sciences* 109 (2012): 7593–7594.

30. Clark and Harris, "Fire and Its Roles in Early Hominid Lifeways"; J. A. J. Gowlett, J. W. K. Harris, D. Walton, and B. A. Wood, "Early Archaeological Sites, Hominid Remains and Traces of Fire from Chesowanja, Kenya," *Nature* 294 (1981): 125–129.

31. R. Wrangham, *Catching Fire: How Cooking Made Us Human* (New York: Basic Books, 2009).

32. Clark and Harris, "Fire and Its Roles in Early Hominid Lifeways," 22.

33. K. L. Kramer and P. T. Ellison, "Pooled Energy Budgets: Resituating Human Energy Allocation Trade-offs," *Evolutionary Anthropology* 19 (2010): 136–147. For discussions of the evolution of the sexual division of labor and how resources became pooled, see R. Bird, "Cooperation and Conflict: The Behavioral Ecology of the Sexual Division of Labor," *Evolutionary Anthropology 8* (1999): 65–75; C. Panter-Brick, "Sexual Division of Labor: Energetic and Evolutionary Scenarios," *American Journal of Human Biology* 14 (2002): 627–640; H. Kaplan, K. Hill, J. Lancaster, and A. M. Hurtado, "A Theory of Human Life History Evolution: Diet, Intelligence, and Longevity," *Evolutionary Anthropology* 9 (2000): 156–185.

34. C. O. Lovejoy, "The Origin of Man," *Science* 211 (1981): 341–350. See also, C. B. Stanford, *Upright: The Evolutionary Key to Becoming Human* (Boston: Houghton Mifflin, 2003).

35. D. L. Leonetti and B. Chabot-Hanowell, "The Foundation of Kinship: Household," *Human Nature* 22 (2011): 16–40.

36. S. Blaffer Hrdy, *Mothers and Others: The Evolutionary Origins of Mutual Understanding* (Cambridge: Belknap Press of Harvard University Press, 2009).

Notes to Chapter 4

1. W. D. Funkhouser and W. S. Webb, *Ancient Life in Kentucky* (Frankfort, KY: The Kentucky Geological Society, 1928).

2. P. Metcalf and R. Huntington, *Celebrations of Death: The Anthropology of Mortuary Ritual,* 2nd ed. (Cambridge: Cambridge University Press, 1991).

3. Discussion based on G. C. Conroy, *Reconstructing Human Origins: A Modern Synthesis* (New York: Norton, 1997), 344–373; C. Stanford, J. S. Allen, and S. Antón, *Biological Anthropology: The Natural History of Humankind* (Upper Saddle River, NJ: Pearson, 2013), 378–413.

4. C. Stringer, "The Status of *Homo heidelbergensis* (Schoetensack 1908)," *Evolutionary Anthropology* 21 (2012): 101–107. More recent analyses of the Mauer mandible suggest that it does not match some of the other transitional specimens that well, which would mean that only a subset of these "archaic *Homo sapiens*" should be included within the taxon. See also M. Street, T. Terberger, and J. Orschiedt, "A Critical Review of the German

Paleolithic Hominin Record," *Journal of Human Evolution* 51 (2006): 551–579.

5. F. C. Howell, "Observations on the Earlier Phases of the European Lower Paleolithic," *American Anthropologist* 68 (1966): 111–140; C. B. Stringer, E. Trinkaus, M. B. Roberts, S. A. Parfitt, and R. I. Macphail, "The Middle Pleistocene Human Tibia from Boxgrove," *Journal of Human Evolution* 34, (1998): 509–547.

6. H. Thieme (1997). "Lower Palaeolithic Hunting Spears from Germany," *Nature* 385 (1997): 807–810.

7. See E. Trinkaus and P. Shipman, *The Neandertals* (New York: Vintage Books, 1992), for a historical overview of Neandertal discoveries and interpretation.

8. Discussion based on Conroy, *Reconstructing Human Origins,* 402–458; Stanford, Allen, and Antón, *Biological Anthropology,* 389–413.

9. Trinkaus and Shipman, *The Neandertals*, 384–410.

10. R. E. Green, J. Krause, A. W. Briggs et al., "A Draft Sequence of the Neandertal Genome," *Science* 328 (2010): 710–722; J. D. Wall, M. A. Yang, F. Jay, S. K. Kim, E. Y. Durand, L. S. Stevison, C. Gignoux, A. Woerner, M. F. Hammer, and M. Slatkin, "Higher Levels of Neanderthal Ancestry in East Asians Than in Europeans," *Genetics* 194 (2013): 199–209. K. Prüfer, F. Racimo, N. Patterson et al., "The Complete Genome Sequence of a Neanderthal from the Altai Mountains," *Nature* 505 (2014): 43–49. For an excellent overview of this topic see T. R. Disotell, "Archaic Human Genomics," *Yearbook of Physical Anthropology* 55 (2012): 24–39.

11. See J. S. Allen, *The Lives of the Brain: Human Evolution and the Organ of Mind* (Cambridge, MA: Belknap Press of Harvard University Press, 2009), 258–259, for discussion and references.

12. The ability to look ahead is a critical one and may be uniquely human. See T. Suddendorf and M. Corballis, "The Evolution of Foresight: What Is Mental Time Travel and Is It Unique to Humans?" *Behavioral and Brain Sciences* 30 (2007): 299–351.

13. S. Mithen, *The Singing Neanderthals* (Cambridge, MA: Harvard University Press, 2006); C. Stringer and C. Gamble, *In Search of the Neanderthals* (New York: Thames and Hudson, 1993).

14. Stanford, Allen, and Antón. *Biological Anthropology,* 436–442; J. J. Shea, "*Homo sapiens* Is as *Homo sapiens* Was: Behavioral Variability Versus 'Behavioral Modernity' in Paleolithic Archaeology," *Current Anthropology* 52 (2011): 1–35.

15. Stanford, Allen, and Antón, *Biological Anthropology,* 385–388; Stringer and Gamble, *In Search of the Neanderthals,* 143–178.

16. Mithen, *The Singing Neanderthals*, 230.

17. Metcalf and Huntington, *Celebrations of Death,* 24.

18. M. D. Russell, "Bone Breakage in the Krapina Hominid Collection," *American Journal of Physical Anthropology* 72 (1987): 373–379; S. Defleur, T. White, P. Valensi, L. Slimak, and E. Crégut-Bonnoure, "Neanderthal Cannibalism at Moula-Guercy, Ardèche, France," *Science* 286 (1999): 128–131.

19. P. Pettitt, *The Palaeolithic Origins of Human Burial* (London: Routledge, 2011).

20. Ibid., 136–138.

21. M. Balter, "Neandertal Champion Defends the Reputation of Our Closest Cousins," *Science* 33 (2011): 642–643; M. Balter, "Did Neandertals Truly Bury Their Dead?" *Science* 33 (2012): 1443–1444.

22. R. H. Gargett, "Grave Shortcomings: The Evidence for Neanderthal Burial," *Current Anthropology* 30 (1989): 155–174; R. H. Gargett, "Middle Palaeolithic Burial Is Not a Dead Issue: The View from Qafzeh, Saint-Césaire, Kebara, Amud, and Dederiyeh," *Journal of Human Evolution* 37 (1999): 27–90.

23. D. M. Sandgathe, H. L. Dibble, P. Goldberg, and S. P. McPherron, "The Roc de Marsal Neandertal Child: A Reassessment of Its Status as a Deliberate Burial," *Journal of Human Evolution* 61 (2011): 243–253.

24. Quoted in Balter, "Did Neandertals Truly Bury Their Dead?" 1444.

25. Stringer and Gamble, *In Search of the Neanderthals,* 160.

26. Pettitt, *The Palaeolithic Origins.*

27. Metcalf and Huntington, *Celebrations of Death,* 24.

28. This topic is comprehensively reviewed in Pettitt, *The Palaeolithic Origins.*

29. J. D. Moore, *The Prehistory of Home* (Berkeley: University of California Press, 2012), 27–28. Moore points out that the standard of proof for some of these houses or living spaces can seem extraordinarily high to outside observers. This is no doubt true, but these two sites in particular, while intriguing, have issues that have prevented them from being generally accepted. Terra Amata was accepted for many years as showing evidence of "beach huts" in the south of France, but reanalysis of the excavation and its records casts doubts on this interpretation.

30. S. McBrearty and A. S. Brooks, "The Revolution That Wasn't: A New Interpretation of the Origin of Modern Human Behavior," *Journal of Human Evolution* 39 (2000), 517.

31. J. E. Yellen, "Behavioural and Taphonomic Patterning at Katanda 9: A Middle Stone Age Site, Kivu Province, Zaire," *Journal of Archaeological Science* 23 (1996), 917.

32. R. G. Klein with B. Edgar, *The Dawn of Human Culture* (New York: John Wiley & Sons, 2002), 191.

33. See, for example, C. P. Beaman, "Working Memory and Working Attention," *Current Anthropology* 51 (2010): S27–S38; M. N. Haidle,

"Working-Memory Capacity and the Evolution of Modern Cognitive Potential," *Current Anthropology* 51 (2010): S149–S166.

Notes to Chapter 5

1. J. S. Allen, *The Omnivorous Mind: Our Evolving Relationship with Food* (Cambridge: Harvard University Press, 2012).

2. R. G. Tugwell, "Human Nature in Economic Theory," *The Journal of Political Economy* 30 (1922), 319. Tugwell appeared on the cover of the June 25, 1934, issue of *Time* magazine.

3. J. Tooby and L. Cosmides, "The Psychological Foundations of Culture," in J. H. Barkow, L. Cosmides, and J. Tooby (eds.), *The Adapted Mind: Evolutionary Psychology and the Generation of Culture*, 19–136 (New York: Oxford University Press, 1992); for Rob Becker on caveman days, see http://defendingthecaveman.com/.

4. G. A. Akerlof and R. J. Shiller, *Animal Spirits: How Human Psychology Drives the Economy, and Why It Matters for Global Capitalism* (Princeton: Princeton University Press, 2009).

5. C. Levallois, J. A. Clithero, P. Wouters, A. Smidts, and S. A. Huettel, "Translating Upwards: Linking the Neural and Social Sciences via Neuroeconomics," *Nature Reviews Neuroscience* 13 (2012): 789–797.

6. G. Loewenstein, S. Rick, and J. D. Cohen, "Neuroeconomics," *Annual Review of Psychology* 59 (2008): 647–672.

7. D. Kahneman, *Thinking, Fast and Slow* (New York: Farrar, Straus and Giroux, 2011).

8. Loewenstein, Rick, and Cohen, "Neuroeconomics"; see also M. L. Platt and S. A. Huettel, "Risky Business: The Neuroeconomics of Decision Making Under Uncertainty," *Nature Neuroscience* 11 (2008): 398–403.

9. S. M. McClure, D. I. Laibson, G. Loewenstein, and J. D. Cohen, "Separate Neural Systems Value Immediate and Delayed Monetary Rewards," *Science* 306 (2004): 503–507.

10. C. Sharp, J. Monterosso, and P. R. Montague, "Neuroeconomics: A Bridge for Translational Research," *Biological Psychiatry* 72 (2012): 87–92.

11. My sources for this section are R. J. Shiller, *Irrational Exuberance,* 2nd ed. (New York, Broadway Books: 2005); T. Sowell, *The Housing Boom and Bust,* revised edition (New York: Basic Books, 2009); F. Islam, *The Default Line: The Inside Story of People, Banks, and Entire Nations on the Edge* (London: Head of Zeus, 2013). I also made use of the online article "Timeline of the United States Housing Bubble," http://en.wikipedia.org/wiki/Timeline_of_the_United_States_housing_bubble.

12. Islam, *The Default Line,* 133.

13. Shiller, *Irrational Exuberance,* 27.

14. See HUD Release 99-131, July 29, 1999, "Cuomo Announces Action to Provide $2.4 Trillion in Mortgages for Affordable Housing for 28.1 Million Families," http://archives.hud.gov/news/1999/pr99-131.html.

15. R. Dawkins, *The Selfish Gene,* new ed. (Oxford: Oxford University Press, 1989).

16. S. Blackmore, "Why I No Longer Believe Religion Is a virus of the Mind," *The Guardian*, September 16, 2010, http://www.theguardian.com /commentisfree/belief/2010/sep/16/why-no-longer-believe-religion-virus-mind.

17. L. J. Vale, *From the Puritans to the Projects: Public Housing and Public Neighbors* (Cambridge: Harvard University Press, 2000), 119.

18. Ibid., 120.

19. Islam, *The Default Line,* 132.

20. Ibid., 142.

21. C. Mayer and K. Pence, "Subprime Mortgages: What, Where, and to Whom?" *Finance and Economics Discussion Series,* Divisions of Research & Statistics and Monetary Affairs, Federal Reserve Board, Washington, DC, 2008, http://www.federalreserve.gov/pubs/feds/2008/200829/200829pap.pdf.

22. Sowell, *The Housing Boom,* 96.

23. Shiller, *Irrational Exuberance,* 56–59.

24. Kahneman, *Thinking, Fast and Slow,* 85–88.

25. Ibid., 87.

26. R. J. Samuelson, "Our Economy's Crisis of Confidence," *Washington Post*, June 14, 2010, http://www.washingtonpost.com/wp-dyn/content/article /2010/06/13/AR2010061303330.html.

27. R. Dunbar, *Grooming, Gossip, and the Evolution of Language* (Cambridge: Harvard University Press, 1996).

28. For a review of language evolution, see J. S. Allen, *The Lives of the Brain: Human Evolution and the Organ of Mind* (Cambridge: Belknap Press of Harvard University Press, 2009), Chapter 9.

29. B. Boyd, *On the Evolution of Stories: Evolution, Cognition, and Fiction* (Cambridge: Belknap Press of Harvard University Press, 2009), 166.

30. R. J. Sternberg, *Love Is a Story: A New Theory of Relationships* (New York: Oxford University Press, 1998).

31. Vale, *From the Puritans to the Projects.*

32. L. J. Vale, *Purging the Poorest: Public Housing and the Design Politics of Twice-Cleared Communities* (Chicago: University of Chicago Press, 2013).

33. Ibid., 17.

34. S. A. Venkatesh, *American Project: The Rise and Fall of a Modern Ghetto* (Cambridge: Harvard University Press, 2000), 243.

35. D. B. Hunt, *Blueprint for Disaster: The Unraveling of Chicago Public Housing* (Chicago: University of Chicago Press, 2009), 142.

36. Venkatesh, *American Project,* 66.

37. Vale, *Purging the Poorest,* 22.

38. There are several videos on YouTube that detail the demolition of these projects.

39. B. Shrader, "State Housing in New Zealand," December 20, 1012, New Zealand Ministry for Culture and Heritage. http://www.nzhistory.net.nz/culture/state-housing-in-nz, updated December 20, 2012.

40. S. Hanlon, "Tax Expenditure of the Week: The Mortgage Interest Deduction," *Center for American Progress*, January 26, 2011, http://www.americanprogress.org/issues/open-government/news/2011/01/26/8866/tax-expenditure-of-the-week-the-mortgage-interest-deduction/.

41. E. Anderson, "Beyond *Homo economicus*: New Developments in Theories of Social Norms," *Philosophy & Public Affairs* 29 (2000): 170–200.

Notes to Chapter 6

1. United Nations, "The Universal Declaration of Human Rights," December 10, 1948, http://www.un.org/en/documents/udhr/.

2. The Scottish Government, "Homelessness," last updated April 20, 2015, http://www.scotland.gov.uk/Topics/Built-Environment/Housing/homeless.

3. United States Department of Housing and Urban Development, "The 2013 Annual Homeless Assessment Report (AHAR) to Congress," 2013, https://www.onecpd.info/resources/documents/ahar-2013-part1.pdf.

4. National Law Center on Homelessness & Poverty, "Homelessness in the United States and the Human Right to Housing," January 14, 2004, http://www.mplp.org/Resources/mplpresource.2006-06-13.0349156065/file0.

5. National Coalition for the Homeless, "How Many People Experience Homelessness?" July 2009, http://www.nationalhomeless.org/publications/facts/How_Many.pdf.

6. European Commission, "Confronting Homelessness in the European Union: Commission Staff Working Document," Brussels, February 20, 2013, http://ec.europa.eu/social/BlobServlet?docId=9770&langId=en.

7. According to United Nations Special Rapporteur on Adequate Housing Miloon Kothari. As quoted in G. Capdivila, "Human Rights: More Than 100 Million Homeless Worldwide," Inter Press Service News Agency, March 30, 2005, http://www.ipsnews.net/2005/03/human-rights-more-than-100-million-homeless-worldwide/.

8. National Coalition for the Homeless, "How Many People Experience Homelessness?"; National Coalition for the Homeless, "Substance Abuse and Homelessness," July 2009, http://www.nationalhomeless.org/factsheets/addiction.pdf.

9. L. C. Ivers and K. A. Cullen, "Food Insecurity: Special Considerations for Women," *American Journal of Clinical Nutrition* 94 (2011): 1740S–1744S; B. A. Laraia, "Food Insecurity and Chronic Disease," *Advances in Nutrition* 4 (2013): 203–212; C. Hadley and C. L. Patil, "Food Insecurity in Rural Tanzania Is Associated with Maternal Anxiety and Depression," *American Journal of Human Biology* 18 (2006): 359–368.

10. Y. Liu, R. S. Njai, K. J. Greenlund, D. P. Chapman, and J. B. Croft, "Relationships Between Housing and Food Insecurity, Frequent Mental Distress, and Insufficient Sleep Among Adults in 12 States, 2009," *Preventing Chronic Disease* 11 (2014): 130334.

11. M. A. Grandner, M. E. Ruiter Petrov, P. Rattanaumpawan, N. Jackson, A. Platt, and N. P. Patel, "Sleep Symptoms, Race/Ethnicity, and Socioeconomic Position," *Journal of Clinical Sleep Medicine* 9, (2013): 897–905.

12. M. B. Kushel, R. Gupta, L. Gee, and J. S. Haas, "Housing Instability and Food Insecurity as Barriers to Health Care Among Low-Income Americans," *Journal of General Internal Medicine* 21 (2005): 71–77.

13. K. A. McLaughlin, A. Nandi, K. M. Keyes, M. Uddin, A. E. Aiello, S. Galea, and K. C. Koenen, "Home Foreclosure and Risk of Psychiatric Morbidity During the Recent Financial Crisis," *Psychological Medicine* 42 (2012): 1441–1448; K. A. Cagney, C. R. Browning, J. Iveniuk, and N. English, "The Onset of Depression During the Great Recession: Foreclosure and Older Adult Mental Health," *American Journal of Public Health* 104 (2014): 498–505.

14. McLaughlin et al., "Home Foreclosure and Risk."

15. Cagney et al., "The Onset of Depression."

16. See, for example, A. Stevens and J. Price, *Evolutionary Psychiatry: A New Beginning* (New York: Routledge, 1996).

17. United States Department of Housing and Urban Development, "The 2013 Annual Homeless Assessment Report."

18. C. Wildeman and N. Emanuel, "Cumulative Risks of Foster Care Placement by Age 18 for U.S. Children, 2000–2011," *PLOS One* 9 (2014): article e92785.

19. Ibid., 6.

20. See several chapters in M. S. Gazzaniga, *The New Cognitive Neurosciences,* 2nd ed. (Cambridge, MA: MIT Press, 2000).

21. B. Jones Harden, "Safety and Stability for Foster Children: A Developmental Perspective," *The Future of Children* 14 (2004): 31–47.

22. Studies referenced here come from the review by R. Grant, D. Gracy, G. Goldsmith, A. Shapiro, and I. E. Redlener, "Twenty-five Years of Child and Family Homelessness: Where Are We Now?" *American Journal of Public Health* 103 (2013): e1–e10.

23. D. B. Cutts, A. F. Meyers, M. M. Black, P. H. Casey, M. Chilton, J. T. Cook, J. Geppert, S. E. de Cuba, T. Heeren, S. Coleman, R. Rose-Jacobs, and D. A. Frank, "US Housing Insecurity and the Health of Very Young Children," *American Journal of Public Health* 101 (2011): 1508–1514.

24. H. Turnbull, K. Loptson, and N. Muhajarine, "Experiences of Housing Insecurity Among Participants of an Early Childhood Intervention Programme," *Child: Care, Health, and Development* 40 (2013): 435–440.

25. American Academy of Pediatrics, "Health Care of Young Children in Foster Care," *Pediatrics* 109 (2002): 536–541.

26. See references in J. J. Doyle, "Child Protection and Child Outcomes: Measuring the Effects of Foster Care," *The American Economic Review* 97 (2007): 1583–1610.

27. Doyle, "Child Protection and Child Outcomes; J. J. Doyle, "Child Protection and Adult Crime: Using Investigator Assignment to Estimate Causal Effects of Foster Care," *Journal of Political Economy* 116 (2008): 746–770.

28. R. R. Newton, A. J. Litrownik, and J. A. Landsverk, "Children and Youth in Foster Care: Disentangling the Relationship Between Problem Behaviors and Number of Placements," *Child Abuse & Neglect* 24 (2000): 1363–1374. See also S. James, J. Landsverk, and D. J. Slymen, "Placement Movement in Out-of-Home Care: Patterns and Predictors," *Children and Youth Services* 26 (2004): 185–206.

29. M. McCoy-Ruth, M. Freundlich, and T. Thorpe-Lubnueski, *Time for Reform: Preventing Youth from Aging Out on Their Own,* Report from The Pew Charitable Trusts, 2008, http://www.pewtrusts.org/uploadedFiles/wwwpewtrustsorg/Reports/Foster_care_reform/Kids_are_Waiting_TimeforReform0307.pdf.

30. L. D. Leve, G. T. Harold, P. Chamerlain, J. A. Landsverk, P. A. Fisher, and P. Vostanis, "Children in Foster Care—Vulnerabilities and Evidence-Based Interventions That Promote Resilience Processes," *Journal of Child Psychology and Psychiatry* 53 (2012): 1197–1211.

31. D. A. Bangasser and R. J. Valentino, "Sex Differences in Stress-Related Psychiatric Disorders: Neurobiological Perspectives," *Frontiers in Neuroendocrinology* 35 (2014): 303–319.

32. M. Dozier, M. Manni, M. K. Gordon, E. Peloso, M. R. Gunnar, K. Chase Stovall-McClough, D. Eldreth, and S. Levine, "Foster Children's Diurnal Production of Cortisol: An Exploratory Study," *Child Maltreatment* 11 (2006): 189–197.

33. Ibid.

34. Ibid., 195.

35. K. Bernard, Z. Butzin-Dozier, J. Rittenhouse, and M. Dozier, "Cortisol Production Patterns in Young Children Living with Birth Parents

vs Children Placed in Foster Care Following Involvement of Child Protective Services," *Archives of Pediatric and Adolescent Medicine* 164 (2010): 438–443.

36. P. A. Fisher, M. J. Van Ryzin, and M. R. Gunnar, "Mitigating HPA Axis Dysregulation Associated with Placement Changes in Foster Care," *Psychoneuroendocrinology* 36 (2011): 531–539; P. A. Fisher, M. Stoolmiller, M. R. Gunnar, and B. O. Burraston, "Effects of a Therapeutic Intervention for Foster Preschoolers on Diurnal Cortisol Activity," *Psychoneuroendocrinology* 32 (2007): 892–905; see also P. A. Fisher and M. Stoolmiller, "Intervention Effects on Foster Stress: Associations with Child Cortisol Level," *Development and Psychopathology* 20 (2008): 1003–1021.

37. H. W. H. van Andel, L. M. C. Jansen, H. Grietens, E. J. Knorth, and R. J. van der Gaag, "Salivary Cortisol: A Possible Biomarker in Evaluating Stress and Effects of Interventions in Young Foster Children?" *European Child & Adolescent Psychiatry* 23 (2014): 3–12.

38. M. R. Gunnar, S. J. Morison, K. Chisholm, and M. Schuder, "Salivary Cortisol Levels in Children Adopted from Romanian Orphanages," *Development and Psychopathology* 13 (2001): 611–628; A. Pesonen, K. Räikkönen, K. Feldt, K. Heinonen, J. G. Eriksson, and E. Kajantie, "Childhood Separation Experience Predicts HPA Axis Hormonal Responses of Late Adulthood: A Natural Experiment of World War II," *Psychoneuroendocrinology* 35 (2010): 758–767; M. Kumari, J. Ead, M. Bartley, S. Stansfeld, and M. Kivimaki, "Maternal Separation in Childhood and Diurnal Cortisol Patterns in Mid-life: Findings from the Whitehall II Study," *Psychological Medicine* 43 (2013): 633–643. Cortisol changes are also seen in nonhuman primates: X. Feng, L. Wang, S. Yang, D. Qin, J. Wang, C. Li, L. Lv, Y. Ma, and X. Hu, "Maternal Separation Produces Lasting Changes in Cortisol and Behavior in Rhesus Monkeys," *Proceedings of the National Academy of Sciences* 108 (2011): 14312–14317.

39. Quote from R. A. C. Hoksbergen. "Psychic Homelessness," in G. J. M. Abbarno (ed.), *The Ethics of Homelessness: Philosophical Perspectives*, 105–121 (Amsterdam: Editions Robopi B.V., 1999), 105. See also, R. Hoksbergen and J. ter Laak, "Psychic Homelessness Related to Attachment Disorder: Dutch Adult Foreign Adoptees Struggling with Their Identity," in R. A. Javier, A. L. Baden, F. A. Biafora, and A. Camach-Gingerich (eds.), *Handbook of Adoption: Implications for Researchers, Practitioners, and Families* (Thousand Oaks, CA: Sage, 2007), 474–490.

40. J. M. Murphy, "Psychiatric Labeling in Cross-Cultural Perspective," *Science* 191 (1976): 1019–1028.

41. Ibid., 1025.

42. S. V. Roberts, "Reagan on Homelessness: Many Choose to Live in the Streets," *New York Times,* December 23, 1988. This is a report of Reagan's interview with David Brinkley of ABC News.

43. S. Fazel, V. Khosla, H. Doll, and J. Geddes, "The Prevalence of Mental Disorders Among the Homeless in Western Countries: Systematic Review and Meta-regression Analysis," *PLoS Medicine* 5 (2008): e225; National Law Center on Homelessness & Poverty, "Homelessness in the United States and the Human Right to Housing"; K. J. Hodgson, K. H. Shelton, M. B. M. van den Bree, and F. J. Los, "Psychopathology in Young People Experiencing Homelessness: A Systematic Review," *American Journal of Public Health* 103 (2013): e24–e37; D. Folsom and D. V. Jeste, "Schizophrenia in Homeless Persons: A Systematic Review of the Literature," *Acta Psychiatrica Scandinavica* 105 (2002): 404–413.

44. J.-H. Baik, "Dopamine Signaling in Reward-Related Behaviors," *Frontiers in Neural Circuits* 7 (2013): 152.

45. E. F. Torrey, *Nowhere to Go: The Tragic Odyssey of the Homeless Mentally Ill* (New York: Harper and Row, 1988), 36; see also E. F. Torrey, "Fifty Years of Failing America's Mentally Ill," *The Wall Street Journal*, February 4, 2013, http://online.wsj.com/news/articles/SB1000142412788732 3539804578260023200841756.

46. S. Sheehan, *Is There No Place on Earth for Me?* (New York: Vintage, 1982). See also S. Sheehan, "The Last Days of Sylvia Frumkin," *The New Yorker*, February 20, 1995, 200–211.

47. S. E. Estroff, *Making It Crazy: An Ethnography of Psychiatric Clients in an American Community* (Berkeley: University of California Press, 1981), 60.

48. Anonymous, "Collyer Brothers," Wikipedia, accessed June 29, 2015, http://en.wikipedia.org/wiki/Collyer_brothers.

49. American Psychiatric Association, "Fact Sheet: Obsessive Compulsive and Related Disorders," 2013, http://www.dsm5.org/Documents/Obsessive %20Compulsive%20Disorders%20Fact%20Sheet.pdf.

50. G. Fleury, L. Gaudette, and P. Moran, "Compulsive Hoarding: Overview and Implications for Community Health Nurses," *Journal of Community Health Nursing* 29 (2012): 154–162; D. F. Tolin, S. A. Meunier, R. O. Frost, and G. Steketee, "Course of Compulsive Hoarding and Its Relationship to Life Events," *Depression and Anxiety* 27 (2010): 829–838; S. R. Woody, K. Kellman-McFarlane, and A. Welsted, "Review of Cognitive Performance in Hoarding Disorder," *Clinical Psychology Review* 34 (2014): 324–336.

51. Of course, the American Psychiatric Association does not define the "intended use" of home, which is reasonable since defining all the normative behaviors referenced in the DSM would be impractical. But it is a basic critique of the DSM that normative behaviors are implicit in their pathological definitions without reference to their cultural variation or evolutionary origins.

52. R. O. Frost and G. Steketee, *Stuff: Compulsive Hoarding and the Meaning of Things* (Boston: Mariner Books, 2010).

53. H. Takahashi, N. Yahata, M. Koeda, T. Matsuda, K. Asai, and Y. Okubo, "Brain Activation Associated with Evaluative Processes of Guilt and Embarrassment: An fMRI Study," *NeuroImage* 23 (2004): 967–974; P. Michl, T. Meindl, F. Meister, C. Born, R. R. Engel, M. Reiser, and K. Hennig-Fast, "Neurobiological Underpinnings of Shame and Guilt: A Pilot fMRI Study," *Social Cognitive and Affective Neuroscience* 9, (2014): 150–157.

54. Michl et al., "Neurobiological Underpinnings of Shame and Guilt," 155.

55. E. Goffman, *Stigma: Notes on the Management of Spoiled Identity* (Englewood Cliffs, NJ: Prentice-Hall, 1963), 3.

56. A. C. Krendl, E. A. Kensinger, and N. Ambady, "How Does the Brain Regulate Negative Bias to Stigma?" *Social Cognitive and Affective Neuroscience* 7 (2012): 715–726.

57. J. Decety, S. Echols, and J. Correll, "The Blame Game: The Effect of Responsibility and Social Stigma on Empathy for Pain," *Journal of Cognitive Neuroscience* 22 (2009): 985–997.

58. R. T. Azevedo, E. Macaluso, V. Viola, G. Sani, and S. M. Aglioti, "Weighing the Stigma of Weight: An fMRI Study of Neural Reactivity to the Pain of Obese Individuals," *NeuroImage* 91 (2014): 109–119.

59. L. T. Harris and S. T. Fiske, "Dehumanizing the Lowest of the Low: Neuroimaging Responses to Extreme Outgroups," *Psychological Science* 17 (2006): 847–853.

Notes to Chapter 7

1. E. Cook and R. Dunifon, "Do Family Meals Really Make a Difference?" Cornell University College of Human Ecology, 2012, http://www.human.cornell.edu/pam/outreach/upload/Family-Mealtimes-2.pdf.

Notes to Epilogue

1. A. Damasio, *The Feeling of What Happens: Body and Emotion in the Making of Consciousness* (New York: Harcourt Brace, 1999).

2. D. L. Algase, "Wandering in Dementia," *Annual Review of Nursing Research* 17 (1999): 185–217.

INDEX

STEPHANIE SHEFFIELD

JOHN ALLEN is a neuroanthropologist and research scientist at the Dornsife Cognitive Neuroscience Imaging Center and the Brain and Creativity Institute at the University of Southern California. The author of several trade books and textbooks, he lives near Lexington, Kentucky.

The Macquarie Dictionary of Australian Colloquialisms

MACQUARIE LIBRARY

Published by The Macquarie Library Pty Limited
43 Victoria Street, McMahons Point, NSW, Australia 2060
First published 1984

Produced in Australia for the Publisher
Phototypeset in Australia by Photoset Computer Service Pty Ltd Sydney

Printed in Australia by
Globe Press Pty Ltd, Brunswick, Vic 3056

National Library of Australia Cataloguing-in-Publication Data

Aussie talk

ISBN 0 949757 22 5.

1. English language — Australia — Terms and phrases
I. Delbridge, Arthur, 1921-.

427'.994

PUBLISHING DIRECTOR: SUSAN BUTLER

Production Manager: Cecille Weldon

A number of words entered in this dictionary are derived from trademarks. However, the absence of this indication of derivation should not be regarded as affecting the legal status of any trademark.

Contents

Editorial Staff

General Editor	Arthur Delbridge
Executive Editor	Susan Butler
Senior Editor	Margaret McPhee Martin
Editors	Pat Kreuiter
	Jane Butler
	Jill Bull
	Susan Lack
	Susan McGrath
	Alison Clark
	Richard Tardif
	Jessie Terry
	William E. Smith
Computer Systems	Pat Kreuiter
	Robert Mannell
Editorial Assistants	Marjorie Atkinson
	Kristine Burnet

Editorial Committee
The Macquarie Dictionary

Editor-in-Chief: A. Delbridge, M.A.(Lond.), B.A., Dip.Ed.(Syd.), Professor of Linguistics and Director of the Speech and Language Research Centre, Macquarie University.

J. R. L. Bernard, B.Sc., B.A., Ph.D., Dip.Ed.(Syd.), Associate Professor in Linguistics, Macquarie University.

D. Blair, M.A.(Syd.), Senior Tutor in Linguistics, Macquarie University.

W. S. Ramson, M.A.(N.Z.), Ph.D.(Syd.), Reader in English, Australian National University, and Editor of the Australian National Dictionary.

Executive Editor: Susan Butler, B.A.

Aussie Talk

A. Delbridge

Some people are shy about revealing their interest in language that is apparently not quite proper. But in writing this dictionary we have had our whispering informants to whom we are grateful: *Have you got* "Seen more tails than Hoffmann", for a racehorse that never wins? *Have you got* "wimps"? *Have you got* "wheelies"? *Have you got* "Drinking pigs down the rubbidy and throwing the duds over the eighteen"? "Pigs"? Well, "pig's ear": beer. "Rubbidy"? "Rub-a-dub": pub. "Duds"? empties, of course. And "eighteen"? That's "eighteen pence", for fence. So: "drinking beer down the pub and throwing the empties over the fence."

Word-play that depends on rhyming is not distinctively Australian, but many Australians engage in it, on occasions, especially men. The game is inexhaustible: no keen player is content with the rhymes already known and used; the invention of new ones and of new combinations marks the good player. Some of these catch on, others don't. Naturally, those that have caught on are the most likely to appear in a dictionary. But there is room here only for a sample, even so.

The particular game is known in the dictionary trade as *rhyming slang.* But in this dictionary, as in its parent *The Macquarie Dictionary,* we otherwise avoid using the word *slang* for a particular sector of the vocabulary. We agree with the Editor of Webster's *Third International,* who wrote "No word is inevitably slang". We prefer to recognise a class of words and phrases which are available for use when the speaker or writer feels free of some of the personal and social constraints which may otherwise affect the choice of words. This class of words and phrases we call *colloquialisms.* They are words and phrases available for use in conversational speech or writing in which the speaker or writer (to quote from the Macquarie Dictionary definition of *colloquial*) "is under no constraint to choose standard, formal, conservative, deferential, polite, or grammatically unchallengeable words, but feels free to choose words as appropriate from the informal, slang, vulgar, or taboo elements of the lexicon." Colloquial expressions may be chosen because they make an utterance appear to be (for example) more personal, more direct, more sincere, more sociable, more blunt, more playful, more amusing — any of these — than other expressions could. It is a matter of rhetoric, of making choices that answer needs. The question of correctness, often anxiously raised, is not a real question at all. Novelists can not avoid appropriate colloquialisms in dialogue when the *personae* of their novels speak colloquially. The facts of language include colloquialism, and this dictionary holds the mirror up to the use of colloquialisms in Australia. It deals with the whole range of colloquialisms in use, and in this respect differs from Professor Wilkes's *Dictionary of Australian Colloquialisms,* which is restricted to Australianisms, and which (with some exceptions) excludes the primary uses of words, like *arse* and *bloke,* which are much the same here as they are in other English-speaking countries.

How was it made?

In making this collection we started with the entries marked *Colloq.* in the Macquarie Dictionary. But we gratefully acknowledge our indebtedness, whether direct or indirect, to others who have worked in this field, among whom Baker, Ramson and Wilkes are pre-eminent. In trying to ensure that our coverage was both full and up-to-date, we used as sources of citations urban and country newspapers, recent fiction, and the language of radio and television and advertising. We have been indebted also to the members of the Macquarie Dictionary Society who have kindly become our informants, often with access to local or otherwise special information about the colloquial words of their communities. We have not restricted our attention to written citations, but have used our ears.

It has not been easy to settle on some criteria for identifying words and phrases as colloquial, and for selecting them for entry in the dictionary. One difficulty concerns certain grey areas: the dividing line, for example, between colloquialism and idiom. Take the very *fall. The Oxford Dictionary of Current Idiomatic English* records *fall out of* as in 'fall out of bed') and *fall out with* (meaning 'disagree' or 'quarrel') as idioms. But they are not colloquial: one could use them in any context or situation without constraint. On the other hand *fall over oneself (to do something)* is both idiomatic and colloquial; *fall over backwards (to do something)* is likewise. And to say of something brand new which has apparently been stolen that it 'fell off the back of a truck' is certainly both idiomatic and colloquial.

Another grey area of interest is in phrases that make a comparison: *tight as a drum, rough as bags, game as Ned Kelly, silly as a square wheel, mad as a cut snake, flat out like a lizard drinking,* etc. In reading, one might come upon a colourful, amusing, or memorable phrase not unlike those just cited, which nevertheless has been used a few times, or perhaps even only once, the product of one particular stimulus. These do not earn a place in this dictionary. There are others which get cited in studies like Baker's, but have not earned a place here either because there is no clear evidence of their being in general use, or because they are individual variants of a well-worn comparison where there are so many possibilities for further playful variation that to attempt to list even a few would be idle:

> *bald as a bandicoot* (or a *stone,* or a *billiard ball,* or an *opera house shell,* or an *egg)*
> *touchy as a Queensland buffalo* (or a *taipan,* or a *scrub bull in a bog*)

An apt or colourful comparison is not inevitably colloquial, nor does it get recognition as a phrase of the language until it achieves some density of usage.

The other grey area deserving mention is with extensions to the meaning of words already in non-colloquial use. One may feel constrained ordinarily to use the word *candidate* only for a person seeking an office, an honour, a status etc. But in more formal usage, one might extend the reference to some inanimate object seen as a suitable subject for a particular status or place: *that idea is a candidate for the wastepaper basket.* This sense of *candidate* is entered in this dictionary as a colloquialism. Similarly *travel* in the sense of 'go very fast' is entered as colloquial: *That souped-up car can certainly travel.* Similarly, *persuader* gets a new sense with its reference to a jockey's whip; and *user,* in the traffic with drugs, comes to mean an addict.

Our other major difficulty in settling criteria for selection of entries comes from the nature of one part of our evidence, in the form of citations from works of fiction. The difficulty is that dialogue in novels and plays presents a re-creation of colloquial speech, speech as the author feels it for the reader or audience, not speech as it really is. In nineteenth century Australian fiction, (and in some of later date) dialogue had to be made self-explanatory, so that British or otherwise polite readers would not be mystified:

> "He was shot by a highway robber?" inquired Devereux, "What you call a bushranger in Australia, don't you?" (Boldrewood).

It is used as a rhetorical device, to provide local colour and spirit:

> "You don't need to swell your head with shaping destiny or interpreting life according to those new-fangled blokes who never baked a damper, or felled a tree, or rode a buck-jumper, or killed a snake or a beast, or tanned a hide, or broke in a team of bullocks, or knocked up a coffin for a mate out of stringybark or drank water out of their hats." (Brent of Bin Bin).

There's an awful falsity in the presentation of Australian talk in much Australian fiction, that comes from a preoccupation with local background. Tom Inglis Moore complained in an article in *Southerly* that in Australian novels the interest shown in displaying the local background has ousted the foreground where the individual characters stand. Even Henry Lawson had to put people right about the linguistic romanticising of Australian experience:

> "No bushman thinks of 'going on the wallaby', or 'walking Matilda', or 'padding the hoof': he goes on the track — when forced to it . . ."

For *Aussie Talk* we have tried to find colloquialisms in their natural state, and to moderate the influence of writers who felt they had to explain them to their audience, or who forced them to a hot-house luxuriance, for display purposes chiefly.

How do colloquialisms emerge?

Many colloquial words and phrases are simple substitutions for words in the non-colloquial part of the vocabulary, chosen for a particular rhetorical effect. Rhyming slang, already mentioned, is simple substitution. One chooses not *shark* but *Noah's ark*; then not *Noah's ark* but *Noah's.* So *Noah's* becomes the word for shark, if you feel like using it.

There are other types of substitution equally available. One of the most productive is *metaphor*, when one refers to something by the name of something else. Thus an old motor car is called a *bomb*, a sentimental film is called a *weepie*; too much of something is a *welter*, a kookaburra is the *bushman's clock*, a dust storm becomes a *Wimmera shower*, a failure a *washout*, a good child a *little vegemite.*

But substitution may be even more mechanical: *utility van* (or *truck*) becomes *utility*, which becomes *ute*, and that is colloquial. Similarly, a stipendiary steward at the races becomes a *stipe.* Or stripping the process down further, *male chauvinist pig* (which may or may not be colloquial) becomes *MCP*, (which is!). Sometimes the substitution is by translation: thus for *go away!* or *buzz off!* one uses a Spanish word (here by way of the States), *vamoose.* Or one makes up a new word on the analogy of an older one: for example, if *millions* doesn't seem enough, say *zillions!* Or one substitutes the sound of an action for the description of it: *Yum-yum* is an expression of delight at the thought of eating "yummy" foods.

But substitution is not the only process. Comparison is another, though many well-worn comparisons, though they may be homely, or witty, or offensive, or colourful, are not therefore colloquial. *Silly as a curlew,* or *wary as a mallee hen guarding her eggs* may be homely bush phrases that once had some currency, but they do not meet the criteria we have suggested for a colloquialism. But *mad as a cut snake* and *mad as a gum-tree full of galahs* qualify because the comparisons are not literal, and their assumed meaning is kept alive only by their repeated use in a colloquial style.

Who will find it useful?
Colloquialisms belong primarily to talk, and to representations of talk in writing. But since the choice of words and phrases is ultimately not constrained, except by laws affecting libel, defamation and obscenity, a close acquaintance with the colloquialisms of Aussie talk is a necessary part of linguistic competence in Australia. *Aussie Talk* is designed to meet the needs of people who are looking for help in this most inscrutable, most unstable, most diversified sector of Australian English. Australians whose mother-tongue is Australian English may otherwise be left in the dark by the talk of another generation or another locality, or just of another individual. Recent migrants, especially those who are learning Australian English as a second language, might like Nino Culotta's archetypal migrant in *They're A Weird Mob* feel the need for a key to popular Australian talk. And the increasing number of readers and even scholars gaining an acquaintance with Australian literature and Australian film from abroad, in America, in Japan, in Europe — they need this broad picture of Australian colloquialism too. This dictionary, like all the others in the Macquarie set, is designed to meet particular needs of particular readers. But the particular readers for this one, as for its parent, *The Macquarie Dictionary* may be found in Peking or Los Angeles, as well as in Melbourne, or Wollongong, or Alice Springs.

A *phr.* **give (someone) the big A,** to reject or rebuff (someone). [standing for *A(rse)*]

Abdul *n.* (*often offensive*) nickname for a Turk.

Abo *n.* **1.** (*often offensive*) an Aborigine. *—adj.* **2.** Aboriginal. *—phr.* **3. give it back to the Abos** or **blacks,** See **blacks.** [shortened form of *Aborigine*]

aboriginality *n.* a colloquial anecdote on Aboriginal or Australian bush subjects in general. [from *Aboriginalities,* the title of a former regular feature in the *Bulletin,* a Sydney weekly journal]

absconder *n.* (formerly) an escaped convict. Also, **absconder, absentee into the woods.**

abso-bloody-lutely *adv.* a jocularly emphatic version of **absolutely.**

Academy Award *phr.* **an Academy Award job,** an exaggerated reaction made to impress or to secure an advantage. [from the awards made by the Motion Picture Academy of America]

accident *n.* **1.** an unplanned pregnancy. *—phr.* **2. to be an accident getting ready to happen, a.** to be unattractive, as in physical features, dress, etc. **b.** to be in a state of confusion and disorder. **c.** to be very clumsy.

accidentally *phr.* **accidentally on purpose,** with a hidden purpose.

ac-dc *adj.* bisexual; attracted to both males and females as sexual partners. [from the sense of *ac-dc* which relates to an electric device which can operate from either an alternating current or a direct current power source]

ace *adj.* **1.** excellent; first in quality; outstanding: *The kids think he's ace.* *—n.* **2.** the anus. *—phr.* **3. on one's ace,** on one's own; alone. [from *ace* a playing card marked with a single spot, in many games counting as the highest in its suit]

acid[1] *phr.*

1. come the acid over, *N.Z.* to act sharply or viciously towards.

2. put the acid on, to ask (something) of (someone) in such a manner that refusal is difficult; pressure (someone).

acid[2] *n.* **1.** LSD. *—phr.* **2. drop acid,** to take LSD.

acid head *n.* one who takes LSD.

ack-ack *n.* **1.** anti-aircraft fire. **2.** anti-aircraft arms. *—adj.* **3.** anti-aircraft. [used by radio operators for A.A. (anti-aircraft) *Ack* represents A]

ack-emma *adv.* a.m. [used by radio operators *Ack* represents A, *Emma* represents M]

Ack-i-Foof *n.* an A.I.F. man. [a humorous variant of the military signal code used in World War I and World War II to denote a member of the Australian Imperial Forces]

ack-willie *adj.* (in military jargon) absent without leave. [from the military signalling code in which *Ack* represented A and *Willie* represented W, signalling A.W. short for *A.W.L.: A(bsent) W(ithout) L(eave)*]

acquire *v.* to steal; obtain illicitly or deviously.

acre *n.* **1.** the buttocks. **2.** the anus.

across *phr.* **come across, a.** to pay up; settle an outstanding debt, etc. **b.** to own up to (something); give information. **c.** (usually of a woman) to grant sexual favours.

act *phr.*

1. act the goat or **angora,** to play the fool; behave in a foolish fashion.

2. act up, a. to play up; take advantage of. **b.** (of a car, etc.) to malfunction.

3. bit of an act, artificial or insincere behaviour.

4. bung or **stack on an act, a.** to display bad temper. **b.** to behave in a manner especially put on for the occasion.

5. get one's act together, to organise or clarify ideas, plans, etc.

6. hard or **difficult act to follow, a.** a performance or performer having such appeal that it is difficult for the next event in the program not to be overshadowed. **b.** any event or person whose brilliance overshadows that of others following.

action *n.* **1.** the focus of interest and activity: *I want to be where the action is.* *—phr.* **2. a piece of the action,** a profitable involvement in the enterprise.

action stations *pl. n.* **1.** positions taken up preparatory to action. **2.** a warning or command to get ready for action. [from military term *action stations* the command to take up stations in readiness for or during battle; the stations themselves]

actor *n.* one whose ordinary behaviour is often ostentatious, bizarre or showy.

Adam *phr.* **not to know (someone) from Adam,** not to know (someone) at all. [from *Adam* the name of the first man]

Adam's ale *n.* water.

add *phr.* **add up, a.** to amount (*to*): *It adds up to murder.* **b.** to make sense, be logically consistent: *The facts don't add up.*

adder *phr.* **have death adders in one's pocket,** to be mean with one's money.

adjective *phr.* **the great Australian adjective,** See **Australian.**

advert *n.* an advertisement. Also, **ad.** [shortened form]

aerial ping-pong *n.* Australian Rules football. [so-called because the frequent kicking and high marking which characterises the game ensures that the ball is in the air most of the time]

Afghan *n.* **1.** (formerly) a camel driver employed in the outback of Australia. **2.** a travelling merchant. [from the camel drivers from Afghanistan who worked in inland Australia in the second half of the 19th century]

Afghan's fly trap *n.* a hole cut in the seat of one's trousers which thus attracts flies away from the face. Also, **Australian fly trap.** See **Bedourie fly veil.**

african *n.* a cigarette. [rhyming slang *African nigger* cigger]

after dinner mint *n.* sexual favours granted at the conclusion of an expensive night out, as in payment.

agates *pl. n.* the testicles. Also, **agotts.** [?from *agate* a child's playing marble]

age *phr.* **act** or **be one's age,** a command to behave sensibly; a command to stop behaving stupidly.

aggie *n.* a child's playing marble made of quartz or of a glass imitation of it; agate. [from *agate* a variety of quartz]

aggro *adj.* **1.** aggressive; dominating. — *n.* **2.** aggression; violence. **3.** aggravation.

agony *phr.* **put/pile/turn on the agony,** to exaggerate a story, misfortunes, etc., for effect.

agony column *n.* a newspaper column of advertisements, especially those arising from personal distress; personal column.

agricultural *n.* (in cricket) a wild stroke without finesse that digs up the turf. Also, **agricultural shot.**

ahead *phr.*

1. be ahead, to be to the good; be winning: *I was well ahead in the deal.*

2. get ahead, to do well, as in one's job, etc.

A.I.F. *adj.* deaf. [rhyming slang]

air *phr.*

1. air and exercise, *Obsolete.* a gaol sentence.

2. clear the air, to eliminate dissension, ambiguity, or tension from a discussion, situation, etc.

3. off the air, a. crazy. **b.** very angry.

4. (up) in the air, undecided or unsettled.

air-raid *v.* to protest volubly; nag; scold.

airs and graces[1] *pl. n.* **1.** faces. **2.** braces. [rhyming slang]

airs and graces[2] *phr.* **putting on airs and graces,** putting on affected mannerisms, speech, etc.

airy-fairy *adj.* of little substance; whimsical; fanciful: *an airy-fairy excuse.*

aisle *phr.* **lay them in the aisles, a.** to amuse greatly. **b.** to impress people favourably.

ajax *v.* to clean (baths, etc.) using a household cleaning powder. [from the Trademark *Ajax* name of such a powder]

akubra *n.* a broad-brimmed hat. [Trademark]

alarm bird *n.* a kookaburra.

Albany doctor *n.* a strong, cool wind blowing after a hot day. [Albany, a town in Western Australia]

albert *n.* a kind of watch chain. Also **Albert Chain.** [named after Prince *Albert* consort of Queen Victoria]

alberts *pl. n.* covering for the feet made from sacking, cloth, etc. worn by swagmen, tramps, etc. Also, **Prince Alberts.** [named after Prince *Albert* because of his alleged poverty before he became Queen Victoria's consort]

albino *n. Northern Australia.* (*often offensive*) a white woman. [from *albino* a person with a pale skin]

Al Capone *n.* telephone. [rhyming slang]

Alderman Lushington *n.* See **lushington.**

alec *n.* See **smart alec.**

alf *n.* **1.** a heterosexual male. **2.** a male whose behaviour shows contempt for cultural pursuits, prejudice towards minority groups, low estimation of women as a class, and a marked preference for male social company. **3.** a dull and ineffectual person.

Alfred, Royal *n.* See **Royal Alfred.**

alibi *n.* an excuse. [from legal sense *alibi* a defence by an accused person that he was elsewhere at the time of the offence]

Alice *phr.* **the Alice, a.** Alice Springs, a town in the Northern Territory. **b.** the train from Sydney to Alice Springs.

alive *phr.* **alive and kicking,** very much alive.

alkie *n.* a heavy drinker; an alcoholic. Also, **alky.**

all *phr.*

1. all aboard!, *N.Z.* a cry given by shearers, when shearing is about to begin.
2. all cush, okay; all right.
3. all in, (of a person or animal) exhausted.
4. all laired up, flashily dressed, as for some special occasion.
5. all over the place, in complete chaos.
6. all over (someone) like a rash, See **rash.**
7. all piss-'n-wind, loquacious, insincere.
8. all serene, okay; all right.
9. all Sir Garnet, okay; all right. Also, **all cigarnette (segarnio) (sogarnio).** [probably after *Sir Garnet* Wolseley, 1833-1913, British field marshal noted for his integrity]
10. all there, shrewd; alert; smart.
11. not all there, a. dull-witted. **b.** crazy; insane.
12. it's all right for you, addressed to someone better off or better placed than oneself.

All Blacks *pl. n.* the New Zealand international Rugby Union Football team. [from the colour of the team's uniform]

all-day sucker *n.* a large flat round sweet, often on a stick.

allergy *n.* a dislike or antipathy; *an allergy to hard work.* [from *allergy* a state of physical hypersensitivity to certain things which are normally harmless]

alley[1] *n.* **1.** a two-up school, run on organised lines and under strict control. **2.** *Horseracing.* position at the barrier, drawn by a horse for a race: *He went to the front from a wide alley. —phr.*

3. alley up, to pay up (a debt, etc.).

4. up one's alley, in the sphere that one knows or likes best.

alley[2] *n.* **1.** a large playing marble. *—phr.*

2. make one's alley good, to curry favour; improve one's standing.

3. toss or **pass in one's alley,** *Obsolete.* to die; give in.

alley[3] *v.* to go. [?French *allez* a command meaning 'go']

alley clerk *n.* a person who arranges bets for a two-up player.

alley loafer *n.* a moneyless two-up player who is never allowed a seat round a ring.

all-fired *adj. U.S.* extreme; excessive.

all-in *adj.* with extras included; inclusive: *at the all-in rate.*

all-overish *adj. Obsolete.* drunk.

all right *adv.* **1.** settled, or agreed on by bribery; *The job is all right. —phr.* **2. make it all right with,** to bribe.

all-set *adj.* ready; arranged in order; comfortable.

all-time *adj.* greatest of all time to date; outstanding: *He's an all-time rogue.*

all-up *adj.* **1.** total; inclusive: *The all-up weight is three tonnes. —phr.* **2. bet all-up,** to place the winnings of a previous race on one or more later races.

almighty *adj.* **1.** great; extreme: *He's in an almighty fix. —adv.* **2.** used to add emphasis to a statment: *too almighty clever.*

almonds *pl. n.* socks. [rhyming slang *almond rocks* (a boiled sweet) socks]

alone *phr.*

1. all alone like a country dunny, all alone; isolated; forlorn.

2. let alone, See **let.**

along *phr.*

1. get along, a. to be on amicable terms. **b.** to manage successfully; cope.

2. get along with you, an exclamation of dismissal or disbelief.

3. go along with, to agree with.

also-ran *n.* **1.** *Horseracing.* an unplaced horse in a race. **2.** a nonentity.

altar *phr.* **lead to the altar,** to marry (someone).

alter *v.* to castrate or spay.

altogether *phr.* **the altogether,** the nude.

amber fluid *n.* beer. Also, **amber liquid.**

ambo *n.* an ambulanceman.

Amen snorter *n.* a clergyman. [from *amen* used after a prayer]

ammo *n.* ammunition.

amp *n.* an amplifier.

ampster *n.* a decoy who works with a sideshow operator, acting as if he were an enthusiastic member of the audience, so as to arouse the interest of others. Also, **amster, Amsterdam.** [from *Amsterdam* rhyming slang for *ram* a trickster's confederate, from *ramp* a swindle]

anarchist *n.* a match, as used to produce a flame.

anatomical *adj.* bawdy; sexual. Also, **anatomic.**

anatomy *n.* body; bodily form; figure: *What a great anatomy that girl has!*

anchor *n.* **1.** (*pl.*) brakes: *hit the anchors.* —*v.* **2.** to take up residence (in a place); settle. —*phr.* **3. swallow the anchor,** (of a sailor) to settle down on shore. [from *anchor* a device for holding boats, vessels, etc., in place]

ancient history *n.* information or events of the recent past which are common knowledge or are no longer relevant.

andy mac *n. Obsolete.* a zack; sixpence; five cents. [rhyming slang]

anecdotage *n.* a state of old age in which a person is given to excessive reminiscence.

angel *n.* a financial backer of a play, campaign, actor, candidate, etc.

angel bruiser *n. Sport.* a very high kick in football, or a high shot in golf or cricket; rain-maker.

angels *n.* See **hell's angels.**

angie *n.* name for cocaine when used as a drug of addiction.

angle[1] *n.* a devious, artful scheme, method, etc. [from *at an angle,* slanting, not upright]

angle[2] *phr.* **angle for,** to try to get something by scheming, using tricks or artful means: *to angle for a compliment.* [from *angle,* to fish with hook and line]

angora *phr.* **act the angora,** to behave in a foolish fashion. [a humorous variation of **act the goat.** from *Angora* goat]

angry young man *n.* a young man, especially an artist or writer, outspokenly disgusted with the existing social order.

animal act *n.* a mean or despicable action.

ankle biter *n.* a young child.

annihilate *v.* to defeat utterly, as in argument, competition, or the like.

annual *n.* a bath; ablutions.

anotherie *n.* another one. Also, **anothery.**

ant *phr.* **have ants in one's pants,** to be restless or impatient.

ante *n.* **1.** a payment, usually monetary, extracted as part of a bargain. —*phr.*

2. ante up, a. to pay one's share as contribution. **b.** to put up the price or amount to be paid or contributed.

3. raise or **up the ante, a.** to increase suddenly the price to be paid for goods or services. **b.** to raise the requirements (for a job, etc.). [from *ante* (in a poker game) the stake put into the pool by each player]

antipodes *phr.* **the Antipodes,** a humorous term for Australia. [from *antipodes* the part of the world diametrically opposite, as Australia is from Britain]

antique *adj.* old-fashioned; antiquated: *an antique idea.*

ants pants *n.* the ultimate in style, novelty or cleverness: *He's really the ants pants.*

any *phr.* **get any,** to have sexual intercourse: *Are you getting any?*

anyrate *adv.* anyway.

anything *phr.* **like anything,** greatly; with great energy or emotion.

Anzac *n.* **1.** a soldier from Australia or New Zealand. **2.** Also, **bronzed Anzac.** (*sometimes ironic*) a healthy and heroic Australian male. [from *Anzac* a member of the *A(ustralian and) N(ew) Z(ealand) A(rmy) C(orps)* during World War I]

Anzac button *n.* a nail used in the place of a missing button.

Anzac wafer *n.* an army wafer, as issued during World War I.

A-OK *adj.* very good; functioning correctly. Also, **A-okay.**

AOT *adv.* upside-down; head over heels. [*A(rse) O(ver) T(it)*]

APC *n.* a quick wash; whore's bath. [*A(rm) P(its)* and *C(rotch)*, also the brandname of a headache powder]

ape *phr.* **go ape (over),** to react with excessive and unrestrained pleasure, excitement, etc.

ape hangers *pl. n.* bicycle handlebars so curved that the handles are above the level of the rider's shoulders.

apeman *n.* (*offensive*) an uncouth, virile man. Also, **ape.**

apoplectic *adj.* bad-tempered; choleric.

apple *phr.* **she's apples** or **she'll be apples,** all is well.

Apple *phr.* **the Big Apple,** New York.

applecart *phr.* **upset the applecart,** to disrupt plans; disturb the existing state of affairs.

applecatchers *pl. n.* loosely fitting short trousers gathered in at the knees; knickerbockers. Also, **poopcatchers.**

Apple Isle *n.* the island of Tasmania, one of the States of Australia. [from the fact that Tasmania was particularly noted for its export of apples]

apple pie *n.* sky. [rhyming slang]

apple-pie *phr.*

1. apple-pie bed, a bed shortsheeted, or in any way made uncomfortable as a joke.

2. apple-pie order, perfect order.

apples *pl. n.* stairs. [rhyming slang *apples and pears*]

apple sauce[1] *interj. British.* nonsense, rubbish.

apple sauce[2] *n.* a horse. [rhyming slang]

appro *n.* **1.** approval. —*phr.* **2. on appro,** for examination, without obligation to buy.

apron-strings *phr.* **tied to the apron-strings,** emotionally dependent on or bound to a person, as a child is to its mother.

Arab *n.* (*often offensive*) a foreigner.

argue *phr.* **argue the toss,** to dispute a decision or command.

argy-bargy *n.* **1.** argumentative talk; wrangling; disputation. —*v.* **2.** to wrangle; argue tediously; bandy words. Also, **argie-bargie, argle-bargle.**

aristotle *n.* a bottle. Also, **aris, aras.** [rhyming slang]

ark *phr.* **out of the ark,** very old.

arm *phr.* **chance one's arm,** to take a risk.

armchair *adj.* stay-at-home; amateur: *an armchair critic, an armchair philosopher.*

armchair ride *n.* (usually in a competitive situation) easy progress, easy success.

armed *phr.* **armed up,** *Prison.* carrying a weapon.

around *phr.* **have been around,** to be experienced, especially worldly-wise.

Arrows, the *pl. n.* the Canberra soccer team in the National Soccer League.

arse *n.* **1.** rump; bottom; buttocks; posterior. **2.** a despised person. **3.** impudence: *What arse!* **4.** a woman considered as a sex object: *She's a nice bit of arse.* *—phr.*

5. arse about, in reverse or illogical order: *He did the exercise completely arse about.*

6. arse about face, changed in direction; back to front.

7. arse about or **around,** to act like a fool; waste time.

8. arse out, to dismiss; reject.

9. arse over tit or **kettle/arse over apex** or **turkey, a.** upside down. **b.** fallen heavily and awkwardly usually in a forward direction.

10. arse up, to spoil; cause to fail.

11. down on one's arse, out of luck; destitute.

12. get one's arse into gear, to become organised and ready for action.

13. give (someone) the arse, a. to reject or rebuff (someone). **b.** to dismiss from employment.

14. have one's arse in one's hand, to be very annoyed.

15. kiss my arse! an expression of derision.

16. know one's arse from one's elbow; know one's arse from a hole in the ground; know one's arse from a hole in a flowerpot, to be aware, be well informed.

17. pain in the arse, See **pain.**

18. smart arse, another name for **smart alec.**

19. to think the sun shines out of one's arse, See **sun.**

20. up Cook's arse, *N.Z.* an expression of disgust.

arse bandit *n.* a homosexual who is sexually aggressive.

arsehole *n.* **1.** the anus. **2. a.** a despised place: *This town is the arsehole of the universe.* **b.** a despised person. *—v.* **3.** to remove a person from a place quickly and without ceremony; to throw someone out. **4.** to dismiss; sack. *—phr.*

5. arsehole about, to fool around.

6. arsehole off, to depart quickly or unobtrusively.

7. as ugly as a bag full of arseholes, unattractive; ugly.

8. from arsehole to breakfast time, completely.

9. have one's arsehole pointing to the ground, to be fit and able; alive and kicking.

arseholes *interj.* an exclamation of disgust or disbelief.

arse-licker *n.* a self-seeking flatterer; a fawning, servile parasite; a sycophant.

arse-up *adj.* **1.** wrong side up; topsy-turvy; incorrect. *—adv.* **2.** in a clumsy fashion.

arsy *adj.* lucky. Also, **arsey, arsie.**

arsy-versy *adj.* back to front; topsy-turvy.

artic *n.* *N.Z.* a semitrailer. [shortened form of *articulated lorry*]

artichoke *n.* **1.** a debauched old woman. **2.** *Building Trades.* an architect.

artist *n.* a person noted or notorious for a reprehensible aspect of his behaviour: *a booze artist; a bullshit artist.*

artiste *n.* **1.** (often mockingly or tongue-in-cheek) a female artist in the more popular theatrical arts, with more pretension than talent. **2.** a female striptease artist.

arty *adj.* **1.** ostentatious in display of artistic interest. —*n.* **2.** an arty person. **3.** *Army.* the artillery.

arty-crafty *adj.* affectedly artistic; artistic but useless, pretentious, or trivial.

arty-farty *adj.* pretentious; precious.

arvo *n.* the afternoon. Also, **aftie, afto.**

a.s.a.p. as soon as possible. [*a(s) s(oon) a(s) p(ossible)*]

Ashes, the *pl. n.* the trophy, an urn containing a cremated cricket stump, kept permanently in England, played for by England and Australia in Test cricket.

ask *phr.* **ask for it** or **trouble,** to behave so as to invite trouble.

aspro[1] *n.* an Associate Professor.

aspro[2] *n.* a male prostitute.

ass *n. U.S.* arse.

at *phr.*

1. at that, as things stand: *Let it go at that.*

2. be at, be engaged in or preoccupied with: *What are you at these days?*

3. be at someone, to be critical.

attaboy *interj.* an exclamation of approbation or exhortation.

aunt *phr.* **my sainted aunt!** an expression of surprise or disbelief.

Auntie Nellie *n.* the belly, especially with reference to indigestion. [rhyming slang]

aunty *n.* **1.** a conservative body or organisation, especially the Australian Broadcasting Commission. **2.** an effeminate or homosexual older male. Also, **auntie.**

aunty's downfall *n.* a humorous term for gin; aunty's ruin, mother's ruin.

Aussie *adj.* **1.** Australian. —*n.* **2.** an Australian. **3.** Australia. Also, **Aus.**

Aussie battler *n.* See **battler.**

Aussieland *n.* Australia.

Aussie Rules *n.* Australian Rules football.

Aussie wuzzy *n. Army.* a member of an Australian carrying party in New Guinea in World War II. [modelled on *fuzzy wuzzy* a native]

Australian *phr.* **the great Australian adjective,** *bloody,* an intensive signifying approval, as in *bloody beauty,* or disapproval, as in *bloody bastard.*

Australian flag *n.* the bottom of a man's shirt when it has come out from his trousers.

Australian fly trap *n.* See **Afghan's fly trap.**

Australian salute *n.* the movement of the hand and arm to brush away flies from one's face. Also, **Barcoo salute.**

auto *phr.* **on auto,** (of a person) not giving full attention; not thinking.

autumn leaf *n. Horseracing.* the name given to a jockey, apprentice, etc., who has often fallen from his mount.

awake *phr.* **awake up,** fully aware: *He's awake up to what's going on.*

away *adj.* **1.** on the move; having started; in full flight. —*phr.*
2. get away, a. to leave in a hurry. **b.** to start in a race: *The horses got away well.*
3. get away (with you), an expression of disbelief, surprise, etc. Also, **get out (with you).**

awkward *phr.* **the awkward squad,** one of a number of groups, as sporting teams, marching squads, etc., to which the clumsiest and most awkward people have been assigned.

axe *n.* **1.** a guitar, usually electric. —*v.* **2.** to cut out; abolish (a project, etc.). **3.** to reduce (expenditure, prices, etc.) sharply. **4.** to dismiss from a position. —*phr.*
5. have an axe to grind, to have a private purpose or selfish end to attain.
6. the axe, a. a drastic cutting down (of expenses). **b.** dismissal from a job, position, or the like; the sack.

axe-breaker *n.* any tree whose timber has a hard, close or interlocking grain.

axe-handle *n.* a rough unit of measurement.

axeman *n.* a personnel officer or the like, one of whose duties is to dismiss staff.

axle-grease *n.* **1.** butter. **2.** money.

ayatollah *n.* any autocratic leader. [a reference to *Ayatollah* Khomeini, a Shiite leader, whose rule in Iran has been marked by strict enforcement of Islamic laws]

B *n.* **1.** a bastard. *—phr.* **2. not to know B from a bull's foot,** to be stupid and ignorant.

baal *adv.* **1.** not. *—interj.* **2.** no. [Aboriginal]

babbling brook *n.* a cook. Also, **bab, babbler.** [rhyming slang]

babe *n.* a girl.

baby *n.* **1.** an invention or creation of which one is particularly proud. **2.** a girl. *—phr.* **3. leave (someone) holding the baby,** to abandon (someone) with a problem or responsibility not rightly his.

baby's cries *n.* eyes. [rhyming slang]

baccy *n.* tobacco. Also, **bacco, bakky.**

bach *n.* **1.** a bachelor. *—v.* **2.** Also, **batch. a.** to keep house alone or with a companion when neither is accustomed to housekeeping: *She was baching with a friend at North Sydney.* **b.** to live alone: *an old hand at baching.*

bachelors' hall *n.* a building on a station for the accommodation of the jackeroos, etc.

back *phr.*

1. back and fill, to vacillate.

2. back up for, to seek more of (something such as money or a further helping of food).

3. be or **get on (someone's) back,** to stand over, to urge constantly to further action: *He's always on my back.*

4. break the back of, a. to deal with or accomplish the most difficult or arduous part of (a task, etc.). **b.** to overburden or overwhelm.

5. get off (someone's) back, to cease to annoy or harrass.

6. get one's back up, to become annoyed.

7. out the back, in the backyard, as an outside toilet.

8. put (someone's) back up, to arouse (someone's) resentment.

9. see the back of, a. to be rid of (a person). **b.** to be finished with (a situation, task, etc.).

backblocks *pl. n.* **1.** remote, sparsely inhabited inland country. **2.** the outer suburbs of a city.

backbone *phr.* **see (someone's) backbone,** (formerly) to flog.

backbreaker *n.* a wave which, in shallow water, instead of breaking evenly from the top, crashes violently down, throwing surfers to the bottom; dumper.

backburner *phr.* **put on the backburner,** to delay immediate action on.

backchat *n.* impertinent talk; answering back.

back door *n.* **1.** the anus (especially used in referring to anal intercourse). *—phr.* **2. by** or **through the back door,** secretly; by hidden, obscure or dishonourable means.

backhander *n.* a bribe: *He slipped the witness a backhander.*

back of beyond *n.* **1.** a remote, inaccessible place. **2.** the far outback. *—adv.* **3.** in the outback. Also, **back o' beyond, beyond set o' sun, back of sunset, back of Bourke.**

backroom boys *pl. n.* people operating behind the scenes, usually in enterprises in which they do not wish their involvement to be known to the public.

back scratcher *n.* one who provides a service to another in expectation of receiving a service in return.

back seat *phr.* **take a back seat,** to retire into obscurity, or into an insignificant or subordinate position.

back-seat driver *n.* **1.** a passenger in a car who offers unsolicited advice to the driver, as though trying to drive the car himself. **2.** one who proffers advice or gives orders in matters which are not his responsibility.

backslapping *n.* **1.** hearty fraternisation, especially between men. *—adj.* **2.** jovial and friendly.

backstop *n.* a person or thing which is relied upon for assistance when all else fails: *I have $500 as a backstop.*

backstreet *adj.* illegal, illicit or improper.

backyarder *n.* **1.** an abortionist without recognised medical training. **2.** a person who carries on business in the backyard of his home.

bacon *phr.*

1. bring home the bacon, to succeed in a specific task.

2. save (someone's) bacon, to save (someone) from a dangerous or awkward situation.

bacon and eggs *pl. n.* legs. [rhyming slang]

bad *phr.*

1. not bad, a. fair; not good. **b.** excellent.

2. in bad with, out of favour with.

3. to the bad, in deficit; out of pocket: *two hundred dollars to the bad.*

bad apple *n.* **1.** a morally reprehensible person. **2.** one of a group, worse than the rest, and likely to corrupt the others.

baddie *n.* a bad person, especially a villain in a story, play or film.

bad dog *n.* Also, **dog, a dog tied up,** a debt: *He didn't realise just how many bad dogs he'd collected, until word got around of his lottery win.*

bad egg *n.* a person of reprehensible character.

badger *n. Obsolete.* (erroneously) **a.** a wombat. **b.** a bandicoot.

badger-box *n. Obsolete.* a badly built house.

bad hat *n.* a person of reprehensible character.

bad job *n.* See **job.**

bad karma *n.* bad consequences of an action; bad luck. [from *Karma,* the Hindu and Buddhist concept of the cosmic operation of retributive justice, according to which a person's status in life is determined by his own deeds in a previous incarnation]

bad lot *n.* a dishonest, disreputable person, usually considered a failure in life.

badmouth *v.* to speak unfavourably of, to criticise with malice.

bad news *n.* See **news.**

bad penny *n.* someone or something unwanted that seems to be constantly returning: *He kept turning up in Darwin like a bad penny.*

bad trot *n.* a period of ill fortune.

bag *n.* **1.** (*pl.*) a lot; an abundance: *bags of money.* **2.** Also, **old bag.** a disagreeable and unattractive woman. **3.** a chosen occupation, hobby, pursuit, etc.: *Golfing is his bag.* **4.** a measure of marijuana, heroin, etc. **5.** the breathalyser. *—v.* **6.** to arrest and put in gaol. **7.** to criticise sarcastically. *—phr.*
8. bag of mystery, a saveloy or sausage. Also, **mystery bag.**
9. bag of tricks, a. a miscellaneous collection of items. **b.** a person never at a loss.
10. bag of wind, a. a loquacious person. **b.** a football.
11. get a bag, *Cricket.* a sarcastic remark made to a cricketer who has dropped a catch.
12. in the bag, secured; certain to be accomplished: *The contract is in the bag.*
13. rough as bags, uncouth.

bagboy *n.* a bookmaker's clerk.

baggage *n.* a pert or impudent young woman.

bagged *adj.* compelled to undergo a breathalyser test.

baggies *pl. n.* trousers, slacks, jeans, etc., with wide legs.

bagging *n.* severe, especially sarcastic, criticism: *He gave me a bagging this morning over my smelly feet.*

baggy *n. Prison.* a low-ranking prison officer.

bagman *n.* **1.** (formerly) **a.** a swagman; tramp. **b.** a travelling pedlar. **2.** a bookmaker. **3.** a person who collects bribes.

bagman's gazette *n.* (formerly) a fictitious journal said to be the source of rumours in the outback.

bagman's two-up *n.* a swagman's blanket made from two chaff or corn sacks cut open and stitched together.

bag of fruit *n.* a suit. [rhyming slang]

bags *interj.* Also, **I bags.** an exclamation by which one establishes right by virtue of making the first claim: *Bags I have first ride.*

bagswinger *n.* **1.** a streetwalking prostitute. **2.** a bookmaker.

bag system *n.* (formerly) the issue of a bag of groceries each dole day to the unemployed. [from the depression of the 1930s]

bag test *n.* examination by the breathalyser.

bail *phr.*
1. bail (someone) out, to help (someone) out of trouble or difficulty.
2. bail up, a. *Bushranging.* to hold the victims under guard, and confine them or tie them up. **b.** to hold up and rob. **c.** to delay (someone) unnecessarily, as in conversation.
3. jump bail, to forfeit one's bail by absconding or failing to appear in court at the appointed time.

bait *v.* (of a line fisherman) to lose one's bait to the fish: *I think I'm baited.*

bait bobber *n.* a fisherman who uses a rod or who fishes with a cork on his line.

baitlayer *n.* a cook.

baker *n.* See **floury baker.**

balancer *n.* a fraudulent bookmaker.

baldheaded *phr.* **go at (something) baldheaded,** to act suddenly with great (and often rash) energy.

baldy *n.* **1.** (sometimes offensive) a term of address for a bald person. **2.** a worn tyre having little or no tread.

Bali belly *n.* diarrhoea, as suffered by travellers.

ball *n.* **1.** a testicle: *She kicked him in the balls.* —*v.* **2.** to have intercourse with. —*phr.*
3. a ball of muscle, a person who is very healthy and in good spirits.
4. balls and all, aggressively and enthusiastically.
5. ball up, *U.S., British.* to bring to a state of confusion.
6. do (one's) balls on, to become infatuated with (someone).
7. grab or **have a ball,** to enjoy oneself; have a good time.
8. have (someone) by the balls, to have (someone) in one's power.
9. have the ball at one's feet, to be in a position of immediate opportunity.
10. have the ball in one's court, to have the opportunity or obligation to act.
11. keep one's eye on the ball, to be alert.
12. keep the ball rolling, to keep something going; to keep up the rate of progress or activity.
15. on the ball, in touch with a situation, reality, etc.; alert; sharp.
16. play ball, to work together (with); cooperate.
17. start the ball rolling, to start an operation; set an activity in motion.
16. that's the way the ball bounces, *Chiefly U.S.* that's how things are.

ball and chain *phr.* **the old ball and chain,** the wife.

ballarat *n.* **1.** a cat. **2.** a homosexual who plays a passive role; cat. [rhyming slang; from *Ballarat,* a town in Victoria]

ball-drainer *n.* a sexually exhausting experience.

ball game *phr.* **a (whole) new ball game,** a new set of circumstances calling for a different approach.

ball-out *v.* to reprimand. [misuse of *bawl* to cry out loudly, scold, + *-out*]

ballpark *adj.* **1.** roughly estimated. *—phr.* **2. off the ballpark,** unofficially; at a guess.

balls *pl. n.* **1.** courage, moral strength. **2.** the testicles. *—interj.* **3.** an exclamation of repudiation, ridicule, etc. *—phr.* **3. balls around,** to deliberately waste time in inconsequential activity.

balls-up *n.* **1.** confusion arising from a mistake; a mess. **2.** the mistake itself. *—v.* **3.** to bring to a state of hopeless confusion or difficulty. [probably from *ball up* to bring to a state of confusion, by confusion with *ball* a testicle]

ball-tearer *n.* something extremely good or dynamic: *a ball-tearer of a book.*

bally *adj.* **1.** confounded (used humorously or for emphasis). *—adv.* **2.** very. [originally euphemism for *bloody*]

baloney *n.* nonsense; insincere or idle talk; eyewash; waffle. Also, **boloney.** [? from *Bologna* sausage]

balt *n.* (*offensive*) an immigrant to Australia from any of the countries of Central or Eastern Europe. [from *Baltic,* of or pertaining to the Baltic Sea or the countries on its shores]

bananabender *n.* a Queenslander.

banana republic *n.* **1.** any small tropical country, especially of South or Central America, considered as backward, politically unstable, etc., and dependent on the trade of rich foreign nations. **2.** any backward country or state. *—phr.* **3. the Banana Republic,** Queensland.

bananas *phr.* **go bananas, a.** to become uncontrollably angry. **b.** to become mentally unbalanced.

bandaid solution *n.* a temporary, expedient, and generally unsatisfactory solution to a problem.

b. and d. *n.* bondage and discipline, as related to sexual practices.

bandicoot *v.* **1.** to dig up (root vegetables, potatoes, etc.) with the fingers, leaving the top of the plant undisturbed. *—phr.*
2. bald as a bandicoot, remarkably bald.
3. barmy as a bandicoot, mad.
4. like a bandicoot on a burnt ridge, lonely and forlorn.
5. lousy as a bandicoot, miserly.
6. miserable or **poor as a bandicoot,** of wretched character. [from *bandicoot,* a small, ratlike Australian marsupial, relatively harmless and sufficiently remote from human contact to be tagged with a variety of pejorative epithets. It is sometimes considered a nuisance because it damages gardens while digging for insects]

bandit *n.* **1.** a poker machine; one-armed bandit. **2.** *Originally U.S.* an aggressive homosexual.

band moll *n.* a young female who serves the sexual needs of the members of a rock group.

band of hope *n.* soap. [rhyming slang, with reference to the name of the Temperance Mission]

band wagon *phr.* **climb** or **jump on the band wagon,** to join the winning side; take advantage of a popular movement or fashion; follow the crowd. [from U.S. *band wagon,* originally a wagon, often elaborately decorated, used to transport musicians, as at the head of a procession or parade]

bandy *phr.* **knock (someone) bandy,** to get the best of (someone); to astonish (someone).

Bandywallop *n.* an imaginary remote town.

bang *n.* **1.** sexual intercourse. **2.** a thrill; excitement. —*v.* **3.** Also, **bang up.** (of a man) to have sexual intercourse with (a woman). —*phr.*

4. bang goes, that's the end of: *Bang goes that idea.*

5. bang like a dunny door in the wind, (of a female) to be very free with sexual favours.

6. bang on, dead-centre; right on mark; correct.

7. bang something over, to accomplish a task, etc. quickly or with little effort: *We'll bang it over in no time.*

8. bang to rights, red-handed.

9. the whole bang lot, everything.

10. with a bang, impressively; successfully: *The party went with a bang.*

bange *v.* **1.** to rest; relax. —*n.* **2.** a rest. Also, **banje.** [British dialect *benge* to lounge around]

banged up *adj.* pregnant.

banger *n.* **1.** a bunger; a type of firework. **2.** *British.* a sausage. —*phr.* **3. three bangers short of a barbie,** See **barbie.**

bango *n.* a bush food of boiled flour and sugar.

bangtail muster *n.* **1.** a round-up of cattle for counting. **2.** a carnival or sports day in a country town. [from *bangtail,* an animal with a docked tail, especially a horse + *muster*]

bang-up *adj.* first-rate; excellent.

banjo *n.* **1.** a shovel or spade. **2.** a frying pan. **3.** a shoulder of mutton. **4.** *Tin Mining.* a device in which tin is washed from dirt. [from similarity in shape to the *banjo,* a musical instrument of the guitar family]

bank[1] *phr.* **bank and bank,** *N.Z.* (of a river) in flood.

bank[2] *phr.* **bank on** or **upon,** to rely or count upon.

banker *phr.* **run a banker,** (of a river) to be flowing up to the top of the banks.

bar[1] *phr.*

1. go off the bars, *Prison.* to commit suicide by hanging.

2. no holds barred, with no restrictions. [wrestling]

3. not to have a bar (of), not to tolerate: *I won't have a bar of it.*

bar[2] *adj.* **1.** (*in children's speech*) inviolate: *Don't touch me, I'm bar.* —*n.* **2.** (*in children's games*) anything which acts as a sanctuary; a position from which one cannot be assailed. [from *barley, British Obsolete,* an appeal for respite]

bar[3] *phr.* **wouldn't know (someone) from a bar of soap,** See **soap.**

barb *n.* a black dog like a kelpie.

barbed wire *phr.* *Qld.* **on the barbed wire,** to be on a beer drinking spree. [*barbed wire* refers to a bottle or can of XXXX beer, with the four X's displayed on the label, resembling barbed wire]

barber *v.* **1.** *Prison.* to steal from residential premises. —*n.* **2.** *N.Z.* a cold, keen, cutting wind.

barbie *n.* **1.** a barbecue. *—phr.* **2. three bangers short of a barbie,** dull-witted; weak of intellect; stupid. Also, **barby.** [shortened form]

barbie doll *n.* moll. [rhyming slang]

Barcoo *n. Qld* **1.** an electric storm. **2.** Also, **Barcoo buster.** a westerly gale in the outback. [from *Barcoo* a river and district in Queensland]

Barcoo challenge *Shearing. n.* **1.** (formerly) a challenge for the highest tally of the day, made by scraping the points of the shears on the floor or wall. **2.** a similar challenge made by throwing the belly wool of the first sheep shorn over another shearer's head. [See **Barcoo**]

Barcoo rot *n.* a chronic streptococcal skin infection. [See **Barcoo**]

Barcoo salute *n.* See **Australian salute.** [See **Barcoo**]

Barcoo sandwich *n.* any outback food considered to be inedible, as (humorously) a goanna between two slabs of bark. Also, **Borroloola sandwich.** [See **Barcoo**]

Barcoo spews *pl. n.* attacks of heat-induced vomiting. [See **Barcoo**]

bardy *phr.* **starve the bardies!** an exclamation of surprise or disgust.Also, **bardi.** [Aboriginal: an edible wood-boring grub or its larvae]

barebelly *n.* a sheep which, due to defective growth or illness, has lost the wool from its belly and inner hindlegs; rosella.

barf *v. U.S.* to vomit.

bar fly *n.* a habitual drinker at bars.

barge *n.* **1.** any old or unwieldy boat. **2.** a cumbersome surfboard. *—phr.*
3. barge in, to intrude clumsily as into a conversation.
4. barge into, to collide clumsily with (someone or something).

barge pole *phr.* **not to touch with a barge pole,** to have nothing to do with; not to go near. [from the long pole used to propel a barge]

bargo *n.* someone who will eat anything.

bark[1] *v.* **1.** to advertise a cheap show at its entrance. **2.** to vomit. *—phr.*
3. bark up the wrong tree, to mistake one's object; assail or pursue the wrong person or purpose.
4. keep a dog and bark oneself, to indulge in unnecessary or duplicated activity.

bark[2] *phr.* **short of a sheet of bark,** weak of intellect; silly; stupid.

Bark *n.* an Irish person.

barker *n.* one who stands before a shop, nightclub, etc., calling passers-by to enter.

barking jackass *n.* a kookaburra.

barleycorn *n.* See **John Barleycorn.**

barmaid's blush[1] *n.* **1.** a drink made from rum and raspberry. **2.** a drink made from port wine and lemonade.

barmaid's blush[2] *n.* a flush in poker. [rhyming slang]

barmy *adj.* mad; stupid; silly.

barn *phr.* **born in a barn/field/tent,** of a person who constantly forgets to close a door.

barney *n.* **1.** an argument; fight. **2.** humbug; cheating. **3.** an unfair contest or prizefight. **4.** a spree. —*v.* **5.** to argue or fight. [British dialect]

Barney's bull *n.* **1.** a clumsy or incompetent person. **2.** an object of no value. —*phr.* **3. all behind like Barney's bull, a.** late; backward. **b.** exhausted.

barossa *n.* a girl. [rhyming slang *Barossa* Pearl, girl, from the Tradename of a wine produced in the Barossa Valley, South Australia]

barrack[1] *v.* to jeer, shout derisively at (a player, team, etc.). [Aboriginal *barak* (a negative)]

barrack[2] *phr.* **barrack for,** to support, shout encouragement and approval.

barracouta *n.* *N.Z.* a long, raised bread loaf. Also, **barracuda.** [from its similarity in shape to the *barracouta* and *barracuda,* both being elongated fish]

barrel[1] *n.* **1.** a theatrical performance to raise money for a worthy cause. —*v.* **2.** to knock over, run into or strike hard: *I'll barrel the bloke.* —*phr.*

3. barrel along, to move along swiftly and confidently.

4. over a barrel, at a disadvantage; in difficulty.

5. scrape the bottom of the barrel, See **scrape.**

barrel[2] *n.* a hat. [rhyming slang, *barrel of fat* hat]

barrier reef *n.* teeth. [rhyming slang]

barrow[1] *phr.*

1. push one's barrow, to campaign vigorously in one's own interest.

2. push (someone's) barrow, to take up (someone's) cause for them.

barrow[2] *v.* (of a shedhand who is learning to become a shearer) to finish shearing a sheep left partly shorn by a shearer at the end of a shearing run.

bart *n.* *Obsolete.* a girl.

base *n.* **1.** the buttocks. —*phr.*

2. not to get to first base with, to fail to establish any contact or rapport.

3. base over apex, Also, **arse over tit.** fallen heavily and awkwardly, usually in a forward direction.

base wallah *n.* a member of the armed services employed on a military base. [from *base* + *wallah* Hindi *-wālā* person]

base walloper *n.* *N.Z.* a member of the armed services employed on a military base. [*base wallah* + *walloper,* pun on *base* arse]

bash *n.* **1. a.** a party. **b.** a drinking spree. —*v.* **2.** to move quickly and noisily through the bush. —*phr.*

3. bash it, an exclamation of contempt, disgust, etc.

4. bash one's brains out, to expend a great deal of effort in intellectual activity.

5. bash the ear, to talk incessantly; harangue. [from *earbasher*]

6. give it a bash, a. to make an attempt. **b.** to go on a drinking spree.

7. have a bash at, make an attempt at.

basher *n.* a man.

bash hat *n.* a safety helmet; hard hat.

bashing *n.* **1.** a hard or arduous job: *spud bashing; bush bashing. See* **spinebashing. 2.** excessive attention, exposure or use: *That song's had a bashing on air lately.* **3.** such attention, etc., evincing hostility: *Canberra bashing is my favourite sport.*

basinful *n.* a superabundance of trouble, distress, etc.

basket *n.* a euphemism for bastard.

basket case *n.* a person in an advanced state of nervous tension or mental instability.

Bass and Flinders *pl. n.* windows. [rhyming slang]

bastard *n.* **1.** an unpleasant or despicable person. **2.** any person (without negative connotations). *—phr.*

3. a bastard of a thing, a terrible thing.

4. don't let the bastards grind you down, an exhortation to keep one's resolve and remain firm and cheerful in the face of difficulties. [often and variously rendered in mock Latin, as *nil carborundum bastardis*]

5. happy or **lucky as a bastard on Father's Day, a.** unhappy, unlucky. **b.** extremely pleased.

bastard from the bush *n.* a person who is ruthlessly overbearing and who cadges shamelessly. [?from a poem 'The Captain of the Push' attributed to Henry Lawson, 1867-1922, Australian poet and short-story writer]

bastardise *v.* to seek to humiliate, as a part of initiation into a regiment, college, etc.

bastardly *adj.* mean; despicable.

bastardry *n.* obnoxious and unpleasant behaviour.

bat[1] *n.* **1.** a blow as with a bat. **2.** rate of motion: *to go at a fair bat.* **3.** a spree; binge: *to go on a bat. —phr.*

4. bat along, to travel at speed.

5. bat or **fan the breeze,** See **breeze.**

6. carry one's bat, to accomplish any difficult, lengthy, or dangerous task. [from the cricket sense, *carry one's bat,* to remain at the wicket as a batsman throughout an innings]

7. go to bat for, to champion; take up the cause of.

8. off one's own bat, independently; without prompting or assistance.

[from *bat* a club, etc., as used in certain sporting games]

bat[2] *n.* **1.** a cranky or silly woman: *She's an old bat. —phr.*

2. blind as a bat, very blind.

3. have bats in the belfry, to have mad notions; to be crazy or peculiar.

4. like a bat out of hell, at speed; quickly.

[from *bat* nocturnal flying mammal]

bat and ball *n.* wall. [rhyming slang]

bathers *n.* swimming costume.

bathplug *n.* bastard. Also, **bathtub.**

bats *adj.* mad; crazy.

battle *phr.*

1. battle on, to continue a struggle; endure.

2. on the battle, working as a prostitute.

battleaxe *n.* a domineering woman.

battler *n.* **1.** one who struggles continually and persistently against heavy odds. **2.** Also, **little Aussie battler.** a typical member of the working class in Australia. **3.** a conscientious worker, especially one living at subsistence level. **4.** an itinerant worker reduced to living as a swagman. **5.** a small time punter who tries to live on his winnings. **6.** a prostitute.

batty *adj.* crazy; silly.

bay *phr.* **away in the bay,** (of a person) lost in thought; preoccupied; paying no attention to what is going on.

Bay, the *n.* **1.** a local name for various towns or areas, such as Byron Bay, Batemans Bay, Moreton Bay, Nelson Bay, Port Phillip Bay, etc. **2.** Long Bay Gaol, Sydney.

Bazza-land *n.* Australia. [from Barry (*Bazza*) McKenzie, a character created by Barry Humphries, Australian satirist]

BBQ *n.* barbecue.

beach *phr.* **on the beach, a.** without a job. **b.** without money.

beach-bash *v.* to lie on a beach, especially at night, and for amorous purposes.

beach bum *n.* one who spends most of his or her life lazing about on a beach.

beak[1] *n.* a person's nose.

beak[2] *n.* **1.** a magistrate; judge. **2.** a schoolmaster.

beam[1] *phr.* **broad in the beam,** (of a person, usually female) very wide across the buttocks.

beam[2] *phr.*

1. off the beam, a. wrong; incorrect; out of touch with the situation. **b.** crazy.

2. on the beam, just right; exact; correct; in touch with the situation.

[from the radio *beam* used to indicate the correct flight course]

bean *n.* **1.** a coin; anything of the least value: *I haven't a bean.* **2.** the head. —*v.* **3.** *U.S.* to hit on the head. —*phr.*

4. a row of beans, anything significant: *It doesn't add up to a row of beans.*

5. full of beans, energetic; vivacious.

6. give (someone) beans, to berate; attack (someone).

7. know how many beans make five, to be aware; be well informed.

8. not to know (or care) beans, not to know (or care about) something.

9. spill the beans, to divulge information, often unintentionally.

bean-ball *n.* *Cricket.* a full toss which goes towards the batsman's head. Also, **beamer.**

bean brain *n.* a fool.

beanpole *n.* a tall, lanky person.

bear *phr.* **like a bear with a sore head,** intensely irritable; grumpy.

beardie *n.* a bearded or frill-necked lizard.

bear's paw *n.* a saw; a saw of any kind used by a workman. [rhyming slang]

Bears, the *pl. n.* **1.** the North Sydney team in the N.S.W. Rugby Football League. **2.** the Caulfield team in the Victorian Football Association.

beat *n.* **1.** a beatnik. **2.** a deadbeat; loafer; sponger. **3.** a sheep or cattle run under the supervision of one man. —*adj.* **4.** exhausted; worn out. **5.** defeated. —*phr.*

6. beat about the bush, See **bush.**

7. beat hollow, to win or beat overwhelmingly.

8. beat it, to go away; depart.

9. beat one's meat, Also, **beat off.** (of a male) to masturbate.

10. beat or **jump the gun,** See **gun.**

11. beat the living daylights out of, to thrash soundly as a punishment; beat violently.

12. beat the skins, to drum.

13. off the beat, out of the usual routine.

beaten track *phr.* **off the beaten track, a.** isolated; remote. **b.** in unfamiliar territory. **c.** unusual.

beat-up *n.* **1.** a media story of small significance which is given spurious importance by an expanded, often sensational treatment. —*adj.* **2.** old; dilapidated: *a beat-up old car.*

beaut *adj.* **1.** fine; good: *a beaut car.* —*interj.* **2.** Also, **you beaut!** an exclamation of approval, delight, enthusiasm, etc. —*n.* **3.** Also, **beauty. a.** something successful or highly valued. **b.** a pleasant, agreeable, trustworthy person.

beautiful people *pl. n.* **1.** a fashionable social set of wealthy, well-groomed, usually young people. **2.** hippies. Also, **Beautiful People.**

beauty *n., interj.* See **beaut.** Also, **bewdy, bewdy bottler.**

beaver *n.* **1.** the pubic area. **2.** the vagina. **3.** a female.

bed *phr.* **get into bed with,** Also, **go to bed with, a.** to have sexual intercourse with. **b.** to be committed to an alliance, deal, etc.: *to get into bed with the unions.*

bed and breakfast *n. Prison.* imprisonment for seven days.

beddy-byes *phr.* (*in children's speech*) **go (to) beddy-byes,** to go to bed.

bedourie *n.* a bushman's camp oven, the lid of which serves as a frying pan. [from *Bedourie,* a town in Queensland]

Bedourie fly veil *n.* the tail of one's shirt, dabbed with excrement, and allowed to hang out so as to attract flies away from the face. See **Afghan fly trap.** [See **bedourie**]

Bedourie shower *n.* a dust storm. [see **bedourie**]

bedroom mug *n.* a chamber-pot.

bee[1] *phr.*

1. bee in one's bonnet, a. an obsession. **b.** a slightly crazy idea, attitude, fad, etc.

2. bees and honey, money. [rhyming slang]

3. the bee's knees, someone or something arousing great admiration.

bee[2] *phr.* **bee eff,** bloody fool. Also, **b fool.** [from the initials *B(loody) F(ool)*]

beef *n.* **1.** weight, as of human flesh. **2.** *Chiefly U.S.* a complaint. —*v.* **3.** *Chiefly U.S.* to complain; grumble. —*phr.*

4. beef up, to increase, enlarge: *the airline will beef up the number of flights.*

5. beef out, to expand unnecessarily.

6. beef to the ankles, overweight.

beef bayonet *n.* an erect penis. Also, **beef bugle.**

beefcake *n.* photographs of men in newspapers, magazines, etc., posed to display their bodies and emphasising their sex appeal. See **cheesecake.**

beefeater *n. U.S.* an Englishman. [from *beefeater* a yeoman of the guard or a warder of the Tower of London]

beer gut *n.* a paunch caused by excessive beer drinking.

beer money *n.* **1.** a gratuity. **2.** any money set aside for spending on pleasure, especially by a husband.

beer-up *n.* a drinking party devoted largely to beer-drinking and talk.

beetle *n.* **1.** a Volkswagen car of the first type produced, so called because of its shape. —*v.* **2.** Also, **beetle off** or **along,** to move swiftly, especially in an aeroplane.

beggar *n.* **1.** one who is remarkably keen on or adept at something: *He's a beggar for work; a beggar at chess.* —*phr.* **2. beggar for punishment,** a person who consistently exerts himself.

beggars-on-the-coals *pl. n.* small thin dampers. Also, **beggars-in-the-pan, buggers-on-the-coals.**

behind *n.* the buttocks.

bejesus *phr.* **knock the bejesus out of,** to destroy the self-confidence of or defeat utterly.

bell *n.* **1.** a telephone call: *to give someone a bell.* —*phr.*

2. ring a bell, to remind one; jog the memory.

3. ring the bell, to excel; surpass all others.

4. with bells on, See **knob.**

bell sheep *n. Shearing.* the last sheep shorn by any shearer before meal or tea intervals or at the end of the day.

belly *phr.* **lower than a snake's belly,** See **snake.**

belly-ache *n.* **1.** a pain in the stomach, especially colic. **2.** a cause of discontent. **3.** a complaint. —*v.* **4.** to complain or grumble.

bellybuster *n.* a badly-judged dive in which one's stomach hits the water first.

bellybutton *n.* the navel. Also. **belly button.**

bellyflop *n.* **1.** Also, **bellyflopper.** a bellybuster. —*v.* **2.** to dive so that one's stomach hits the water first.

bellyful *n.* enough or more than enough.

Belsen horror *n.* one who is unusually thin or emaciated. [from *Belsen* German concentration camp in World War II]

belt *v.* **1.** to give a thwack or blow to. **2.** to eat or drink quickly: *Belt that food into you.* **3.** to move quickly: *to belt along.* —*phr.*

4. below the belt, against the rules; unfairly. [from boxing sense, i.e. against the Queensberry Rules]

5. belt into, to begin with speed and vigour.

6. belt up, be quiet; shut up.

7. tighten one's belt, to live frugally.

8. have under one's belt, to achieve or experience: *She's got hours of study under her belt.*

belting *n.* a thrashing or beating.

bend *phr.*

1. bend the elbow, to drink alcoholic liquor usually to excess.

2. round the bend, mad.

bender *n.* **1.** a drinking spree. **2.** something very good of its kind; corker. **3.** a homosexual.

benjo *n.* a toilet. [probably from Japanese impolite term *benjo* toilet]

benny *n.* a benzedrine pill.

bent *adj.* **1.** stolen: *to sell bent goods.* **2.** thievish; having little or no regard for the law; dishonest. **3.** diverging from what is considered to be normal or conservative behaviour, as by taking illegal drugs, practising homosexuality, etc. —*n.* **4.** one who diverges from the orthodox patterns of society as a homosexual, a taker of illegal drugs, etc.

bent cop *n.* a dishonest policeman, especially one who takes bribes.

beresk *adj.* a humorous term for **berserk.**

bergoo *n.* See **burgoo.**

berk *n.* (*offensive*) an unpleasant or despicable person. Also, **birk, burk, burke.** [rhyming slang, *Berkshire Hunt* cunt]

berko *adj.* berserk.

berley *n.* **1.** vomit, resulting from seasickness. **2.** leg-pulling; good humoured deceit. Also, **burley, birley.**

Berries, the *pl. n.* the Canterbury-Bankstown team in the N.S.W. Rugby Football League.

berth *n.* **1.** job; position. —*phr.* **2. give a wide berth to,** to avoid; keep away from.

best *phr.*

1. get or **have the best of,** to defeat.

2. give (someone or **something) best,** to admit defeat to (someone or something).

bester *n.* a fraudulent bookmaker.

bet *n.* **1.** a chance of success: *Your best bet is to catch the earlier train.* **2.** a thing, person, etc., or eventuality on which to gamble or stake one's hopes: *He's a bad bet.*

—*phr* **3. bet London to a brick,** See **brick.**

4. bet Sydney to the bush, to wager recklessly.

5. bet your boots, See **boot.**

6. bet your life, to stake all upon the truth of a statement, etc.

7. you bet, you may be sure; certainly.

betcha *interj.* Also, **betcher.** an assertion: *Betcha she'll win.* [*(you) bet your (life)*]

better half *n.* **1.** one's wife. **2.** one's spouse or partner.

Betty Martin *phr.* **(all) my eye and Betty Martin,** See **eye.**

between *phr.* **between you, me and the gatepost,** in confidence.

bewdy! *n., interj.* very good! terrific! (a humorous variant of **beauty.)**

b.f. *n.* bloody fool. [initials]

bi *adj.* **1.** bisexual. —*n.* **2.** a bisexual.

bib *phr.*

1. keep one's bib out, to refrain from interfering with or inquiring into (the affairs of another).

2. put or **stick one's bib in, a.** to interfere. **b.** to intervene.

bib and tucker *n.* clothes, especially one's best: *For the party she put on her best bib and tucker.*

bibful *n.* an unexpected and sometimes embarrassing disclosure: *Yesterday the Prime Minister spilt a bibful.*

bible-basher *n.* a person of excessive religious, especially evangelical, zeal often relying much on narrow biblical interpretations. Also, **bible-banger, bible-puncher.**

Bible Belt *n.* an area noted for its fundamentalist religious beliefs. [originally of southern U.S.A.]

bickie *n.* **1.** money. *—phr.* **2. big bickies,** a lot of money. Also, **bikkie.**

biddy *n.* **1.** old woman. **2.** chicken; young woman.

biff *n.* **1.** a blow; punch. *—v.* **2.** to punch.

big *adj.* **1.** strong; powerful. *—adv.* **2.** boastfully: *to talk big.* **3.** on a grand scale; liberally: *to think big.* *—phr.*

4. be in big on, to be an important confidant or partner in, frequently, a commercial venture.

5. be in big with, to be highly favoured by (someone).

6. big on, knowledgeable and enthusiastic about: *big on wine.*

big A *n.* **1.** arse. *—phr.* **2. give (someone) the big A,** to get rid of (someone).

big boys *pl.n.* the most powerful and influential members of a group.

Big Brother *n.* a dictator or figure of authority, especially one who tries to control people's lives and thoughts. [from a character in the novel '1984', by George Orwell (1903-1950)]

big C *n.* cancer.

big dollars *pl.n.* a lot of money: *You don't need big dollars to holiday in Tasmania.*

big fellow *phr.* **make a big fellow of (one's self),** to increase (one's) standing by some action.

big fish *n.* an important, powerful person. [from the fable of the big fish in the little pond]

biggie *n.* (*especially in children's speech*) a big person or thing.

big gun *n.* **1.** a powerful or influential person. **2.** *Surfing.* a long heavy surfboard.

big-headed *adj.* conceited; vain.

big jobs *pl. n.* See **job.**

big league *n.* the top level in any business or pursuit: *He's in the big league now.*

big lunch *n.* **1.** (*in children's speech*) the lunch period in the school day. **2.** the food eaten then.

big men *pl. n.* Australian Rules football players.

big mouth *n.* **1.** a garrulous person. *—phr.* **2. shut one's big mouth,** to cease disclosing information; stop talking.

big-mouth *v.* to boast; skite.

big noise *n.* a very important person.

big-note *v.* **1.** to boast of or promote (oneself): *He big-notes himself at every committee meeting.* *—phr.* **2. big-note man,** a wealthy man, especially one who is ostentatious with his money.

big shot *n.* a very important person.

big smoke *n.* See **smoke.**

big spit *n.* vomit: *to go for the big spit.*

big stick *n.* **1.** threat of punishment or disadvantage. *—phr.* **2. break eggs with a big stick,** See **eggs.**

big sticks *pl. n.* (in Australian Rules) the goal posts: *He's dobbed it through the big sticks.*

big-time *adj.* **1.** at the top level in any business or pursuit: *big-time boys.* —*n.* **2.** the top level, especially in business or society. —*phr.* **3. the big-time,** the social milieu of those who have achieved fame and success: *She has hit the big-time now.*

bigwig *n.* a very important person.

bike *n.* **1.** a woman who will have sexual intercourse with any man who asks her. —*phr.* **2. get off one's bike,** to get angry; lose control of oneself.

bikie *n.* a member of a gang of motorcycle riders. Also, **biker.**

bikkie *n.* See **bickie.**

bilge *n.* Also, **bilge water.** nonsense; rubbish.

bilge artist *n.* a person notorious for speaking or writing specious nonsense.

billabong *phr.* **on the billabong,** out of work.

Bill Harris *n. World War I Military* bilharzia, a type of parasitic blood fluke.

billion *n.* a large amount.

Bill Masseys *pl.n. N.Z.* boots, especially military boots. [named after William *(Bill)* F. *Massey,* Prime Minister of New Zealand, 1912-1925]

billposter *phr.* **as busy as a one-armed billposter in a high wind,** in a state of frantic activity; busy; harassed.

billy[1] *n.* **1.** a tin with a lid and a wire handle used for boiling water, making tea, etc. over an open fire. —*phr.* **2. boil the billy, a.** to make tea, not necessarily with a billy can. **b.** to stop for refreshments.

billy[2] *n.* a male goat. Also, **billygoat.**

billygoat *n.* one who is incompetent.

billygoat rider *n.* a Freemason. [from the goat used in certain ceremonies of the Freemasons]

billylid *n.* child; kid. [rhyming slang]

billyo *phr.*
1. go to billyo, get lost!
2. like billyo, a. with gusto: *We laughed like billyo.* **b.** with great speed: *He rode like billyo.*
3. off to billyo, a. off course; astray; in error. **b.** a long way ahead. Also, **billyoh.**

bimbo *n.* **1.** a homosexual. **2.** (formerly) a homeless young man in the depression of the 1930s, usually with implication of homosexuality. [from Italian *bambino* a baby boy]

bin *n.* **1.** a pocket in a garment. **2.** a mental asylum. [short for *loony bin*] **3.** a gaol.

bind *n.* **1.** a nuisance; bore. —*phr.* **2. in a bind,** in a dilemma.

binder *n.* a solid meal; a feed.

bine *n. W.A.* a newly-arrived Englishman. [from their supposed habit of smoking *Woodbines,* a cheap brand of cigarettes originating in England]

binge *n.* a spree; a period of excessive indulgence, as in eating or drinking. [Lincolnshire dialect *binge* to soak]

binghi *n.* an Aboriginal. [Aboriginal: brother]

bingie *n.* the stomach; belly. Also, **bingy, bingey, bingee, binjy.** [Aboriginal]

bingle *n.* **1.** a dent or fracture in a motor vehicle, surfboard, etc. —*v.* **2.** to damage (a motor vehicle, etc.).

bingo *interj.* **1.** an exclamation expressing triumph at a discovery or achievement. [from the call by the winning player of a *bingo* game] —*phr.* **2. like bingo, a.** with gusto. **b.** with great speed.

bint *n.* a girl. [Arabic]

bird[1] *n.* **1.** a person, especially one having some peculiarity: *a strange bird.* **2.** a sound of derision, especially hissing: *to give someone the bird.* **3.** *Horseracing.* a certainty to win, a cert. **4.** a girl; a girlfriend. **5.** a secret source of information: *A little bird told me.* —*v.* **6.** *Theatre.* to boo an actor. —*phr.*
7. bird in the hand, that which is sure though perhaps not entirely satisfactory. [from proverb, *a bird in the hand is worth two in the bush*]
8. birds of a feather, people of similar character or like tastes.
9. dead bird, a certainty.
10. get the (big) bird, a. *Theatre.* (of an actor) to be booed and hissed. **b.** to be dismissed from employment.
11. give (someone) the bird, to ridicule (someone).
12. like a bird, easily, swiftly.
13. make a bird of it, complete a task, project, etc.
14. (strictly) for the birds, of little importance; trivial; not worthy of consideration.

bird[2] *n.* **1.** a prison sentence. [rhyming slang, *birdlime* time] **2.** a prison. **3.** *World War II.* a military prisoner.

bird brain *n.* a frivolous or scatter-brained person.

birdcage *n.* **1.** an enclosure on a racecourse where horses are paraded before a race and to which they return after it. **2.** *N.Z.* a used-car dealer's display yard, enclosed by a wire fence.

birdcage boy *n.* *N.Z.* a used-car dealer.

birder *n.* a hunter of mutton-birds.

birdlime *n.* time, as a prison sentence or the time of day. Also, **bird.** [rhyming slang]

birdman *n.* an aviator.

birdwatcher *n.* (*humorous*) one who displays a keen interest in looking at beautiful young women.

birl *n., v.* See **burl.**

birthday suit *n.* the naked skin; the state of nakedness.

biscuit *phr.* **take the biscuit,** See **cake.**

biscuit bombers *pl. n.* *World War II Air Force.* planes which dropped supplies into remote areas of New Guinea.

biscuits and cheese *n.* *World War II.* knees. Also, **biscuits.** [rhyming slang]

bish *v.* to throw.

bishop *phr.* **bury the bishop,** See **bury.**

Bishop Barker *n.* beer served in a tall glass. [named after Frederick *Barker,* Bishop of Sydney 1845-81, who was unusually tall]

bit *n.* **1.** a girl. **2.** sexual intercourse: *Did you get a bit?* —*phr.*
3. a bit, a sum of money: *Reg had a bit on the winner of the last race; Ann was left a bit when her grandmother died.*
4. a bit of all right, something or someone exciting admiration.
5. a bit on the side, a. something beyond the usual arrangement. **b.** an extra-marital affair.

6. have a bit both ways, to attempt to cover oneself against any eventuality.

bitch *n.* **1.** a woman, especially a disagreeable or malicious one. **2.** a complaint. *—v.* **3.** to complain. **4.** to spoil; bungle.

bite *v.* **1.** to trouble; worry; disturb: *What's biting him?* **2.** to react angrily: *Don't tease her, she bites.* *—n.* **3.** a person from whom one anticipates borrowing money: *He'd be a good bite.* **4.** a reaction: *Did you get a bite from Robin?* **5.** *N.Z.* a nagging person. *—phr.*
6. bite off more than one can chew, to attempt more than one can handle.
7. bite (someone's) ear, to harangue.
8. bite the dust, See **dust.**
9. put the bite on, to cadge.
10. raise a bite, to tease until one gets a reaction.

bitser *n.* **1.** a mongrel. **2.** any contrivance the parts of which come from miscellaneous sources, as a billycart. **3.** an animal or person of mixed stock. Also, **bitzer.**

bitsy *adj.* small; tiny. Also, **little bitsy, itsy bitsy.**

bitten *adj.* tricked; cheated.

bitumen *phr.*
1. end of the bitumen, the outskirts of town, often with reference to a shanty settlement.
2. the bitumen, a. a tarred or sealed road. **b.** any bituminised area.

bitumen blonde *n.* **1.** a brunette. **2.** an Aboriginal woman.

bitzer *n.* See **bitser.**

bivvy *n. Military.* bivouac; any field exercise which extends over at least two days.

biz *n.* **1.** business: *big biz; show biz.* *—phr.* **2. what's the biz?** what is the latest news? [short form of *business*]

bizzo *n.* **1.** worthless or irrelevant ideas, talk, writing, etc.: *politics, and all that bizzo.* **2.** any object or device for which one does not know the correct name. [shortened form of *business* + *-o*]

blabbermouth *n.* one who talks too much or who talks indiscreetly.

black *phr.* **in the black, a.** financially solvent. **b.** of betting odds, any bet above or including even money.

black and tan *n.* a drink made by mixing beer and stout.

blackbird *n.* (formerly) a Kanaka kidnapped and transported to Australia as slave labour. [Kanakas were Pacific Islanders, especially from the New Hebrides (Vanuatu), kidnapped as labour for the cotton and sugar plantations of Fiji, Samoa and Queensland during the period 1850-1900]

blackbirder *n.* one engaged in kidnapping and transporting Kanakas to Australia.

black book *n.* **1.** a book of names of people liable to censure or punishment. **2.** a small book containing a private list of names and addresses of associates in business, partners in sex, crime, etc. *—phr.*
3. be in someone's black books, to be in disfavour.

black box *n.* any device, invention, etc., the workings of which are mysterious or kept secret.

blackboy *n. Obsolete.* an Aboriginal man or youth.

blackfellow *n.* (*especially in Aboriginal pidgin*) an Aboriginal. Also, **blackfella, blackfeller.**

blackfellow's button *n.* a small glasslike body (australite) believed to be of meteoric origin, found across the Nullarbor Plain. Also, **emu button.**

black Friday *n.* any Friday which falls on the 13th of the month, thought to be generally unlucky.

black gold *n.* **1.** coal. **2.** oil. **3.** rutile.

black hat *n.* (formerly) an immigrant; new chum, i.e. one still wearing city clothes.

black hole *n.* any small, overcrowded room. [from *Black Hole of Calcutta,* a small cell in Fort William, Calcutta, into which, in 1756, 146 Europeans were thrust for a night, only 23 of whom were alive in the morning]

blackjack *n.* **1.** a thick black bituminous waterproofing compound. **2.** caramel, burnt sugar, treacle, black toffee, liquorice, or any other dark sweet substance. **3.** a pot for boiling water or for cooking.

blackjack merchant *n.* a bad plumber; one who disguises poor work under much blackjack.

blackleg *n.* **1.** one who continues to work during a strike, takes a striker's place or refuses to join a union; scab. **2.** a swindler, especially in racing or gambling.

blackletter day *n.* an unlucky day.

black lightning *n.* an Aboriginal cooking fire, signal smoke fire, ceremonial fire, etc. regarded as the cause of a bushfire.

black list *n.* **1.** a list of persons under suspicion, disfavour, censure, etc. or a list of fraudulent or unreliable customers or firms. —*v.* **2.** to put on a black list, as being under suspicion, disfavour, censure, etc.

blackman's potatoes *pl. n.* early Nancy, a small Australian wildflower.

black maria *n.* a closed vehicle used for conveying prisoners to and from gaol. Also, **Black Maria.**

black peter *n.* a punishment cell in a prison, usually without light, where a prisoner is put in solitary confinement.

black prince *n.* a large slim-bodied cicada.

blacks *phr.* **give it back to the blacks,** an expression of annoyance or disgust at any inhospitable feature of Australia.

blackstrap *n.* treacle; golden syrup.

black stump *n.* **1.** an outer limit of mythical distance. —*phr.*

2. back of or **beyond the black stump,** in the far outback; in any remote region.

3. this side of the black stump, a loose measure of comprehensiveness or distance: *I make the best pumpkin soup this side of the black stump; He is the biggest bore this side of the black stump.*

black taxi *n.* a chauffeur-driven Commonwealth Government car, which provides transport for politicians and high public servants.

black velvet *n.* **1.** a drink made from a mixture of stout and champagne. **2.** a black girl or woman, considered as a sex object. See **yellow satin.**

bladder *phr.* **have a Japanese bladder,** See **Japanese.**

blade *n.* a knife (as a weapon): *to carry a blade.*

blah *n.* high-sounding empty talk; eloquent rubbish.

Blake, Joe *n.* See **Joe Blake.**

Blamey, Lady *n.* See **Lady Blamey.**

blank *adj.* **1.** a euphemism for any vulgar or taboo word. *—phr.*
2. draw (a) blank, to get no results; be unsuccessful; fail.
3. fire blanks, to be impotent.

blank cheque *n.* a free hand; carte blanche.

blanket *n.* a sexual partner.

blankety *adj.* Also, **blanketty, blanky.** a euphemism for any obscene language.

blarney *phr.* **have kissed the blarney,** to be an entertaining and charming talker. [from the *Blarney* stone, a stone in Blarney Castle near Cork, Ireland, said to confer skill in flattery to anyone who kisses it]

blast *interj.* **1.** an exclamation of anger or irritation. *—v.* **2.** to criticise someone abusively. *—n.* **3.** a wild, noisy party or gathering.

blasted *adj.* **1.** intensifier expressing anger, annoyance, disgust, etc. **2.** intoxicated.

blast-off *n.* a rapid departure.

blatt *n.* **1.** a printed sheet as a daily communique, a set of instructions, etc. *—phr.* **2. the daily blatts,** the daily newspapers. [German *Blatt* leaf]

blazer *n.* anything intensely bright or hot.

blazes *phr.*
1. go to blazes, an exclamation of dismissal, contempt, anger, etc.
2. the blazes, a word used to add emphasis to a statement, question, etc.: *What the blazes is going on here?*

bleat *v.* to complain; moan.

bleed *v.* **1.** to obtain, as in excessive amount, or extort money from. *—phr.* **2. bleed dry,** to extort money from.

bleeder *n.* a person; fellow.

bleeding *adj. British.* bloody.

bless *phr.* **not a penny to bless oneself with,** destitute.

blighter *n.* **1.** a person; fellow. **2.** a despicable person; cad.

Blighty *n. British Military.* **1.** England. **2.** *Chiefly in World War I.* a wound serious enough to get one sent back to England. Also, **blight.** [Hindu *bilāyatī* foreign, European]

blimey *interj.* an exclamation expressing surprise or amazement. Also, **blimy, bli'me.** [shortened form of *gorblimey*]

blind *adj.* **1.** drunk. **2.** made without knowledge in advance: *a blind date.* *—n.* **3.** a bout of excessive drinking; drinking spree: *to go on a blind.* *—phr.* **4. to talk (someone) blind,** to talk to excess.

blind alley *n.* a position or situation offering no hope of progress or improvement. [from *blind alley,* a road, street, etc. closed at one end]

blinder *n.* a dazzling display of skill, especially at sport.

blind Freddy *n.* **1.** an imaginary person representing the lowest level in perception or competence. *—phr.* **2. even blind Freddy**

could see that, an expression used to indicate that something needs no explanation or defence. Also, **blind Freddie.**

blind mullet *n.* faeces.

blind spot *n.* a matter about which one is ignorant, unintelligent or prejudiced, despite knowledge of related things.

blind-stab *v. Prison.* to break and enter without ascertaining beforehand whether valuable goods are present.

blink *phr.* **on the blink,** not working properly.

Blinky Bill *n.* a fool. [rhyming slang *Blinky Bill* dill]

Bliss, Johnny *n.* See **Johnny Bliss.**

blister[1] *n.* **1.** a summons in law. **2.** a summary demand for payment of a debt. **3.** a scathing communication: *The boss sent me a blister.* **4.** *British.* an unpleasant person. —*v.* **5.** (of a policeman) to record a person's name for an alleged offence.

blister[2] *n.* sister. Also, **skin and blister.** [rhyming slang]

blob *n.* **1.** (in cricket) nought; no runs: *out for a blob.* **2.** a fool.

block *n.* **1.** the head. —*phr.*

2. lose or **do one's block,** to become very angry; lose one's temper.

3. off one's block, insane.

blockbuster *n.* anything large and spectacular, as a lavish theatrical production, impressive political campaign, etc. [from *blockbuster* an aerial bomb used in World War II as a large-scale demolition bomb]

blockbusting *adj.* spectacular.

blocked *adj.* dazed, stupefied as a result of drugs.

blocker *n.* **1.** the owner of an irrigation block. **2.** *S.A.* the owner of a vineyard.

blockie *n.* a small farmer.

Block, the *phr.* **do the Block,** to promenade in the fashionable area of town. [originally the section of early Melbourne bounded by Collins, Elizabeth and Swanston Streets]

bloke *n.* **1.** man; fellow; guy. —*phr.* **2. the Bloke,** the boss, the person in charge. [from Shelta (a tinkers' jargon of Ireland and parts of Britain)]

blood *phr.*

1. get blood from or **out of a stone,** to attempt something that is impossible (often used with reference to someone who is mean or hard-hearted).

3. (one's) blood is worth bottling, (one) is exceptionally meritorious or praiseworthy.

blood alley *n.* a red marble.

blood-and-guts *adj.* violent.

blood and thunder *n.* **1.** violence; sensationalism; bombast; extravagant anger. —*adj.* **2.** (of films, plays, etc.) characterised by violence and noise.

blood donor *n.* a wharf labourer who supplements a gang as required.

bloodhound *n.* a detective; sleuth. [from *bloodhound,* one of a breed of dogs with a very acute sense of smell, used for tracking game, human fugitives, etc.]

bloodhouse *n.* a particularly rough hotel, uncomfortable and disorderly.

blood money *n.* small remuneration earned by great effort.

blood oath *interj.* an exclamation usually expressing agreement, affirmation, etc. Also, **bloody oath!, my bloody oath!**

Bloods *pl. n.* **1.** (formerly) the South Melbourne team in the Victorian Football League (now the Sydney Swans). **2.** the Mordialloc team in the Victorian Football Association. **3.** the West Adelaide team in the South Australian National Football League.

Bloodstained Angels *pl. n.* Also, **Bloodstained Niggers.** the Essendon team in the Victorian Football League.

bloodsucker *n.* **1.** an extortioner. **2.** a person who sponges on others. **3.** a leech.

bloody *adj.* **1.** a word used to add emphasis in signifying approval, as in *bloody beauty,* or disapproval, as in *bloody bastard.* **2.** (of people) difficult; obstinate; cruel. **3.** (of events) cruel; unjust; unbearable. —*adv.* **4.** very; extremely: *a bloody wonderful game; a bloody awful thing to happen.*

bloody-minded *adj.* **1.** obstructive; unhelpful; difficult. **2.** deliberately cruel or unpleasant.

blooey *phr.* **go blooey,** to go wrong, awry, amiss.

bloomer *n.* an embarrassingly foolish mistake; laughable blunder.

blooming *adj.* (*euphemism*) bloody.

blooper *n.* a slip of the tongue, especially of a broadcaster, resulting in a humorous or indecorous misreading.

blot *n.* **1.** the anus. —*phr.*

2. blot one's copybook, to blemish one's character or reputation by some action.

3. on the blot, sitting down.

blotting paper *n.* food taken while drinking alcoholic beverages, to mitigate the effects of the alcohol. [from *blotting paper,* soft, absorbent paper]

blotto *adj.* under the influence of drink.

blow¹ *n.* **1.** the first and longest stroke made in shearing a sheep. —*phr.* **2. strike a blow,** to begin or resume work. [from *blow* a sudden stroke with hand, fist or weapon]

blow² *v.* **1.** to play on any musical instrument, or sing, usually with other musicians at a jam. **2.** to waste; squander: *to blow one's money.* **3.** to boast; brag. **4.** to depart. **5.** to fail in something: *to blow an exam.* **6.** to ejaculate; experience orgasm. **7.** Also, **blow out.** *Horseracing.* (of odds on a horse offered by bookmakers) to lengthen. **8.** (pp. **blowed**), (*euphemism*) to damn: *Blow that!* —*n.* **9.** a musical performance, usually with other musicians, and often improvised; gig; jam. **10.** a walk in the fresh air; airing. **11.** boasting or bragging. **12.** a short rest, especially during hard physical labour: *We'll have a blow now.* —*phr.*

13. blow in, to make an unexpected visit; drop in; call.

14. blow (one) out, to put into a state of surprise or delight.

15. blow one's cool, to lose one's temper.

16. **blow one's mind,** to achieve a state of euphoria, as with drugs.
17. **blow one's own trumpet,** to boast.
18. **blow one's top** or **stack/blow one's boiler,** to lose one's temper.
19. **blow over, a.** to cease; subside. **b.** to be forgotten.
20. **blow up,** to scold or abuse.
21. **blow the whistle on,** to inform upon; report to authority.
22. **blow through, a.** to depart. **b.** to evade a responsibility.
23. **blow up, a.** to lose one's temper. **b.** *N.Z. Sport.* to halt play for an infringement by blowing on a whistle.

blowed *v. euphemism.* damned: *Well I'll be blowed.*

blower *n.* **1.** a supercharger in an engine. **2.** a telephone.

blow-hard *n.* a garrulous person.

blowie *n.* a blowfly.

blow-in *n.* an unexpected visitor or newcomer to an area.

blow job *n.* fellatio.

blown-out *adj.* Also, **blown-away.** amazed, usually from delight.

blow-out *n.* a big meal or lavish entertainment; spree.

blow-up *n.* a violent outburst of temper or scolding.

blub *v.* to weep; cry noisily. [short for *blubber*]

bludge *v.* **1.** to evade responsibilities. **2.** to pimp; live on the earnings of a prostitute. **3.** to cadge. —*n.* **4.** a job which entails next to no work. —*phr.*
5. **bludge on,** to impose on others.
6. **on the bludge,** imposing on others.

bludger *n.* **1.** someone living on the earnings of a prostitute. **2.** someone who imposes on others, evades responsibilities, does not do a fair share of the work etc.

blue *adj.* **1.** dismal: *I'm feeling blue.* **2.** obscene or pertaining to obscenity: *a blue joke.* —*n.* **3.** a fight; dispute. **4.** a summons, especially for a traffic offence. [from colour of paper] **5.** a mistake. **6.** a nickname for a red-headed person. —*v.* **7.** to spend wastefully; squander. **8.** to fight. —*phr.*
9. **bet on the blue/Mary Lou/nod,** *Horseracing.* to bet on credit.
10. **cop the blue,** to take the blame.
11. **once in a blue moon,** rarely and exceptionally.
12. **scream blue murder, a.** to scream loudly. **b.** to protest vociferously.
13. **true blue,** loyal; faithful; genuine.

blue-arsed fly *n.* **1.** the fly which causes sheep strike, the blue blowfly. —*phr.* **2. buzz around like a blue-arsed fly,** to be very busy or active, especially in an erratic and frenzied fashion.

Bluebags *pl. n.* See **Blues, the.**

blue balloon *n.* the bag used to measure the alcohol content of exhaled breath; breathalyser; balloon.

bluebird *n.* a police van used for transporting prisoners; Black Maria; paddy wagon.

Bluebirds *pl. n.* See **Blues.**

bluebottle *n.* a stinging jellyfish.

blue devils *pl. n.* **1.** low spirits. **2.** delirium tremens.

blue duck *n.* **1.** *Chiefly Military.* a baseless rumour; a failure. **2.** something which does not perform as expected; a dud.

blue-eyed *adj.* **1.** darling; favourite. **2.** ingenuous; innocent.

blue funk *n.* a state of extreme fear.

blue light *n.* Also, **blue lighter.** a highway patrol-car.

blue movie *n.* a pornographic film.

blue-rinse *adj.* (*offensive*) of or pertaining to well off, middle-aged women of conservative and trivial outlook: *the blue-rinse set.*

blue ruin *n. British.* gin.

blues *pl. n.* despondency; melancholy. [short for *blue devils*]

blue-singlet *adj.* Also, **blue-singletted.** working-class.

Blues, the *pl. n.* **1.** the New South Wales State representative team. **2.** Also, **Bluebirds.** the Carlton team in the Victorian Football League. **3.** the Manly team in the N.S.W. Rugby Football Union. **4.** Also, **Bluebags.** the Newtown team in the N.S.W. Rugby Football League.

blue streak *n.* something moving very fast.

blue-tongue *n.* a handyman on a station, in a hotel, etc.; rouseabout.

bluey *n.* **1.** a rolled blanket (originally blue) containing the possessions carried by a traveller through the bush; swag; shiralee. **2.** a beachworm; slimy; bungum worm (so called because of the blueish-black spots on the front part of the worm which emerges first from the sand). **3.** a type of overcoat. **4.** a lorikeet. **5.** a red dog. **6.** a nickname for a red-headed person. **7.** a summons in law. *—phr.*

8. hump the bluey or **humping bluey, a.** to carry a swag in the outback. **b.** to be unemployed and on the road, as an itinerant worker.

9. put a knot in one's bluey, a. to prepare for travel. **b.** to prepare for action.

bluff *phr.* **call (someone's) bluff,** to challenge or expose (someone's) bluff.

blunt *n. Obsolete.* money.

blurter *n.* the anus.

BO *n.* body odour, especially due to excessive perspiration. [*B(ody) O(dour)*]

board *n.* **1.** Also, **shearing board. a.** the floor of a woolshed. **b.** the shearers employed in a woolshed. *—adj.* **2.** (often plural) relating to the playing of card games for money. *—phr.*

3. across the board, in a comprehensive fashion.

4. board out, to discharge or retire (a serviceman) on medical grounds.

5. go by the board, to be discarded, neglected or destroyed.

6. sweep the board, (of gambling, especially cards) to win all.

boardboy *n. Shearing.* a shedhand.

boardie *n.* an enthusiast for the sport of surfboard riding.

boat *phr.*

1. burn (one's) boats, See **burn.**

2. miss the boat, See **miss.**

3. push out the boat, to pay for a round of drinks; begin a drinking bout.

4. rock the boat, to make difficulties.

boatie *n.* a person who owns and runs a small craft.

boat people *pl. n.* refugees from South-East Asia, setting out for Malaysia, Hong Kong or Australia by boat.

boatrace *n.* **1.** a competition between teams of beer drinkers to see which team can drink its beer the fastest. **2.** *Horseracing.* a race in which the result is predetermined or rigged. **3.** any situation in which the outcome has been secretly secured.

bob[1] *n.* **1.** (formerly) a shilling. *—phr.* **2. bob in,** (formerly) a subscription of one shilling to a common fund.

bob[2] *n. Convict Obsolete.* a flogging of fifty lashes. [from *bob* shilling. See *tester, bull, canary*]

bob[3] *n.* altered form of 'God' used in oaths, etc.: *So help me bob!*

Bob *phr.* **staggering Bob,** a newly born calf. [British dialect]

bobberie *n.* commotion; fuss.

bobby[1] *n.* **1.** a policeman. **2.** *Prison.* a prison officer; screw. [special use of *bobby* for Sir *Robert* Peel, who as British Home Secretary, organised the Metropolitan Police Force (1828)]

bobby[2] *n. Chiefly W.A.* **1.** a small beer glass. **2.** the contents of such a glass. [British dialect *bobby* small]

bobby-dazzler *n.* an excellent thing or person: *you little bobby-dazzler. Also,* **ruby-dazzler.**

bobbysocks *pl.n. Chiefly U.S.* ankle-socks, especially as worn by young girls. Also, **bobbysox.**

bobbysoxer *n. Chiefly U.S.* an adolescent or teenage girl.

Bob Hope *n.* **1.** soap. **2.** dope. [rhyming slang]

Bob Munro *phr.* **in you go says Bob Munro,** *N.Z.* an expression of encouragement addressed to someone entering upon some enterprise; a toast.

bobsy-die *n.* **1.** a fuss; panic. **2.** boisterous merriment: *kick up bobsy-die.* [British dialect *bobs-a-dying*]

bob's your uncle *interj.* **1.** (also as a response in conversation expressing compliance) it's all right; there you are. **2.** (as a response to a statement which proves nothing) so what?

boche *n.* (*offensive*) a German. Also, **Boche.** [alteration of French *caboche* head, pate, noodle]

boco *n.* See **boko**[1]**.**

bod *n.* a person: *an odd bod.* [short for *body*]

bodger *n.* **1.** *World War II Army.* a worthless person. *—adj.* **2.** inferior; false; second-rate (as a name, receipt, etc.)

bodgie[1] *adj.* **1.** inferior; worthless. **2.** (of names, etc.) false; assumed. *—n.* **3.** a worthless person. **4.** a person who has assumed an alias or who is in some way acting under false pretences. *—v.* **5.** Also, **bodgie up.** to repair superficially; to remove temporarily any obvious defects. Also, **bodger.** [Obsolete British *bodge* to patch or mend clumsily]

bodgie[2] *n.* especially in the 1950s, one of a group of young men usually dressed in an extreme American-influenced fashion and given to wild or exuberant behaviour. See **widgie.**

body *n.* **1.** a person. —*phr.* **2. over one's dead body,** under no circumstances; never.

body hire *n.* recruitment of casual labour for building jobs from a central point.

body shop *n.* an agency for computer programmers.

body-snatching *n.* taking members of one organisation into another.

boffin *n.* a person who is enthusiastic for and knowledgeable in any pursuit, activity, study, etc., especially a research scientist.

bog *n.* **1.** defecation. **2.** a lavatory or latrine. —*v.* **3.** to defecate. —*phr.*
4. bog down, to sink in or as in a bog: *I'm bogged down with work today.*
5. bog in, to eat voraciously.

Bogan gate *n.* a gate roughly constructed from droppers and fencing wire. [?from *Bogan Gate* in central N.S.W.]

Bogan shower *n.* a dust storm. [from *Bogan* River and district in western N.S.W.]

bogart *v.* to smoke more than one's fair share of a marijuana cigarette: *Don't bogart that joint!*

bogeyhole *n.* a swimming hole.

boggabri *n.* any of several rather weedy herbs of inland Australia, as fat hen, a kind of wild spinach.

bogghi *n. Shearing.* **1.** (formerly) the handpiece of clippers used in shearing sheep. **2.** the shears themselves. Also, **boggi, bog-eye.**

boggle *n.* bungle; botch.

boghouse *n.* a toilet, especially an outside toilet.

bogie[1] *n.* **1.** a swim or bath. **2.** a swimming hole. Also, **bogey, bogeyhole.** [Aboriginal]

bogie[2] *n.* snot. Also, **bogey.**

bog-Irish *n.* an uneducated, ignorant Irishman.

bogtrotter *n.* **1.** (*offensive*) a rural Irishman. **2.** an itinerant labourer.

bogy *n. Military.* an unidentified aircraft. Also, **bogey, bogie.** [from obsolete *bog* variant of *bug* a bogey, hobgoblin]

bogyman *n.* **1.** a hobgoblin; evil spirit. **2.** anything that frightens or annoys one. Also, **bogieman.**

bohunk *n. U. S.* (*offensive*) an unskilled or semiskilled foreign-born labourer, specifically a Bohemian, Magyar, Slovak, or Croatian.

boil *v.* **1.** to feel very hot. —*phr.*
2. boil the billy, make tea; have a break.
3. boil up, to make tea.

boiled shirt *n.* a pompous person.

boiler *n.* **1.** an old woman. —*phr.*
2. blow one's boiler, See **blow.**
3. burst one's boiler, to fail; expend all one's energies and resources in an unsuccessful attempt.

boilermaker *n.* whisky followed by a beer chaser.

boilover *n.* **1.** a sudden conflict between persons. **2.** a win by a long-priced entrant in horse or greyhound race.

boil-up *n.* **1.** Also, **brew-up.** a break for tea. **2.** a sudden excitement or conflict.

boko[1] *n.* the nose. Also, **boco.**

boko[2] *n.* an animal which has lost one eye. Also, **boco.**

bollix *v.* to bungle; make a mess of. [variant of *bollocks* nonsense]

bollock *v.* to bungle.

bollocks *pl.n.* **1.** rubbish; nonsense. **2.** the testicles. Also, **bollicks.**

bollocky *adj.* **1.** naked: *stark bollocky.* —*phr.* **2. in the bollocky,** naked. Also, **bollicky, bols.**

boloney *n.* See **baloney.**

bolshie *n.* **1.** (*sometimes cap.*) Bolshevik. —*adj.* **2.** bolshevistic. **3.** obstinate; difficult; tiresome. Also, **bolshy.** [short for *Bolshevik*]

bolt *phr.*

1. bolt (it) in, *Horseracing.* to win easily.

2. bolt out of or **from the blue,** a sudden and entirely unexpected occurrence.

3. shoot one's bolt, See **shoot.**

4. have shot one's bolt, to have reached the limit of one's endurance or effort.

bolter *n.* **1.** *Horseracing.* a horse which wins a race unexpectedly, or by an unexpectedly large margin. **2.** (formerly) an escaped convict.

bomb *n.* **1.** an old car. **2.** a failure, as in an examination. **3.** a drug. —*phr.* **4. go like a bomb, a.** to go successfully. **b.** (of a motor vehicle, etc.) to go rapidly.

Bombay bloomers *pl.n.* loose-fitting shorts for summer or physical education wear. [from the baggy shorts worn as part of the uniform of servicemen during World War II]

bombed *adj.* heavily under the influence of drugs. Also, **bombed out.**

Bombers, the *pl. n.* the Essendon team in the Victorian Football League. Also, **the Dons.** [from its proximity to the Essendon aerodrome]

bomb-happy *adj. Military.* shell-shocked.

bombo *n.* cheap wine.

bombshell *n.* **1.** a sudden or devastating action or effect: *His resignation was a bombshell.* **2.** a woman who is physically well-endowed. —*phr.* **3. drop a bombshell,** to make a startling or unexpected announcement.

Bondi *phr.*

1. give (someone) Bondi, to give (someone) a thrashing.

2. shoot through like a Bondi tram, a. to depart in haste. **b.** World War II army expression meaning to go A.W.L. (absent without leave). [from *Bondi,* a beach and suburb of Sydney]

bone *v.* **1.** to steal. —*n.* **2.** See **hambone.** —*phr.*

3. have a bone to pick with (someone), to confont (someone) with a (usually minor) cause for dissatisfaction with them.

4. not to make old bones, (of a person or animal) to be unlikely to reach a great age.

5. point the bone at, a. to bring or wish bad luck upon. **b.** to indicate (a guilty person).

bonecrusher *n. Rugby Football.* a heavy tackle.

boned out *adj.* exhausted.

bonehead *n.* a stupid, obstinate person; a blockhead.

bone-idle *adj.* extremely lazy.

boner *n.* a mistake: *He made a boner that time.*

bones *pl.n.* the trombone section of a band.

boneshaker *n.* any ancient and rickety bicycle or other vehicle.

boneyard *n.* a cemetery.

bong[1] *v.* **1.** to hit, especially on the head. —*n.* **2.** a blow.

bong[2] *n.* **1.** the act of smoking hashish or marijuana, using a type of water-cooled pipe or hookah. —*phr.* **2. bong on,** to take part in a long smoking session.

bonkers *adj.* **1.** crazy. **2.** *Navy.* slightly drunk.

bonk wagon *n.* *Cycling.* a vehicle which accompanies cyclists taking part in road races and carries spare parts, spare clothing and food supplies. Also, **sag wagon.**

bonus *n.* a bribe. [from *bonus* something given or paid over and above what is due]

bonzer *adj.* excellent; attractive; pleasing. Also, **bonza, boshter, bosker.**

boo[1] *n.* *Chiefly U.S.* marijuana.

boo[2] *interj.* **1.** an exclamation used to express contempt, disapprobation, etc., or to frighten. —*phr.* **2. wouldn't say boo to a goose,** (of a person) to be timid or shy.

booay *n.* **1.** a remote country district. —*phr.* **2. up the booay, a.** in the backblocks. **b.** in difficulties; in a predicament. **c.** completely wrong. Also, **boo-eye, boo-ai, boohai.** [Maori *puhoi*]

boob[1] *n.* **1.** a fool; a dunce. **2.** a foolish mistake. **3.** a woman's breast.

boob[2] *n.* prison.

boob happy *adj.* *Prison.* suffering from a form of neurosis brought about by the strain of gaol routine.

boobhead *n.* *Prison.* a recidivist.

boo-boo *n.* an error, usually of judgment: *He made a classic boo-boo.*

boob tube *n.* **1.** a television set. **2.** a strapless elasticised upper garment worn by women.

booby *n.* the worst student, player, etc., of a group. [probably from Spanish *bobo* fool]

booby hatch *n.* *U.S.* **1.** a lunatic asylum. **2.** gaol.

boodle *n.* *Chiefly U.S.* **1.** (*often offensive*) the lot, pack, or crowd: *the whole boodle.* **2.** a bribe or other illicit gain in politics. **3.** anything dishonestly and ruthlessly appropriated; loot. —*v.* **4.** to obtain money dishonestly, as by corrupt bargains. [Dutch: stock, lot]

boofhead *n.* a fool.

boohai *n.* See **booay.**

book *phr.*

1. by the book, a. formally. **b.** authoritatively; correctly.

2. clean up the books, *Prison.* to confess to numerous unsolved crimes.

3. in (someone's) good books, in favour.

4. in (someone's) bad books, out of favour.

5. it doesn't suit one's book, it doesn't fit in with one's arrangements. [originally with reference to a betting book]

6. make a book, *Horseracing.* to lay and receive bets at such odds that whichever horse wins, a profit is made.

7. take a leaf out of (someone's) book, to emulate (someone).

8. throw the book at, a. to bring all possible charges against (an offender). **b.** sentence (an offender) to the maximum penalties. **c.** punish severely.

bookie *n.* a bookmaker.

bookie's runner *n. Horseracing.* a bookmaker's assistant engaged in collecting prices and laying off bets with other bookmakers, etc.

Booligal *phr.* **Hay, Hell and Booligal,** See **Hay.**

boom *phr.* **lower the boom on, a.** to prohibit; refuse: *He lowered the boom on further discussion.* **b.** to criticise severely.

boom alley *n. Aeronautics.* a flight path for supersonic aircraft.

boomer *n.* **1.** something large, as a surfing wave. **2.** something successful or popular, as a party or song; bottler. **3.** a mature male kangaroo. [Warwickshire dialect]

boomerang *n.* **1.** a scheme, plan, argument, etc., which recoils upon the user. **2.** a small fish that is thrown back into the water. **3.** that which is expected to be returned by a borrower. **4.** a dishonoured cheque. *—v.* **5.** to return or recoil upon the originator: *The argument boomeranged.* *—adj.* **6.** returning; rebounding: *a boomerang decision.* *—phr.*

7. able to sell boomerangs to the blacks, to have a persuasive personality, manner, etc.

8. to swear on a bag of boomerangs, to affirm. [from *boomerang* a bent or curved piece of hardwood used as a missile by Australian Aboriginals, one form of which can be thrown so as to return to the thrower]

boomerang bender *n.* a teller of tall stories.

boom galloper *n. Horseracing.* a horse whose previous performance has been outstanding and which thus is tipped to win his next races easily.

boondocks *n. Chiefly U.S.* **1.** an uninhabited and densely overgrown area, as a swamp, forest, etc. **2.** a remote suburb or rural area.

boondoggle *v.* **1.** to fritter away one's time on work that is unnecessary. *—n.* **2.** pointless and time-wasting activities. [from the plaited leather cord made by Boy Scouts to be worn around the neck, considered to be a time-wasting piece of handicraft; coined by R.H. Link, an American Scoutmaster]

boondy *n. W.A.* a stone or pebble. [Aboriginal]

boong *n.* (*offensive*) **1.** an Aboriginal. **2.** a black man.

boot *n.* **1.** an electric shock. *—v.* **2.** to kick; drive by kicking. **3.** to dismiss; expel; discharge: *booted out of the club.* *—phr.*

4. be too big for one's boots, to hold too high an opinion of oneself; be conceited.

5. bet your boots, to be certain.

6. boot home, a. *Horseracing.* to ride to win, kicking the horse to greater speed. **b.** to emphasise strongly. **c.** to push into position forcibly.

7. boot off, a. to leave; to go. **b.** *Prison.* to escape.

8. boots and all, completely; with all one's strength or resources: *Go in boots and all.*

9. get the boot, to be discharged.

10. order of the boot, a dismissal; an order to leave.
11. put in the boot, See **put.**
12. the boot's on the other foot, the true position is the reverse.

bootlace *phr.* **not be somone's bootlace,** to be of an inferior standard, to lack ability, etc.: *He's not a footballer's bootlace.*

bootlaces *pl. n. Shearing.* strips of skin cut off from sheep by rough shearers.

bootleg *adj. Chiefly U.S.* **1.** made, sold, or transported unlawfully. **2.** unlawful; clandestine. [from the practice of concealing illegal spirits in the bootleg]

bootlicker *n.* one who curries favour; a flatterer.

bootstraps *phr.* **pull oneself up by the bootstraps,** to advance to success solely by one's own efforts.

booze *n.* **1.** alcoholic drink. **2.** a drinking bout; spree. *—v.* **3.** to drink immoderately. *—phr.* **4. on the booze,** drinking immoderately.

booze artist *n.* a person noted for being a heavy drinker.

booze bus *n. N.S.W.* a mobile police unit used for random breath tests.

booze hound *n.* a heavy drinker.

boozer *n.* **1.** one who drinks immoderately. **2.** a hotel.

boozeroo *n. N.Z.* **1.** a drinking spree. **2.** a public house; hotel.

booze-up *n.* a drinking party, bout or spree.

boozington *n.* a drunkard.

boozy *adj.* **1.** drunken. **2.** addicted to alcohol.

bop[1] *v.* **1.** to hit or punch: *Jack bopped him on the head.* **2.** to fight.

bop[2] *v.* to dance to pop or rock music. [from *bebop,* a type of jazz music]

bo-peep *n.* a peep, view: *Have a bo-peep at that!*

borak *phr.* **to poke borak at someone,** to ridicule. Also, **borac, borack, borax.** [Aboriginal]

bore *phr.*
1. bore up, to upbraid; rebuke vehemently.
2. bore it up you, an offensive retort expressing contempt, dismissal.

Boroughs, the *pl. n.* the Port Melbourne team in the Victorian Football Association.

Borroloola sandwich *n.* See **Barcoo sandwich.**

bosh *n.* complete nonsense; absurd or foolish talk or opinions. [Turkish: empty, vain]

bosie *n. Cricket.* a delivery bowled by a wristspinner which looks as if it will break one way but in fact goes the other way; googly; wrong'un. [named after the inventor of the googly, B.J.T. *Bosanquet,* 1877-1936, English cricketer]

bosker *adj.* excellent; delightful. Also, **boshter.**

boss *n.* **1.** one who employs or superintends workmen; a foreman or manager. **2.** anyone who asserts mastery, especially one who controls a political or other body. **3.** the headmaster or headmistress of a school. **4.** an informal mode of address, not necessarily implying difference in status. *—v.* **5.** to be master of or over; manage; control. —*adj.* **6.** chief; master: *boss cook.* [Dutch *baas* master]

boss cocky *n.* **1.** a farmer who employs labour. **2.** a boss. [*boss* master + *cocky* a farmer]

boss-eyed *adj.* lacking in perception; based on false perception: *a boss-eyed attempt.* [from *boss-eyed* having a squint]

boss of the board *n.* **1.** *Shearing.* the owner, manager or contractor who hires shearers. **2.** a boss. Also, **boss over the board.**

bot *n.* **1.** a person who cadges persistently. **2.** an insect larva infecting the skin, sinuses, nose, eye, stomach, or other parts of animals or man. **3.** *N.Z.* a minor ailment, as a bad cold; the wog. **4.** bottom; the buttocks. —*v.* **5.** to cadge. —*phr.*

6. how're the bots biting, *N.Z.* a humorous greeting.

7. on the bot, cadging.

8. the bot, tuberculosis, or other chest complaint. Also, **bott.**

Botany Bay barfly *n* a Sydney drunk. [from *Botany Bay* formerly the name for New South Wales]

Botany Bay dozen *n. Convict Obsolete.* a punishment consisting of twenty-five lashes. [from *Botany Bay,* N.S.W., formerly a place of detention and punishment]

bottle *v.* **1.** to knock over people as though they were bottles: *Bottle 'im!* —*phr.*

2. be on the bottle, to be on a drinking bout; be intoxicated.

3. hit the bottle, to drink heavily; become an alcoholic.

4. the bottle, a. intoxicating drink. **b.** bottled milk for babies (opposed to *the breast*): *raised on the bottle.*

bottle-holder *n.* **1.** a boxer's or wrestler's second. **2.** any supporter or assistant.

bottle-oh *n.* one who collects empty bottles for sale or re-use. Also, **bottle-o.** [from the collector's cry *bottle-oh*]

bottler *n.* **1.** something exciting admiration or approval: *You little bottler.* **2.** Also, **bottley.** a marble made of glass.

bottle-washer *phr.* **chief cook and bottle-washer,** a person who, as well as being responsible for some enterprise, also does much of the work, especially manual work for it.

bottley *n.* a clear-glass marble, originally the stopper in an obsolete type of aerated-water bottle.

bottom *phr.*

1. bottoms up! an exclamation used as an encouragement to finish a drink.

2. the bottom line, a. the final cost. **b.** the end result of any given action.

bottom drawer *n.* a store of clothes, linen, etc., kept by young women in expectation of being married; glory-box; hope chest.

bottomless cup *n.* (in cafes) a coffee cup that may be refilled without extra charge to the customer.

bottom-of-the-harbour *adj.* **1.** relating to a tax-evasion scheme whereby a company with a tax liability is sold and then resold, so that the records, etc., for that company seemingly disappear, and the tax is not paid. **2.** concealed for illegal purposes; disposed of.

bounce *v.* **1.** (of cheques) to be dishonoured; to be returned unpaid. **2.** to eject or discharge summarily. **3.** to arrest. **4.** to bully; be overbearing. *—n.* **5.** *U.S.* expulsion; discharge; dismissal. *—phr.* **6. the bounce,** *Australian Rules Football.* the start of the game, when the umpire bounces the ball to begin play.

bouncer *n.* one employed in a place of public resort to eject disorderly persons.

bounder *n.* an obtrusive, ill-bred person; a vulgar upstart.

bouquet *n.* an offensive smell.

Bourke *phr.* **back of Bourke,** any remote, unsettled outback area. [from *Bourke* a town in north-west N.S.W.]

Bourke shower *n.* a dust storm. [See **Bourke**]

Bourke Street *phr.* **not to know whether it's Tuesday or Bourke Street,** to be stupid, ignorant or confused. [from *Bourke Street* in Melbourne]

bow and arrow *n.* sparrow. [rhyming slang]

bower bird *n.* a person who collects trivia and useless objects. [from the *bower bird* a bird species, the males of which build bowerlike structures and decorate them with objects such as shell, bone, pieces of glass, etc.]

bowl over *v.* to dumbfound.

box *n.* **1.** the vagina. **2.** a coffin. *—phr.*

3. be a box of birds, *N.Z.* be lively, energetic, in good health; be in good spirits.

4. box on, a. to argue; fight. **b.** to show no sign of giving in.

6. box up, to bring about a state of confusion as a result of mismanagement.

7. make a box of, to muddle.

8. nothing out of the box, not remarkable; mediocre.

9. one out of the box, an outstanding person or thing.

10. the box, a television set.

11. the whole box and dice, the whole; the lot.

boxer *n.* one who runs a two-up school and receives a set proportion, usually twenty percent, of the spinner's earnings; ringie.

box-on *n.* a dispute; fight.

box seat *n.* **1.** any position of vantage. *—phr.* **2. be in the box seat,** to have reached a peak of success; be in the most favourable position.

box-up *n.* a confusion or muddle. [originally of sheep mixed together]

boy *phr.*

1. boys in blue, the police force.

2. the boys, a group of male friends.

3. the old boy, a. one's father. **b.** one's boss. **c.** the penis.

boyo *n.* a disorderly or rowdy young man, especially from the country. [originally Anglo-Irish]

boy scout *n.* an earnest, serious-minded person. [from the *Boy Scouts* organisation of boys]

brace *phr.*

1. in a brace of shakes, immediately.

2. brace up, to raise one's strength or vigour.

bracelets *pl.n.* handcuffs. [from *bracelet* an ornamental band or circlet for the wrist or arm]

bracer *n.* a stimulating drink; tonic.

braces *n.* horse or dog races. [rhyming slang]

bracket *n.* the nose.

braid *phr.* **pull the braid,** *World War II Military.* to invoke the rights due to superior rank; pull rank.

braille *phr.* **a bit of braille,** a racing tip.

brain *n.* **1.** a highly intelligent or well-informed person. *—v.* **2.** to hit someone hard, especially about the head; cuff: *If you do that again I'll brain you! —phr.*

3. have (something) on the brain, to have as an obsession; be preoccupied with (something).

4. go off one's brain, to become frenzied with worry, anger, etc.

5. pick (someone's) brains, to use (another person's) work or ideas to one's own advantage.

brain drain *n.* the steady flow of young scientists, artists, etc., emigrating to work in countries other than the country of their birth.

brainpan *n.* the skull or cranium.

brainstorm *n.* a sudden inspiration, idea, etc.

brainwave *n.* a sudden idea or inspiration.

brasco *n.* a toilet.

brass *n.* **1.** excessive assurance; impudence; effrontery. **2.** money. **3.** Also, **top brass. a.** high ranking military officers. **b.** the people in the most senior positions in an organisation. *—v.* **4.** to cheat or defraud. *—phr.* **5. part brass rags with,** *Military.* to have an argument with.

brassed *adj.* **1.** cheated; defrauded. *—phr.* **2. brassed off,** bad-tempered; disenchanted; disillusioned.

brass hat *n.* a high-ranking army, navy, or airforce officer.

brass monkey *adj.* cold: *brass monkey weather.* [from the saying 'cold enough to freeze the balls of a *brass monkey*']

brass tacks *pl.n.* **1.** basic facts; realities. *—phr.* **2. get down to brass tacks** or **tintacks,** discuss something seriously.

brassy *adj.* brazen.

bread *n.* **1.** money; earnings. *—phr.* **2. know which side one's bread is buttered,** See **know.**

bread-and-butter *adj.* concerned with necessities or the means of living: *Tax is a bread-and-butter issue.*

breadbasket *n.* the stomach.

break *n.* **1.** an opportunity; chance. **2.** a social error or slip; an unfortunate remark. *—phr.*

3. break even or **square,** to have one's credits or profits equal one's debits or losses.

4. break it down, a. stop it. **b.** calm down; be reasonable.

6. break one's duck, *Cricket.* (of a batsman) to score his first run in an innings.

7. break out, to bring out, especially refreshments: *Let's break out the beer.*

8. break up, to collapse with laughter.

9. give (someone) a break, to give (someone) a fair chance.

10. those are the breaks, that is how life is.

breaker breaker! *interj. C.B. Radio.* a call used by an operator to request access to a channel.

breakfast *phr.*

1. all over the place like a madwoman's breakfast, See **madwoman.**

2. bushman's/dingo's breakfast. Also **dog's/pelican's breakfast,** attention to various basic needs upon waking, as a drink of water, a good look around, urination, a wash, a scratch, etc. but not usually including eating.

3. See **dog's breakfast.**

breath *phr.* **save (one's) breath to cool (one's) porridge,** a command to keep (one's) temper; to reserve (one's) energy for a more worthwhile cause.

breath test *n.* a test to measure the alcohol content of exhaled breath.

breeze *n.* **1.** an easy task: *It's a breeze.* **2.** a rest: *Give it a breeze.* *—v.* **3.** Also, **breeze along.** to move or proceed in a casual, quick, gay manner. *—phr.*

4. bat or **fan the breeze,** to engage in idle conversation.

5. breeze through, to perform without effort.

6. have the breeze up, to be afraid.

7. put the breeze up, to make afraid.

breezer *n.* **1.** a fart. **2.** the anus.

brekkie *n.* Also, **brekky.** (originally children's speech) breakfast.

Brethren, the *pl. n.* the Past Brothers team in the Queensland Rugby Football League. Also, **Brothers.**

brew *phr.* **make a brew,** to make a pot of tea; prepare any drink.

brewer's droop *n.* alcoholically-induced sexual impotence.

brick *n.* **1.** a good fellow. **2. a.** (formerly) the sum of £10. **b.** the sum of $10. **3.** *Music.* an amplifier. **4.** *Prison.* prison sentence of ten year's duration. *—v.* **5.** to falsify evidence. *—phr.*

6. a brick short (of a load), simple-minded.

7. bet London to a brick, a. to bet on a certainty. **b.** to be firmly convinced. [use of *brick* a sum of money]

8. built like a brick (shithouse), (of a person) well-built; stocky.

9. to drop a brick, to make a social blunder or solecism.

brickbat *n.* an unkind remark; caustic criticism.

brickfielder *n.* **1.** a dry, dusty wind. **2.** (in Sydney) a southerly buster. [from *Brickfield* Hill in Sydney from which winds blew dust]

bride *phr.* **off like a bride's nightie,** See **off.**

bridge *v.* **1.** to show or display (something), usually inadvertently. *—phr.* **2. burn (one's) bridges,** See **burn. 3. chuck a bridge,** (of a female) to sit with underpants visible.

4. cross (one's) bridges, See **cross.**

bridle path *n.* the Milky Way. [Northern Territory usage]

brief *n. Prison.* a barrister.

brigalow *phr.* **the brigalow,** the outback; country where brigalow, a species of acacia, is the main vegetation.

bring *phr.*
1. bring down the house, to receive rapturous applause.
2. bring in, *N.Z.* to bring (land) into cultivation.
3. bring off, to induce an orgasm in.
4. bring on, to excite sexually, so as to induce orgasm.
5. bring out, to induce (workers, etc.) to leave work and go on strike.
6. bring up, to vomit.
7. bring up with a jolt, to cause to stop suddenly, especially for reappraisal.

brinkmanship *n.* the practice of courting disaster, especially nuclear war, to gain one's ends.

brinnie *n.* a small stone. Also, **brinny.**

briny *n.* the sea. [from *brine* salted water; the sea or ocean]

brisket *n.* the human chest. [from *brisket* the breast of an animal]

Brissy *n.* Brisbane. Also, **Brizzie.**

Bristols *pl.n.* breasts. [rhyming slang, *Bristol City* (a British Soccer Team), titty]

Brit *n.* a person from Britain. [short for *British*]

britches *pl.n.* **1.** trousers. *—phr.* **2. too big for one's britches,** conceited; arrogant. [variant of *breeches* trousers]

British *phr.* **best of British luck,** an expression of goodwill sometimes implying that the recipient has little chance of success or lucky outcome.

Brits, Jimmy *pl. n.* See **Jimmy Brits.** Also, **Jimmie Britt, Edgar Britt.**

broad *n.* a woman.

broadie *n.* a deliberate sideways skid by which a car is turned to travel in the opposite direction.

broke *adj.* **1.** out of money. *—phr.*
2. broke to the wide, out of money; bankrupt.
3. flat or **dead broke,** completely out of money.
4. go for broke, a. (gambling, investment, etc.) to risk all one's capital in the hope of a very large gain. **b.** to take a major risk in pursuing an activity, objective, etc., to its extreme.

brolly *n.* **1.** an umbrella. **2.** *British Military.* a parachute.

bromide *n.* **1.** a sedative. **2.** a person who is platitudinous and boring. **3.** a tiresome platitude.

bronze *n.* **1.** Also, **bronzo,** the anus. **2.** the buttocks.

bronzer *n.* a male homosexual.

bronze wing *n.* *N.T., W.A.* a half-caste Aborigine.

brothel *n.* any room in a disorderly state.

brothel boots *pl.n.* soft-soled footwear. Also, **brothel creepers.**

brown *Obsolete.* a copper coin, especially a penny.

brown bomber *n.* (formerly) an officer employed to enforce parking and other associated traffic regulations. See **grey ghost.** [from the colour of the uniform]

browned off *adj.* bored; discontented; fed up. Also, **browned-off.**

brown-eye *n.* the anus.

brownie *n.* a bottle of beer.

brownnose *v.* **1.** to flatter servilely. —*n.* **2.** a self-seeking servile flatterer; one who curries favour; bootlicker.

Brown's cows *phr.* **like Brown's cows,** in a disorderly progression; all over the place.

bruiser *n.* a tough fellow; bully.

brumby *n.* a wild horse, especially one descended from runaway stock. [Aboriginal]

brummy *adj.* **1.** shoddy; cheap. —*n.* **2.** anything shoddy or cheap. Also, **brum.** [from *Brummagem* an altered form of Birmingham where cheap articles were manufactured]

brush *n.* **1. a.** a girl or a woman. **b.** girls or women collectively. **2.** the female pubic area. —*phr.* **3. brush up, a.** to polish up; smarten. **b.** to revise and renew or improve one's skill in.

brusher *n.* **1.** a person who moves quickly. [from *brusher* a small wallaby noted for its brisk movements] —*phr.*

2. to get brusher, to be rejected; dismissed: *He's gone, she gave him the brusher last week.*

3. to give brusher, a. to leave with debts unpaid. **b.** to abandon a task.

brush-off *n.* an abrupt or final dismissal or refusal.

brute *n.* a selfish or unsympathetic person.

btm *n.* a euphemism for bottom or buttocks.

bub *n.* **1.** Also, **bubba,** a baby. **2.** one of the breasts of a woman.

bubble and squeak *n.* **1.** a beak (a magistrate). **2.** Greek. —*v.* **3.** Also, **bubble.** to speak (inform). [rhyming slang]

bubblegum music *n.* pop music designed to appeal to pre-teenagers; cradle rock.

bubbly *n.* champagne.

buck[1] *n.* **1.** *U.S.* (*offensive*) a male Indian or Negro. —*adj.* **2.** (*offensive*) male: *a buck nigger. [from buck* the male of certain animals]

buck[2] *v.* **1.** to resist obstinately; object strongly: *to buck at improvements.* —*phr.*

2. buck up, a. to hurry. **b.** to become more cheerful, vigorous, etc. **c.** to force or urge (someone) to hurry. **d.** to make more cheerful, vigorous, etc. **3. give it a buck,** to make an attempt; chance.

buck[3] *phr.*

1. pass the buck, to shift the responsibility or blame to another person.

2. the buck stops here, the acceptance of final responsibility. [from *buck, (Poker)* any object in the kitty which reminds the winner that he has some privilege or duty when his turn to deal next comes]

buck[4] *n. Originally. U.S.* **1.** a dollar. —*phr.* **2. a fast buck,** money earned with little effort, often by dishonest means. [shortened form of *buckskin,* the skin of a buck or deer, formerly an accepted form of exchange on the U.S. frontier]

bucked *adj.* cheered; elated; encouraged.

bucket *v.* **1.** to criticise or make scandalous accusations about (someone). —*phr.*

2. **empty/tip/drop the bucket on,** to make scandalous accusations or revelations about (someone); criticise strongly.

3. **kick the bucket,** See **kick.**

4. **scrape the bottom of the bucket,** See **scrape.**

buckjumper *n.* **1.** a tram. **2.** a small damper or scone.

buckle *v.* **1.** to arrest. *—n.* **2.** the condition (degree of fatness) of a horse. *—phr.*

3. **buckle down,** to set to work with vigour.

4. **buckle under,** to give up; despair.

5. **buckle under to,** to yield; give way.

6. **get buckled,** *Prison.* to be arrested.

Buckley's chance *phr.*

1. Also, **Buckley's, Buckley's hope,** a very slim chance; forlorn hope.

2. **Buckley's and none,** (*humorous*) two chances amounting to next to no chance. [probably a pun on *Buckley and Nunn,* a Melbourne store]

bucko *n.* **1.** a young man. **2.** a blustering bully.

buck-passing *n.* the avoidance of responsibility by passing it on to another.

buck's fizz *n.* a drink made with champagne and orange juice.

buckshee *adj.* free of charge. [variant of *baksheesh,* a charitable gift or gratuity (used in Middle Eastern countries)]

bucks party *n.* a party held on the eve of a wedding for the bridegroom by his male friends. Also, **bucks' party, bucks ding.**

bud[1] *phr.* **nip in the bud,** to stop (something) before it gets under way.

bud[2] *n. U.S.* **1.** brother. **2.** man or boy (as a term of address). [alteration of *brother*]

buddy *n.* comrade; mate. [from *bud,* alteration of *brother*]

buddy-buddy (*often offensive*) *n.* **1.** a close friend or associate, especially one in a relationship based on a common interest or ambition. *—adj.* **2.** sycophantic: *buddy-buddy tactics.* *—v.* **3.** to behave in a sycophantic manner: *to buddy-buddy an official.*

budgeree *adj. Obsolete.* good; fine. [Aboriginal]

budgie *n.* a budgerigar.

buff *n.* **1.** an enthusiast; an expert (sometimes self-proclaimed): *a wine buff.* [from the buff uniforms worn by New York volunteer firemen in the 1820s] **2.** the bare skin.

Buffaloes, the *pl. n.* the Darwin team in the Northern Territory Football League.

buffer *n.* a foolish man, especially one who is elderly and pompous.

bug *n.* **1.** a malady, especially a virus infection. **2.** an idea or belief with which one is obsessed. **3.** a microphone hidden in a room to tap conversation. *—v.* **4.** to install a bug in (a room, etc.). **5.** to cause annoyance or distress to (a person). **6.** to equip (a building) with burglar alarms.

bugaboo *n.* some imaginary thing that causes fear or worry; a bugbear; a bogy.

bug-eyed *adj.* popeyed; with eyes protruding like those of a bug, as a sign of surprise, tiredness, etc.

bugger *n.* **1.** (*humorous*)a person; child: *Come on, you old bugger.* **2.** a foul, contemptible person. **3.** a nuisance, a difficulty; something unpleasant or nasty: *That recipe is a real bugger; It's a bugger of a day.* —*v.* **4.** to damn or curse, as an indication of contempt or dismissal: *Bugger him, I'm going home; Bugger it!* **5.** to cause inconvenience to someone; delay. — *interj.* **6.** a strong exclamation of annoyance, disgust, etc.: *Oh, bugger!* —*phr.*

7. bugger about or **around,** to mess about; fiddle around.

8. bugger me dead or **Charlie,** an exclamation of surprise.

9. bugger off, to remove oneself; depart.

10. bugger up, to cause damage, frustration or inconvenience to.

11. play silly buggers, to engage in time-wasting activities and frivolous behaviour. [from *bugger* one who practices bestiality or sodomy, from Middle Latin *Bulgarus* a Bulgarian, a heretic; certain Bulgarian heretics being charged with this activity]

bugger-all *n.* very little; nothing: *He's done bugger-all all day.* Also, **bugger all.**

buggered *adj.* **1.** tired out; exhausted. **2.** broken; wrecked. **3.** damned: *I'm buggered if I'll do that.*

buggerise *phr.* **buggerise about** or **around,** to behave aimlessly or ineffectually.

buggerlugs *n.* a mock abusive term, used affectionately.

buggery *phr.*

1. go to buggery!, go away; leave me alone.

2. like buggery, considerably: *It hurts like buggery.*

3. off to buggery, a. greatly off course; in error; astray. **b.** a long way away.

buggy *n.* a motor car. [a jocular reference to *buggy* a two-wheeled horse-drawn carriage]

buggy-ride *phr.* **thanks for the buggy-ride,** *U.S.* an expression of gratitude.

bug house *n.* a picture theatre; cinema.

bug-hunter *n.* a collecting entomologist. [*bug* any insect + *hunter*]

bugle *n.* **1.** the nose. —*phr.* **2. on the bugle,** smelly.

bug rake *n.* a comb.

bugs *n.* money. Also, **bugs bunny.** [rhyming slang *Bugs Bunny* money]

built *adj.* *Originally U.S.* of a woman, well-developed.

bulk *adj.* a great many: *We've got bulk people staying at our place.*

bull[1] *n.* **1.** *Military.* the polishing and cleaning of equipment. **2.** a police officer. —*v.* **3.** to have intercourse with (a woman). —*phr.*

4. bull at a gate, an impatient and headstrong person.

5. bull in a china shop, an inept or clumsy person in a situation requiring care or tact.

6. take the bull by the horns, to tackle a difficult situation.

bull[2] *n.* *Convict Obsolete.* a flogging of seventy-five lashes. [from obsolete British slang *bull,* a crown piece or five shillings. See **bob, canary, tester**]

bull[3] *n.* **1.** nonsense. **2.** trivial or boastful talk. —*v.* **3.** to deceive; dupe. **4.** to boast; exaggerate. —*interj.* **5.** Also, **bulls.** an exclamation implying that what has been said is nonsensical or wrong. [shortened form of *bullshit*]

Bullamakanka *n.* an imaginary remote town; any remote place.

bullamakau *n.* bully beef. [New Guinea term from World War II]

Bullants, the *pl. n.* the Preston team in the Victorian Football Association.

bull artist *n.* one notorious for his excessive talk which is usually boastful, exaggerated and unreliable. Also, **bullshit artist.**

bull car *n.* a police car.

Bulldogs, the *pl. n.* **1.** the Canterbury-Bankstown team in the N.S.W. Rugby Football League. **2.** the Central Districts team in the South Australian National Football League. **3.** the East Sydney Australian Rules Football team. **4.** the Footscray team in the Victorian Football League. **5.** the South Fremantle team in the West Australian Football League. **6.** the Western Districts team in the Queensland Rugby Football Union. **7.** the Western Suburbs team in the Queensland Rugby Football League.

bulldoze *v.* **1.** to put pressure on; to coerce or intimidate, often with violence or threats. **2.** to push legislation, a motion, etc., through the process of a parliament, meeting, etc., with undue haste.

bulldozer *n.* a person who intimidates.

bulldust *n.* **1.** fine dust on outback roads. **2.** nonsense; bullshit. —*v.* **3.** to boast, to exaggerate.

bullet *n.* **1.** a recording which is moving rapidly up the popularity chart. **2.** the symbol placed beside the name of such a recording on the chart. —*phr.*

3. to get or **have the bullet,** to be dismissed from employment.

4. bite (on) the bullet, to face up bravely; resign oneself to an unavoidable ordeal.

bullet-head *n.* an obstinate or stupid person.

bullhead *n.* a stupid or obstinate person.

bull-headed *phr.* **go bull-headed at,** to act aggressively, blunderingly.

bull Joe *n.* a bull ant. Also, **bulljoe.**

bullo *n.* nonsense; rubbish.

bullock *v.* **1.** to force: *to bullock one's way through.* **2.** to do heavy, strenuous manual work.

bullock horn *v.* to pawn. [rhyming slang]

bullock-puncher *n.* the driver of a bullock team. Also, **bull-puncher.**

bullocky *n.* violent language. [from *bullocky* the driver of a bullock team. Bullock drivers were well known for their colourful language]

bullocky's joy *n.* **1.** golden syrup. **2.** treacle. Also, **cocky's joy.**

bullpen *n.* *U.S.* a place for the temporary confinement of prisoners or suspects.

bullseye *interj.* an exclamation indicating triumph, success, etc.

bullshit *n.* **1.** nonsense. —*v.* **2.** to deceive; outwit. —*interj.* **3.** an expression of disgust, disbelief, etc. Also, **bullsh.**

bullshit artist *n.* See **bull artist.**

bull's roar *phr.* **not get within a bull's roar,** to be not close at all.

Bulls, the *pl. n.* the Wide Bay team in the Queensland Rugby Football League.

bullswool *n.* **1.** nonsense. **2.** a tall tale. —*interj.* **2.** an exclamation of disbelief, disgust, etc.

bully *adj.* **1.** fine; excellent; very good. —*interj.* **2.** an exclamation indicating approval.

bully beef *n.* chief. [rhyming slang]

bully beef bomber *n. World War II Air Force.* Also, **biscuit bomber.** a supply or transport plane in New Guinea.

bum *n.* **1.** the rump; buttocks. **2.** a shiftless or dissolute person. **3.** a habitual loafer and tramp. —*v.* **4.** to get for nothing; borrow without expectation of returning: *to bum a cigarette.* **5.** to sponge on others for a living; lead an idle or dissolute life. —*adj.* **6.** of poor, wretched, or miserable quality; bad: *a bum deal.* —*phr.*

7. go bite your bum, an impolite dismissal indicating the speaker's wish to end the conversation.

8. bum a ride, to appeal successfully for a free ride in a car, etc.

bumble *v.* **1.** to proceed clumsily or inefficiently: *to bumble along.* **2.** to mismanage: *The government bumbled its way through crisis after crisis.*

bumboy *n.* (*offensive*) a follower or employee who is obsequious and servile.

bumf *n.* See **bum fodder. Also, bumph.**

bumfluff *n.* light hair growing on the face of an adolescent male.

bum fodder *n.* **1.** toilet paper. **2.** Also, **bumf.** (*offensive*) written or printed matter judged suitable only for use as toilet paper.

bumfreezer *n. Chiefly British.* a short coat.

bummer *n.* a fiasco; failure; disappointment.

bump *phr.*

1. bump up, to increase (in extent, etc.).

2. bump off, to kill.

bump cap *n.* a type of safety helmet of light construction, as used in the meat industry.

bumper *n.* **1.** something unusually large or full. **2.** a cigarette end; a discarded cigarette, partly smoked. **3.** a pickpocket's accomplice. —*phr.* **4. not worth a bumper,** worthless.

bumper-to-bumper *adj.* (of traffic) dense, and moving slowly.

bum's rush *n.* **1.** the peremptory dismissal or bodily removal of an unwanted person. **2.** the peremptory rejection of an idea or proposal.

bum steer *n.* incorrect information or advice.

bumsucker *n.* a sycophant. Also, **bumlicker, arse-licker.**

bumzack *n. W.A.* a cadger, especially of drinks.

bun *phr.*

1. have a bun in the oven, to be pregnant.

2. do one's bun, *N.Z.* to lose one's temper.

3. to take the bun or **cake,** See **cake.**

bunce *n.* **1.** profit, especially unexpected. **2.** a share or commission. [origin obscure]

bunch *n.* a group of human beings: *a fine bunch of boys.*

bunch of fives *n.* the fist.

Bundaberg blanket *n.* rum. [from *Bundaberg,* Queensland, centre of sugar-growing district and famed for the production of rum]

Bundaberg honey *n.* golden syrup (a sweet syrup derived from sugar processing). [See **Bundaberg blanket**]

bundle *n.* **1.** a lot of money. *—phr.*

2. bundle off or **out,** to leave or send away hurriedly or unceremoniously.

3. drop a bundle, a. to give birth. **b.** to reveal startling news, information **c.** to suffer heavy monetary loss as the result of gambling, etc.

4. drop one's bundle, to give up, especially out of a sense of despair or inadequacy.

bundy *phr.*

1. bundy on, to begin work.

2. bundy off, a. to finish work. **b.** *Prison.* to die, especially from a drug overdose.

3. punch the bundy, a. to begin work. **b.** to be in regular employment. [from *bundy* a time clock which is used to record the arrival and departure times of employees. Trademark]

bunfight *n.* any noisy or disorganised gathering of people, as at a crowded party. Also, **bun fight.**

bung *n.* **1.** a memo to an employee, especially of a government department, calling him to account for a breach of regulations on his part. *—v.* **2.** to put: *Bung it in the cupboard.* **3.** to toss to another person; throw. *—adj.* **4.** not in good working order; impaired; injured. *—phr.*

5. bung it in, Gunga Din!, a direction to proceed, as from a building labourer to someone pouring concrete or tipping 'fill' from a truck.

6. bung it on, to behave temperamentally.

7. bung on, a. stage; put on: *bung on airs and graces.* **b.** to prepare or arrange, especially at short notice: *We'll bung on a party.*

8. bung on side, to behave in a pompous and overbearing manner.

9. go bung, a. to break down; cease to function. **b.** to fail; become bankrupt.

10. put a bung in it, *Military.* be quiet; close the door.

bunger *n.* **1.** a firework which produces a loud bang. **2.** a thickly-rolled cigarette.

bung-eye *n.* an infection of the eyes, as sandy blight.

bunghole *n.* cheese.

bungum worm *n. S.A.* a beachworm found on the southern and eastern coasts of Australia, giving off mucus when handled, and commonly used for bait; slimy.

bun hat *n. N.Z.* a bowler hat.

bunk[1] *n.* **1.** any bed. *—v.* **2.** to occupy a bunk; sleep, especially in rough quarters. *—phr.* **3. give someone a bunk-up,** to give someone a leg up. [from *bunk* a built-in platform bed]

bunk[2] *n.* humbug; nonsense; bunkum. [short for *bunkum*]

bunk[3] *phr.* **do a bunk,** to run away; take flight.

bunker *v.* to place in a situation from which it is difficult to extricate oneself; stymie. [originally golfing term]

bunkum *n.* **1.** insincere talk; claptrap; humbug. **2.** *Chiefly U.S.* insincere speechmaking intended merely to please voters. Also, *U.S.* **buncombe.** [alteration of *Buncombe,* a county in the U.S., in North Carolina, from its Congressional representative's phrase, 'talking for Buncombe']

bunny *n.* **1.** a rabbit. **2.** a fool. **3.** one who accepts the responsibility for a situation, sometimes willingly: *to be the bunny.* **4.** a nightclub hostess attired in a costume with simulated rabbit's ears and tail.

bun rush *n.* the movement of a disorganised crowd of people, all attempting to do something at the same time.

bun wagon *n.* a closed police vehicle used for conveying prisoners; black maria.

bunyip *n.* an imposter. [from *bunyip* an imaginary creature of Aboriginal legend]

bunyip aristocracy *n.* an often offensive term for Australians who consider themselves to be aristocrats. [originally derogatory title for suggested Australian peerage]

Burdekin duck *n.* (formerly) a meat fritter. [from *Burdekin,* a river and district in Queensland]

Burdekin vomit *n.* a gastric ailment. [See **Burdekin duck**]

burg *n. U.S.* a city or town. [variant of *burgh,* a borough, from Middle English, a town]

burgoo *n.* any porridge-like food, as prison hominy. Also, **bergoo.** [Persian *burghul,* a bruised grain]

burl *n.* **1.** an attempt. **2.** a gamble as in two-up: *have a burl.* *—v.* **3.** to move quickly: *to burl along.* **4.** to cause (a coin, etc.) to spin. **5.** to taunt and jeer at: *to burl the science master.* *—phr.* **6. give it a burl,** make an attempt. Also, **birl.** [northern British dialect *birl* to spin]

burn *v.* **1.** Also, **burn off.** to race, on a motorcycle or in a car. **2.** *U.S.* to be electrocuted in an electric chair. *—n.* **3. a.** an unofficial speed trial, on a motorcycle or in a car. **b.** a fast run. **4.** a cigarette. *—phr.*

5. burn one's bridges or **boats,** to put oneself in a position from which there is no retreat; to make an irrevocable decision.

6. burn one's fingers, See **finger.**

7. burn out, to become exhausted; to become deficient in energy or drive: *Many Olympic swimmers are burnt out before they are 20.*

8. burn the midnight oil, to stay up late studying, etc.

9. slow burn, a display of slowly mounting anger.

burnt offering *n.* over-cooked food.

burnt stick *phr.* **better than a poke in the eye with a burnt stick, a.** an expression of pleasure at some event to one's advantage. **b.** an ironic situation which could be worse.

burp *v.* **1.** to belch. **2.** to cause (a baby) to belch, especially to relieve flatulence after feeding. [imitative]

burry *n.* (*usually offensive*) an Aboriginal.

burst *phr.* **on the burst, a.** on a drinking bout. **b.** with a sudden dash, as a player in a football game.

burton *phr.* **go for a burton,** to be killed or destroyed; to disappear.

bury *phr.*

1. bury the hatchet, to be reconciled after hostilities.

2. bury the bishop, of a man, to have sexual intercourse.

bus *n.* **1.** a motor car or aeroplane. **2.** *Prison.* a large police or prison van used to transport prisoners, as from court to gaol, etc. *—phr.* **3. miss the bus,** See **miss.**

bush *v.* **1.** to rough it; camp out: *bushing it under the stars. —phr.*

2. beat about the bush, to fail to come to the point; prevaricate.

3. go bush, a. to turn one's back on civilisation; adopt a way of life close to nature. **b.** Also, **take to the bush.** to disappear suddenly from one's normal surroundings or circle of friends.

4. in the bush, *Surfing.* beyond the line of the breakers.

5. Sydney or the bush, See **Sydney.**

bush Baptist *n.* a person of vague but strong religious beliefs, not necessarily associated with a particular denomination.

bush bashing *n.* **1.** clearing virgin bush. **2.** (in bush walking) making a path through virgin bush.

bush bellows *pl. n.* hats (as used to fan camp fires).

bush boss *n. N.Z. Timber Industry.* an overseer of logging in the bush.

bush capital, the *n.* Canberra.

bush carpenter *n.* a rough, amateur carpenter.

bush champagne *n.* a drink reputedly made from a saline and methylated spirits.

bush cure *n.* a household remedy; traditional medicine, especially as practised in bush communities.

bushed *adj.* **1.** lost. **2.** exhausted. **3.** confused.

bushel o' coke *n.* bloke, any man, particularly a young man. [rhyming slang]

bushfire *phr.* **get on like a bushfire,** (of a friendship, project, etc.) to develop or proceed rapidly. [variant of *wildfire*]

bushfire blonde *n.* **1.** Also, **strawberry blonde,** a redhead, usually female. **2.** a drink, consisting usually of cherry brandy and lemonade.

bushie *n.* a person, usually unsophisticated, who lives in the bush. See **townie.** Also, **bushy.**

bush-lawyer *n.* **1.** a person with a good knowledge of the law, but without legal qualifications. **2.** an argumentative person who scores by making legal points. Also, **bush lawyer.**

bush liar *n.* a teller of tall tales.

bushman's bible *n.* the Bulletin. [a widely read weekly journal, founded in 1880, which in its first 30 years was a focus of radical Australian nationalism]

bushman's breakfast *n.* See **breakfast.**

bushman's clock *n.* a kookaburra.

bush miles *pl.n.* distance roughly calculated to include the twists, turns and difficulties of a route winding through the bush.

bush oyster *n.* (*humorous*) a testicle, usually of a sheep.

bushranger *n.* **1.** (*offensive*) any man or business enterprise seen as unprincipled and extorting. **2.** a con-man. [from *bushranger* (formerly) a bandit or criminal (often an escaped convict) who hid in the bush and led a predatory life]

bush reckoning *n.* measurement of time or distance usually underestimated or unrealistic: *It takes five days to get there, two days by bush reckoning.*

bush refrigerator *n.* a cabinet, for the storage of perishable foodstuffs, which allows a breeze to blow through wet fabric, such as hessian, thus reducing the temperature inside; cool safe; Coolgardie safe.

bush telegraph *n.* an unofficial chain of communication by which information is conveyed and rumour spread, by word of mouth. Also, **bush telegram, bush wire, bush wireless.**

bush tucker *n.* simple fare, as eaten by one living in or off the bush.

bush week *n.* **1.** a fictitious week when country people come to town. **2.** circumstances in which unsuspecting people are imposed upon: *What do you think this is — bush week?* Also, **bushweek.**

bushwhack *v.* **1.** to live as a bushwhacker. **2.** to ambush. **3.** *N.Z.* to clear land of timber. Also, **bushwack.** [back formation from *bushwhacker*]

bushwhacked *adj.* **1.** extremely fatigued; beaten; exhausted. **2.** astonished; annoyed. Also, **bushwacked.**

bushwhacker *n.* **1.** one who lives in the bush; a bushie. **2.** *N.Z.* one who clears the land of bush, especially an axeman in cutting timber. [*bush,* + *whacker* one who strikes with a smart resounding blow or blows, as in clearing land of timber]

bush wire *n.* See **bush telegraph.** Also, **bush wireless.**

bushytailed *phr.* **bright-eyed and bushytailed,** full of health and good spirits.

business *n.* **1.** defecation. **2.** prostitution. *—phr.*
3. business as usual, proceeding normally.
4. to mean business, to be in earnest.

busman's holiday *n.* a holiday on which one does one's regular work, or some similar activity.

bust *v.* **1.** to burst. **2.** to (go) bankrupt. **3.** to reduce in rank or grade; demote. **4.** to subdue; break the spirits of (a brumby, etc.). **5.** to break and enter with intent to steal. **6.** (of police) to raid or arrest. *—n.* **7.** the act of breaking and entering. **8.** a police raid. **9.** a drunken party or spree; brawl. *—adj.* **10.** broken; ruined. **11.** bankrupt. *—phr.*
12. bust one's gut or **boiler,** to overdo anything: *Don't bust your gut over that job.*
13. bust up, a. to part finally; quarrel and part. **b.** to smash. **c.** to interrupt violently a political meeting or other gathering.

buster *n.* **1.** a cold, violent southerly wind, often after a heatwave: *a southerly buster.* **2.** *Chiefly U.S.* a small boy. **3.** a term of address to a man or boy, either casually friendly or covertly aggressive. **4.** something very big or unusual for its kind. **5.** *U.S.* a frolic; a spree. —*phr.* **6. come a buster,** to fail, usually because of a misfortune.

bust-up *n.* **1.** a disruption; disturbance or commotion, as one which brings a meeting to a sudden end. **2.** a final parting, often with ill-feeling. **3.** a financial or commercial failure. **4.** a riotous party.

but *adv.* **1.** (*in children's speech*) however; though (used at the end of a sentence): *A picnic was planned, it rained but.* —*phr.* **2. no buts about it,** without restriction or objection.

butch *n.* **1.** a lesbian or woman exhibiting extravagantly masculine characteristics. **2.** a man, especially one of notable physical strength. —*adj.* **3.** of a man or woman, exhibiting masculine characteristics.

butcher *n. S.A.* a small glass, used primarily for serving beer.

butchers[1] *adj.* **1.** ill. —*phr.* **2. go butchers (hook) at,** to become angry with. [rhyming slang, *butcher's hook* crook]

butchers[2] *n.* a look. [rhyming slang *butcher's hook* look]

butcher's canary *n.* (*humorous*) a blowfly.

butt *phr.* **butt in,** to interrupt; interfere; intrude.

butter *v.* to flatter grossly. Also, **butter up.**

butterball *n.* a fat, round person.

butter-fingers *n.* a person who fails to catch or who drops things easily.

butterfly *n.* (in two-up) a coin which when tossed fails to spin; floater.

buttery *adj.* grossly flattering.

buttinski *n.* **1.** a stickybeak; an interfering person. **2.** a telephone device enabling a technician to cut into a local cable and speak to a subscriber, as when a fault is being repaired. [a mock-foreign coinage from *butt* + *in*]

button *n.* **1.** the nose: *He was hit on the button.* —*phr.*
3. button down, to restrain.
4. button one's lip or **face,** to be silent; stop talking.
5. button up, a. stop talking. **b.** (of a scheme, etc.) to arrange successfully; bring to a successful conclusion.

buttonhole *v.* to seize by or as by the buttonhole in the lapel of the jacket and detain in conversation.

butty *n. British.* **1.** a workmate, especially in a colliery. **2.** a thick slice of buttered bread.

buy *n.* **1.** a bargain: *a good buy.* —*v.* **2.** to accept: *Do you think he'll buy the idea?* —*phr.*
3. buy back the farm, See **farm.**
4. buy it, to die: *He bought it at Bathurst.*
5. buy in, to join in; become involved.
6. buy into, to choose to become involved in: *buy into an argument.*
7. buy into trouble, to undertake a course of action against the better judgment of oneself or others.

buzz *n.* **1.** a rumour or report. **2.** a telephone call. **3.** a feeling of exhilaration or pleasure, especially as induced by drugs. —*v.* **4.** to make a telephone call to. **5.** *Aeronautics.* **a.** to fly an aeroplane very low over: *to buzz a field.* **b.** to signal or greet (someone) by flying an aeroplane low and slowing the motor spasmodically. —*phr.* **6. buzz around like a blue-arsed blowfly,** See **blue-arsed fly.** **7. buzz off** or **along,** to go; leave.

buzz word *n.* a word used for its emotive value or its ability to impress the listener.

bye-bye *interj.* **1.** goodbye. —*n.* **2.** (*pl.*) (*in children's speech*) sleep: *Go to bye-byes.*

B.Y.O. *adj.* **1.** of a dinner, party, restaurant, etc., to which one brings one's own supply of liquor. Also, **B.Y.O.G.** —*n.* **2.** an unlicensed restaurant which allows clients to bring their own liquor. [short for *B(ring) Y(our) O(wn) G(rog)*]

cab *n.* **1.** a taxicab. —*phr.* **2. first cab off the rank,** the first person to do something. [short for *taximeter cab*]

cabbage *n.* **1.** paper money. —*phr.* **2. the cabbage garden, land** or **patch,** Victoria.

cabbage-gardener *n.* a Victorian. Also, **cabbage-patcher, cabbage-lander.**

cabbie *n.* a cab driver. Also, **cabby.**

caboodle *n.* the whole lot/pack/crowd. [unexplained variant of *boodle* the lot, pack or crowd]

cack *n.* **1.** muck; filth. **2.** faeces. —*v.* **3. a.** to defecate. **b.** to laugh uncontrollably. **4.** to soil with excrement: *He cacked his corduroys.* Also, **kack.**

cack-handed *adj.* **1.** clumsy with the hands; maladroit. **2.** left-handed. Also, **cacky-handed.**

cackle *v.* **1.** to chatter noisily. —*n.* **2.** idle talk. —*phr.* **3. cut the cackle,** to be quiet. [from *cackle* a shrill cry, as uttered by a hen after laying an egg]

cackle berry *n.* an egg. Also, **cackle fruit.**

cactus *n.* **1.** the backblocks. —*adj.* **2.** ruined; useless. —*phr.* **3. in the cactus,** in difficulties, in trouble. [from *cactus* a plant common in hot dry inland regions, especially of America]

cady *n.* **1.** a hard straw hat formerly worn by men, especially on recreational outings; boater. **2.** a man's hat, of any style or material. Also, **caddie.** [origin unknown]

caf *n.* a cafeteria. Also, **caff.**

cages *pl. n. Prison.* (in a maximum security gaol) the exercise yards for solitary confinement prisoners.

cahoot *phr.* **in cahoot** or **cahoots,** in partnership; in league. [from French *cahute* hut, cabin]

Cain *n.* **1.** a murderer. *—phr.* **2. raise Cain, a.** to make a commotion. **b.** to complain or remonstrate vociferously. [from *Cain* first son of Adam and Eve, who murdered his brother Abel (Gen. 4)]

Cain and Abel *n.* a table. [rhyming slang]

cake *phr.*

1. a slice of the cake, See **slice.**

2. cakes and ale, the good things and pleasures of life.

3. hot cake, See **hot.**

4. piece of cake, something easily accomplished or obtained.

5. take the cake or **bun, a.** to win the prize. **b.** to surpass all others; excel.

6. the cake, a profit or reward to be shared.

cakehole *n.* the mouth.

calaboose *n. U.S.* a prison cell; gaol. [from Spanish *calabozo* dungeon]

calf *n.* **1.** an awkward, silly boy or man. *—phr.* **2. in calf,** pregnant.

calicos *phr.* **raise the twin white calicos,** *Australian Rules.* to score a goal.

call *phr.*

1. call it a day, to bring an activity to a close whether temporarily or permanently.

2. call of nature, the need to urinate or defecate.

3. the call, *Two-up,* the right to call, i.e. to nominate either heads or tails to win the toss.

calling card *n.* See **visiting card.**

camel driver *n. Horseracing.* an unsuccessful jockey.

camp[1] *v.* to sleep: *You take that bed, I'll camp here.*

camp[2] *adj.* **1.** exaggerated and often amusing or effeminate in style. **2.** effeminate; given to acting and speaking with exaggerated mannerisms. **3.** homosexual. *—n.* **4.** an exaggerated, often amusing or effeminate style, mannerism, or the like. **5.** a person or his work displaying this quality. *—v.* **6.** to act in a camp manner. **7.** to be a homosexual. **8.** to perform or imbue (something) with a camp quality. *—phr.*

9. camp as a row of tents, homosexual.

10. camp it up, a. to make an ostentatious or affected display. **b.** to flaunt homosexuality. [? from *camp* impetuous, uncouth person, hence objectionable, effeminate; in some senses, probably special use of *camp* brothel]

can[1] *phr.* **can do!** an exclamation indicating acquiescence in a request.

can[2] *n.* **1.** the blame for something: *to carry the can.* **2.** (*pl.*) a set of earphones. **3.** *U.S.* dismissal. **4.** *U.S.* the toilet or bathroom. **5.** *U.S.* the buttocks. *—v.* **6.** *U.S.* to dismiss; fire. *—phr.*

7. can it, to be or become silent.

8. can of worms, a situation, problem, etc., bristling with difficulties.

9. carry the can, See **carry.**

10. in the can, completed; made final. [from *in the can* (of a film) ready for distribution; filmed, developed, and edited]

11. the can, gaol.

canary[1] *n.* Also, **canary bird.** (formerly) a convict. [named for the black and yellow colour of the prisoner's clothing]

canary[2] *n. Convict Obsolete.* a flogging of one hundred lashes. [shortened form of *canary*-bird, obsolete British thieves' slang for guinea (21 shillings); by extension great, large, etc. See **bob, bull, tester**]

cancer stick *n.* a cigarette.

candidate *n.* a suitable object: *That idea is a candidate for the waste-paper basket.*

candle *phr.*
1. burn the candle at both ends, to lead a too strenuous existence; attempt to do too much, as by making an excessive demand on one's available energy, rising early and retiring late, etc.
2. can't hold a candle to, to be totally inferior to.
3. not worth the candle, not worth the effort or expense.

candy man *n. U.S.* a drug pusher.

canned *adj.* **1.** recorded: *canned laughter.* **2.** prepared in advance. **3.** drunk.

cannon *n.* a revolver.

canoe *phr.* **paddle one's own canoe,** to be independent; manage on one's own.

canoodle *v.* to indulge in fondling and petting.

canter *phr.* **win at a canter,** to win easily.

Canuck *n.* a Canadian, especially a French Canadian.

cap[1] *phr.* **if the cap fits, wear it,** if the judgment applies, accept it.

cap[2] *n.* a capsule.

cap[3] *n.* a caption.

Cape *phr.* **run via the Cape,** *Horseracing.* to race wide. [from the long sea route to England, via the Cape of Good Hope]

caper *n.* a trick; dodge. [from *caper* a playful leap or skip; a prank]

capital *phr.* **with a capital,** used to add emphasis to a statement: *It's upmarket, with a capital 'u'.*

capper *phr.* **put the capper on,** *N.Z.* finish; give the final touches to.

Capras, the *pl. n.* the Central Queensland team in the Queensland Rugby Football League.

captain *v.* (of women) to accept drinks from a stranger in a hotel, in return for the promise of sexual intercourse.

Captain Cook *n.* a look. Also, **Captain, captain.** [rhyming slang]

Captain Cooker *n. N.Z.* **1.** a wild boar. **2.** any pig, especially a poor, mangy specimen. Also, **cooker.**

carbie *n.* a carburettor.

carcass *n.* **1.** (*humorous*) a living body. *—phr.* **2. move** or **shift one's carcass,** to move away; get out of the way. Also, **carcase.** [from *carcass* the body of a slaughtered animal after removal of the offal, etc.]

card *n.* **1.** a resource, plan, idea, approach to a problem or proposition, etc.: *That's his best card; to have a card up one's sleeve.* **2.** a person of some indicated characteristic: *a queer card.* **3.** a likeable, amusing, or facetious person. *—phr.*
5. on the cards, likely to happen.
6. pick the card, *Horseracing.* to choose the winners in every race on the program.
7. put one's cards on the table, to speak plainly, candidly; disclose all information in one's possession.

cardie *n.* a cardigan.

careful *adj.* mean; parsimonious.

cark *v.* **1.** to collapse; die. *—phr.* **2. cark it, a.** to collapse; die. **b.** (of a machine) to fail; break down. [? shortened form of *carcass* body of a slaughtered animal]

carn *interj.* come on! (a sporting barracker's cry): *Carn the Blues.* [altered form of *come on*]

carnal car *n.* See **shaggin' wagon.**

carnie[1] *n.* a nubile girl under the age for legal sexual intercourse. [*carn(al),* in legal phrase, unlawful carnal knowledge, + *-ie*]

carnie[2] *n.* **1.** a person who works in a carnival. *—adj.* **2.** in the style of or suitable for a carnival. [*carn(ival)* + *-ie*]

carpet *v.* **1.** to reprimand. *—phr.* **2. on the carpet, a.** under consideration or discussion. **b.** before an authority for a reprimand.

carrot *n.* **1.** an incentive: *to dangle a carrot.* **2.** a migrant.

carry *phr.*

1. carry on, to behave in an excited, foolish, or improper manner; flirt.

2. carry the can, to do the dirty work; bear the responsibility; take the blame.

3. carry the day, to succeed; triumph.

cart *phr.* **in the cart,** in an awkward or unpleasant predicament.

carve *phr.* **carve up, a.** to slash (a person) with a knife or razor. **b.** to distribute profits, a legacy, illegal gain, an estate, etc. **c.** to defeat, as in a match.

cas *n.* a casualty ward.

casanova *n.* any man noted for his amorous adventures. [from Giovanni Jacopo *Casanova,* 1725-98, Italian adventurer and writer]

case[1] *n.* a peculiar or unusual person: *He's a case.*

case[2] *v.* to examine or survey (a house, bank, etc.) as in planning a crime.

caser *n. Obsolete.* (formerly) a five shilling piece; a crown.

cash *phr.*

1. cash in, a. to obtain an advantage. **b.** *U.S.* to die.

2. cash in on, a. to gain a return from. **b.** to turn to one's advantage.

3. cash in one's chips, a. (in poker, etc.) to hand in one's counters, etc., and get cash for them. **b.** to die.

4. cash up, *Obsolete.* to pay in settlement of a debt or fine: *He had to cash up a fiver.*

cashed-up *adj.* (especially of seasonal workers) recently paid and therefore with ready money: *The cashed-up shearers spent their cheques in the Menindee pub.* Also, **chequed-up**

casket *phr.* **the (Golden) Casket,** a lottery in Queensland.

cast *n. N.Z.* drunk.

casting couch *n.* the couch in a film or stage director's office, supposedly for the seduction of those auditioning. [to succeed by means of the casting couch is to gain success in return for sexual favours]

castor *adj.* pleasing; excellent: *She'll be castor!*

cat[1] *n.* **1.** a person, especially a young jazz musician or devotee of jazz. **2.** a homosexual who plays a passive role. **3.** the cat-o'-nine-tails; a whip with nine tails. *—phr.*

4. cat on a hot tin roof, See **hot.**

5. go like a cut cat, See **cut.**

6. kick/whip/flog the cat, See **kick.**

7. let the cat out of the bag, to disclose information, usually unintentionally.

8. put or **set the cat among the pigeons,** to induce discord into a peaceful situation.

9. the cat's pyjamas/miaow, an excellent person, proposal, etc.

10. too much of what the cat licks itself with, very talkative (too much tongue).

cat² *n.* a catamaran.

cat-and-dog *adj.* **1.** characterised by constant hostility and frequent quarrels: *They lead a cat-and-dog life together.* *—phr.* **2. raining cats and dogs,** to be raining very heavily.

catch *n.* **1.** a person of either sex regarded as a desirable matrimonial prospect: *The young doctor was the best catch in town.* **2.** a difficulty, usually unseen: *What's the catch?* *—phr.*

3. catch it, to get a scolding or a beating.

4. catch on, a. to become popular. **b.** to grasp mentally; understand.

5. catch out, a. to trap somebody, as into revealing a secret or displaying ignorance. **b.** to surprise.

6. catch up on, to make up a deficiency: *She caught up on her sleep.*

catch-22 *n.* a situation or rule which prevents the completion of an operation and may establish a futile self-perpetuating cycle. [from the title of the novel (1961) by J. Heller, American novelist]

catfight *n.* a fight between two women, especially two prostitutes.

cathouse *n.* a brothel.

Cats, the *pl. n.* the Geelong team in the Victorian Football League.

cat's whisker *n.* **1.** a small amount, often of distance: *He won the race by a cat's whisker.* *—phr.* **2. the cat's whiskers,** an excellent proposal, person, etc.: *He thinks he's the cat's whiskers.*

caulie *n.* a cauliflower.

caution *n.* a person or thing that is unusual, odd, amazing, etc.

caveman *n.* a man who behaves in a rough, primitive manner, especially towards women.

Cazaly *phr.* **up there Cazaly,** (a cry of encouragement). Also, **up there Cazzer.** [from Roy *Cazaly*, 1893-1963, an Australian Rules Football player]

ceiling *phr.* **hit the ceiling,** See **hit.**

Centralia *n.* the inland region of continental Australia.

centre *n.* **1.** *Two-up.* the one who holds all bets in a game of two-up made by the spinner. *—phr.* **2. the (red) centre,** the remote interior of Australia.

cert *n.* something regarded as certain to happen, to achieve a desired result as winning a race, etc.; a certainty.: *That horse is a dead cert.*

chaff *n.* money.

chaffcutter *n.* a motor vehicle, aeroplane, etc., which has a very noisy engine, especially an air-cooled engine, as in an early model Volkswagen.

chai *n.* tea. [Hindi *chā* from Peking Chinese *ch'a*]

chain *phr.* **drag the chain,** to shirk or fall behind in one's share of work or responsibilities.

chair *phr.* **be in the chair,** be the person in a group of drinkers whose turn it is to buy drinks.

chairborne *adj.* having a desk or office job (opposed to a more active one).

chalk *phr.*
1. chalk up, a. to score: *They chalked up 360 runs in the first innings.* **b.** to ascribe to: *It may be chalked up to experience.*
2. by a long chalk, by far; by a considerable extent or degree.
3. know chalk from cheese, See **know.**
4. like chalk and cheese, complete opposites.

chalkface *phr.* **work at the chalkface,** (*humorous*) (of teachers) to be actively engaged in teaching in the classroom. [variation of *work at the coalface* See **coalface**]

chalkie *n.* **1.** a schoolteacher. **2.** *Stock Exchange.* the person who records transactions on the board.

champ[1] *phr.* **champ at the bit,** to be anxious to begin.

champ[2] *n.* a champion.

champagne taste *n.* extravagant desires, usually beyond the means of one who has them: *a champagne taste and a beer income.*

champers *n.* champagne.

champion *adj.* **1.** first-rate. —*adv.* **2.** in a first-rate manner.

chance *phr.* **chance one's arm,** to make an attempt, often in spite of a strong possibility of failure.

chancy *adj.* uncertain; risky. Also, **chancey.**

change *n.* **1.** information of advantage: *get no change out of someone.* —*phr.* **2. change one's tune,** to assume a different, usually humbler, attitude.

chaps *n.* a fellow; man or boy. Also, **chappie, chappy.**

char *n.* **1.** a charwoman. —*v.* **2.** to do housework by the hour or day for money.

character *n.* an odd or interesting person: *John is really quite a character!*

charge *n.* **1.** an alcoholic drink. **2.** a thrill; a kick. —*phr.* **3. charge like a wounded bull,** to ask prices that are excessively high.

charity moll *n.* **1.** a woman who gives sexual favours in return for expensive meals, theatre tickets, etc. **2.** an amateur prostitute, who does not charge the going fee, and thus deprives the professional prostitute of custom.

charlie[1] *n.* **1.** a fool; a silly person: *a right charlie.* **2.** (*pl.*) a woman's breasts. **3.** menstruation; periods: *Charlie is visiting.* Also, **charley.**

charlie[2] *n.* a girl. [rhyming slang *Charlie Wheeler* sheila]

Charlie *n. Mil.* an enemy Asian soldier, especially a Vietnamese soldier. Also, **Charley.** [from military signals code *Victor* representing V, *Charlie* representing C, for Viet Cong]

charming *interj.* an expression of surprised disapproval.

chartbound *adj.* (of a pop-music record) likely to achieve a place in the charts of best-selling records each week.

chase *phr.*
1. chase up, to give attention to anything which is dilatory, running behind schedule, etc.: *Chase up the kids, their dinner is ready.*
2. go chase yourself, go away! (an impolite dismissal).

chaser *n.* a drink of water, beer, or other mild beverage taken after a drink of spirits: *He ordered a whisky and a beer chaser.*

chat *v.* **1.** to reprove. *—n.* **2.** a louse; similar vermin. **3.** a dirty or slovenly person. *—phr.* **4. chat up,** to talk persuasively to or flirt with: *to chat up a girl.*

chatty *adj.* **1.** dirty. **2.** busy.

cheap *adj.* **1.** embarrassed; sheepish: *He felt cheap about his mistake.* **2.** (usually of a woman) free with sexual favours. *—phr.*
3. cheap drunk, one who easily becomes intoxicated.
4. on the cheap, a. at a low price. **b.** done in a substandard manner.

cheapie *n.* a cheap product.

cheap shot *n.* a mean or despicable act or remark.

cheapskate *n.* a term of contempt for a stingy person.

cheat *phr.* **cheat on, a.** to deceive. **b.** to be sexually unfaithful to (one's spouse or lover).

cheat sheet *n.* a time sheet; a sheet or card recording the hours worked by an employee.

check *v.* **1.** to pay special attention (to something): *Check the dress on that redhead! —phr.* **2. check out,** investigate.

checkout chick *n.* a young girl serving at a checkout or cash counter in a supermarket.

cheddar *phr.* **hard** or **stiff chedder,** See **cheese.**

cheek *n.* **1.** impudence or effrontery: *She had the cheek to ask me my age! —v.* **2.** to address saucily; to be impudent. *—phr.*
3. cheek by jowl, close together; adjacent; in close intimacy.
4. turn the other cheek, to accept provocation pacifistically.
5. with one's tongue in one's cheek, mockingly; insincerely.

cheerio *interj.* goodbye.

cheers *interj.* to your health! (a toast, as in drinking to someone's health).

cheese *n.* **1.** an attractive girl or young woman. [from *cheesecake*] *—phr.* **2. hard** or **stiff cheese** or **cheddar, a.** bad luck. **b.** an offhand expression of sympathy. **c.** a rebuff to an appeal for sympathy.

cheesecake *n.* *Originally U.S.* **1.** photographs of pretty girls in newspapers, magazines, etc., posed to display their bodies, and emphasising their sex appeal. **2.** an often pejorative term for sex appeal.

cheesed-off *adj.* **1.** bored; fed up. **2.** irritated; annoyed. Also, **cheesed off, cheesed.**

cheesy[1] *adj.* smelly.

cheesy[2] *adj.* artificial: *a cheesy grin.* [from the request by a photographer to say 'cheese', to produce an apparent smile in his subject]

chequed-up *adj.* See **cashed-up.**

cherry *n.* **1.** *Cricket.* a cricket ball, especially a new one. **2.** virginity. **3.** a virgin woman. *—phr.* **4. two bites of** or **at the cherry,** two attempts.

cherry picker *n.* **1.** a crane, especially one mounted on a truck, with an enclosed platform at one end, designed to lift people to a height where they can perform such functions as changing streetlights, checking stores, lopping trees, etc. **2.** *Film Industry.* an extension arm with a hydraulic movable platform which is used instead of a camera boom.

chest *phr.* **get (something) off one's chest,** to bring into the open (a pressing worry).

chestnut *n.* an old or stale joke, anecdote, etc.: *Not that hoary old chestnut again!*

chevoo *n.* a party or spree; shivoo.

chew *n.* **1.** *N.Z.* (*in children's speech*) a sweet, a lolly. —*phr.*

2. chew (someone's) ear, to talk to (someone) insistently and at length.

3. chew the fat or **rag,** to gossip.

chewie *n.* **1.** chewing gum. —*phr.* **2. chewie on your boot! a.** *Australian Rules.* a cat-call, intended to disconcert a football player kicking for a goal. **b.** an exclamation of contempt for the tardiness of another contestant in a race.

chew-'n'-spew *n.* **1.** *Prison.* prison stew. **2.** any ready cooked snack or meal, such as hamburgers, fish and chips, etc. Also, **chew and spew.**

chiack *v.* **1.** to jeer; taunt; deride; tease. —*n.* **2.** jeering cheek. Also, **chyack.** [British dialect *chi-hike* a salute, exclamation]

chick *n.* **1.** a young girl. —*phr.* **2. pull a chick,** to seduce a girl or woman.

chicken *n.* **1. a.** a young person, especially a young girl. **b.** (*usually with a negative*) See **spring chicken. 2.** a coward. —*adj.* **3.** cowardly. —*phr.*

4. chicken out, to withdraw because of cowardice, tiredness, etc.

5. count one's chickens before they are hatched, to act on an expectation which has not yet been fulfilled.

6. play chicken, a. to perform a dangerous dare. **b.** (of a person) to stand in the path of an approaching vehicle daring the driver to run him down. **c.** (of the drivers of two vehicles) to proceed along a collision course, as a test of courage.

chickenfeed *n.* **1.** a meagre or insignificant sum of money. **2.** anything or anybody insignificant.

chicken-hearted *adj.* timid; cowardly.

chief *n.* a boss.

chillybin *n.* a portable icebox; an esky. Also, **chillibin.**

chime *phr.* **chime in,** to break suddenly into a conversation, especially to express agreement.

chimneypot *n.* a top hat.

chin *v.* **1.** to talk. —*phr.*

2. keep one's chin up, to remain cheerful, especially under stress.

3. take it on the chin, to take suffering or punishment stalwartly.

china *n.* a mate; friend. [rhyming slang *china plate* mate]

chinaman *phr.* **kill a chinaman, a.** to put a jinx on oneself, as by killing a Chinaman. **b.** to commit a petty sin or offence.

Chinaman's luck *n.* infallible good luck.

chin-chin *n.* **1.** goodbye. **2.** (a toast, as in drinking to someone's health).

Chinee *n.* a Chinese. [by back-formation]

Chink *n.* (*sometimes l.c.*) (*offensive*) a Chinese. Also, **Chinkie.**

chinwag *n.* **1.** a chat; a conversation. —*v.* **2.** to gossip: *chinwagging over the back fence.*

chip *v.* **1.** to reprimand: *He chipped me for being late.* —*phr.*
2. cash in one's chips, See **cash.**
3. chip in, a. to contribute money, help, etc. **b.** to interrupt; enter uninvited into a debate or argument being conducted by others.
4. chip off the old block, a person inheriting marked family characteristics.
5. chip on the shoulder, a grudge.
6. have had one's chips, to have lost one's opportunity.
7. spit chips, See **spit.**
8. the chips are down, the moment of decision has been reached.

chipper *adj.* lively; cheerful. [? British dialect *kipper* frisky]

chippie *n.* a carpenter.

chips *pl.n.* money.

Chips, the *n.* a daytime train running between Sydney and the Blue Mountains. See **Fish.**

chisel *v.* **1.** to cheat; swindle. **2.** to get by cheating or trickery.

chizz *n. British.* a swindle; an unfair or unlucky event.

Chloe *phr.* **drunk as Chloe,** very drunk.

chock *phr.* **chock of,** fed up with: *I'm chock of this party, let's go home.*

chock-a-block *adj.* **1.** full; overcrowded. —*adv.* **2.** in a jammed or crowded condition.

chocker *adj.* **1.** full; replete. **2.** intoxicated. Also, **chockers.**

chockie *n.* chocolate. Also, **choc, chokkie.**

choco[1] *n.* a member of the Australian militia or a conscripted soldier in World War II. Also, **chocko.** [abbreviation *chocolate soldier,* a reluctant soldier, from the operetta 'The Chocolate Soldier' by Oscar Strauss]

choco[2] *n.* (*offensive*) a member of a dark-skinned race. Also, **chocko.**

chocolate-box *adj.* gaudily pretty.

chocolate drop *n. British.* a negro.

chocolate frog[1] *n. Prison.* an informer. [rhyming slang *chocolate frog* dog: an informer]

chocolate frog[2] *n.* wog. [rhyming slang]

choice *adj.* **1.** fantastic; wonderful. —*phr.* **2. choice language,** colourfully vulgar language.

chokeman *n. Prison.* a thug.

choker *n.* a cravat or high collar.

choky *n.* prison.

chompers *pl. n.* the teeth.

choof *phr.* **choof off,** See **chuff off.**

chook *n.* **1.** a woman: *silly old chook.* **2.** a silly person. —*phr.*
3. See **headless chook.**
4. hope your chooks turn to emus and kick your dunny down, a general term of abuse or derision.

5. **how are your mother's chooks?** a general expression of greeting. Also, **duck.** [from *chook* a domestic fowl]

chook's bum *n.* the mouth.

chop[1] *n.* **1.** a share, cut: *in for one's chop.* **2.** an accented strum on a guitar. **3.** one's deathblow: *He got the chop.* **4.** the sack; dismissal. *—v.* **5.** to dismiss; give the sack to; fire. **6.** to criticise. *—phr.* **7. chop off,** to finish suddenly.

chop[2] *phr.* **in the chops,** in the mouth. Also, **chap.**

chop[3] *phr.* **not much chop,** no good. [Hindi *chhāp* impression, stamp]

chop chop *interj.* hurry up. [Pidgin English *chop* quickly]

chopper[1] *n.* **1.** *Tennis.* a type of handgrip on a racket used when serving a ball with a spin. **2.** a helicopter.

chopper[2] *n.* a Roman Catholic. [See **rock chopper**]

chop-up *n.* a dividing of goods.

chow *n.* food.

Chow *n.* (*offensive*) a Chinaman. [Pidgin English *chow-chow*; origin uncertain]

Chrissie *n.* Christmas.

Christ *interj.* an exclamation indicating surprise, indignation, etc. Also, **Christ Almighty.**

christen *v.* to make use of for the first time: *We can christen the new cups tonight.*

christening *phr.* **like a moll/prostitute/gin at a christening,** out of place; ill at ease.

Christmas *phr.*

1. **as regular as Christmas,** regularly.

2. **done up like a Christmas tree,** overdressed; garish.

3. (*pl.*) **have all one's Christmases come at once,** to have extreme good fortune.

4. **think one is Christmas,** to be pleased with oneself; be elated.

Christmas hold *n.* *Wrestling, etc.* a hold in which one grabs the opponent's testicles. [from *nuts* testicles; a Christmas hold is a handful of nuts]

chromo *n.* a prostitute.

chronic *adj.* very bad; deplorable. Also, **chronical.**

chrysie *n.* a chrysanthemum.

chuck *v.* **1.** to do; perform, usually with some flamboyance: *chuck a U-ie, chuck a willy, chuck a mental.* *—phr.*

2. **chuck in,** to resign from: *He's chucked in his job.*

3. **chuck it in,** to desist; give up (something begun) without finishing: *There were so many problems with the job that finally she just chucked it in.*

4. **chuck off at,** to speak sarcastically or critically about.

5. **chuck one's hand in,** to give up; refuse to go on.

6. **chuck one's weight about,** to be overbearing; interfere forcefully and unwelcomely.

7. **chuck out,** to eject: *They chucked him out of the nightclub.*

8. **chuck up,** to vomit.

9. **the chuck,** dismissal.

chucker-out *n.* one employed at a place of public entertainment to eject undesirable persons; bouncer. Also, **chucker-outer.**

chucklehead *n.* a blockhead; fool.

chuddy *n.* chewing gum. Also, **chuttie.**

chuff[1] *phr.* **chuff off,** to go away. Also, **choof.**

chuff[2] *n.* the buttocks: *Get off your chuff, mate.*

chuffed *adj.* pleased; delighted: *She was really chuffed about her exam results.*

chug-a-lug *v.* **1.** to down a drink quickly without pause. —*n.* **2. a.** one long drink swallowed without pause, often at a special social occasion. **b.** a drinking bout. —*interj.* **3.** a term used as an encouragement to drink in such a manner.

chum *phr.*
1. chum up to, to behave obsequiously towards.
2. chum up with, to become friendly with.
[from *chum* companion; friend]

chump *n.* **1.** a blockhead or dolt. **2.** the head.

chunder *v.* **1.** to vomit. —*n.* **2.** the act of vomiting. **3.** the substance vomited.

chunderous *adj.* revolting, unpleasant.

churn *phr.* **churn out,** to produce constantly and copiously.

churned up *adj.* upset; agitated: *churned up about the accident.*

chute *n.* **1.** a parachute. —*phr.* **2. in the chute** or **pipeline,** on the way; in preparation.

chutney barrel *n.* a toilet, especially a sanitary can.

chuttie *n.* chewing gum. Also, **chutty.**

ciao *interj.* **1.** goodbye. **2.** hello. [Italian, alteration of *schiavo* at your service]

cigarette swag *n.* a swag rolled into a long, thin shape like a cigarette.

ciggie *n.* a cigarette. Also, **cigger, cig.**

cinch *n.* something certain or easy.

circs *pl.n.* circumstances.

cissy *n.* See **sissy. Also, cissie.**

city slicker *n.* an often flashily dressed and superficially knowing person who shows considerable adroitness in dealing with a city environment.

civvies *pl. n.* civilian clothes (opposed to *military dress*).

civvy *n.* a civilian (as opposed to a member of the armed forces).

civvy street *n.* civilian life (as opposed to life as a member of the armed forces).

clackers *pl. n.* false teeth.

clag *n. Aeronautics.* clouds.

clam *n.* **1.** a secretive or silent person. —*phr.*
2. clam up, to be silent.
3. clam up on, to refuse to talk to (someone).
[from *clam* a bivalve mollusc with a shell which clamps tightly shut]

clambake *n.* **1.** any social gathering, especially a very cheerful one. **2.** a bungled or unsuccessful performance or rehearsal.

clancy *n.* **1.** (in the petrol industry) an overflow or spillage of petrol, as when filling a tanker. **2.** an overflow, as water, etc. [from the poem *Clancy of the Overflow* by A.B. (Banjo) Paterson]

clanger *n.* a glaring error or mistake, as an embarrassing remark: *to drop a clanger.*

clap[1] *phr.*
1. clap eyes on, to catch sight of.
2. clap on, to increase: *to clap on speed.*
3. clap out, to break down; cease to function: *The car has clapped out.*

clap[2] *n.* gonorrhoea, or any other venereal disease. [Middle English *clapier* brothel]

clapped-out *adj.* **1.** exhausted; weary. **2.** broken; in a state of disrepair.

clapper *n.* the tongue.

claret *n.* blood. [from *claret* red wine]

class *n.* acceptable style in dress or manner: *a person of class.*

classy *adj.* of high class, rank, or grade; stylish; fine: *classy lady.*

claw-hammer coat *n.* a tail coat.

claytons *adj., adv.* an imitation or substitute: *This is a claytons government. Their love affair went claytons, but then picked up again.* [from *Claytons* tradename of a non-alcoholic drink which was advertised as 'the drink you have when you're not having a drink']

clean *adj.* **1.** free from addiction to drugs. *—phr.*
2. clean out, a. to use up; exhaust. **b.** to take all money from, especially illegally: *to clean out the bank.*
3. clean up, a. to make (money, or the like) as profit, gain, etc.: *to clean up at the races.* **b.** *Sport, etc.* to defeat crushingly: *Carlton cleaned up Richmond last Saturday.*
4. come clean, to make a full confession.
5. keep it clean, to stop using obscene language or making indecent jokes.

cleaner *phr.* **take (someone) to the cleaners,** to strip (someone) of all assets, money, etc., usually in gambling.

cleaner upper *n.* (*humorous*) the person responsible for cleaning or tidying up.

clean potato *n.* **1.** (formerly) a convict set free. **2.** a law-abiding person.

cleanskin *n.* one who is free from blame, or has no record of police conviction.

cleanskin nobby *n.* an opal found in an abandoned mine, which has been overlooked by previous miners. Also, **cleanskin.**

clean-up *n.* a very large profit.

clear *phr.*
1. clear as mud, not clear; confused.
2. clear off, to disappear; vanish.
3. clear out, to go away.
4. clear the air, a. to relieve tension. **b.** to remove misunderstanding.
5. clear the decks, to make room or prepare for action of any kind
6. clear with, to obtain approval: *to clear with the boss.*
7. in the clear, free from the imputation of blame, censure, or the like: *The verdict put her in the clear.*

clever *adj.* **1.** superficially smart or bright; facile: *Don't get clever with me, mate.* **2.** sly; cunning: *Watch him, he's a clever piece of work.*

clever dick *n.* a conceited, smug person, who displays his prowess at the expense of others.

click *n.* **1.** (*pl.*) Also, **klick.** kilometres. — *v.* **2.** to make a success; make a hit. **3.** to fall into place or be understood: *His story suddenly clicked.* **4.** to establish an immediate affinity, usually with a member of the opposite sex: *They clicked right away.*

clicker *n.* **1.** a foreman in a printing works. **2.** a foreman shoemaker.

climb *phr.* **climb down,** to withdraw from an untenable position; retract an indefensible argument.

clinah *n. Obsolete.* girl friend. Also, **cliner.** [German *kleine* little]

clinch *n.* an embrace or passionate kiss.

clink *n.* a prison; gaol. [apparently from *Clink* Prison in Clink Street, Southwark, London]

clinker *n.* something first-rate or worthy of admiration.

clip *v.* **1.** to hit with a quick, sharp blow. **2.** to defraud. — *n.* **3.** a quick, sharp blow or punch. **4.** rate; pace: *at a rapid clip.* **5.** general appearance; looks.

clipjoint *n.* a nightclub or restaurant, etc., where prices are exorbitant and customers are swindled.

clipper *n.* a first-rate person or thing.

clippie *n. Chiefly British.* one whose job involves clipping tickets, as a tram conductress, a railway employee at a platform barrier, etc.

clipping *adj.* swift: *a clipping pace.*

cliquey *adj.* snobbishly exclusive; cliquish. Also, **cliquy.**

clit *n.* the clitoris.

clobber[1] *v.* to batter severely; maul: *Say that again and I'll clobber you.* [? blend *club* and *slobber*]

clobber[2] *n.* clothes or gear: *Sunday clobber.* [alteration of *clothes*]

clock[1] *n. Prison.* a prison sentence of twelve months' duration.

clock[2] *v.* to hit; strike. Also, **clonk.** [?echoic]

clock bird *n.* the kookaburra.

clocker *n. Horseracing.* a timekeeper.

clock-watcher *n.* an employee who spends a lot of time thinking about the end of the working day.

clod *n.* a stupid person; a blockhead; dolt.

clonk *v.* **1.** to hit; punch. — *n.* **2.** a hit; punch.

close call *n.* a narrow escape. Also, **close shave, close thing.**

closed doors *phr.* **behind closed doors,** in secret.

closed shop *n.* a restricted operation or group of people. [from *closed shop,* a workshop, factory, etc., where only trade union members may be employed]

close shave *n.* a narrow escape.

closet *adj.* secret: *a closet drinker, a closet queen.*

clot *n.* a stupid person.

clothes horse *n.* a person who pays particular attention to dress and who wears clothes well, especially a model or mannequin.

clottish *adj.* foolish; silly.

cloud *phr.*

1. have one's head in the clouds, to be divorced from reality; be in a dreamlike state.

2. on cloud nine, in a state of bliss.

cloud-cuckoo-land *n.* a fanciful place of unrealistic notions. [from Aristophones' *The Birds*]

cloudland *n.* a region of unreality, imagination, etc.; dreamland.

clout *n.* **1.** a blow, especially with the hand; a cuff. **2.** power; force; effectiveness: *The committee has no political clout.* —*v.* **3.** to strike, especially with the hand; cuff. **4.** to cheat by evading payment, especially of a gambling debt; to welsh. —*phr.*

5. clout down on, to clamp down on.

6. clout on, to steal, or take without permission.

cloven hoof *phr.* **show the cloven hoof** or **foot,** exhibit a base motive. [a reference to the cloven-footed devil]

clover *phr.* **in clover,** in comfort or luxury.

clubbie *n.* a member of an organised surfboard riding club.

clucky *adj.* **1.** feeling disposed to have children. **2.** fussy and overprotective of children. [from *clucky* (of a hen) broody]

clue *phr.*

1. clue (someone) up, to give (someone) the facts.

2. not to have a clue, to be ignorant, inexperienced.

clued-up *adj.* well-informed.

clueless *adj.* helpless; stupid; ignorant.

cluey *adj.* **1.** well-informed. **2.** showing good sense and keen awareness.

clump *n.* **1.** *British.* a blow; a clout. —*v.* **2.** *British.* to strike; punch.

clumsy duck *n.* *Law.* an application made to an officer of the court, usually the attorney general, that a bill of indictment not be signed; no bill application.

clutch *n.* a group; a bunch.

Clydesdales, the *pl. n.* the Toowoomba team in the Queensland Rugby Football League.

coal *phr.*

1. coals to Newcastle, anything supplied unnecessarily.

2. add coals to the fire, to make a bad situation, dissension, etc., worse.

3. take/haul/rake/etc., over the coals, to scold; reprimand.

coalface *phr.* **work at the coalface,** to play an active rather than an administrative part in a project, organisation, etc.

coast *v.* **1.** (formerly) to loaf; move about the country as a swagman. —*phr.*

2. coast along, to act or perform with minimal effort.

3. the coast is clear, the danger has gone.

coaster *n.* **1.** a lazy person; loafer. **2.** *Obsolete.* a sundowner. **3.** *N.Z.* a person from the west coast.

coat *phr.*

1. on the coat, *Prison.* ostracised.

2. pull the coat, a signal to avoid (someone).

Coat-Hanger, the *n.* the Sydney Harbour Bridge.

cobar *n.* copper coinage. [named after *Cobar* a copper-mining town in western N.S.W.]

Cobar shower *n.* a dust storm. [named after *Cobar* a town in western N.S.W.]

cobber *n.* **1.** a mate; friend. —*phr.* **2. cobber up with,** *N.Z.* to make friends with. [origin uncertain, probably related to British dialect *cob* to form a friendship with]

cobblers[1] *pl. n.* **1.** balls, testicles. **2.** nonsense; rubbish. *—phr.* **3. a load of old cobblers,** a lot of nonsense. [rhyming slang *cobbler's awls,* balls]

cobblers[2] *pl. n.* legs. [rhyming slang *cobblers pegs*]

Cobras, the *pl. n.* the Camberwell team in the Victorian Football Association.

coca cola *n. Cricket.* a bowler. [rhyming slang, from the name of the drink]

cock[1] *n.* **1.** a mate, friend or fellow. **2.** the penis. *—phr.*

3. cock a snook or **snoot,** to put a thumb to the nose, in a contemptuous gesture.

4. cock of the walk, one who asserts himself domineeringly, as the leader of a gang.

5. hot cock, nonsense.

6. cock up, to make a mess of; ruin: *You really cocked that up.*

cock[2] *n.* nonsense. [short for *poppycock*]

cockabully *n. N.Z.* (*especially in children's speech*) any of several kinds of small freshwater fish. [? Maori *kokopu*]

cock-a-hoop *adj.* in a state of unrestrained joy.

cock-and-bull story *n.* an absurd, improbable story told as true.

cockatoo[1] *n.* one who keeps watch during a two-up game, or other illegal activity. [from *cockatoo* an Australian crested parrot, known for its habit of posting 'sentries' to noisily warn the feeding flock of any approaching danger]

cockatoo[2] *n.* (formerly) a convict from Cockatoo Island, Port Jackson. Also, **Cockatoo Islander.**

cockatoo farmer *n.* a farmer, especially one who farms in a small way. Also, **cocky.**

cockatoo fence *n.* a rough fence made of logs and branches. Also, **cockatoo gate.**

cocked hat *phr.* **knock into a cocked hat,** to damage or destroy completely; outdo, overcome, or defeat utterly. [a *cocked hat* is a hat with the brim turned up on two or three sides, common in the 18th century]

cockeye bob *n.* a sudden storm or squall. Also, **cockeyed bob.**

cockeyed *adj.* **1.** twisted or slanted to one side. **2.** foolish; absurd. **3.** drunk. [from *cockeyed* having a squinting eye; cross-eyed]

cockie *n.* a cockroach.

cock sparrow *n.* a conceited little man. [from *cock sparrow* a male sparrow]

cocksucker *n. Chiefly U.S.* a homosexual.

cocksucking *adj. Chiefly U.S.* awful; disgusting (an intensive indicating extreme disapproval).

cockteaser *n.* See **prickteaser.**

cock-up *n.* a mess; a tangle.

cocky[1] *adj.* arrogantly smart; pertly self-assertive; conceited: *a cocky fellow, air, answer.*

cocky[2] *n.* **1.** a cockatoo, or other parrot. **2.** See **cockatoo farmer.** *—v.* **3.** to follow the occupation of a farmer. [abbreviation of *cockatoo*]

cocky boss *n.* a figure of petty authority.

cocky's clock *n.* the kookaburra.

cocky's friend *n.* fencing wire.

cocky's joy *n.* **1.** golden syrup. **2.** treacle. Also, **cocky's delight, bullocky's joy.**

coconut *n. N.Z.* (*offensive*) a Pacific Islander.

cod *n.* (*sometimes pl.*) the scrotum.

codger *n.* **1.** a mean, miserly person. **2.** an odd or peculiar (old) person: *a lovable old codger.* **3.** a fellow; a chap.

codswallop *phr.* **a load of old codswallop,** rubbish or nonsense.

coffin nail *n.* a cigarette.

coin *v.* **1.** to make or gain (money) rapidly. *—phr.*

2. coin a phrase, (*humorous*) to use an acknowledged cliché.

3. coin it, to make a lot of money.

coit *n.* **1.** the anus. **2.** the buttocks. Also, **quoit.**

coke *n.* cocaine.

cold *adj.* **1.** unconscious because of a severe blow, shock, etc.: *knocked cold.* **2.** *Prison.* innocent of a charge laid against one. *—phr.*

3. cold feet, loss of courage or confidence for carrying out some undertaking.

4. (someone) couldn't catch a cold, *Sport.* (a player, etc.) is incompetent or in poor form.

5. in the cold, neglected; ignored.

6. leave one cold, See **leave. 7. throw cold water on,** to dampen enthusiasm of (a person), or for (a thing); discourage: *She threw cold water on the idea.*

coldie *n.* a bottle or can of cold beer.

cold storage *n.* abeyance; indefinite postponement: *to put an idea into cold storage.*

cold-turkey *n.* **1.** unrelieved, blunt, matter-of-fact statement or procedure. **2.** the sudden and complete withdrawal of narcotics as a treatment of drug addiction. *—v.* **3.** to treat (an addict) in this manner.

coldwater man *n. Obsolete.* a teetotaller.

collar *v.* **1. a.** to lay hold of, seize, or take: *They chased and collared the bag-snatcher.* **b.** to get the better of, as in cricket: *to collar the bowling.* **2.** to gain a monopoly over: *to collar the market in wool.* *—phr.* **3. soft collar,** an easy job.

collect *n.* **1.** a winning bet. *—v.* **2.** to run into or collide with, especially in a motor vehicle.

college *n.* a prison.

Collins Street cocky *n.* (in Melbourne) one who owns a country property, often for purposes of tax avoidance, but who lives and works in Melbourne. See also **Pitt Street farmer, Queen Street bushie.** [from *Collins Street* a major street in the city of Melbourne]

collywobble *n.* any small insect.

collywobbles *n.* **1.** stomach-ache. **2.** diarrhoea. [*colic* + *wobbles*]

colonial *phr.*

1. long colonial, a very cold glass of beer.

2. my colonial oath, (*euphemism*) my bloody oath.

combo *n.* **1.** *Northern Australia.* a white man who lives with an Aboriginal woman. *—phr.* **2. go combo,** to begin such a relationship. Also, **wambo, wombo.**

come *v.* **1.** to have an orgasm. **2.** to play the part of: *Don't come the* *great lady with me.* —*n.* **3.** semen. —*phr.*

4. come across, a. to pay or give. **b.** to communicate successfully; to be understood. **c.** (of a woman) to give sexual favours.

5. come again? a request to repeat, expand or explain (a statement, etc.).

6. come clean, to confess.

7. come down, a. to lose wealth, rank, etc.: *He has really come down in the world.* **b.** to travel, especially from a town.

8. come down on, to scold; blame.

9. come in, (of odds on a horse, dog, etc.) to become lower.

10. come in, spinner! (in two-up) toss the coins.

11. come nothing, *Prison.* to make no admissions.

12. come off it, a command to stop; lay aside (a pretentious attitude) etc.): *Come off it, mate!*

13. come one's guts, *Prison.* to confess.

14. come out, a. to admit openly one's homosexuality. **b.** *Obsolete.* to leave Europe or America to make a home in Australia.

15. come out on, to declare support for: *to come out on abortion.*

16. come out with, to blurt out.

17. come over, to happen to; affect: *What's come over him?*

18. come round, a. to relent. **b.** to change (an opinion, direction, etc.).

19. come that on, to attempt to hoodwink (someone) with an argument, device, etc., which is blatantly a deception: *Don't come that on me.*

20. come the raw prawn, See **prawn.**

21. come to light with, to produce; supply: *He came to light with the beer at the right moment.*

22. come undone or **unstuck,** to break down; collapse.

23. come up against, to meet difficulties or opposition.

24. come up trumps, to be successful.

comeallyers *pl.n.* songs suitable to sing around a campfire. [from the opening words of many such songs *Come all ye*]

come-at-able *adj.* accessible.

comeback *n.* **1.** a retort; repartee. **2.** a ground for complaint.

comedown *n.* a letdown; disappointment.

come-lately *n.* a new-comer to a situation, social group, area, employment, etc. Also, **Johnny-come-lately, Jill-come-lately, Shirley-come-lately.**

come-on *n.* an inducement; lure: *The free drinks were just a come-on.*

comer *n.* one who or something that is coming on or promising well.

comeuppance *n.* a well-deserved punishment or retribution; one's just deserts. Also, **comeupance.**

comfy *adj.* comfortable.

comic cuts *pl. n.* guts. [rhyming slang]

common-or-garden *adj.* ordinary.

compo[1] *n.* **1.** compensation for injury at or in connection with a person's work; workers' compensation. —*phr.* **2. on compo,** in receipt of such payment. Also, **comp.** [short for *compensation*]

compo[2] *n.* any of various combined substances such as plaster or mortar, made by mixing ingredients. [short for *composition*]

con[1] *adj.* **1.** confidence: *con game, con man.* —*n.* **2.** a confidence trick; swindle. —*v.* **3.** to swindle; defraud. **4.** to deceive with intent to gain some advantage.

con[2] *n.* a convict.

concern *n.* any material object or contrivance: *fed up with the whole concern.*

concertina *n.* **1.** a sheep with a very wrinkled skin. **2.** a side of lamb.

conchie *n.* **1.** a conscientious objector. **2.** one who is overconscientious. —*adj.* **3.** overconscientious. Also, **conch, conchy, conshie, conshy.** [short for *conscientious*]

confab *n.* **1.** confabulation; discussion. —*v.* **2.** to confabulate; converse.

confetti *phr.* **cowyard/ farmyard/Flemington confetti,** rubbish, nonsense: *talking cowyard confetti.*

confounded *adj.* damned: *a confounded lie.*

conk *n.* **1.** the nose. **2.** a blow; a violent stroke. —*v.* **3.** to hit or strike, especially on the head. —*phr.* **4. conk out, a.** (of an engine) to break down. **b.** to faint; collapse. **c.** to die. Also, **konk.** [probably an alteration of *conch* a spiral shell]

con man *n.* one who swindles by gaining the victim's confidence and then inducing the victim to part with property or money; confidence man.

connect *v.* to understand or see a connection: *I didn't connect until I saw her photo in the newspaper.*

connie *n.* **1.** a conductress, usually on trams. **2.** a type of marble.

conniptions *pl. n.* a fit of hysteria or anger: *Don't have conniptions, it's only a game.*

constabulary *n.* (*in jocular use*) the police.

contemplate *phr.* **contemplate one's navel,** to indulge in introspection; daydream.

continental *n.* the least bit; a damn: *I don't give a continental.* [from *continental* a piece of paper money issued by the Continental Congress during the War of American Independence]

contract *n.* **1.** an agreement to kill a nominated person or persons: *There is a contract out on Slippery Sam.* **2.** any difficult job.

cooee *phr.*
1. not within cooee, far from achieving a given goal: *He is not within cooee of finishing the job by Friday.*
2. within cooee, a. within calling distance. **b.** close to achieving a given goal. [from *cooee* a prolonged clear call used most frequently in the bush as a signal to attract attention; Aboriginal]

cook[1] *v.* **1.** to concoct; invent falsely; falsify: *cook the books.* **2.** to ruin; spoil: *that's cooked it.* —*phr.* **3. cook one's goose,** to frustrate or spoil one's plans.

cook[2] *n.* a look. [rhyming slang *Captain Cook* look]

cookie *phr.*
1. a smart cookie, a particularly intelligent person.
2. that's the way the cookie crumbles, that's how things are. Also, **cooky.**

cooking *phr.*
1. be cooking, in full swing: *That band is really cooking now.*

2. be cooking with oil or **gas,** to be in control of the situation; to begin to achieve some results or understanding: *Now we are cooking with gas!*

3. what's cooking?, a request to be told what is happening.

cool *adj.* **1.** (of a number or sum) without exaggeration or qualification: *a cool thousand.* **2.** smart; up-to-date; fashionable. **3.** all right; okay: *Don't worry, it's cool.* **4.** attractive; excellent. *—phr.*

5. cool it, stop doing (something); take it easy.

6. cool off or **down,** to become calmer; to become more reasonable.

7. cool one's heels, See **heel.**

8. keep one's cool, to remain calm and self-controlled.

9. lose or **blow one's cool,** to become angry.

cooler *n.* a prison.

coon *n.* (*offensive*) a dark-skinned person.

coop *n.* **1.** a prison. *—phr.* **2. fly the coop,** to escape from a prison, etc. [from *coop* an enclosure, cage or pen]

coot *n.* **1.** a fool; simpleton. **2.** a man.

cootie *n.* the head-louse.

cop *n.* **1.** a policeman. **2.** a silent cop; traffic dome. **3.** an arrest; a state of being caught. **4.** something advantageous, or profitable. *—v.* **5.** to steal. **6.** to receive in payment. **7.** to accept resignedly; put up with: *Would you cop a deal like that?* **8.** to be allotted, receive: *He copped more than his fair share. —phr.*

9. a sure cop, something certain.

10. a sweet cop, an easy job.

11. cop a load, to contract venereal disease.

12. cop it, to be punished.

13. cop it sweet, a. to endure or put up with an unpleasant situation. **b.** to have a lucky break.

14. cop out, a. to opt out of (something). **b.** to fail completely.

15. cop the lot, to bear the brunt of some misfortune.

16. cop this! Also, **cop this** or **that lot!** look at this! (implying **a.** admiration. **b.** astonishment. **c.** contempt.) **17. not much cop,** not worthwhile.

[Old English *coppian,* lop, steal]

cop-out *n.* a way out of a situation of embarrassment or responsibility: *Her going overseas was a bit of a cop-out.*

copper *n.* a policeman. [from *cop*]

copper-bottomed *adj.* sound, especially financially sound.

copper's nark *n.* a police informer.

coppertail *n.* a working-class person (as opposed to a *silvertail*).

coppertop *n.* a red-headed person. Also, **coppernob.**

cop shop *n.* a police station.

copter *n.* a helicopter.

copybook *phr.* **blot one's copybook,** to spoil, damage or destroy one's reputation or record.

cor *interj.* an exclamation of surprise.

cords *pl. n.* corduroy trousers.

cork *phr.* **put a cork in it,** to be quiet; cease to talk. [from *cork* a stopper for a bottle, etc.]

corked *adj.* drunk.

corker *n.* **1.** something striking or astonishing. **2.** something very good of its kind.

corn[1] *n.* a trite or sentimental writing or style.

corn[2] *phr.* **on the corn,** in prison. [a reference to the type of food served in prison]

cornball *n.* one who is corny; a sentimentalist.

corner *n.* **1.** an awkward or embarrassing position, especially one from which escape is impossible. *—phr.*

2. corner a wave, *Surfing.* to stay on the shoulder of the wave.

3. cut corners, a. to take short cuts habitually. **b.** to bypass an official procedure, or the like.

4. get one's corner, to obtain one's share.

5. the Corner. Also, **Corner country.** the land where the borders of Queensland, South Australia and New South Wales meet.

6. turn the corner, to begin to get well; improve.

corns *phr.* **tread on someone's corns,** to hurt someone's feelings. [from *corn* a painful callosity on the toes or feet]

cornstalk *n.* **1.** a person native to or resident in New South Wales. **2.** a tall, thin person. **3.** *Obsolete.* a native-born Australian, being generally taller and thinner than the immigrant.

corny *adj.* **1.** old-fashioned; lacking subtlety. **2.** sentimental; mawkish and of poor quality.

corporation *n.* the abdomen, especially when large and prominent.

corpse *v.* **1.** to collapse or fall asleep from exhaustion, alcohol, drugs, etc. *—phr.* **2. corpse reviver,** See **reviver.** [from *corpse* a dead body, usually of a human being]

corpus delicti *n.* (*humorous*) a shapely young woman.

corroboree *n.* a meeting or gathering, especially if large and noisy. [Aboriginal; an assembly]

cossie *n.* a swimming costume. Also, **cozzie.**

cot *phr.* **hit the cot,** to go to bed.

cot case *n.* (*humorous*) someone who is exhausted, drunk, or in some way incapacitated, and fit only for bed.

cottage *n.* **1.** a public lavatory frequented by male homosexuals. *—v.* **2.** (of a male homosexual) to frequent public lavatories in search of casual sex: *go cottaging.*

cottager *n.* a male homosexual who frequents public lavatories.

cotton *v.* **1.** to make friends. **2.** to get on together; agree. *—phr.*

3. cotton on, to understand; perceive a meaning or purpose.

4. cotton to or **with,** to become attached or friendly.

cotton-picking *adj. Originally U.S.* unworthy; simple: *out of one's cotton-picking mind.*

cottonwool *n.* **1.** a protected and comfortable state or existence. *—adj.* **2.** protected and comfortable.

Cougars, the *pl. n.* the Kilsyth team in the Victorian Football Association.

cough *phr.* **cough up,** to give; hand over: *to cough up (with) the money.*

cough drop *n. British.* a fool, simpleton.

count *phr.*

1. count in, to include.

2. count out, to exclude: *Count me out.*

3. out for the count, a. completely exhausted. **b.** unable to continue an activity.

counter *phr.* **under the counter, a.** clandestine or reserved for favoured customers. **b.** in a manner other than that of an open and honest business transaction; clandestinely and often illegally. [from *counter* a table or board on which money is counted, business is transacted, or goods are laid for examination]

counterjumper *n.* a salesman at a counter.

country *n. Sport.* any part of the ground on which a sporting event takes place which is far from the main area of activity, as the outfield in cricket, or the part of the course away from the stands in horseracing.

country cousin *n.* a person, sometimes a relative, from the country to whom the sights and activities of a large city are novel and bewildering.

country dunny *phr.* **all alone like a country dunny,** See **alone.**

court *phr.* **out of court,** out of the question; not to be considered.

Cousin Jack *n.* **1.** a Cornishman. **2.** (formerly) a Cornish miner in South Australia.

couth *adj.* civilised, well-mannered. [back formation from *uncouth* uncivilised]

cove *n.* **1.** a man: *a rum sort of cove.* **2.** a boss, especially the manager of a sheep station. **3.** (formerly) the master of a house or shop. **4.** (formerly) a convict overseer. [said to be from Romany *kova* creature]

cover-up *n.* an attempt at concealment: *A government cover-up on the deteriorating state of the economy.*

cow *n.* **1.** an ugly or bed-tempered woman: *mean old cow.* **2.** a term of abuse: *You miserable cow. —phr.*

3. a cow of a (something or **someone),** a difficult, unpleasant, disagreeable (thing or person): *That's a cow of a thing to say.*

4. a fair cow, anything regarded as disagreeable or difficult.

5. chase or **hunt up a cow,** to find a dry spot outdoors, usually with sexual intentions.

6. poor cow, (expressing sympathy) unfortunate person.

7. till the cows come home, for a long time; for ever: *to talk till the cows come home.*

cowal *n.* a swampy depression in the outback, especially in red sand country, usually shaded by low growing trees or shrubs.

coward's castle *n.* parliament when used as an arena in which to vilify and abuse others while under parliamentary privilege.

cowardy custard *n.* (*in children's speech*) a coward: *cowardy, cowardy custard.*

cowbang *v.* to run a dairy farm. Also, **cowspank.**

cow cake *n.* a cow pat; cow dung.

cow cocky *n.* a dairy farmer. Also, **cowbanger, cowspanker.**

cow juice *n.* milk.

cowpuncher *n. U.S.* a cowboy.

cow shot *n.* (in cricket) a stroke made without style or discrimination.

cowspank *v.* See **cowbang.**

cowyard confetti *n.* See **confetti.**

coz *n.* a cousin.

cozzie *n.* See **cossie.**

crab *v.* **1.** to find fault with. **2.** to spoil. **3.** to impede; impair. *—phr.* **4. draw (the) crabs,** to attract the unwelcome attention of an authority, as the police.

crabby *adj.* irritable.

crabs *pl. n.* body lice, particularly of the pubic area, which cause severe itching.

crabstick *n.* an ill-tempered person.

crack *v.* **1.** to break into (a safe, vault, etc.). **2.** to solve (a mystery, etc.). **3.** to obtain: *crack an invite.* **4.** to open and drink (a bottle of wine, etc.): *Let's crack a bottle on that news!* **5.** to give way to pressure. *— n.* **6.** (of a prostitute) the act of having sexual intercourse with a client. **7.** a try; an opportunity or chance: *I'd like a crack at that job.* **8.** the anus. **9.** a joke; gibe. **10.** a moment; instant: *He was on his feet again in a crack.* **11.** a burglary. **12.** the female pudendum. **13.** a prostitute's customer. *—adj.* **14.** of superior excellence; first-rate: *a crack rider.* *—phr.*

15. crack a fat, to have an erection.

16. crack a fruity or **shitty,** to lose one's temper.

17. crack hardy or **hearty,** to endure with patience; put on a brave front.

18. crack it, a. to have or offer sexual intercourse, especially as a prostitute. **b.** to be successful.

19. crack onto, to establish a close relationship with (someone); chat up.

20. crack up, a. to suffer a physical, mental or moral breakdown. **b.** to praise; extoll. **c.** to break into uncontrollable laughter.

21. cracked up to be, (something) as described; as it is supposed to be: *It isn't all it's cracked up to be.*

22. fair crack of the whip, a. a chance. **b.** an appeal for fairness.

crackbrain *n.* an insane person.

cracked *adj.* mentally unsound.

cracker *n.* **1.** something which has a particular quality in a high degree: *This pace is a cracker.* **2.** a prostitute. *—phr.*

3. not to have a cracker, to be without money.

4. not worth a cracker, (something) of little worth.

crackerjack *n.* **1.** a person of marked ability; something exceptionally fine. *—adj.* **2.** of marked ability; exceptionally fine. Also, **crackajack.**

cracker night *n.* formerly the night of Empire Day (24 May) which was traditionally celebrated with fireworks, now generally held on the Queen's Birthday holiday weekend; bonfire night.

crackers *adj.* insane; crazy.

cracking *adj.* **1.** first-rate; fine; excellent. **2.** fast; vigorous: *a cracking pace.* *—phr.*

3. cracking it, working as a prostitute.

4. get cracking, to start an activity, especially energetically.

crackpot *n.* **1.** an eccentric or insane person. *—adj.* **2.** eccentric; insane; impractical.

cracksman *n.* a burglar.

crack-up *n.* a breakdown in health.

cradle snatcher *n.* one who shows romantic or sexual interest in a much younger person.

crammer *n.* a lie.

cramp *phr.* **cramp one's style,** to be hindered from showing one's best abilities, etc.

crane *v.* to hesitate at danger, difficulty, etc.

crank *n.* **1.** an eccentric person, or one who holds stubbornly to eccentric views. **2.** an eccentric notion. *—phr.* **3. crank up,** to build up; put into motion.

crap *n.* **1.** excrement. **2.** nonsense; rubbish. **3.** junk; odds and ends. *—v.* **4.** to defecate. **5.** to make a mess of; bungle. *—phr.*

6. crap off, to annoy; disgust.

7. crap on, to talk nonsense.

8. crap out, to fail.

crapper *n.* a toilet.

crash *v.* **1.** to come uninvited to (a party, etc.). **2.** to enter without buying a ticket: *to crash the gate.* **3.** to defecate. **4.** to collapse or fall asleep with exhaustion. *—n.* **5.** the act of defecating. *—adj.* **6.** characterised by all-out, intensive effort, especially to meet an emergency: *a crash program; a crash diet.* *—phr.* **7. crash through or crash,** make a do-or-die attempt.

crash cart *n. Medicine.* a mobile trolley fitted for emergencies (especially cardiac arrest) and wheeled to hospital wards as needed.

crash-hot *adj.* excellent. Also, **crash hot.**

crashing *adj.* complete and utter: *a crashing bore.*

crate *n.* a motor vehicle, aeroplane, or the like, especially a dilapidated one.

craw *phr.* **stick in one's craw,** to be irritating or annoying.

crawler *n.* **1.** an abject flatterer. **2.** a person who is slow, or lazy, or unfit to work.

craze *n.* a mania; a popular fashion, usually short-lived.

crazy *adj.* **1.** intensely enthusiastic or excited: *I'm absolutely crazy about the idea.* *—n.* **2.** an insane person: *Don't mind her, she's a crazy.*

cream *v.* **1.** to beat up in a fight. *—n.* **2.** semen. *—phr.*

3. cream one's jeans, a. to have an orgasm while dressed. **b.** to become extremely excited.

4. cream on the bun, something extra; a luxury item.

creamed *adj.* **1.** beaten up in a fight. **2.** *Surfing.* wiped off one's surfboard by a wave.

creamer *n.* a coward.

creamie *n.* (*offensive*) a quarter-caste aborigine.

cream puff *n.* a weak person; an effeminate person; a cissy.

crease *v.* to strike (a person).

creek *phr.*

1. up the creek or **up shit creek,** in a predicament; in trouble. Also, **up the** or **shit creek in a barbed wire canoe; up the** or **shit creek without a paddle.**

2. up the creek without a paddle, unmarried and pregnant.

Creek, the *n.* Albion Park racecourse, Brisbane.

creep *n.* **1.** an unpleasant, obnoxious, or insignificant person. *—phr.*

2. make (someone's) flesh creep, to frighten; repel.

3. the creeps, the thrill resulting from an undefined dread.

creepers *pl.n.* thick rubber-soled shoes.

creeping Jesus *n.* a slinking, fawning person.

creepy *adj.* (of a person) unpleasant, obnoxious or insignificant.

creepy-crawly *n.* **1.** an insect. *—adj.* **2.** having or causing a creeping sensation.

crew *n.* a company; crowd: *He is hanging about with a strange crew these days.*

cricket *n.* fair play: *His behaviour was not cricket.* [from the game of *cricket*]

cricket score *n. Prison.* a very long gaol sentence.

crikey *interj.* an expression of surprise; a mild oath. Also, **cricky, crickey.**

crim *n.* **1.** a criminal. *—adj.* **2.** of or pertaining to a crime.

crime *n.* a foolish or senseless act: *It's a crime to have to work so hard.*

cripes *interj.* an expression of amazement, disgust, or the like. [euphemistic variant of *Christ*]

critter *n. U.S.* a creature.

croak *v.* **1.** to die. **2.** to kill.

crobby *adj.* cross; irritated.

crock *n.* **1.** a worn-out, decrepit old person. **2.** an old motor car. [from *crock* an old ewe; an old horse]

crocodile *n.* **1.** a double file of persons, as schoolchildren out for a walk. *— v.* **2.** to swim with another person holding on to one's shoulders. *—phr.* **3. crocodile tears,** false or insincere tears; a hypocritical show of sorrow. [crocodiles are said to shed tears over those they devour]

crook *n.* **1.** a dishonest person; a swindler. *—adj.* **2.** fraudulent. **3.** sick; disabled. **4.** bad; inferior. **5.** unpleasant; difficult: *a crook job.* *—phr.*

6. crook steer, See **steer.**

7. go crook at or **on,** to upbraid noisily.

8. put (someone) crook, to give wrong or bad advice.

crooked *phr.* **crooked on,** angry with.

crookie *n.* any thing or person that proves a failure.

crool *v.* See **cruel.**

cropper *phr.* **come a cropper, a.** to fall heavily, especially from a horse. **b.** to fail; collapse, or be struck by misfortune.

cross *adj.* **1.** illegal; dishonest. *—phr.*

2. cross a bridge when one comes to it, face one's problems when one comes to them.

3. cross as two sticks, very peevish or annoyed. [a pun on the cross formed by two sticks]

4. cross one's t's and dot one's i's, See **dot.**

5. on the cross, dishonestly.

6. when the cross turns over, at the end of a drover's watch, as indicated by the tilting of the Southern Cross from an eastern to a western position in the night sky: *Call me when the cross turns over.*

crosspatch *n.* a cross person.

crotchety *adj.* irritable, difficult, or cross.

crow *n.* **1.** an unattractive woman. *—phr.*

2. as the crow flies, in a straight line.

3. eat crow, to be forced to do or say something very unpleasant or humiliating.

4. land where the crow flies backwards, the remote outback.

5. starve/stiffen/stone the crows, an exclamation of astonishment.

6. the day the crow or **eagle shits,** payday. Also, **crow** or **eagle shit day.**

crowd *v. U.S.* to urge; press by solicitation; annoy by urging: *to crowd a debtor for immediate payment.*

croweater *n.* a South Australian.

crown *v.* to hit on the top of the head: *The ball flew into the crowd and crowned him.*

cruddy *adj.* inferior; unworthy.

cruel *v.* **1.** Also, **crool.** to impair, spoil: *to cruel one's chances. —phr.* **2. cruel one's pitch,** to spoil one's opportunity.

cruet *n.* **1.** the head. *—phr.* **2. do one's cruet,** lose one's head.

cruise *v.* **1.** to go out with a view to picking up a sexual partner. *—phr.* **2. cruise on** or **along,** to maintain a moderate level of activity; lead an uneventful life.

cruiser *n.* **1.** a very large beer glass. **2.** the contents of such a glass.

crumb *n.* **1.** a stupid person: *He's a real crumb. —phr.* **2. crumb bum,** a stupid or worthless person.

crumb act *phr.* **put on the crumb act,** to impose or bludge on someone.

crummy *adj.* very inferior, mean, or shabby.

crumpet *n.* **1.** a girl or woman considered as a sexual object: *a nice bit of crumpet.* **2.** (of a male) sexual intercourse with a woman: *Had any crumpet lately?* **3.** the head: *soft in the crumpet. —phr.* **4. not worth a crumpet,** worthless; of little or no value.

crunch *n.* a moment of crisis. Also, **cruncher.**

crunchie *n.* a pimple.

crush *n.* **1.** an infatuation: *He had a schoolboy crush on his new neighbour. —phr.* **2. crush the price,** to lay out a bookmaker's money on a horse so that the odds shorten.

crusher *n.* a person employed by a bookmaker to crush the price.

crust *n.* **1.** impertinence. **2.** a livelihood: *What do you do for a crust? —phr.*

3. down to the last crust, destitute.

4. the crust, *Prison.* a sentence for vagrancy: *to do the crust.*

crustacean *phr.* **come the uncooked crustacean,** See **prawn.**

crutch *n.* anything relied on, trusted: *Grog is his crutch.* [from *crutch* a staff or support]

cry *phr.*

1. a far cry, only remotely related; very different: *Life in the outback was a far cry from her London childhood.*

2. cry off, to break a promise, agreement, etc.

cry-baby *n.* one given to crying like a baby, or to weak display of injured feeling.

crypto *n.* a person who conceals his allegience to a political group. [from *crypto-* a word element meaning 'hidden']

crystal highway *n.* the name applied to any long stretch of road where windscreen breakages are frequent, as the Sale-Bairnsdale section of the Princes Highway in Victoria, or the highway from Marlborough to Sarina in Queensland. [from the glittering appearance of the shattered glass along the side of the road]

cub *n.* (*humorous*) an awkward or uncouth youth.

cuckoo *adj.* crazy; silly; foolish.

cud *phr.* **chew the cud,** to reflect; meditate.

cudgel *phr.*

1. cudgel one's brains, to think hard.

2. take up the cudgels, to engage in a contest, usually on behalf of another.

[from *cudgel* a club; weapon]

cuff *phr.*

1. off the cuff, impromptu; on the spur of the moment: *to speak off the cuff.*

2. on the cuff, a. on credit. **b.** *N.Z.* excessive; unfair.

cuie *n.* a cucumber. Also, **cu, cuey, cukie.**

cultural cringe *n.* a sense of the inferiority of the culture of one's own country. [a phrase coined by A.A. Phillips to describe the attitude of Australians to the culture of Australia]

culture-vulture *n.* a derisive or humorous term for one who takes excessive interest in the arts.

Cunnamulla gun *n.* a shearer who claims expertise he does not have. [from *Cunnamulla* a town in Queensland]

cunning kick *n.* a secret reserve of money. [*cunning* + *kick* trouser pocket]

cunt *n.* **1.** the female pudendum. **2.** (*offensive*) a woman considered as a sexual object. **3.** (*offensive*) any person. **4.** sexual intercourse.

cunt-hair *n.* *Chiefly Military.* (used in communicating measurements) a very small fraction of the smallest graduation on a scale: *59.2mm and a cunt-hair.*

cunthook *n.* *N.Z.* (*offensive*) an unpleasant or despicable man.

cunt-struck *adj.* infatuated with women.

cup *phr.*

1. in one's cups, intoxicated; tipsy. *—phr.*

2. one's cup of tea, See **tea.**

3. the Cup, the Melbourne Cup (Australia's best-known horse race).

cupboard love *n.* affection inspired by considerations of gain.

cuppa *n.* a cup of tea: *It's time for a cuppa!*

curate's egg *n.* an object or event of mixed quality, or which inspires mixed feelings: *bad (or good) in spots, like the curate's egg.*

curl *v.* **1.** to shrink away, as in horror or disgust: *The sight of blood always makes me curl. —phr.*

2. curl the mo, a. to succeed brilliantly. **b.** an exclamation indicating surprised admiration.

3. curl up one's toes, a. to give in. **b.** to die.
4. make (someone's) hair curl, to cause (someone) to be astonished.

currency *n.* **1.** (formerly) one born in Australia. —*adj.* **2.** (formerly) born in Australia, as opposed to *sterling,* one born in Britain or Ireland. —*phr.* **3. currency lad (lass),** (formerly) a man (woman) born in Australia.

curry *phr.* **give (someone) curry,** to abuse (someone) angrily. [from *curry* a hot tasting Indian mixture of spices]

Curry, the *n.* Cloncurry, a town in north-western Queensland.

curse *phr.* **the curse,** menstruation.

curtain lecture *n.* a private scolding, especially of a man by his wife.

curtains *pl. n.* the end, especially of life.

cush *phr.* **all cush,** *Obsolete.* See **all.** Also, **cush-n-all, cush-n-andy.**

cushy *adj.* easy; pleasant.

cuss *originally U.S.* —*n.* **1.** a curse. **2.** a person or animal: *a queer but likeable cuss.* —*v.* **3.** to curse.

cussed *adj.* **1.** cursed. **2.** obstinate; perverse.

custard-pie *adj.* of or pertaining to slapstick comedy. [from the convention in this type of comedy of throwing pies]

customer *n.* a person one has to deal with; a fellow: *a queer customer.*

cut *v.* **1.** to renounce; give up: *to cut out grog.* **2.** to refuse to recognise socially. **3.** to absent oneself from. **4.** to be hurt or offended by another's actions: *She was really cut because he didn't phone.* —*adj.* **5.** drunk. **6.** diluted; adulterated; impure: *cut heroin.* —*n.* **7.** a share: *His cut was 20 per cent.* **8.** a refusal to recognise an acquaintance. —*phr.*
9. cut a dash, to make an impression by one's ostentatious or flamboyant behaviour or dress.
10. cut and dried, fixed or settled in advance.
11. cut and run, to leave unceremoniously and in great haste.
12. cut both ways, to be beneficial in some ways and disadvantageous in others.
13. cut corners, See **corner.**
14. cut down or **back to size,** to reduce, especially a self-important person, to a status, level, or frame of mind more in keeping with that person's position.
15. cut in, a. to allow oneself (or someone else) a share: *He cut his brother in on the deal.* **b.** to begin to shear sheep. **c.** to join a cardgame by taking the place of someone who is leaving.
16. cut it, make the grade; be competent: *She couldn't cut it in her new job.*
17. cut no ice, See **ice.**
18. cut loose, to free oneself from restraint.
19. cut one's losses, to abandon a project in which one has already invested some part of one's capital, either material or emotional, for no return, so as not to incur more losses.
20. cut one's stick, to be off. [supposedly one cuts a stick to help one's journey along]
21. cut one's teeth on, See **teeth.**
22. cut out, a. to spend all the money represented by (a cheque). **b.** to stop; cease: *The shearing cut out on Friday.* **c.** to leave a cardgame.
23. cut out for, fit or suitable for.

24. cut (someone) dead, to refuse to greet or acknowledge (someone).
25. cut up, a. to criticise severely. **b.** to upset or cause distress to.
26. cut up rough or **nasty,** to behave badly; become unpleasant.
27. go like a cut cat, to go very fast.
28. in for one's cut, participating in the expectation of a share in the spoils or profit.
29. mad as a cut snake, See **mad.**
30. the cuts, a caning.

cute *adj.* pleasingly pretty or dainty.

cutie *n.* a pleasing person; one with a winning personality, especially female. Also, **cutey.**

cut-lunch *phr.* **carry a cut-lunch,** to have an honest job.

cut-lunch commando *n.* **1.** *Military.* a person who has a desk job but who is officious or overbearing in manner. **2.** an amateur or part-time soldier. **3.** an amateur hunter who makes up in bravado for what he lacks in skill.

cut-lunch revolutionary *n.* one who professes revolutionary ideals from the safety of his position in the establishment.

cut-out *n.* **1.** *Shearing.* the completion of shearing: *After the cut-out the cashed-up shearers headed for town.* **2.** the last of the flock to be shorn.

dab hand *n.* a person particularly skilled: *She's a dab hand at breaking in horses.*

dabs *pl. n.* fingerprints.

dad *n.* a form of address to an older man.

dad and mum *n.* rum. [rhyming slang]

daddy *phr.*

1. sugar daddy, a rich, usually old man, who lavishes money and gifts on a young woman or boy in return for sexual favours or companionship.

2. the daddy of them all, the biggest, most powerful, most impressive, etc.

Dad'n'Dave *n.* **1.** a shave. [rhyming slang] —*adj.* **2.** of or pertaining to anything amusingly or ludicrously countrified. [from two bush characters in the writings of Steele Rudd]

Dad's army *n.* a body of retired or unfit men or mothballed plant or machinery brought back into service: *a Dad's army of old power stations.*

daffy *adj.* silly; weak-minded; crazy. [Scottish dialect *daff* a fool, akin to *daft*]

dag[1] *phr.* **rattle your dags,** a command to hurry up.

dag[2] *v. Horseracing.* to follow (a jockey) offering assistance in the expectation of receiving racing information.

dag[3] *n.* an odd, eccentric or amusing person.

dag[4] *n.* **1.** an untidy, slovenly person. **2.** a person who, while neat in appearance and conservative in manners, lacks style or panache. [back formation from *daggy*]

dagger[1] *phr.*

1. at or **with daggers drawn,** on very bad terms.

2. **look daggers,** to cast angry, threatening, or vengeful glances.

dagger[2] *n.* a follower; hanger on. [from *dag*[2]]

daggy[1] *adj.* **1.** dirty; slovenly; unpleasant. **2.** conservative and lacking in style, especially in appearance.

daggy[2] *adj.* stupid; idiotic; eccentric.

dago *n.* **1.** (*offensive*) an Italian. **2.** any person of Latin race. [Spanish *Diego* James]

dagwood *n.* a large sandwich, with a number of different fillings, and sometimes more than two slices of bread. [from *Dagwood* Bumstead, the hero of a U.S. comic strip 'Blondie' who was given to making sandwiches of this type]

daisy *n.* **1.** something fine or first-rate. *—adj.* **2.** fine; first-class; excellent; first-rate. *—phr.* **3. push up daisies,** to be dead and buried.

daisy chain *n.* a group (chain) of homosexual males engaged in lovemaking.

daisy-cutter *n.* (in cricket, football, tennis, etc.) a ball which, after being struck or kicked skims near the ground.

daks *pl. n.* trousers [Trademark]

damage *n.* the cost; expense: *What's the damage sport?*

dame *n.* a woman.

damn *interj.* **1.** an expression of anger, annoyance, or emphasis. *—phr.*

2. as near as damn it, as near as conceivably possible.

3. not to give a damn, not to care; be unconcerned.

4. not worth a damn, worth a negligible amount.

damn-all *n.* **1.** a negligible amount: *He's done damn-all today. —adj.* **2.** very few.

damnedest *n.* the limit of personal effort, or an object's or element's natural function: *to do one's damnedest.*

damper[1] *phr.* **put a damper on,** to discourage.

damper[2] *n.* bush bread, made from a simple flour and water dough and cooked on the coals.

Dan *n.* a sanitary man. [from the jingle *Dan, Dan the dunny man*]

dance *phr.*

1. dance on air, to be hanged.

2. **lead someone a merry dance,** to frustrate someone, as by constantly changing one's moods, intentions, attitudes, etc.

dancer *phr.* **disco** or **Spanish dancer,** cancer. [rhyming slang]

dander *phr.* **get one's dander up,** show anger or temper. [?figurative use of *dander* dandruff; or figurative use of *dander* ferment]

d and m *phr.* deep and meaningful (as a discussion, etc.). [Initials]

dandruff *n. Tennis.* the puff of chalk which rises into the air when the ball strikes the line.

dandy[1] *n.* **1.** something very fine or first rate. *—adj.* **2.** fine, first-rate. [?special use of *Dandy,* variant of *Andy* (Andrew)]

dandy[2] *n.* a carton of ice cream. Also, **dixie.**

dangle *v.* to be hanged.

Daphne *n.* the uncultivated Australian female; female version of **ocker.**

Dapto Dog *n.* a wog. [rhyming slang]

darbies *pl. n. Prison* handcuffs.

Darby and Joan *n.* the typical 'old married couple' happily leading a life of placid domesticity.

dark *phr.*

1. in the dark, in ignorance.

2. keep it dark, to keep a secret.

dark horse *n.* a person whose capabilities may be greater than they are known to be. [from the racehorse or competitor about whom little is known, and who may unexpectedly win]

darkie *n.* **1.** (*offensive*) a dark-skinned person. **2.** waste matter from the intestines; faeces. **3.** Also, **darky, darkey.** *—phr.* **4. drop** or **choke a darkie,** to defecate.

dark'un *n.* a shift of twenty-four hours on the wharves.

Darling pea *phr.* **get the Darling pea,** to be insane or behave strangely. [from toxic species of the *Darling Pea,* plants which, when eaten by cattle, cause a disease of the brain]

Darling shower *n.* a dust storm. [from *Darling* River and region, in western N.S.W.]

darn *interj.* **1.** a mild expletive. *—phr.* **2. not give a darn,** to be utterly indifferent. [variant of *damn*]

darnedest *phr.* **do one's darnedest,** to exert oneself to the utmost; try very hard.

dart¹ *n.* a cigarette.

dart² *n.* **1.** a plan or idea; scheme. **2.** a particular fancy or favourite.

Dart *phr.* **the Old Dart,** England.

Darwin stubby *n.* (formerly) an 80oz bottle of beer.

dash *phr.*

1. cut a dash, to create a brilliant impression.

2. do one's dash, to exhaust one's energies or opportunities.

dasher *n.* a spirited person.

date¹ *n.* **1.** a person, usually of the opposite sex, with whom one has a social appointment. *—v.* **2.** to go out with a person or persons of the opposite sex.

date² *n.* **1.** the anus. **2.** the vagina. *—v.* **3.** to have sexual intercourse with. **4.** to poke or prod in the buttocks.

Davy Jones *phr.* **Davy Jones's locker, a.** the ocean's bottom, especially as the grave of all who perish at sea. **b.** the ocean.

day *phr.*

1. call it a day, to bring an activity to a close, either temporarily or permanently.

2. have had one's day, to be old.

3. that'll be the day, an expression indicating disbelief, cynicism, etc.

daylight robbery *n.* a shameless attempt to rob, overcharge or cheat someone. Also, **highway robbery.**

daylights *phr.*

1. beat the living daylights out of (someone), to give (someone) a sound beating.

2. scare the living daylights out of (someone), to give (someone) a fright.

dazzler *n.* something brilliant and showy: *The new play was a dazzler.*

D.C.M. *n.* a discharge from employment; dismissal. [abbreviation of *D(on't) C(ome) M(onday)*]

dead *adj.* **1.** very tired; exhausted. **2.** quiet: *Business is dead today.* **3.** **used to add emphasis to a statement:** *dead right, a dead shot, dead stupid.* *—phr.*
4. a dead give-away, See **give-away.**
5. a dead loss, See **loss.**
6. dead and won't lie down, refusing to give in.
7. dead as a doornail, completely useless.
8. dead as mutton/a mutton chop/a meataxe, a. dead. **b.** out of fashion.
9. dead from the neck up, lacking intelligence; stupid.
10. dead on, exactly right.
11. dead on one's feet, exhausted.
12. dead set, really true.
13. dead to the world, asleep.
14. enough or **fit to wake the dead,** very loud.
15. left for dead, left behind; surpassed; outstripped.
16. play dead, to lie low waiting for an opportunity.
17. run dead, *Horseracing.* (of a horse) to be deliberately pulled up so that it does not run at its best.
18. wouldn't be seen dead with, to refuse to have any association with.

dead-and-buried *adj.* (of an issue) no longer of interest; finished.

deadbeat *n.* **1.** a man down on his luck; vagrant. *—adj.* **2.** of or pertaining to the life of a vagrant.

dead bird *n.* a certainty.

dead centre *phr.* **dead centre of town,** a humorous term for a cemetery.

dead cert *n.* a certainty: *He's a dead cert to win.*

dead duck *n.* **1.** a person lacking good prospects; a failure. **2.** something useless, or worthless, or utterly without promise. *—phr.* **3. to look like a dead duck in a thunderstorm,** to be unattractive, untidy, etc.

dead-end *adj.* **1.** offering no future: *a dead-end job.* **2.** having no apparent hopes or future, as a juvenile delinquent: *dead-end kid.*

dead eye *phr.* **dead eye and horse,** a pie and sauce. [rhyming slang]

dead hand *n.* (formerly) an expert at doing something.

deadhead *n.* **1.** a dull and ineffectual person. **2.** a spent match.

dead heart *n.* the arid central regions of Australia. Also, **Dead Heart.**

dead horse[1] *n.* tomato sauce. [rhyming slang]

dead horse[2] *n.* a debt.

deadhouse *n.* **1.** a morgue. **2.** a room in a hotel to which drunks are removed to sleep off their intoxication.

deadly *adv.* excessively: *deadly dull.*

dead marine *n.* a bottle which had contained beer, whisky, etc., but is now empty. See **marine.**

dead meat ticket *n.* an identity tag worn at all times by Australian soldiers for identification, especially in the event of death or injury in battle.

dead men *pl. n.* empty bottles. [playing on the idea that the spirit has left them]

dead ringer *n.* a person or thing that closely resembles another: *He was* *a dead ringer for the local policeman.* Also, **dead ring.**

deadshit *n.* **1.** a mean and contemptible person. **2.** a no-hoper; a dullard. *—adj.* **3.** despicable; mean: *It was a deadshit thing to do.*

dead spit *n.* the image, likeness, or counterpart of a person, etc.

dead-wood *n.* a person or thing regarded as useless; a hindrance or impediment.

deal *n.* **1.** a measured quantity, as of marijuana. *—phr.*
2. a big deal, an important event; a serious matter.
3. big deal!, an ironic exclamation indicating contempt, disbelief, etc.

dear John *n.* a letter to a lover ending a relationship. [from such letters sent to frontline soldiers by girlfriends at home during World War II]

death *phr.*
1. at death's door, in danger of death; gravely ill.
2. do to death, to repeat until hackneyed; overdo.
3. fate worse than death, a. (*usually humorous*) a circumstance regarded as particularly horrible. **b.** rape. **c.** (*humorous*) sexual intercourse.
4. in at the death, present at the climax or conclusion of a series of events or a situation.
5. like death (warmed up), appearing or feeling extremely ill or exhausted.
6. like grim death, tenaciously, firmly: *He hung on like grim death.*
7. pack death, See **pack.**

death adder *n.* **1.** a mean, avaricious person. *—phr.* **2. have a death adder in one's pocket,** to be mean or parsimonious.

death-knock *n.* the end; the last minute.

death seat *n.* **1.** the front passenger seat in a motor car. **2.** (in a trotting race) a position on the outside of the leader, from which it is very hard to win.

debag *v.* to remove the trousers of, as a joke or punishment.

debil-debil *n.* a devil. [Aboriginal pidgin]

debil-debil country *n.* harsh terrain not easily traversed.

debug *v.* **1.** to detect and remove faults in (an electronic system). **2.** to detect and remove electronic listening devices from (a room or the like).

debunk *v.* to strip of false sentiment, etc.; to make fun of.

deck *v.* **1.** to knock (someone) to the ground. *—n.* **2.** a portion of a narcotic drug. See **bag.** *—phr.*
3. clear the decks, to prepare for action of any kind.
4. hit the deck, a. to fall to the ground or floor. **b.** to rise from bed.
5. not playing with a full deck, mentally inadequate; insane.
6. on deck, a. on duty; on hand; present at the time. **b.** alive: *Is he still on deck?*

decker *n.* a hat.

dee *n.* a detective or policeman.

deejay *n.* one who comperes radio programs of gramophone records; disc jockey. Also, **D.J.**

deener *n. Obsolete.* a shilling. Also, **deenah, deaner.**

deep *phr.*
1. go off the deep end, a. to get into a dither; to become hysterical. **b.** to go to extremes.

2. **in deep water,** in trouble or difficulties.

deep freeze *phr.* **give (someone) the deep freeze,** to ignore or act coldly towards.

deep north *n.* (*sometimes caps*) Queensland, often so called on analogy with the deep south (of the United States), because of supposed conservative and racially intolerant attitudes.

dekko *n.* a look or view. Also, **dek.**

deknacker *v.* to castrate.

Delhi belly *n.* diarrhoea, as suffered by travellers.

deli *n.* a delicatessen. Also, **dellie.**

Delilah *n.* a seductive and treacherous woman. [named after the woman who betrayed Samson to the Philistines]

delish *adj.* delicious.

deliver *v.* **1.** to perform a task competently and professionally; come up to expectations: *He seems to have the qualifications, but can he deliver?* —*phr.*

2. deliver a serve, to make an opening statement as part of a planned debate.

3. deliver the goods, come up with the goods.

delo *n.* a delegate.

demo *n.* a demonstration.

demob *n.* **1.** demobilisation. —*v.* **2.** to discharge (a soldier) from the army. [short for *demobilise*]

demolish *v.* to eat or drink greedily.

demon *n.* **1.** (formerly) a convict; bushranger. **2.** (formerly) a trooper. **3.** a detective; a policeman, especially a motorcycle policeman.

Demons, the *pl. n.* **1.** the Blacktown City team in the National Soccer League. **2.** the Melbourne team in the Victorian Football League. **3.** the North Hobart team in the Tasmanian Australian National Football League. **4.** the Perth team in the West Australian Football League. **5.** the Redcliffe team in the Queensland Rugby Football Union.

derby *phr.*

1. local derby, any sporting contest between teams from the same area.

2. soapbox derby, a race for billycarts.

dermo *n.* dermatitis.

dero *n.* a vagrant, especially one with a weird or sinister appearance. Also, **derro.** [shortened form of *derelict*]

derrière *n.* (*humorous*) the buttocks; the bottom.

derry *n.* **1.** an alarm; pursuit, especially by the police. —*phr.* **2. have a derry on,** to be prejudiced against.

Derwent duck *n.* (formerly) a convict at Hobart on the river Derwent, Tasmania.

Derwenter *n.* an ex-convict. [from the convict settlement on the Derwent River in Tasmania]

desert rat *n.* *Military.* a member of the various Allied forces who fought in the desert campaigns in North Africa, 1941-42, especially one who assisted in the defence of Tobruk. [named after the divisional sign, a jerboa (desert rat), of the British 7th Armoured Division]

desperation *n.* extreme keenness or urgency, especially in sport: *a desperation tackle.*

destructo *n.* **1.** a person whose behaviour is so wild or outlandish that he is regarded as being suicidal. **2.** a wild, brawling party. Also, **destroy party.**

deuce *n.* the devil (used in mild oaths and exclamations).

deuced *adj.* **1.** confounded; excessive. *—adv.* **2.** Also, **deucedly.** confoundedly; excessively.

devil *n.* **1.** a person, usually one in unfortunate circumstances: *He has not a cent to his name, poor devil.* **2.** the errand boy, or the youngest apprentice in a printing office, as a printer's devil. **3.** fighting spirit. *—phr.*

4. as the devil loves holy water, not at all.

5. between the devil and the deep blue sea, faced with two equally distasteful alternatives.

6. devil of a, extremely difficult: *The unexpected news put him in a devil of a situation.*

7. give the devil his due, to do justice to or give deserved credit to an unpleasant or disliked person.

8. go to the devil, a. to fail completely; be ruined. **b.** to become depraved. **c.** an expletive expressing annoyance, disgust, impatience, etc.

9. raise the devil, to make a commotion.

10. speak or **talk of the devil,** here comes the person who has been the subject of conversation.

11. the devil to pay, serious trouble to be faced.

12. the devil! a mild oath used to express disgust, anger, astonishment, negation, etc.

devilish *adj.* **1.** excessive; very great. *—adv.* **2.** excessively; extremely.

devil-may-care *adj.* reckless; careless; happy-go-lucky.

devil-on-the-coals *n.* a small, thin, quickly made damper.

devil's darning needle *n.* a dragon fly.

devil's punch bowl *n.* a deep hollow in a hillside.

Devils, the *pl. n.* the Northern Suburbs team in the Queensland Rugby Football League.

devil take the hindmost *n.* a cycling race over four to eight kilometres in which the rider finishing last in a lap drops out of the race. [from the saying *the devil take the hindmost,* the least fortunate being abandoned to their fate]

devo *n.* a term of insult or abuse. [shortened form of *deviant*]

dewdrop *n.* a drop at the end of a person's nose.

dial *n.* the human face.

diamond[1] *n.* mate. [rhyming slang *diamond plate* mate]

diamond[2] *phr.* **a rough diamond,** a person without refinement of manner but having an essentially good or likeable personality. [from a diamond which is crude, unwrought, undressed, or unprepared]

dibs *pl.n.* **1.** marbles. *—phr.*

2. dibs in, taking part; included, especially in a game.

3. play for dibs, to play with the object of keeping what has been won.

[from *dibstones,* (formerly) a children's game played with small stones]

dice *v.* **1.** to throw away, reject. *—phr.*
2. dice with death, to act dangerously or take a risk.
3. no dice, of no use; unsuccessful; out of luck.

dicey *adj.* dangerous; risky; tricky.

dick¹ *n.* a detective. [shortened form of *detective*; influenced by dick²]

dick² *n.* **1.** the penis. Also, **dickie. 2.** a foolish, unattractive person; dickhead. *—phr.*
3. clever dick, a smart aleck.
4. have had the dick, to be finished or ruined.

dicken *interj.* **1.** (a questioning interjection) is it true?; really? **2.** (an affirmative interjection) it is true!; really! Also, **dickon.**

dickens *phr.* **(what) the dickens,** a mild oath.

dickhead *n.* **1.** a fool; idiot: *Country people are scornful of Sydney dickheads. —adj.* **2.** Also, **dickheaded.** stupid; foolish: *It was a dickhead thing to do.*

dickie *n.* (*in children's speech*) a penis.

Dickless Tracy *n.* a policewoman. [a humorous reference to the comic strip character *Dick Tracy,* + *dick* penis]

dicky¹ *n.* a donkey, especially a male. Also, **dickey, dickie.**

dicky² *adj.* **1.** unsteady, shaky; in bad health; in poor condition. **2.** difficult; untenable: *a dicky position.*

dickybird *n.* (*in children's speech*) a bird.

did *n.* a toilet. Also, **diddy, didee.** [(in children's speech) modification of *diaper*]

diddle *v.* to cheat; swindle; victimise. [origin uncertain]

diddums *interj.* an exclamation indicating that the speaker thinks that the person addressed is being childish and petulant. [in speech to children *did he* (*she*)]

diddy *n.* a woman's breast or nipple. [variant of *titty*]

die *v.* **1.** to desire or want keenly or greatly: *I'm dying for a drink. —phr.*
2. die in harness, to die while still working at a job.
3. die on (someone), a. to fall asleep while in the company of (someone). **b.** to let (someone) down; fail to keep a promise.
4. die on one's feet, to fail; cease to function.
5. die with one's boots on, a. to die by violent means. **b.** to die before one is incapacitated and bedridden.

Diehards, the *pl. n.* the Valleys team in the Queensland Rugby Football League.

difference *phr.* **split the difference, a.** to pay equal shares. **b.** to compromise.

dig *v.* **1.** to understand or find to one's taste. **2.** to take notice of; pay attention to. *— n.* **3.** a digger (a common form of address among men). **4.** a cutting, sarcastic remark. *—phr.*
5. dig down or **deep,** pay, as from scant funds.
6. dig in, a. to maintain one's position or opinion firmly. **b.** to apply oneself vigorously.
7. dig one's heels or **toes in,** See **heel.**
8. dig out or **up,** to discover, find, reveal.

digger *n.* **1.** an Australian soldier, especially one who served in World War I. **2.** a common form of address in Australia and New Zealand. [originating on the goldfields] **3.** *N.Z. Prison.* a punishment cell.

diggings *n.* living quarters; lodgings. Also, **digs.**

dill *n.* a fool; an incompetent. Also, **dillpot.**

dilly[1] *adj.* queer; mad; crazy. [*d(aft)* + *(s)illy*]

dilly[2] *n.* something remarkable of its kind; a delightful or excellent person or thing: *That's a dilly of an idea.* [*del(ightful)* or *del(icious)* + *y*]

dillybag *n.* any small bag for carrying food or personal belongings. [Aboriginal *dilly* + *bag*]

dillydally *v.* to waste time, especially by indecision; trifle; loiter.

dilutee *n.* an inadequately trained or otherwise incompetent tradesman. [from *dilutee* a man, ineligible for service in World War II, who was trained at a trade by means of a crash course and certificated after the war]

dim *adj.* **1.** (of a person) stupid; lacking in intelligence. *—phr.* **2. take a dim view,** to have a disparaging, adverse opinion of an action, a project, etc.

dime *phr.* **a dime a dozen,** See **dozen.**

dimwit *n.* a stupid or slow-thinking person.

din-din *n.* (*in children's speech and humorous*) dinner. Also, **din-dins.**

dine *phr.*

1. dine at the Y, to engage in cunnilingus.

2. dine out on, to entertain with (a particular joke, anecdote, etc.) at a dinner or other social occasion.

ding[1] *v.* **1.** to throw away. **2.** to smash; damage. *—n.* **3.** a damaged section on a car, bike, surfboard, etc. **4.** a minor accident involving a car, bike, surfboard, etc. **5.** an argument. **6.** the penis. **7.** (*offensive*) a foreigner, especially an Italian or Greek.

ding[2] *n.* a party, especially a wild or successful one. [short for *wing-ding*]

ding-a-ling *n.* a fool; idiot; eccentric person.

dingbat *n.* an eccentric or peculiar person.

dingbats *adj.* **1.** peculiar; odd; crazy. *—phr.*

2. the dingbats, a. delirium tremens. **b.** a fit of madness or rage.

3. give (someone) the dingbats, a. to cause (someone) to worry. **b.** to make (someone) frightened or nervous.

ding-dong *adj.* **1.** vigorously fought with alternating success: *a ding-dong contest.* *—n.* **2.** a loud and vigorous argument.

dinger *n.* **1.** Also, **ding.** the buttocks; arse. **2.** a shanghai; catapult.

dingie *n.* **1.** Also, **dingy, dingi.** a faked illness. *—phr.* **2. throw a dingie,** to pretend to be ill in order to avoid an uncongenial task.

dingo *n.* **1. a.** a contemptible person; coward. **b.** one who shirks responsibility or evades difficult situations. *—v.* **2.** to act in a cowardly manner. **3.** to shirk, evade, or avoid; to spoil or ruin. *—phr.*

4. a dingo's breakfast, See **breakfast.**
5. dingo on (someone), to betray (someone). Also, **turn dingo.**
6. dry as a dingo's tail, See **dry.**
7. put on a dingo act, act in a cowardly way.
[from *dingo* the Australian wild dog which is said to be treacherous and cowardly]

dink *v.* **1.** to carry or convey a second person on a horse, bicycle or motorcycle. Also, **double-dink.** —*n.* **2.** a minor injury (on a surfboard or in a car).

dinkum *adj.* **1.** Also, **dinky-di, dinky.** true; honest; genuine: *dinkum Aussie.* **2.** seriously interested in a proposed deal, scheme, etc. —*adv.* **3.** truly. See **fair dinkum.** —*n.* **4.** an excellent or remarkable example of its kind: *You little dinkum.* —*phr.* **5. dinkum oil,** See **oil.**

dinkus *n.* a small drawing used to decorate a page, or to break up a block of type. Also, *U.S.* **dingus.** [coined by an artist in the Bulletin c.1920; from *dinky*]

dinky[1] *adj.* **1.** of small size. **2.** neat; dainty; smart. —*n.* **3.** a small tricycle. **4.** a small rowing or sailing boat, sometimes inflatable; dinghy.

dinky[2] *adj., adv., n.* dinkum.

dinky-di *adj.* true; honest; genuine: *He's a dinky-di Aussie.* [alteration of *dinkum*]

dinner *phr.* **done like a dinner,** completely defeated or outwitted.

dinnies *pl. n.* dinner. Also, **din din, din dins.**

dinnyhayser *n.* anything superior or excellent. [from *Dinny Hayes,* pugilist]

dip *n.* **1.** *Prison.* a pickpocket. —*phr.* **2. dip into one's pocket,** to spend money.
3. dip in willy, to steal from a person's pocket.
4. dip one's lid to (someone), to pay honour to or congratulate (someone).
5. dip out, a. to miss out, **b.** to fail: *He dipped out in his exams.*
6. dip out on, to remain uninvolved; avoid.
8. dip south, to reach into one's pocket.
9. dip the wick, to have sexual intercourse.

dippy *adj.* stupid; foolish.

dipso *n.* one who suffers from an insatiable craving for alcoholic drink; a dipsomaniac.

dirt *n.* **1.** abusive or scurrilous language. **2.** unsavoury or malicious gossip. —*phr.*
3. do (someone) dirt, to behave unfairly or wrongly towards (someone). Also, **do the dirty on.**
4. eat dirt, to accept insult without complaint.
5. treat (someone) like dirt, to show little liking or respect for (someone).
[from *dirt* which consists especially of any foul or filthy substance, as excrement, mud, etc.]

dirt-cheap *adj.* very inexpensive.

dirty *adj.* **1.** morally unclean; indecent. **2.** angry. —*adv.* **3.** very; extremely: *dirty big.* —*phr.*
4. be dirty on, to be angry with.
5. do the dirty on, See **dirt.**

dirty ditty *n.* a rude song or joke.

dirty dog *n.* an unprincipled person.

dirty half mile *n.* King's Cross, Sydney.

dirty joke *n.* a joke on the taboo subject of sex.

dirty linen *phr.* **wash one's dirty linen in public,** reveal or discuss matters of a scandalous nature from one's private life.

dirty look *n.* an angry or sullen expression: *She gave him a dirty look.*

dirty old man *n.* a man, usually of mature years, who is considered to have an unhealthy interest in sexual matters.

dirty tricks *pl. n.* underhand activities designed to discredit or smear a political opponent.

dirty weekend *n.* a weekend spent with a lover in sensual delights.

dirty word *n.* **1.** a vulgar word. **2.** something one doesn't mention because it is as objectionable as if it were a vulgar word: *Work is a dirty word round here.*

dirty work *phr.* **do (someone's) dirty work,** to take over an unpleasant task for (another person) usually without reward or thanks.

disaster *n.* a total failure, as of a person, machine, plan, etc.

disaster area *n.* **1.** a person who seems overwhelmed by misfortune: *She is a walking disaster area.* **2.** an untidy place.

discombobulate *v.* to upset, confuse (a person). [mock Latin formation, from *discompose* or *discomfort*]

dish *n.* **1.** an attractive girl or man. —*v.* **2.** to abandon; discard; sack. —*phr.*

3. dish it out, to punish (someone).

4. dish out, to distribute; share out. **5. dish up, a.** to defeat; frustrate; cheat. **b.** to provide as entertainment, information, etc.

dished *adj.* defeated; frustrated; cheated.

dishwater *n.* **1.** any weak drink. —*phr.* **2. dull as dishwater,** very boring.

dishy *adj.* (of persons or things) attractive.

dismals *pl. n.* gloom; melancholy; dumps.

disso *n.* a wharf labourer suffering from disability or advanced age, who is fit only for work on the wharf, not on the ship.

distance *phr.*

1. go the distance, to complete a proposed undertaking, as of a horse able to complete a course, a student able to gain his award, etc.

2. keep one's distance, to maintain a proper degree of reserve or aloofness.

ditch *n.* **1.** the sea. —*v.* **2.** to get rid of; get away from. **3.** to crash-land (an aeroplane), especially in the sea. —*phr.* **4. last ditch,** the last defence; utmost extremity.

ditto *n.* a duplicate or copy.

dive *n.* a disreputable place, as for drinking, gambling, etc., especially a cellar or basement.

dive-bomb *v.* **1.** to jump into water, as a swimming pool, creek, etc., with the knees tucked under the chin. —*n.* **2.** the action of dive-bombing.

divvy[1] *n.* **1.** a dividend. **2.** (*pl.*) rewards; profits; gains. —*phr.*

3. divvy up, to share: *divvy up the proceeds.*

4. divvy with, to share with.

[shortened form of *dividend*]

divvy[2] *n. Golf, Cricket, etc.* a piece of turf cut out with a club or bat making a stroke; divot.

DIY *adj.* do-it-yourself.

dizzy *adj.* **1.** foolish or stupid: *a dizzy dame.* *—phr.* **2. dizzy limit,** See **limit.**

do *n.* **1.** a swindle. **2.** a festivity or treat: *We're having a big do next week.* *—v.* **3.** to serve; suffice for: *This will do us for the present.* **4.** to provide; prepare: *This pub does lunches.* **5.** to cheat or swindle. **6.** to commit burglary; steal. **7.** to use up; expend: *He did his money at the races.* **8.** to treat violently; beat up. **9.** to have sexual intercourse with. *—phr.*

10. do a Melba, See **Melba.**

11. do a (someone), to imitate (someone): *He did a Whitlam.*

12. do away with, to kill.

13. do down, to get the better of; cheat.

14. do for, a. to accomplish the defeat, ruin, death, etc., of. **b.** to cook and keep house for. **c.** to charge with a certain offence: *I've been done for speeding again.*

15. do in, a. to kill; murder. **b.** to exhaust; tire out. **c.** to ruin.

16. do one's thing, to act according to one's own self-image.

17. do or die, of supreme effort: *a do or die attempt at the record.*

18. do out of, to deprive, cheat, or swindle of.

19. do over, a. to redecorate; renovate: *to do a room over.* **b.** to assault. **c.** to have sexual intercourse with.

dob *v.* **1.** *Football.* to kick, usually accurately, especially in shooting for goal: *He's dobbed another goal.* *—phr.*

2. dob in, a. to contribute money to a common fund: *We'll all dob in and buy him a present.* **b.** to betray, report (someone) as for a misdemeanour. **c.** to nominate (someone absent) for an unpleasant task.

5. dob on, to inform against; betray.

dobber *n.* an informer; telltale.

doc *n.* a doctor. [shortened form]

dock *phr.* **in dock, a.** (of equipment, etc.) out of order and being fixed. **b.** (of a person) ill; laid up.

doco *n.* a documentary.

doctor *n.* **1.** an expert; one who makes the final decision: *You're the doctor.* **2. a.** a ship's cook. **b.** a sheep-station cook. **3.** a strong fresh breeze blowing at the end of a hot day, as the Albany, Esperence, Fremantle doctor, etc. *—v.* **4.** to repair or mend. **5.** to tamper with; falsify; adulterate. **6.** to castrate or spay. *—phr.*

7. go for the doctor, a. to bet all one's money on a race. **b.** to make an effort that consumes all one's resources.

8. just what the doctor ordered, something agreeable; the perfect solution.

dodge *n.* an ingenious expedient or contrivance; a shifty trick.

dodger *n.* food, especially bread. [British dialect]

dodgy *adj.* difficult; awkward; tricky.

dodo *n.* **1.** a silly or slow-witted person. *—phr.* **2. dead as a dodo,** inanimate and beyond possibility of revival. [from the clumsy, flightless *dodo,* a bird now extinct]

doer *n.* **1.** an amusing or odd person. —*phr.* **2. hard doer,** a stronger version of a **doer.**

dog *n.* **1.** a despicable fellow. **2.** *Prison.* an informer. **3.** a dingo. —*phr.*
4. a dog tied up, an outstanding account. See **bad dog.**
5. every dog has his day, everyone achieves some success or happiness eventually.
6. go to the dogs, to go to ruin.
7. to have a dog's chance, to have no chance.
8. lame dog, an unfortunate person; a helpless person. Also, **lame duck.**
9. lead a dog's life, to have a harassed existence; to be continuously unhappy.
10. let sleeping dogs lie, to refrain from action which might alter the existing situation.
11. on the dog list, to be barred from drinking in a hotel.
12. put on (the) dog, to behave pretentiously; put on airs.
13. try it out on the dog, to make a sample or trial of a dubious project, procedure, etc., in order to test its feasibility.

dog and bone *n.* the telephone. [rhyming slang]

dogbox *n.* **1.** *N.Z.* cramped quarters; a kennel-like room. —*phr.* **2. in the dogbox,** *N.Z.* in disfavour. [from *dogbox* a carriage on a passenger train in which there is no access by corridor from one compartment to another]

dog-collar *n.* a clerical collar.

dog-eat-dog *adj.* extremely competitive: *dog-eat-dog society.*

doggie bag *n.* a bag provided by a restaurant or the like, for carrying home leftovers. Also, **doggy bag.**

doggo *adv.* **1.** out of sight. —*phr.* **2. lie doggo,** to hide; remain in concealment.

doggone *U.S.* *iadj.* **1.** darned; damn. —*adv.* **2.** very; extremely: *He's doggone rich.*

doghouse *phr.* **in the doghouse,** in disgrace; unpopular.

dog licence *n.* (formerly) citizenship papers issued to an Aborigine permitting the consumption of alcohol.

do-gooder *n.* a well-intentioned, but often clumsy social reformer.

dogs *pl. n.* **1.** greyhound racing. **2.** the feet.

dogsbody *n.* an overworked drudge.

dog's breakfast *n.* **1.** a mess; a confused state of affairs. Also, **dog's dinner. 2.** See **breakfast.**

dog's cock *n.* an exclamation mark (!).

dog's disease *n.* influenza.

dog's eye *n.* a meat pie. [rhyming slang]

dog squad *n.* *Prison.* undercover police.

dog tag *n.* an identity disc. Also, **meat tag, dead meat ticket.**

dog-tired *adj.* very tired.

dog-tucker *adj.* (of a sheep) fit only to be used as dog food.

dog watch *n.* See **graveyard shift.**

doing *n.* **1.** a scolding. —*phr.*
2. doing over, a beating.
3. drop of the doings, alcohol; grog.
4. nothing doing, an exclamation indicating refusal.
5. the doings, materials or ingredients for a meal.

doldrums *phr.* **in the doldrums,** in a period of dullness, depression, etc. [from the region of relatively calm winds near the equator where sailing ships are often becalmed]

dole bludger *n.* (*offensive*) one who is unemployed and lives on social security payments without making proper attempts to find employment.

doley *n.* a person in receipt of the dole.

doll[1] *n.* **1.** a pretty, but expressionless or unintelligent woman. Also, **dolly bird.** *—phr.* **2. doll up,** to dress (oneself or another) rather too smartly or too much. [from *Doll, Dolly,* for *Dorothy,* woman's name]

doll[2] *phr.* **knock over a doll,** to incur the consequences of one's actions. [from the contest of throwing missiles at dolls at sideshows]

dollar *n.* **1.** (formerly) the sum of five shillings. *—phr.*

2. bottom dollar, the last of a person's money: *down to his bottom dollar.*

3. not the full dollar, intellectually inadequate.

dollop *n.* a lump; a mass.

dolly *v.* to falsify evidence against: *The police dollied Joe.*

dolly catch *n. Cricket.* a simple catch.

dolly's wax *phr.* **(full) up to dolly's wax,** (full) up to one's neck: *I couldn't eat another crumb, I'm full up to dolly's wax.* [a reference to dolls of the 18th and 19th centuries with heads made of wax]

Dolphins, the *pl. n.* **1.** the Frankston team in the Victorian Football Association. **2.** the Redcliffe team in the Queensland Rugby Football League.

d.o.m. *n.* dirty old man. [abbreviation]

domain *phr.*

1. the Domain, a place in Sydney noted for the speakers who address passers-by, usually on Sundays.

2. domain squatter, a loafer or vagrant frequenting the Sydney Domain.

dome *n.* a person's head.

domestic *n.* an argument with one's spouse.

domino *n.* **1.** *Obsolete.* the last in a succession of things, pleasant or unpleasant. **2.** *Obsolete.* the last lash in a flogging.

donah *n.* (formerly) a girl, sweetheart. [Spanish, from Latin: *domina* mistress, lady]

donald duck *n.* sexual intercourse. [rhyming slang *donald duck* fuck]

done *interj.* **1.** agreed; settled. *—phr.*

2. done for, a. dead. **b.** close to death. **c.** utterly exhausted. **d.** deprived of one's means of livelihood, etc.; ruined.

3. done in, very tired; exhausted.

4. done out of, cheated; tricked.

5. done up, a. dressed smartly. **b.** finished; ruined.

6. get done, to be completely defeated.

dong *v.* **1.** to hit, punch. *—n.* **2.** a heavy blow. **3.** Also, **donger,** the penis.

donga *n.* **1.** a shallow gully or dried-out water course. **2.** *Chiefly Papua New Guinea.* a house. *—v.* **3.** to loaf; bludge. *—phr.* **4. the donga,**

the bush; the outback. [Afrikaans: water-course, from Zulu *udonga*]

donk[1] *n.* **1.** a small, usually subsidiary, steam engine; donkey engine. **2.** any engine.

donk[2] *n.* **1.** a donkey. **2.** a racehorse.

donkey[1] *n.* **1.** a stupid, silly or obstinate person. —*phr.*
2. donkey's years, a long time.
3. talk the hind legs off a donkey, to talk too much.

donkey[2] *S.A. v.* **1.** to carry a second person on a bicycle, horse, etc.; double. —*n.* **2.** a ride where one is carried as a second person on a bicycle, horse, etc.; a double.

donkey drop *n.* (in cricket) a slow, high ball which looks easy to hit. Also, **donkey shot.**

donkey lick *n.* golden syrup.

donkey-lick *v. Horseracing, etc.* to defeat (another contestant in a race) with ease.

donkey vote *n.* a vote in which the voter simply lists the candidates in the order they appear on the ballot paper.

donkey work *n.* **1.** drudgery; hard, tedious work. **2.** groundwork.

donnybrook *n.* a fight or argument; a brawl. Also, **donneybrook,** *N.Z.* **donny.** [originally with reference to a fair held annually until 1855 at *Donnybrook*, Dublin, famous for rioting and dissipation]

Dons, the *pl. n.* the Essendon team in the Victorian Football League.

doob *n.* a dope; ignoramus.

doodackie *n. N.Z.* (an indefinite name for a thing or person which a speaker cannot or does not designate more precisely). Also, **doodah, doohickie.**

doodad *n.* any trifling ornament or bit of decorative finery.

doodie *n.* a pipe used for smoking tobacco, opium, etc.

doodlebug *n.* a type of self-steering aerial bomb, used by the Germans in World War II over England; buzzbomb.

doolan *n. N.Z.* an Irish Catholic; a Roman Catholic. Also, **dooly, Mickey Doolan.**

Dooley, Larry *phr.* **give some Larry Dooley,** See **Larry Dooley.**

door *phr.*
1. lay at the door of, to attribute to; to impute blame for.
2. show one the door, to dismiss from the house; turn out.

doormat *n.* an uncomplaining person who meekly accepts ill-treatment or bullying.

doornail *phr.* **dead as a doornail,** dead beyond any doubt.

doorstep *n.* an extremely thick slice of bread.

doover *n.* any object (often used jocularly in place of the usual name). Also, **doovah, doofer, dooverlackie, doovahlackie.** [alteration of *do for* in such phrases as *that will do for now*]

doozey *n.* anything especially pleasing: *It is a doozey of an idea.*

dope *n.* **1.** a molasses-like preparation of opium used for smoking. **2.** any drug, especially a narcotic. **3.** marijuana. **4.** a stimulating drug, as one illegally given to a racehorse to induce greater speed. **5.** information or data. **6.** a stupid person. —*v.* **7.** to affect with dope or drugs.

dope fiend *n.* a person addicted to drugs, especially narcotics.

dopey *adj.* **1.** affected by or as by a stupefying drug. **2.** slow-witted; stupid. Also, **dopy.**

dorothy dixer *n.* a question asked in parliament specifically to allow a propagandist reply by a minister. [from *Dorothy Dix* a well-known column of advice to people with emotional problems, the questions are said to have been devised by the columnist, and not sent in by readers]

dose *n.* **1.** venereal disease. *—phr.* **2. go through like a dose** or **packet of salts,** See **salt.**

dosh *n.* money.

doss *n.* **1.** a place to sleep, especially in a cheap lodging house. **2.** sleep. *—v.* **3.** to sleep in a dosshouse. **4.** to make a temporary sleeping place for oneself.

dosser *n.* one who sleeps in a dosshouse.

dosshouse *n.* a cheap lodging house, usually for men only; flophouse.

dot *v.* **1.** to hit; punch. *—phr.*

2. dot and carry one, to walk with a limp.

3. dot one's i's and cross one's t's, to be meticulous; to particularise minutely; pay punctilious attention to detail.

4. hawking the dot, engaging in prostitution.

5. on the dot, punctual; exactly on time.

6. the dots, *Jazz.* musical notation.

7. the year dot, a long time ago: *They've been here since the year dot.*

dotty *adj.* **1.** crazy; eccentric. **2.** feeble or unsteady in gait.

double *phr.*

1. at or **on the double,** fast; quickly; at a run.

2. come the double on, to deceive; doublecross.

3. double or quits, a bet in which a debtor stands to double his debt if he loses or be excused if he wins.

Double Blues *pl. n.* the Sturt team in the South Australian National Football League.

double-dink *v.* to carry or convey a second person on a horse, bicycle, or motorcycle. Also, **double bank, dink.**

double drummer *n.* a variety of cicada with a particularly loud call.

double-dutch *n.* nonsense; gibberish; incomprehensible speech.

double-header *n.* **1.** a double-headed coin. **2.** two attractions, events, etc., featured on the one program, day, etc.

double-take *n.* a second look, either figuratively or literally, given to a person, event, etc., whose significance is suddenly understood.

doubletalk *n.* evasive or ambiguous language.

double time *phr.* **in double time,** with speed; quickly.

double-tongued *adj.* deceitful.

dough *n.* **1.** money. *—phr.* **2. do one's dough, a.** to lose one's money, especially in some speculation or gamble. **b.** to throw away a last chance.

doughboy *n.* **1.** *U.S.* an infantryman. **2.** a rounded lump of dough boiled or steamed as a dumpling; a kind of damper.

doughie *n.* a stupid person.

doughnut *n.* the vulva.

dover *n.* **1.** a bush knife. *—phr.*
2. flash one's dover, to prepare to eat.
3. the run of one's dover, as much to eat as one wants. [originally from brand name on blade of both knife and shears]

down *adv.* **1.** in a prostrate, depressed, or degraded condition. *—adj.* **2.** in prison: *He is down for a few months. —phr.*
3. down and out, without friends, money, or prospects.
4. down at heel, See **heel.**
5. down in the mouth, See **mouth.**
6. down on, over-severe; unnecessarily ready to detect faults and punish harshly.
7. down on one's luck, See **luck.**
8. down south, See **south.**
9. down the plughole/drain/gurgler, wasted (as effort, money, etc.).
10. down time, time lost.
11. down tools, (of workers) to cease to work, as in starting a strike.
12. down with, towards a lower position or total abolition: *Down with school. Down with work.*
13. go down on (someone), to practise fellatio.
14. have a down on, to hold a grudge against.
15. put a down on (someone), to inform against.
16. sent down, a. *Chiefly British.* expelled from university: *sent down from Oxford.* **b.** sent to prison: *sent down for three years.*

downer *n.* **1.** a depressant or tranquilliser. **2.** a depressing experience.

downgrade *phr.* **on the downgrade,** heading for poverty, ruin, etc.

down-under *n.* Australia, New Zealand, and adjacent Pacific Islands (viewed from or as from the Northern Hemisphere). Also, **Down Under.**

downy *phr.* **downy bird,** a canny person.

dozen *phr.*
1. daily dozen, daily physical exercises.
2. a dime a dozen, very cheap; in plentiful supply.

drab *n.* **1.** a dirty, untidy woman; a slattern. **2.** a prostitute.

drack *adj.* unattractive; dressed in a slovenly manner: *a drack sort.* [from *Dracula's Daughter*]

draft dodger *n.* one who evades or attempts to evade conscription into the armed forces.

drag *n.* **1.** somebody or something that is extremely boring. **2.** a puff or a pull on a cigarette. **3.** women's clothes, worn by men; transvestite costume. **4.** a prison sentence of three month's or less duration. **5.** a drag-race. **6.** a road or street: *the main drag. —phr.*
7. drag one's feet, to hang back deliberately; be recalcitrant.
8. drag the chain, to hinder others by doing something slowly.
9. drag up, a. (of a child) to bring up or raise in a socially unacceptable way. **b.** to raise an unpopular or disturbing topic.
10. have a drag with, *U.S.* influence: *He has a drag with the managing director.*

dragon *n.* **1.** a fierce, violent person. **2.** a severely watchful woman.

Dragons, the *pl. n.* **1.** the St George team in the N.S.W. Rugby Football League. Also, **the Saints. 2.** the Northcote team in the Victorian Football Association.

drag-race *n.* a race between cars starting from a standstill, the winner being the car that can accelerate fastest. Also, **drag.**

drain *n.* **1.** a small drink. *—phr.*
2. drain the dragon, to urinate.
3. go down the drain, a. to be wasted. **b.** to become worthless.
4. laugh like a drain, to laugh loudly.

drainpipes *pl. n.* tight, narrow trousers.

drat *interj.* a mild exclamation of vexation. [alteration of *God rot*]

dratted *adj.* confounded.

draught *phr.* **feel the draught,** to feel or be harmed by conditions becoming unfavourable.

draw *phr.*
1. draw a bead on, to take aim at.
2. draw a blank, See **blank.**
3. draw the line, a. to fix a limit. **b.** to decline.
4. draw the teeth of, to render harmless.

drawing board *phr.*
1. back to the drawing board, back to the basic essentials; back to the start.
2. on the drawing board, in preparation.

drawn *phr.* **hung, drawn and quartered,** punished severely.

dreaded lurgi *n.* See **lurgi.**

dreadful *n.* **1.** See **penny-dreadful.** **2.** a periodical given to highly sensational matter.

dreadnought *n.* **1.** a heavyweight boxer. **2.** (formerly) a heavy type of tramcar used in Sydney. **3.** a champion shearer.

dream *n.* **1.** *Prison.* a period of imprisonment of six months. *—phr.*
2. dream up, to invent; to form or plan an idea in the imagination.
3. go like a dream, to succeed or progress very well.

dreamboat *n.* an overwhelmingly attractive member of the opposite sex.

dreamy *adj.* marvellous; extremely pleasing.

dress *phr.*
1. dress down, to scold severely; upbraid and rebuke.
2. dress up, a. to put on best clothes. **b.** to put on fancy dress, costume, or guise.

dressing-down *n.* **1.** a severe reprimand; scolding. **2.** a thrashing; beating.

dribs and drabs *n.* small and often irregular amounts.

drill *n.* the correct procedure; routine: *to know the drill.*

drillion *n.* a great number.

drink *phr.*
1. drink with the flies, See **flies.**
2. the drink, the sea or a large lake.

drip *n.* an insipid or colourless person; a fool.

drippy *adj.* (of a person) colourless; insipid; stupid.

drive *phr.*
1. drive a coach and four (through), to go through easily, without opposition or hindrance: *You could drive a coach and four through that argument.*

2. drive (someone) up the wall, to exasperate (someone).

drongo *n.* **1.** a slow-witted or stupid person. **2.** a raw recruit. [perhaps from *Drongo* the name of a racehorse in the early 1920s which never won a race]

droob *n.* **1.** a minute portion of anything: *He didn't get a droob.* **2.** a weak, ineffectual or slow-witted person.

droog *n.* a slow-witted person; fool.

drool *phr.* **drool over,** to show excessive pleasure at an object or at the prospect of enjoying something.

droopy drawers *n.* a sluggish, apathetic person.

drop *n.* **1.** the gallows. **2.** *Cricket.* the fall of a wicket. **3.** stolen goods. **4.** a place of temporary storage for stolen goods, drugs, etc. —*v.* **5.** to take (drugs) in tablet or capsule form. **6.** to stop, terminate (a topic of conversation, etc.): *Drop it!* **7.** to receive stolen goods. **8.** to fell (a tree, etc.). **9.** *Prison.* to inform on (someone). —*phr.*

10. a drop in the bucket or **ocean,** something very small in comparison with what is required.

11. at the drop of a hat, See **hat.**

12. drop a bundle, See **bundle.**

13. drop like flies, See **flies.**

14. drop off, a. to decrease; decline: *sales have dropped off.* **b.** to fall asleep.

15. drop one's bundle, See **bundle.**

16. drop one's gear, See **gear.**

17. drop or **dump one's load,** (of a male) to ejaculate.

18. get or **have the drop on,** to get or have at a disadvantage.

19. long drop, Also, **high jump.** execution by hanging.

drop-out *n.* one who decides to opt out of conventional society, a given social group or an educational institution: *Student drop-outs became more numerous.*

droppie *n.* *Football.* a drop kick.

drop test *n.* (*humorous*) a wry attempt to fix a faulty piece of equipment, usually electronic, by dropping it.

drover's dog *n.* **1.** (derogatory) a person of no importance, anybody at all. —*phr.* **2. work like a drover's dog,** to work hard and in a determined way.

drown *phr.* **drown one's sorrows,** to get drunk so as to forget one's troubles or problems.

druggie *n.* one who takes drugs habitually.

drum *n.* **1.** a swagman's rolled blanket and the belongings it contains. **2.** a brothel. **3.** information. **4.** lodgings, usually very cheap. —*phr.*

5. drum an idea into (someone), to force understanding by persistent repetition.

6. drum out, to expel or dismiss in disgrace: *to drum out of the army.*

7. drum up business, to organise or promote new business activity.

8. give (someone) the drum, to give (someone) information or advice, usually confidential or profitable.

9. hump one's drum, to carry a swag.

drummer *n.* **1.** a swagman. **2.** a learner, or the slowest shearer, in a shearing gang.

drunk *n.* **1.** a drunken person. **2.** a spree; a drinking party. *—phr.* **3. drunk as Chloe/a piss ant/a skunk/a fowl/etc.,** very drunk.

druthers *pl. n.* choice; preference: *If I'd had my druthers I'd be in France.*

dry *n.* **1.** dry ginger ale: *brandy and dry.* **2.** *U.S.* a prohibitionist. **3.** Also, **Dry.** one who opposes government intervention in the economic affairs of a country; a supporter of free market policies. *—phr.*

4. declare dry, to declare sheep to be sufficiently unaffected by rain for shearing to continue.

5. dry as a dingo's tail; dry as a nun's nasty; dry as a pommy's towel, a. very dry. **b.** very thirsty.

6. dry out, a. to subject (an alcoholic or other drug addict) to a systematic process of detoxification. **b.** (of alcoholics, and other drug addicts) to rid the body of the drug of dependence.

7. dry up, a. to become intellectually barren. **b.** to stop talking.

8. the Dry, the dry season.

dry pan *n.* a sanitary can for an outside toilet.

dry run *n.* a test exercise or rehearsal.

dubbo *adj.* **1.** stupid; imbecilic. *— n.* **2.** an idiot or imbecile. [from *Dubbo* a town in N.S.W.]

duchess¹ *n.* **1.** a woman of showy demeanour or appearance. *— v.* **2.** to treat in an obsequious fashion in order to improve one's social or political standing.

duchess² *n. Chiefly Qld.* a dressing table. [from *duchesse* a dressing table, chest of drawers, etc., with a swing mirror]

duck¹ *n.* **1.** a darling; pet. **2.** a woman: *an old duck.* *—phr.*

3. how are your father's ducks? a general expression of greeting.

4. lame duck, an unfortunate or helpless person.

5. like water off a duck's back, having no effect.

6. take to (something) like a duck to water, to adapt easily to a new situation.

duck² *phr.* **duck out** or **up,** to go away, often in the expectation of an early return: *I just ducked out to the shops for a few minutes.*

duck³ *n. Cricket.* **1.** Also **duck's egg.** a batsman's score of nought. *—phr.* **2. golden duck,** a dismissal with the first ball.

ducks and drakes *pl. n.* **1.** the shakes (as induced by excessive consumption of alcohol). [rhyming slang] *—phr.* **2. make ducks and drakes of, play (at) ducks and drakes with,** to handle recklessly; squander.

ducks and geese *pl. n.* the police. [rhyming slang]

duck's dinner *n.* a drink of water without anything to eat, especially if taken at a normal mealtime.

duck's disease *n.* a short person; a person whose tail is close to the ground.

duckshove *v.* **1.** to use unfair methods; to be unscrupulous in dealings. **2.** (of a taxi driver) to solicit passengers along the roadside, rather than waiting in turn at a rank.

ducky *adj.* **1.** dear; darling. *— n.* **2.** darling; dear; pet. Also, **duckie.**

dud *n.* **1.** any thing or person that proves a failure. **2.** an empty bottle. —*v.* **3.** to swindle; deceive. —*adj.* **4.** useless; defective: *a dud cheque.* —*phr.* **5. dud up, a.** to cause someone to fail. **b.** to misinform someone deliberately.

dud-dropper *n.* one who sells inferior goods as superior goods, in a manner suggesting that they are high-priced goods being sold cheaply because they are stolen.

dude *n.* **1.** *Chiefly U.S.* an affected or fastidious man; fop. **2.** a person brought up in a large city. **3.** a city-dweller holidaying on a ranch. **4.** an adult male; fellow.

duds *pl. n.* **1.** trousers. **2.** clothes, especially old or ragged clothes. **3.** belongings in general.

duff[1] *v.* **1.** to make pregnant. —*phr.* **2. up the duff, a.** pregnant. **b.** ruined; broken.

duff[2] *v.* **1.** to steal (cattle, sheep, etc.) usually altering brands in the process. **2.** to recondition or change the appearance of (especially old or stolen goods). [back formation from *duffer*]

duffer[1] *n.* **1.** *Brit.* a pedlar, especially one who sells cheap, flashy goods as valuable items under false pretences. **2.** a cattle duffer. [British thieves' slang]

duffer[2] *n.* **1.** a plodding, stupid, or incompetent person. **2.** *Mining.* **a.** a shaft yielding no payable ore. **b.** a mine, claim, etc., producing little or no payable ore. **3.** anything inferior or useless. —*phr.* **4. duffer out,** to peter out, as a gold claim.

D.U.I. *n.* a charge made by police against a person who is observed to be driving under the influence of alcohol. [*D(riving) U(nder the) I(nfluence)*]

Duke of Kent *n.* rent. [rhyming slang]

dukes *pl. n.* the hands or fists.

dullsville *n. U.S.* an activity or place which is boring or tedious.

dumbbell *n.* an idiot.

dumbcluck *n.* a fool; idiot.

dumb Dora *n.* a dull or stupid girl or woman.

dumdum *n.* (*especially in children's speech*) a stupid person.

dummy *n.* **1.** a stupid person; dolt. **2.** (especially in buying land) one put forward to act for others while ostensibly acting for himself. **3.** *N.Z. Prison.* a punishment cell; digger. —*phr.* **4. sell a dummy,** *Football.* to make a feigned or pretended manoeuvre; make a dummy pass.

dummy run *n.* a trial; an exploratory attempt.

dump[1] *n.* **1.** a place, house, or town that is poorly kept up, and generally of wretched appearance. —*phr.* **2. sell the dump to, a.** *Football.* to pass the ball to (another player) to avoid being tackled. **b.** to pass on something worthless, or the object of some dispute to (someone else) to avoid trouble.

dump[2] *n.* **1.** (formerly) a round piece cut from the centre of a silver dollar, and used as a coin in early New South Wales. —*phr.* **2. not give a twopenny dump,** not to care. [origin uncertain]

dumper *n.* **1.** a tip-truck. **2.** *Surfing.* a wave which, in shallow water, instead of breaking evenly from the top, crashes violently down, throwing surfers to the bottom.

dumps *pl. n.* **1.** a dull, gloomy state of mind. *—phr.* **2. down in the dumps,** feeling miserable.

dumpty *n.* a toilet.

dungaree settler *n.* a small battling farmer in the early Australian colonial period. Also, **dungaree farmer.**

dungas *pl. n.* dungarees (a coarse cotton fabric of East Indian origin worn by early settlers).

dung puncher *n.* an active male homosexual.

dunk *v.* to dip (biscuits, etc.) into coffee, milk, etc.

dunlop cheque *n.* a dishonoured cheque; rubber cheque. [*Dunlop* is a trademark for the manufacturers of a range of rubber goods, as tyres, etc.]

dunlop overcoat *n.* a condom.

dunno *v.* contraction of *don't know.*

dunny *n.* **1.** an outside toilet, found in unsewered areas, usually at some distance from the house it serves and consisting of a small shed furnished with a lavatory seat placed over a sanitary can. **2.** a toilet. **3.** a sanitary can. *—phr.* **4. all alone like a country dunny,** See **alone.**

dunny budgie *n.* a fly.

dunny can *n.* a sanitary can.

dunny cart *n.* a sanitary cart.

dunny man *n.* the man who removes full sanitary cans and replaces them with empty ones.

durry *n.* a cigarette.

dust *n.* **1.** *Obsolete.* flour. *—phr.*
2. allow the dust to settle, wait until all the problems are solved before proceeding.
3. bite the dust, a. to be killed or wounded. **b.** to fall. **c.** to fail.
4. dust off, to take from storage and make ready for use again; begin to use or practise again.
5. lick the dust, a. to be killed or wounded. **b.** to grovel.
6. raise the dust, kick up the dust, make a fuss; cause a disturbance.
7. shake the dust off one's feet, to depart with scorn.
8. throw dust in (someone's) eyes, to mislead.

dusting *n.* a beating; thrashing.

dust-up *n.* a commotion; fight; scuffle.

dusty *phr.* **not so dusty,** not too bad; quite good.

dutch *n. British.* a wife. [shortened form of *duchess*]

Dutch *phr.*
1. go Dutch, to have each person pay his or her own expenses.
2. in dutch with, in trouble, in disfavour with.

Dutch courage *n.* courage inspired by alcoholic drink.

Dutch treat *n.* a meal or entertainment in which each person pays for himself. Also, **Dutch shout.**

Dutch uncle *n.* a person who criticises or reproves with unsparing severity and frankness.

dyed-in-the-wool *adj.* through and through; complete; inveterate.

dyke[1] *n.* a lavatory.

dyke[2] *n.* a lesbian.

dynamite *n.* **1.** anything or anyone potentially dangerous and liable to cause trouble. **2.** anything or anyone exceptional.

dynamo *n.* an energetic person.

EA *n.* a woman who is free with sexual favours. [*E(nthusiastic) A(mateur),* prostitute's slang]

each *phr.* **bet each way,** to be undecided or neutral.

eager beaver *n.* **1.** a diligent and zealous person. **2.** one over-eager for work.

eagle *phr.* **the day the eagle,** or **crow shits, a.** payday. Also, **eagle shit day. b.** an imaginary time of good fortune.

eaglehawk *v.* to pluck wool from a dead sheep.

Eagles, the *pl. n.* **1.** the Footscray team in the National Soccer League. **2.** the Manly-Warringah team in the N.S.W. Rugby Football League. Also, **Sea Eagles. 3.** the New Norfolk team in the Tasmanian Australian National Football League. **4.** the Wanderers team in the Northern Territory Football League. **5.** the West Torrens team in the South Australian National Football League. **6.** the Yarraville team in the Victorian Football Association.

ear *phr.*

1. be all ears, to listen attentively.

2. bend (someone's) ear, to harangue (someone).

3. do on one's ear, do without difficulty.

4. go in one ear and out the other, to be heard but ignored; to make no impression.

5. have a flea in one's ear, See **flea.**

6. have an ear to the ground, to be well informed about gossip or trends.

7. have one's ears burn, to be embarrassed when one overhears comments, usually unflattering, about oneself.

8. in a pig's ear, See **pig.**

9. on one's ear, drunk.

10. out on one's ear, dismissed summarily.

11. play it by ear, to improvise; to handle a situation without a set plan.

12. set by the ears, to cause to disagree or quarrel.

13. turn a deaf ear, to refuse to help or consider helping.

14. up to one's ears, deeply involved; extremely busy.

15. wet behind the ears, naive; immature.

earbash *v.* **1.** to harangue (someone). **2.** to talk insistently and for a long time.

earbasher *n.* a persistent talker; a bore.

earbite *v.* to borrow habitually.

earful *n.* **1.** a quantity of oral advice, especially unsolicited advice. **2.** a stern rebuke, especially lengthy or abusive.

early *phr.* **early doors,** an early start; leaving the house at an early hour: *Put the children to bed by eight o'clock, because it is early doors tomorrow.*

early bird *n.* **1.** a person who gets up early. **2.** a person who arrives before others. *—phr.* **3. the early bird catches the worm,** those who make an early start on a project, etc., are likely to be the most successful.

early mark *phr.* **take an early mark,** to leave any place of work early.

early opener *n.* a hotel which opens for bar trading before normal hours.

early-risers *pl. n.* thin blankets or rugs, as carried by swagmen.

earn *n. Prison.* proceeds of a robbery.

earth *phr.*

1. on earth, used to emphasise a statement: *What on earth are you doing?*

2. run to earth, to hunt down; track down.

3. the earth, everything; a great deal: *cost the earth; want the earth.*

earthly *adj.* **1.** possible or conceivable: *no earthly use. —phr.* **2. not have an earthly,** to have no idea.

earthman *n.* (used chiefly in science fiction) a native of the planet earth.

earthshaking *adj.* of the greatest importance; tending to cause great upheaval. Also, **earth-shattering.**

earthworm *n.* a mean or grovelling person.

earwig *n.* an eavesdropper; stickybeak.

easies *pl. n. N.Z.* a woman's elasticised foundation garment, without fastenings; step-ins; roll-ons.

Eastern Stater *n. W.A.* (*often offensive*) a person who comes from any of the eastern states of Australia.

Eastern States *pl. n. W.A.* those states of Australia which are east of the Nullarbor Plain, especially South Australia, Victoria, and New South Wales.

easy *adj.* **1.** having no firm preferences in a particular matter: *I'm easy.* **2.** promiscuous; free with sexual favours. **3.** susceptible, as to a loan or a swindle: *easy take. —adv.* **4.** in an easy manner; comfortably: *Take it easy. —phr.*

5. easy come, easy go, (usually with reference to money) that which is easily obtained is often easily lost or spent.

6. easy on the eyes, attractive; good looking.

7. **go easy on, a.** to be lenient with. **b.** to use sparingly: *Go easy on the honey.*
8. **go easy with,** to handle carefully: *Go easy with that valuable jar.*

easy rider *n.* **1.** *U.S.* a lover who is sexually satisfying. **2.** an easyrider motorbike, with a seat sloping up at the back and large enough to seat two people.

easy street *phr.* **on easy street,** a lifestyle characterised by wealth and luxury.

easy wicket *n.* an easy task; a comfortable position in life: *The manager is on an easy wicket here.* [from *easy wicket* (in cricket) a pitch of slow pace which favours the batsmen]

eat *v.* **1.** to cause to worry; trouble: *What's eating you?* **2.** to perform fellatio or cunnilingus with. *—n.* **3.** (*pl.*) food. *—phr.*
4. eat fit to bust, to eat with good appetite.
5. eat one's heart out, to pine.
6. eat one's words, to take back what one has said.
7. eat out of someone's hand, to be uncritically compliant and trusting, often in a servile or sycophantic manner.
8. eat shit, to be subservient or submissive.

eatery *n.* a cafe or restaurant, especially a cheap one. Also, **eating house.**

eating irons *pl. n.* eating utensils.

eau de cologne *n.* the telephone. [rhyming slang]

echo *n.* one who reflects or imitates another. [from *echo* a repetition of a sound]

eco-nut *n.* a person who is deeply concerned with ecology and environment issues. Also, **eco-freak.**

Edgar Britt *n.* **1.** the act of defecating. **2.** (*pl.*) an attack of diarrhoea. Also, Jimmy Britt. [rhyming slang *Edgar Britt* shit]

edge *phr.*
1. have the edge (over), to have the advantage.
2. over the edge, a. mentally disordered or deranged; unbalanced. **b.** unreasonable; beyond the limits of common sense, decency or justice.

edgeways *phr.* **get a word in edgeways,** to succeed in forcing one's way into an animated conversation or in making a remark when a voluble person is for a moment silent. Also, **edgewise.**

edition *n.* any version of anything, especially one resembling an earlier version.

Eels, the *pl. n.* the Parramatta team in the N.S.W. Rugby Football League. from the Aboriginal meaning of Parramatta, 'the place where *eels* lie down']

eff *v.* a euphemism for the word *fuck*: *Eff off.*

egg *phr.*
1. bad egg, a person of reprehensible character.
2. break eggs with a big stick, to act in an ostentatious or flamboyant manner, usually well beyond the requirements of the given situation.
3. have egg on one's face, to be exposed in an embarrassing situation.
4. in the egg, in the planning stages.
5. lay an egg, See **lay.**

6. **one's eggs are cooked,** *N.Z.* one is in serious trouble.
7. **put all one's eggs in one basket,** to devote all one's resources to or risk all one's possessions, etc., on a single undertaking.
8. **teach one's grandmother to suck eggs,** to give explanations, etc. to someone more experienced than oneself.
9. **tread on eggs,** to be very cautious.

egg-beater *n.* a helicopter. [from supposed resemblance of the blades to those of a kitchen *egg-beater*]

eggflip *n. Horseracing, Gambling, etc.* a tip. [rhyming slang]

egghead *n.* an intellectual; highbrow.

egg roll *n.* (*offensive*) a stupid person; idiot.

eggshell blonde *n.* a bald person.

ego trip *n.* behaviour intended to attract attention and admiration, for the sake of boosting one's own ego.

eh *interj.* an expression of surprise or doubt: *Wasn't it lucky, eh?*

eight ball *phr.* **behind the eight ball,** in an awkward or disadvantageous position.

eighteen *n.* an eighteen-gallon keg.

eighteen pence *n.* Also, **eighteen.** a fence. [rhyming slang]

eighty *phr.* **eighty cents in the dollar,** weak of intellect.

Ekka *n.* the Royal National Association Show held at the Brisbane Exhibition Ground.

elbow *phr.*
1. **bend** or **raise one's elbow,** to drink (especially beer).
2. **up to the elbows,** very busy; wholly engaged or engrossed.

elbow grease *n.* vigorous, continuous exertion; hard physical labour.

elbow-room *n.* sufficient room or scope.

el cheapo *adj.* **1.** of or relating to that which is cheap and inferior, as a restaurant, amplifier, record, etc. — *n.* **2.** any such article. **3.** a cheap restaurant.

elegant *phr.* **elegant sufficiency,** a mock-polite way of saying that one has eaten enough.

elephant *n.* See **pink elephant.**

elephant gun *n.* a surfboard used for riding big waves.

elephants *adj.* drunk. [rhyming slang *elephant's trunk* drunk]

elevator *phr.* **(someone's) elevator** or **lift doesn't go to the top floor,** (someone) is intellectually weak or stupid.

elevenses *n. British.* a light mid-morning snack, usually taken at about 11 o'clock.

em *pl. pron.* them: *sock it to em.* Also, **'em.**

embroider *v.* to adorn or embellish rhetorically, especially with fictitious additions.

embuggerance *n. Military.* **1.** an unnecessary or irrelevant interruption in the completion of a task. **2.** an insignificant or irksome factor which will not prevent the achievement of the overall objective.

emma chisit *phr.* the question 'how much is it?' as rendered in strine. See **strine.**

empire *n.* a large and powerful enterprise or group of enterprises controlled by a single person or group of people.

empire builder *n.* one who sets out to increase his influence, area of control and reputation to the greatest possible extent.

empty *adj.* **1.** hungry. **2.** drained of emotion; spent. —*n.* **3.** something empty, as a bottle, can, or the like.

emu *n. Racing.* one who frequents racecourses, trotting and dog tracks, and TAB branches, and collects discarded betting tickets in the hope of finding some which will pay. [from the Australian bird the *emu* which has the habit of picking up things from the ground]

emu-bobber *n.* a person employed to pick up sticks lying around on land which has been cleared or burnt off; stick-picker. [See **emu**]

emu parade *n.* **1.** *Military.* a parade to clean up an area by emu-bobbing. **2.** the picking up of litter in a camping area, school playground, etc., by a group of people organised for this purpose. [See **emu**]

end *phr.*

1. at a loose end, Also, **at loose ends.** unoccupied; with nothing to do: *She asked him to dinner because he seemed to be at a loose end.*

2. get one's end in, (of a male) to have sexual intercourse.

3. go off the deep end, to become violently agitated; to lose control of the emotions.

4. keep one's end up, to see that one's contribution to a joint undertaking is adequately performed.

5. make (both) ends meet, to keep expenditure within one's means.

6. no end, very much; greatly: *The news of the lottery win thrilled them no end.*

7. the (living) end, a. the worst possible. **b.** a person who is incompetent or insufferable in every way.

endless belt *n.* **1.** a girl or woman who is free with sexual favours. **2.** a prostitute. [?from U.S. *band* a woman]

Englander *phr.* **little Englander, a.** an opponent of imperialism, especially one advocating isolationist policies in the late 19th century. **b.** an opponent of Britain's joining the common market, preferring to see Britain less influential but independent.

Enzed *n., adj.* New Zealand. [from the initials *N.Z.*]

Enzedder *n.* a New Zealander.

equaliser *n.* a gun.

esky *n.* a portable icebox; chillybin. [Trademark]

Esperance doctor *n.* a strong cool wind blowing after a hot day. [from *Esperance,* a town in Western Australia]

ethnic *adj.* **1.** (*sometimes offensive*) **a.** of those who seek an older and more simple life style, usually involving the practice of handicrafts and supposed folk ways. **b.** of the life style itself. **c.** odd; quaint. —*n.* **2.** Also, **ethno.** (*offensive*) a migrant.

euchre *v.* to outwit; get the better of, as by scheming. [from the card game *euchre*]

euchred *adj.* beaten; exhausted.

eugarie *n. Qld.* a pipi. Also, **yugarie.**

euphemism *n.* the toilet.

eureka *interj.* an exclamation of truimph at a discovery or supposed discovery. [the reputed exclamation of Archimedes, 287?-212 B.C., mathematician, physicist and inventor, when, after long study, he discovered a method of detecting the amount of alloy in the crown of the king of Syracuse]

even stephen *adj.* (often plural) equal, as shares, etc.

evergreen *adj.* **1.** retaining youthful characteristics in maturity: *an evergreen tennis player.* **2.** retaining popularity from an earlier period: *an evergreen song.*

every *phr.* **every man and his dog,** a lot of people; the general public.

evil eye *n.* the power superstitiously attributed to certain persons of inflicting injury or bad luck by a look: *to put the evil eye on (something)*

evil-minded *adj.* (*humorous*) excessively preoccupied with sex.

ex *n.* **1.** one's former husband or wife. **2.** one's former boy friend or girl friend.

exchequer *n.* funds; finances. [from *exchequer* a treasury, as of a state or nation]

excuse *n.* an inferior or inadequate example of something specified: *She was shabbily dressed and wearing a poor excuse for a hat.*

executive *v.* to murder or assassinate, often through the agency of a hired killer.

ex-govie *n. A.C.T.* a house built by the Commonwealth Government but now privately owned.

expect *v.* **1.** to be pregnant: *My wife is expecting again. —phr.* **2. expect me when you see me,** an expression indicating that one's movements, plans, etc., are uncertain.

expert *n. Shearing.* the man who sharpens the shearers' cutters; the one in charge of machinery in a shearing shed.

ex-pug *n.* a retired boxer. [from *pug* short for pugilist, a boxer]

extra *adj.* **1.** extraordinarily good. *—n.* **2.** *Teaching.* an additional period taught by a teacher because of the absence of a colleague.

eye *v.* **1.** to look or gaze at. *—n.* **2.** a look, glance or gaze: *to cast one's eye on a thing. —phr.*

3. all my eye, nonsense.

4. (all) my eye and Betty Martin, an expression of disbelief or scepticism.

5. cry one's eyes out, to weep copiously.

6. do in the eye, to take advantage of; cheat; swindle.

7. easy on the eye, attractive to look at.

8. eye off, to watch or look at with interest, attention, etc.

9. get one's eye in, to adapt oneself to a situation; become accustomed. [from the sense in cricket and other ball games of being able, through practice, to follow the movement of the ball]

10. give the glad eye, to look amorously at.

11. go eyes out, to work very hard.

12. keep one's eyes open/skinned/peeled, to be especially watchful.

13. lay/clap/set eyes on, to catch sight of; see.

14. make eyes at, to gaze flirtatiously at.
15. pick the eyes out of, to select the best parts, pieces, etc., of (a collection).
16. sight for sore eyes, a welcome sight; an agreeable surprise.
17. slap in the eye, See **slap.**
18. turn a blind eye on or **to,** to pretend not to see; to avoid noticing that which one should oppose or condemn.
19. up to the eyes in, very busy with; deeply involved in.

eyeball *v.* **1.** to look at; *to eyeball a room.* *—phr.*
2. eyeball to eyeball, aggressively face to face; in confrontation.
3. greasy eyeball, a disdainful look: *He gave her the greasy eyeball.*

eyeful *n.* a person of striking appearance, especially a beautiful woman.

eye-opener *n.* **1.** an enlightening or startling disclosure or experience. **2. a.** an alcoholic drink, especially one taken early in the day; heart-starter. **b.** any strong drink, as coffee, taken before one begins the day's activities. **c.** a drug addict's first injection of the day.

Eyetalian *n.* (*humorous*) an Italian. Also, **eyetie, eytie, Itie.**

eyeteeth *phr.*
1. cut one's eyeteeth, to become old and experienced enough to understand things.
2. cut one's eyeteeth on (something) to be brought up on, to gain experience with or reach maturity through contact with (something).
3. give one's eyeteeth for, to desire greatly.

eyewash *n.* **1.** a deception intended to mislead a person into thinking something is good or correct. **2.** nonsense.

f *v.* **1.** to swear at by using the word *fuck*: *he f'ed all and sundry. Also,* **eff.** *—phr.* **2. f, f and f,** full, free and frank (as a discussion, etc.).

3. the big f, (*euphemism*) any of various vulgar, taboo, or obscene words beginning with f, especially *fuck.*

fabulous *adj.* wonderful; exceptionally pleasing. Also, **fab.**

face *n.* **1.** boldness; impudence: *to have the face to ask. —phr.*

2. a face as long as a wet weekend or **a month of Sundays,** a long face; a dismal or disappointed expression.

3. face it out, to ignore or defy blame, hostility, etc.

4. face (someone) out, a. to stare out. **b.** to cause (another) to concede by adhering consistently to a particular version.

5. off one's face, a. mad; insane. **b.** incapacitated as a result of taking drugs, alcohol, etc.

6. put on one's face, to put on make-up.

face-ache *n.* an extremely ugly or irritating person.

face fungus *n.* facial hair, as a moustache, beard, etc. Also, **face fittings.**

faceless men *pl. n.* men who exercise political power without having to take personal or public responsibility for their actions. [term especially applied in the 1963 election campaign to the non-parliamentary members of the ALP Federal Executive, who held power over the parliamentary representatives]

facer *n.* **1.** a sudden and severe check; a disconcerting difficulty, problem, etc. **2.** a blow in the face.

face-saver *n.* one who or that which saves one's prestige: *The team's fine performance after their run of losses was a real face-saver for the coach.*

facial *phr.* **give (someone) a facial,** *Prison.* to slash (someone's) face with a knife, razor, etc.

facts of life *n.* **1.** the details concerning sexual behaviour and reproduction, especially as explained to children. **2.** the harsh realities of life.

fade *v.* **1.** to change the direction of a surfboard towards the breaking part of a wave. *—phr.* **2. do a fade,** to ease oneself out of a difficult situation.

fag *n.* **1.** a cigarette. **2.** a homosexual. *—v.* **3.** to smoke a cigarette.

faggot *n.* a male homosexual.

fair *adv.* **1.** completely: *I was fair flabbergasted.* *—phr.*

2. a fair cop, a. justice. **b.** the discovery of a wrongdoer in the act or with his guilt apparent.

3. a fair treat, excessively.

4. fair and square, honest; just; straightforward.

5. fair crack of the whip, See **whip.**

6. fair shake of the dice, an appeal for fairness or reason.

7. fair suck of the sav or **saveloy,** See **saveloy.** Also, **fair suck of the sauce bottle.**

8. fair to middling, tolerably good; so-so.

9. in a fair way to, likely to; on the way to: *You're in a fair way to becoming an alcoholic, the amount you drink.*

fair buck *n., interj. N.Z.* See **fair go.**

fair dinkum *adj.* **1.** true, genuine, dinkum: *Are you fair dinkum?* *—interj.* **2.** Also, **fair dink.** an assertion of truth or genuineness: *It's true, mate, fair dinkum.*

fair do *n., interj. N.Z.* See **fair go.**

fair enough *adj.* **1.** acceptable; passable. *—interj.* **2.** a statement of acquiescence, or agreement.

fair go *n.* **1.** a fair or reasonable course of action: *Do you think that's a fair go?* **2.** an inadequate opportunity: *He never had a fair go.* *—interj.* **3.** Also, *N.Z.* **fair buck, fair do,** an appeal for fairness or reason: *Fair go, mate!* [from the call in a two-up game signalling that all requirements have been met and requesting that the spinner be allowed to toss the coins without influence]

fair-haired *adj. U.S.* blue-eyed; favourite: *He was the fair-haired boy of the high society matrons.*

fairy *n.* **1.** an effeminate male, usually a homosexual. *—phr.*

2. away with the fairies, a. daydreaming. **b.** mentally unsound or eccentric.

3. shoot a fairy, to fart.

fake *v.* **1.** to get up, prepare, or make (something specious, deceptive, or fraudulent). **2.** to conceal the defects of, usually in order to deceive. **3.** *Theatre, Jazz.* to improvise. *—n.* **4.** Also, **faker.** one who fakes.

faker *n.* **1.** *U.S.* a petty swindler. **2.** *U.S.* a pedlar or street vendor.

Falcons, the *pl. n.* the West Perth team in the West Australian Football League.

fall *v.* **1.** to become pregnant: *They were living in Dubbo when she fell with Jane.* *—phr.*

2. fall about, to laugh immoderately.

3. fall down, to fail: *to fall down on the job.*

4. fall flat, to fail to have a desired effect: *His jokes fell flat.*

5. fall for, a. to be deceived by: *Don't fall for that old story.* **b.** to fall in love with.

6. fall foul, to come into conflict; have trouble: *fall foul of the law.*

7. fall in, a. to agree. **b.** to make a mistake, especially when tricked into doing so: *He really fell into that one!*

8. fall in with, a. to meet and become acquainted with: *She fell in with a bad lot.* **b.** to agree to: *I'll fall in with anything you decide.*

9. fall off the back of a truck, to be obtained by questionable or illegal means.

10. fall on or **upon,** to light upon; chance upon.

11. fall or **land on one's feet, a.** to emerge from a difficult or adverse situation without serious harm. **b.** to be lucky.

12. fall out, a. to disagree; quarrel. **b.** to occur; happen; turn out.

13. fall over oneself, a. to become confused in attempting to take some action. **b.** to be excessively enthusiastic.

14. fall short, to prove insufficient; give out.

15. fall through, to come to naught; fail; miscarry.

fall guy *n.* an easy victim; scapegoat.

fallout *n.* the indirect effects of a decision, event, etc.

falsies *pl. n.* **1.** pads, usually of foam rubber, worn to enlarge the outline of the breasts. **2.** false eyelashes. **3.** false teeth.

family *n.* **1.** (formerly) the criminal fraternity. *—phr.* **2. in the family way,** pregnant.

fan *phr.* **fan the breeze,** to engage in idle talk.

fancy *phr.* **fancy oneself,** to hold an excessively good opinion of one's own merits.

fancy man *n.* **1.** a pimp. **2.** a lover.

fancy woman *n.* **1.** a mistress. **2.** a prostitute.

fandangle *n. Originally U.S.* nonsense; tomfoolery. [perhaps alteration (modelled on *newfangle(d)*) of *fandango* which was occasionally used earlier in this sense]

fang[1] *n.* **1.** a tooth. **2.** the act of eating. **3.** the penis. *—v.* **4.** to borrow. *—phr.* **5. put the fangs** or **nips into,** to attempt to borrow from; to make demands on.

fang[2] *phr.* **fang around,** to drive one's car at a very great speed. [from Juan *Fangio* (born 1911) Argentinian racing-car driver]

fang carpenter *n.* a dentist.

fanny *n.* **1.** the buttocks. **2.** the female pudenda.

fantabulous *adj.* marvellous; wonderful. [blend *fantastic* + *fabulous*]

fantastic *adj.* very good; fine; wonderful: *a fantastic pop song. Also,* **fantastical.**

fanzine *n.* a magazine publishing photographs and adulatory articles about a film star, pop star, etc. [blend *fan* + *(maga)zine*]

F.A.Q. *adj.* of reasonable or average quality. Also, **f.a.q.** [initials *F(air) A(verage) Q(uality)*]

far *phr.*

1. a far cry, very different.

2. far gone, a. extremely mad. **b.** extremely drunk.

farm *phr.* **1. buy back the farm,** to reverse a trend towards excessive foreign ownership of Australian-based companies, by means of legislation or other government control. **2. the Farm, a.** Monash University, Melbourne. **b.** Warwick Farm Racecourse, Sydney. **c.** Eagle Farm Racecourse, Brisbane.

farmer *phr.* **Pitt Street farmer,** See **Pitt Street.**

Farmer Giles *n.* piles; haemorrhoids. [rhyming slang]

farmyard confetti *n.* See **confetti.**

far-out *adj.* **1.** fantastic; wonderful. **2.** extremely unconventional.

fart *n.* **1.** an emission of wind from the anus, especially an audible one. **2.** a foolish or ineffectual person: *stupid old fart.* —*v.* **3.** to emit wind from the anus. —*phr.* **4. fart about** or **around,** to behave stupidly or waste time. [Middle English *ferten*]

fart-arse *n.* **1.** an ineffectual person. —*v.* **2.** to waste time; to idle.

farting *adj. N.Z.* trivial.

fart machine *n.* a small motorised over-snow vehicle; skidoo.

fart sack *n.* a bed or a sleeping bag. Also, **farter.**

fascist *n.* **1.** a dictatorial person: *The new headmaster is a real fascist.* **2.** anyone with extreme right-wing views, especially with regard to race.

fast *phr.*

1. pull a fast one, to act unfairly or deceitfully.

2. play fast and loose with, to behave in an inconsiderate, inconstant, or irresponsible manner towards.

fast buck *n.* money made quickly and often irresponsibly.

fastie *n.* **1.** a deceitful practice; a cunning act. —*phr.* **2. pull** or **put over a fastie,** to deceive; to take an unfair advantage.

fast lane *phr.* **in the fast lane,** (of a lifestyle, etc.) conducted at a hectic pace. [from the *fast lane* on an expressway, motorway, etc.]

fat *n.* **1.** a marble. **2.** an erection. **3.** overtime. —*phr.*

4. a fat chance, little or no chance.

5. a fat lot, little or nothing.

6. chew the fat, See **chew.**

7. crack a fat, See **crack.**

8. live on one's fat, to consume reserves; live on capital.

9. stick fats, not to shift from one's position.

10. the fat is in the fire, an irrevocable (and often disastrous) step has been taken, resulting in dire consequences.

fat cat *n.* a person who expects special comforts and privileges because of his position or wealth: *the fat cats of the Public Service.*

father *phr.* **the father and mother of a (something),** the biggest; a very big (something): *They were having the father and mother of a row, and the shouting could be heard from the end of the street.*

Father's Day *phr.* **happy as a bastard on Father's Day,** See **bastard.**

fatty *n.* a fat person: *He's a real fatty.* Also, **fatso.**

faze *v.* to disturb; discomfit; daunt. [variation of obsolete *feeze* disturb, worry]

fazzo *adj.* (*especially in children's speech*) fabulous; wonderful.

fear *phr.* **no fear!** certainly not.

feather *phr.*
1. a feather in one's cap, a mark of distinction; an honour.
2. feather one's nest, to provide for or enrich oneself.
3. make feathers fly, to cause confusion; create disharmony.
4. show the white feather, to show cowardice.

featherbed *n. Cricket.* a 'dead' pitch which does not give the bowler any assistance.

featherbedder *n.* a person accustomed to comfort.

featherbrain *n.* an irresponsible or weak-minded person.

feather duster *phr.* **a rooster one day and a feather duster the next,** See **rooster.**

feature *phr.* **feature with,** to have sexual intercourse with.

fed[1] *phr.* **fed up (to the back teeth** or **gills),** annoyed; frustrated.

fed[2] *n.* **1.** a federal police officer. **2.** any policeman.

feed *n.* **1.** a meal. **2.** *Theatre* a cue to an actor, especially a comedian.

feedbag *phr.* **put on the feedbag,** to have a meal.

feel *phr.*
1. feel (someone) out, to make exploratory moves in order to assess possible reaction.
2. feel (someone) up, (usually of a man) to attempt the manual sexual stimulation of a woman.
4. feel up to, to be well enough to be capable of; to be able to cope with.

feet *n.* See **foot.**

feller *n.* a form of address, usually aggressive, amongst men. Also, **fella.**

fence *n.* **1. a.** a person who receives and disposes of stolen goods. **b.** the place of business of such a person. *—v.* **2.** to receive stolen goods. *—phr.*
3. over the fence, not reasonable; immoderate.
4. rush one's fences, to act precipitately.
5. sit on the fence, to remain neutral; to avoid a conflict.

fencing *n.* the receiving of stolen goods.

Fernleaf *n.* a New Zealander. [from the leaf of the silver fern, emblem of New Zealand]

ferret *n.* **1.** the penis. **2.** *Cricket.* a tail-end batsman. *—phr.*
3. give the ferret a run, (of a male) to urinate.
4. run the ferret up the drainpipe, (of a male) to have sexual intercourse.

fetch *phr.*
1. fetch and carry, to do minor menial jobs.
2. fetch up, a. to reach as a goal or final state; end up: *You'll fetch up in prison.* **b.** to vomit. **c.** to come to a sudden stop, as a ship running aground, or a walker suddenly pausing.

fib *v.* to beat; give (someone) a rapid succession of blows, as in boxing.

fiddle *n.* **1.** an illegal or underhand transaction or contrivance. *—v.* **2.** to trifle. **3.** to contrive by illegal or underhand means: *fiddle the books. —phr.*
4. fiddle or **fiddlearse about,** to waste time or effort.

5. **fit as a fiddle,** in excellent health.
6. **have a face as long as a fiddle,** to look dismal.
7. **on the fiddle,** manipulating or covering up illicit money-making schemes.
8. **play second fiddle,** to take a minor or secondary part.

fiddle-faddle *n.* **1.** nonsense; something trivial. *—v.* **2.** to fuss with trifles. [reduplication of *fiddle* to trifle]

fiddler[1] *n.* a cheat or rogue.

fiddler[2] *phr.* **in and out like a fiddler's elbow,** engaged in senseless frenzied activity.

fiddley *n.* (formerly) a one pound note (£1); quid. Also, **fid, fiddley-did.** [rhyming slang]

field *phr.* **field a book,** to be a bookmaker.

field day *n.* an occasion of unrestricted enjoyment, amusement, etc.

fielder *n. Horseracing.* a bookmaker.

fierce *adj.* extreme; unreasonable: *The prices are fierce.*

fifth wheel *n.* any extra or superfluous person or thing.

fifty-fifty *n.* **1.** Also, **fifty.** a glass of beer, half old and half new. **2.** a dance, usually held in a country or suburban hall, in which the dancing and music is half old-time and half modern.

fight *phr.*
1. **fight shy of,** to keep carefully aloof from (a person, affair, etc.).
2. **fight like Kilkenny cats,** See **Kilkenny.**

figure *v.* **1.** to be in accordance with expectations or reasonable likelihood: *That figures. —phr.* **2. figure on, a.** to count or rely on. **b.** to take into consideration.

Fiji uncle *n.* See **uncle.**

fill *phr.*
1. **fill in, a.** to act as a substitute; replace. **b.** to give all necessary information, etc.
2. **fill out,** to complete the details of (a plan, design, etc.).
3. **fill** or **fit the bill,** to satisfy the requirements of the case; be or do what is wanted.

filly *n.* a girl. [from *filly* a female foal]

filter *n.* a filter tip cigarette.

filth *n. Originally British.* the police.

filthy *adj.* **1.** (as a general epithet of strong condemnation) highly offensive or objectionable: *filthy lucre.* **2.** very unpleasant: *filthy weather. —phr.* **3. filthy rich,** very rich.

fin *n.* the arm or hand. [from *fin* the paddle-like organ on the body of a fish, etc.]

finagle *v.* **1.** to practise deception or fraud. **2.** to trick or cheat (a person). **3.** to wangle: *to finagle free tickets.*

financial *adj.* having ready money: *to be financial.*

finder *phr.* **finders keepers,** an expression said by someone who has found something, claiming his right to keep it. [from children's rhyme '*finders keepers* losers weepers']

finger *v.* **1.** to point out; accuse. *—phr.*
2. **burn one's fingers,** to get hurt or suffer loss from meddling with or engaging in anything.
3. **finger up,** to stimulate erotically with the fingers.

4. **have a finger in the pie,** to have a share in the doing of something.
5. **have pie on one's fingers,** to make gains from illegal business activities.
6. **pull one's finger out,** to become active; hurry.
7. **put the finger on, a.** to inform against or identify a criminal to the police. **b.** to designate a victim, as of murder or other crime.
8. **slip through one's fingers,** to elude one, as a missed opportunity.
9. **twist round one's little finger,** to dominate; influence easily.

finickity *adj.* fussy. [blend *finick(y)* + *(pernick)ety*]

fink *n.* **1.** a strike-breaker or blackleg. **2.** a contemptible or undesirable person, especially one who reneges on an understanding.

fire *phr.*
1. **fire away,** to begin speaking.
2. **go through fire and water,** to face any hardship or danger.
3. **hang fire,** to be irresolute, postponed or delayed.
4. **play with fire,** to meddle carelessly or lightly with a dangerous matter.
5. **under fire,** under criticism or attack.

firebug *n.* one who maliciously sets fire to buildings or other property, or to bushland.

firefighter *n.* a troubleshooter.

Firm, the *n.* (formerly) J.C. Williamson Ltd, theatrical entrepreneurs.

first *phr.*
1. **draw first blood,** in non-physical competition, to gain the initial advantage.
2. **first and last,** altogether; in all.
3. **first cab off the rank,** See **cab.**
4. **first up,** at the first attempt.

first base *phr.* **get to first base,** to make a slight amount of progress. [from *first base* in the game of baseball]

First Fleeter *n.* a person whose family can be traced back to someone who came to Australia with the First Fleet in 1788.

first string *phr.* **the first string in one's bow,** one's chief skill, asset, etc.

first-timer *n.* *Prison.* a first time prisoner; cleanskin.

fish *n.* **1.** (with an adjective) a person: *a queer fish, a poor fish. —phr.*
2. **a fish out of water,** out of one's proper environment; ill at ease in unfamiliar surroundings.
3. **a fine** or **pretty kettle of fish,** trouble; confusion.
4. **drink like a fish,** to drink alcoholic liquors to excess.
5. **feed the fishes. a.** to be seasick. **b.** to drown.
6. **fish in troubled waters,** to take advantage of uncertain conditions; profit from the difficulties of others.
7. **fish out,** to obtain by careful search or by artifice.
8. **neither fish nor fowl,** Also, **neither fish/flesh/fowl, nor good red herring.** neither one thing nor the other.
9. **other fish to fry,** other matters requiring attention.

Fish, the *n.* a commuter train running between Sydney and the Blue Mountains. See **Chips.**

fishy *adj.* **1.** improbable, as a story. **2.** of questionable character.

fist *n.* **1.** the hand. **2.** a person's handwriting. *—phr.* **3. make a (good, poor) fist of,** to perform a task (well, badly).

fit[1] *v.* **1.** to bring a person before the law on a trivial or trumped-up charge while really intending to victimise him: *He had been trying to fit Chilla for years.* *—n.* **2.** the equipment used to prepare and inject drugs. *—phr.*

3. fit like a glove, to be a perfect fit.

4. fit the bill, to suit; be what is required.

5. fit to be tied, very angry.

fit[2] *phr.*

1. by or **in fits (and starts),** by irregular spells; fitfully; intermittently.

2. throw a fit, to become very excited or angry.

five *phr.* **take five,** to take a break, originally of five minutes, for rest or refreshment, especially as of a performing group in rehearsal.

five-finger discount *n.* stealing; shoplifting.

fiver *n.* **1.** (formerly) a five-pound note. **2.** anything that counts as five.

five-to-two *n.* Jew. [rhyming slang]

fix *v.* **1.** to arrange matters, especially privately or dishonestly, so as to secure favourable action: *to fix a jury or a game.* **2.** to put in a condition or position to make no further trouble. **3.** to get even with; get revenge upon. *—n.* **4.** a position from which it is difficult to escape; a predicament. **5.** a shot of heroin or some other drug. **6.** a bribe. **7.** any dishonest device or trick.

fixed *adj.* arranged with, or arranged, privately or dishonestly.

fixings *pl. n.* appliances; trimmings.

fix-up *n. Prison.* a court case in which evidence has been fabricated or arranged so as to ensure a conviction; frame-up.

fizz *n.* **1.** an informer; stool pigeon. *—phr.* **2. fizz on,** to act as a stool pigeon.

fizzer *n.* **1.** a firecracker which fails to explode. **2.** a failure; fiasco.

fizz-gig *n.* a police informer.

fizzle *n.* **1.** a fiasco; a failure. *—phr.* **2. fizzle out,** to fail ignominiously after a good start.

flab *n.* bodily fat; flabbiness.

flack *n. Originally U.S.* a publicity agent; press secretary; public relations officer. Also, **flak.**

flag *n.* **1.** Also, **Australian flag.** part of a shirt, which has come untucked and hangs out over the trousers. *—phr.*

2. have the flags out, a. to celebrate or welcome. **b.** to be menstruating.

3. hoist the flag, *Prison.* to appeal against a conviction or the severity of a sentence.

4. keep the flag flying, to appear courageous and cheerful in the face of difficulty.

5. raise the flag, *Prison.* to lodge an appeal against a conviction.

6. raise the flags, a. *Rugby, Australian Rules.* to score a conversion. **b.** to accomplish something.

7. show the flag, to put in an appearance.

8. strike or **lower the flag,** to submit or surrender.

9. take the flag, *Australian Rules.* to win the premiership pennant.

flaggie *n.* a flagman; one who signals with a flag.

flak *n.* reactions, repercussions or publicity, usually poor, to some action, decision, etc.: *to cop a lot of flak.* Also, **flack.** [from *flak,* anti-aircraft fire]

flake *v.* **1.** Also, **flake out.** to collapse, faint, or fall asleep, especially as a result of complete exhaustion, or influence of alcohol, drugs, etc. *—n.* **2.** one who is strange or eccentric.

flakers *adj.* unconscious; dead drunk.

flaky *adj.* out of sorts; moody.

flaming *adj.* **1.** used to add emphasis to a statement: *a flaming bore.* **2.** a euphemism for various taboo words: *Stone the flaming crows.*

flannel *n.* **1.** *British.* evasive or flattering talk. *—v.* **2.** *British.* to flatter or talk evasively to.

flap *v.* **1.** to panic, become flustered. *—n.* **2.** a state of panic or nervous excitement: *in a flap.* **3.** a cheque. *—phr.* **4. flap one's gums,** to talk at length; gossip.

flapdoodle *n.* nonsense; bosh.

flapping *phr.* **with one's ears flapping,** with keen interest or astonishment, especially when listening to a conversation not meant to be overheard.

flare-up *n.* a sudden outburst of anger.

flash *n.* **1.** a momentary sensation of pleasure following the injection of certain narcotic drugs. **2.** *Obsolete.* the cant or jargon of thieves. *—v.* **3.** (of a man) to expose the genitals in public. **4.** to make a sudden or ostentatious display of: *to flash one's diamonds.* *—adj.* **5.** pretentious. *—phr.*

6. flash in the pan, something which begins promisingly but has no lasting significance.

7. flash one's dover, see **dover.**

flasher *n.* a man who briefly exposes his private parts in public.

flat *adj.* **1.** (formerly) honest (opposed to *sharp*). *—n.* **2.** (formerly) an honest man. **3.** a policeman. [?from *flatfoot*] **4.** *Prison.* tobacco other than that officially issued. *—phr.*

5. fall flat, to fail completely; fail to succeed in attracting interest, etc.

6. flat as a tack/strap/the Nullarbor, a. very flat. **b.** uninteresting.

7. flat out, a. as fast as possible. **b.** very busy. **c.** lying prone. **d.** exhausted; unable to proceed.

8. flat out like a lizard drinking or **on a rock,** See **lizard.**

flat chat *adv.* at full speed: *He drove flat chat down the road.* Also, **full chat, flat strap, full strap.**

flat-chat *v.* to travel at full speed, especially in a motor vehicle. Also, **flat-strap.**

flatfoot *n.* a policeman.

flat-footed *adj.* **1.** clumsy and tactless. **2.** unprepared, unable to react quickly: *to be caught flat-footed.*

flathead *n.* a fool or simpleton.

flat spin *n.* a state of great confusion.

flatten *v.* **1.** to knock (someone) out. **2.** to crush or disconcert.

flattie *n.* a flat-bottomed dinghy.

flatties *pl. n.* a pair of shoes with low heels.

flaxie *n. N.Z.* a person who works in a flax mill. Also, **flax, flaxy.**

flea *phr.* **flea in one's ear, a.** a discomforting rebuke or rebuff; a sharp hint. **b.** a blow to the ear; a cuff.

fleabag *n.* **1.** a sleeping bag. **2.** a dog; any worthless creature ridden with fleas. **3.** an old hag. **4.** a person, especially a child, who fidgets.

flea-bitten *adj.* shabby; dirty.

fleapit *n.* a shabby, dirty room or building, especially a cinema. Also, **flea house.**

flea rake *n.* a comb.

fleas-and-itches *n.* pictures; the cinema. [rhyming slang]

Flemington confetti *n.* See **confetti.**

flesh *phr.*
1. flesh out, to explain, amplify.
2. in the flesh, in bodily form; in person.
3. pound of flesh, a person's right or due, insisted on mercilessly with a total disregard for others (from Shakespeare's 'The Merchant of Venice').

flick[1] *n.* **1.** a movie film. **2.** (*usually pl.*) a picture theatre. Also, **flickers.**

flick[2] *phr.* **give (someone) the flick,** to reject or dismiss; give up an activity, job, etc. [?from pest exterminator advertisement *One Flick and They're Gone*]

flies *pl. n.* See **fly**[2]**.**

flim-flam *n.* **1.** a piece of nonsense; mere nonsense. **2.** a trick or deception; humbug. —*v.* **3.** to trick; delude; humbug; cheat. —*adj.* **4.** nonsensical; worthless.

fling *phr.*
1. at full fling, at full speed; with reckless abandon.
2. give (something) a fling, to make an attempt at (something), often when not confident of success.

flip[1] *n.* **1.** short flight, especially for pleasure, in an aeroplane. **2.** an irresponsible person. —*v.* **3.** to amaze: *The results will flip you.* **4.** to anger; enrage; annoy. —*phr.*
5. flip one's lid, to become angry.
6. flip oneself off, to masturbate.

flip[2] *adj.* pert; flippant.

flipper[1] *n.* the hand. [from *flipper* a broad flat limb, as of a seal, whale, etc.]

flipper[2] *n.* a fool.

flipping *adj.* used to express disgust, annoyance, etc., often used as a euphemism for *fucking*: *a flipping headache.*

flip side *n.* **1.** the reverse of a gramophone record, usually carrying a song, etc., of less interest, or popularity. **2.** the other side of an argument; situation, etc.

flit *v.* **1.** to change one's residence, especially quickly and surreptitiously. **2.** to elope. —*n.* **3.** a removal, especially a surreptitious one; an elopement: *a moonlight flit.* **4.** *Chiefly U.S.* a male homosexual.

flivver *n.* **1.** *Originally U.S.* something of unsatisfactory quality or inferior grade. **2.** an old cheap car. [originally meaning a failure; blend *flopper* (from *flop*) and *fizzler* (from *fizzer*)]

floater *n.* **1.** a cheque which is not honoured. **2.** *Two-Up.* Also, **butterfly.** a coin which when tossed fails to spin. **3.** *Prison.* a book, magazine, etc., the property of a prisoner, circulated throughout the gaol. **4.** one who often changes his job; a temporary employee; one of the floating population. **5.** *Originally S.A.* a meat pie served in pea soup.

flog *v.* **1.** to sell or attempt to sell. **2.** to steal. *—phr.*
3. flog a dead horse, to make useless efforts, as in attempting to raise interest in a dead issue.
4. flog to death, to overuse.

flogger *n. Australian Rules.* a short stick with a bunch of crepe paper streamers (in team colours) attached, waved to approve a score, or to distract a member of the opposing team. Also, **Patti Duke.**

floor *v.* **1.** to beat or defeat. **2.** to punch or hit (someone) to the ground: *Say that again and I'll floor you!* **3.** to confound or nonplus: *to be floored by a problem. —phr.* **4. wipe the floor with,** to overcome or vanquish totally.

floorer *n.* something that beats, overwhelms, or confounds.

floozy *n.* a worthless woman; harlot. Also, **floosy, floosie, floozie.**

flop *v.* **1.** to fail. **2.** to sit or lie down; rest: *You look exhausted, why don't you flop here for the night? —n.* **3.** a failure. **4.** a place to bed down.

flophouse *n.* a dosshouse.

floppy *adj.* tending to flop.

floury baker *n.* a common cicada of eastern Australia, dark brown with orange markings and a superficial whitish covering resembling flour. Also, **baker, floury miller.**

flower *phr.* **see the flowers,** to be menstruating.

flu *n.* influenza.

fluff *n.* **1.** a blunder or error in execution, performance, etc. **2.** a fart. *—v.* **3.** to fail to perform properly: *to fluff a golf stroke, an examination, lines of a play.* **4.** to blunder; fail in performance or execution. **5.** to break wind. *—phr.* **6. bit of fluff,** a girl, especially one who is superficially attractive.

fluke *v.* to hit, make, or gain by a fluke, lucky chance or accidental advantage.

fluky *adj.* obtained by chance rather than skill.

flummox *v.* to make or become bewildered, confused or unsure of what to do.

flump *v.* **1.** to plump down suddenly or heavily; flop. *—n.* **2.** act or sound of flumping. [blend of *fall* and *plump*]

flunk *v.* **1.** to fail, as a student in an examination. **2.** Also, **flunk out.** to give up; back out. **3.** to remove (a student) as unqualified from a school, course, etc. *—n.* **4.** a failure, as in an examination.

flute *n.* **1.** a topic; subject. **2.** *Horseracing.* a jockey's whip. *—phr.*
3. have the flute, to monopolise a conversation.
4. pass the flute or **kip,** to allow someone else an opportunity to speak.

fluter *n.* a garrulous person.

flutter *v.* **1.** to wager (a small amount). *—n.* **2.** a small wager or bet.

flutter-by *n.* a butterfly.

fly[1] *n.* **1.** an attempt: *Give it a fly. —v.* **2.** *Australian Rules.* to leap high to catch the ball. *—phr.*

3. **fly high. a.** to be ambitious. **b.** to be in a state of euphoria, as induced by drugs.
4. **fly in the face of,** to defy insultingly.
5. **fly off the handle,** See **handle.**
6. **fly out,** to lose one's temper; become suddenly violently angry.
7. **fly a kite, a.** to attempt to obtain reactions to a proposal for a course of action by allowing it to be circulated as a rumour or unconfirmed report. **b.** to undertake some other activity: *Go fly a kite.* **c.** to cash a worthless cheque, or one stolen or forged.
8. **let fly,** See **let.**
9. **on the fly,** *U.S.* **a.** while still in flight; on the volley. **b.** hurriedly.
[from *fly* to move or be borne through the air]

fly[2] *phr.*
1. **a shut mouth catches no flies,** See **mouth.**
2. **drink with the flies,** to drink alone in a pub.
3. **drop like flies,** (of numbers of people or animals) to become sick or die.
4. **fly in the ointment,** a slight flaw that greatly diminishes the value or pleasure of something.
5. **no flies on (someone), a.** (someone) is not easily tricked. **b.** *Obsolete.* honest.
[from *fly* a small two-winged insect, as the housefly, etc.]

fly[3] *adj.* knowing; sharp. [? special use of *fly*[1]]

flyblown *adj.* broke; penniless.

flybog *n.* jam. [World War I; shortened in World War II to **bog**]

fly-by-night *adj.* **1.** irresponsible; not to be trusted; unreliable. —*n.* **2.** a person who leaves secretly at night, as in order to avoid paying his debts. **3.** one who leads a gay night-life.

flying circus *n.* a group of aircraft operating together performing aerobatic manoeuvres.

Flynn *phr.* **in like Flynn,** successful in a particular enterprise, especially sexual. [?from the popular belief in the sexual prowess of Errol *Flynn,* Australian-born U.S. movie actor]

flypaper *phr.* **flypaper for a stickybeak,** a response to any silly or unwarranted question; a put-down; snub

F.O. *n.* a foreign order; a job or repair, not part of the regular work authorised by an employer.

foam *phr.* **foam at the mouth,** to be speechless with some emotion, especially with rage, etc.

foggiest *phr.* **not have the foggiest,** not to have the least idea. [from *foggy* resembling fog, obscure]

foghorn *n.* a deep, loud voice. [from *foghorn* a loud horn for sounding warning signals, as to vessels in foggy weather]

fold *phr.* **fold up,** to fail in business.

folding money *n.* banknotes. Also, **folding stuff, folding variety.**

folkie *n.* one who likes or performs folk music.

folks *pl. n.* the persons of one's own family; one's relatives: *I'm going to see the folks tonight.*

folksy *adj.* **1.** rustic or imitative of the rustic. **2.** *U.S.* sociable; friendly; unceremonious.

foot *v.* **1.** to pay or settle, as a bill. **2.** to walk. —*phr.*
3. fall on one's feet, See **fall.**

4. **feet first, a.** dead. **b.** thoughtlessly; impetuously.
5. **feet of clay,** an imperfection or blemish on what would otherwise have been perfection.
6. **get off on the right (wrong) foot,** to have a good (bad) start.
7. **get one's feet wet,** to obtain first-hand and practical experience.
8. **have one foot in the grave,** to be near death.
9. **keep one's feet on the ground,** to retain a sensible and practical outlook.
10. **my foot!** nonsense!
11. **put one's best foot forward, a.** to make as good an impression as possible. **b.** to do one's very best.
12. **put one's foot down,** to take a firm stand.
13. **put one's foot in one's mouth** or **in it,** to make an embarrassing blunder.
14. **stand on one's own feet,** to be self-sufficient.
15. **sweep off one's feet, a.** to cause one to lose a footing, as a wave, etc. **b.** to impress or overwhelm.
16. **two left feet,** See **left.**

football *n.* an amphetamine pill.

footballer *n.* a person who fights with his feet.

footie *n.* football. Also, **footy, footer.**

footless *adj. U.S.* awkward, helpless, or inefficient.

footling *adj.* foolish; silly; trifling. [from *footle* to talk or act in a foolish manner]

Footscray Alps *n.* any imaginary or impossible place. [from *Footscray* a flat suburb of Melbourne]

footslog *v.* to march or tramp; slog on foot.

footslogger *n.* one who footslogs, but especially an infantryman.

footstep *phr.* **follow in (someone's) footsteps,** to succeed or imitate (someone).

footsy *n.* **1.** (*especially in children's speech*) a foot. *—adj.* **2.** closely, but not publicly, associated with a person or organisation: *He is footsy with the National Party.* *—phr.* **3. play footsies, a.** to touch in secret a person's feet, knees, etc. with one's feet, especially as part of amorous play. **b.** to flirt: *He is playing footsies with the vicar's wife.*

footy *n.* football.

for *phr.* **for it,** about to suffer some punishment, injury, setback, or the like: *You'll be for it if you don't behave!*

forelock *phr.*
1. **take time by the forelock,** to seize an opportunity.
2. **touch/tug/pull one's forelock,** to act in a servile manner.

foreman *phr.* **get the foreman's job,** to take a white collar job as in management, politics, etc., such usually regarded as a betrayal of a working class background. [from the rhyme 'the working class can kiss my arse, I've got the *foreman's job* at last', often sung to the tune of 'The Red Flag']

forget *phr.* **forget it,** to drop the subject.

fork *n.* **1.** a jockey. **2.** the vulva. *—phr.* **3. fork over** or **out,** to hand over; to pay.

form *n.* **1.** a person's record or reputation. **2.** a prison record.

fort *phr.* **hold the fort,** to maintain the existing position or state of affairs.

forty *n.* a thief; scoundrel. [from the Fitzroy *Forty* a Melbourne gang of thieves named after Ali Baba and the Forty Thieves, in 'The Arabian Nights']

fossil *n.* an outdated or old-fashioned person or thing.

foul *phr.* **foul up,** to bungle or spoil; to cause confusion.

four-by-two *n.* **1.** a Jew. **2.** a prison warder; screw. Also, **four-by, fourbie, forbie.** [rhyming slang]

four-eyes *n.* a person who wears glasses.

four-flush *n.* a bluff.

four-flusher *n.* one who makes pretensions that he cannot or does not bear out.

four-legged kangaroo *n.* **1.** any beast, especially a cow. **2.** a dog.

four-legged lottery *n.* a horse race.

four-letter word *n.* anything which is distasteful or unpleasant: *Housework is a four-letter word.*

fourpenny dark *n.* **1.** a small glass of cheap fortified wine. **2.** any cheap wine. [from the time when wine saloons served drinks in miniature mugs, for fourpence a drink]

four-wheeler *n.* the driver of a four-wheel drive vehicle.

fox *v.* **1.** to deceive or trick. **2.** to fetch (often said to a dog as a command): *Fox it!*

foxie *n.* a fox terrier. Also, **foxy.**

foxy lady *n.* a sexy, sophisticated woman of independent spirit.

fraidy-cat *n.* (*especially in children's speech*) a coward; scaredy-cat.

frame *n.* **1.** (in baseball) an inning. **2.** (of livestock) an emaciated animal. **3.** a frame-up. —*v.* **4.** to incriminate unjustly by a plot, as a person. —*phr.* **5. frame a book,** *Horseracing.* to set oneself up as a bookmaker.

frame-up *n.* that which is framed, as a plot, or a contest whose result is fraudulently prearranged; fix-up. Also, **frame.**

franger *n.* a contraceptive sheath.

franglais *n.* **1.** (*offensive*) that type of contemporary French which makes use of imported English words. **2.** a humorous mixture of French and English. [blend French *français* French + *anglais* English]

freak *n.* **1.** a person who is enthused about a particular thing: *a Jesus freak, a drug freak.* —*v.* **2.** Also, **freak out. a.** to have an extreme reaction, either favourable or adverse, to something, especially a drug-induced experience. **b.** to panic. **3.** to frighten: *Spiders really freak her.* —*phr.* **4. freak (someone) out of (their) mind,** to frighten or worry greatly.

freak-out *n.* **1.** an extreme experience, usually terrifying, produced by hallucinogenic drugs. **2.** any unusual or unexpected experience.

freckle *n.* the anus.

Fred *n.* the average, non-discerning Australian male. Also, **Freddie.** [?coined by Max Harris in 'The Angry Eye' (1973)]

Fred Astaire *n.* **1.** a chair. **2.** hair. [rhyming slang]

Fredland *n.* suburban Australia. Also, **Fredsville.** [See **Fred**]

Fred Nerk *n.* an imaginary person regarded as the archetypal fool or simpleton.

free *n.* **1.** (in football) a free kick awarded to a team. *—phr.* **2. make free with, a.** to treat or use too familiarly. **b.** to take sexual liberties with.

freebie *n.* a service or item provided without charge.

freewheel *v.* to act independently, particularly in personal, social matters.

freeze *v.* **1.** to ignore (someone). *—phr.* **2. do a freeze, a.** to be very cold. **b.** to be ignored.

freight *n.* **1.** fare. **2.** money, especially money owed: *He couldn't front the freight to the bookmaker.*

Fremantle doctor *n.* a strong, cool wind blowing after a hot day. [*Fremantle,* a town in Western Australia, + *doctor,* a strong fresh, reviving breeze]

French *n.* **1.** a humorous term for mild swear words: *Pardon my French.* *—adj.* **2.** Also, **French way.** relating to oral sex.

French kiss *n.* a kiss in which the tongue enters the partner's mouth.

French letter *n.* a contraceptive sheath; condom. Also, **Frenchy, Frenchie.**

fresh *adj.* **1.** cheeky. *—phr.* **2. fresh out of,** *Chiefly U.S.* completely lacking in (a commodity, idea, etc.): *We are fresh out of coffee.*

fridge *n.* a frigid girl or woman. Also, **fridgie.**

fried egg *n.* a traffic dome. Also, **silent cop.**

friends *pl. n.* menstrual periods.

frig *v.* **1.** to masturbate. **2.** to have sexual intercourse. *—phr.*
3. frig around, to behave in a stupid or aimless manner.
4. frig up, to confuse; muddle.
[Latin *fricāre* rub]

frigging *adj., adv.* fucking.

frig-up *n.* a confusion; muddle; mess.

frisk *v.* **1.** to search (a person) for concealed weapons, etc., by feeling his clothing. **2.** to steal something from (someone) in this way.

fritz *phr.* **on the fritz,** broken; not in working order.

Fritz *n.* **1.** a German, especially a German soldier. **2.** Germans or a German army collectively.

frog *n.* **1.** (*usually cap*) a Frenchman. [?from *frog-eater*] **2.** a French letter; contraceptive sheath.

frog and toad *n.* a road. [rhyming slang]

frog's eggs *pl. n.* boiled sago or tapioca, especially as served in institutions.

front *v.* **1.** to appear before a court on a charge. **2.** Also, **front up.** to arrive; turn up. **3.** to act as a front (for an illegal activity, etc.). *—n.* **4.** someone or something which serves as a cover for another activity, especially an illegal or disreputable one. *—phr.*
5. more front than Myers/Foy and Gibsons/Mark Foys, hide; cheek; impudence. [from the large frontages and facades of the big department stores, *Myers* in Melbourne, *Foy and Gibsons* in Adelaide and *Mark Foys* in Sydney]
6. up front, in advance: *They paid a thousand dollars up front.*

frost *n.* **1.** something which is received coldly; a failure. **2.** a swindle.

frostie *n.* a cold bottle or can of beer. Also, **frosty.**

froth *phr.* **blow the froth off,** an instruction to stop flattering someone or making exaggerated claims.

froth and bubble *n.* **1.** trouble. **2.** *Horseracing. Gambling.* the double. [rhyming slang]

fruit *n.* **1.** a male homosexual. *—phr.* **2. fruit for** or **on the sideboard, a.** Also, **cream on the bun.** something extra; a luxury item. **b.** an additional source of income.

fruitcake *n.* **1.** a ratbag. *—adj.* **2.** mad; insane. *—phr.*
3. fruitcake factory, a lunatic asylum.
4. nutty as a fruitcake, foolish; very eccentric.

fruit cocktail *n.* a tampon. [?from similar appearance to a cylindrical lolly wrapped in white paper]

fruit factory *n.* a mental hospital.

fruit machine *n. Chiefly British. and U.S.* a poker machine. [the score was originally displayed in the form of replicas of various fruits]

fruit salad *n. Military.* a large collection of medal ribbons.

fruity *adj.* **1.** homosexual. *—phr.* **2. crack a fruity,** See **crack.**

fry *phr.* **small fry, a.** unimportant or insignificant people. **b.** young children. [from *fry* the young of fishes or other creatures]

frying pan *adj.* **1.** of no account; small-time: *just a frying pan thief. —phr.* **2. out of the frying pan and into the fire,** the exchange of one difficult situation for another equally or more difficult.

fryingpan brand *n.* a large brand used by cattle thieves to cover the rightful owner's brand.

fubsy *adj.* short and fat; stumpy.

fuck *v.* **1.** to have sexual intercourse with. *— n.* **2.** a person, as the object of the sexual act: *a good fuck.* **3.** the act of sexual intercourse. *— interj.* **4.** an offensive exclamation of disgust or annoyance, often used as a mere intensive. **5.** an exclamation of wonder or delight. *—phr.*
6. fuck about or **around, a.** to treat (someone) unfairly; deceive, or cause inconvenience, distress, etc., to. **b.** to behave stupidly or insanely.
7. fuck up, to make a mess of; ruin.
8. fuck off, (*often offensive*) to go away; depart.
9. not give a fuck, not to care at all.
10. the fuck, used to add emphasis: *Who the fuck are you?*
11. what the fuck! an exclamation of contempt, dismissal, or the like.

fuckable *adj.* sexually desirable.

fuck-all *n.* very little; nothing: *They've done fuck-all all day.*

fucked *adj.* **1.** exhausted. **2.** ruined; done for. **3.** broken; out of order. *—phr.* **4. get fucked,** (*offensive*) go away; leave me alone.

fucker *n.* **1.** one who fucks; one much given to fucking. **2.** a contemptible person or thing. **3.** (*not necessarily offensive*) any person.

fucking *adj.* **1.** an intensive signifying approval, as in *It's a fucking marvel* or disapproval, as in *fuck-*

ing bastard. —*adv.* **2.** very; extremely: *fucking ridiculous.*

fuck-up *n.* confusion; ruin; miscalculation; mistake.

fuckwit *n.* a nincompoop.

fuckwitted *adj.* foolish; stupid: *a fuckwitted suggestion.*

fuel *phr.* **add fuel to the fire,** to aggravate or exacerbate a situation.

full *adj.* **1.** intoxicated. —*phr.*
2. full as a boot/bull/tick/State school, very drunk.
3. full as a goog, a. very drunk. **b.** full of food.
4. full of oneself, conceited; egoistic.
5. full up, a. (of a person) replete; having eaten enough. **b.** *N.Z.* exasperated; disgruntled.
6. full up to dolly's wax, See **dolly.** Also, **full up to pussy's bow.**

full bore *phr.* **the full bore,** the maximum: *Give it the full bore.*

full chat *adv.* See **flat chat.**

full-frontal *adj.* **1.** giving a complete view of a naked person from the front. **2.** with every detail exposed. —*n.* **3.** a full-frontal picture.

full hand *phr.* **have a full hand,** to be infected with both gonorrhoea and syphilis.

full-on *adj.* **1.** enthusiastic; full of energy; unrestrained. **2.** requiring complete involvement or total commitment.

full points *pl. n.* **1.** (in Australian Rules) a goal. **2.** a full score.

full strap *adv.* to the utmost capacity: *to work at full strap.*

full tilt *adj.* at top speed: *The bus was going full tilt for the station.*

fumble *v.* to touch, fondle, embrace, etc., the body of another as a means to sexual stimulation.

fun *v.* **1.** to make fun; joke: *I was just funning.* —*adj.* **2.** relating to fun; amusing: *a fun thing to do.* **3.** entertaining; lively: *Perth is a fun city.* —*phr.*
4. like fun, not at all.

fun and games *pl. n.* **1.** inconsequential activity. **2.** (*ironic*) amatory play. **3.** (*ironic*) problems and difficulties: *There will be fun and games when the police find out.*

fun bags *pl. n.* female breasts.

fundament *n.* the anus.

funeral *n.* business; worry; concern: *That's his funeral.*

fungus face *n.* a person with a beard or other facial hair.

funk *n.* **1.** cowering fear; state of fright or terror. **2.** one who funks; a coward. —*v.* **3.** to be afraid of. **4.** to frighten. **5.** to shrink from; try to shirk.

funk-hole *n.* a place of refuge from something feared.

funnies *pl. n.* **1.** comic strips. **2.** the section of a newspaper containing them.

funny *adj.* **1.** insolent: *Are you trying to be funny, mate?* —*n.* **2.** a joke: *Tell us a funny.*

funny business *n.* **1.** foolish behaviour. **2.** underhand, dubious, or dishonest dealings. **3.** sexual intercourse or any amorous behaviour.

funny farm *n.* a lunatic asylum; psychiatric hospital.

funny money *n.* **1.** money made by dubious or dishonest means. **2.** counterfeit money.

fur *phr.* **make the fur fly,** to quarrel noisily; make a scene or disturbance.

furburger *n.* **1.** simultaneous fellatio and cunnilingus; sixty-nine. **2.** *Originally U.S.* the vulva.

furphy *n.* a rumour; a false story. [from John *Furphy* manufacturer in Victoria of water and sanitation carts, which during World War I were centres of gossip]

FURTB *adj.* very full (of food, etc.). Also, **F.U.R.T.B.** [from initials *F(ull) U(p) R(eady) T(o) B(urst)*]

fuse *phr.* **blow a fuse,** to lose one's temper.

fusspot *n.* a fussy person; one who is over-particular.

fut *interj.* an exclamation of annoyance, etc. Also, **phut.**

fuzz *n.* **1.** a blur. **2.** frizzy hair. **3.** the police force or a policeman.

fuzzy wuzzy *n.* **1.** *World War II.* native inhabitant of Papua New Guinea. **2.** (*offensive*) a native, especially of Africa. [used originally of Sudanese tribesmen]

fuzzy wuzzy angel *n.* any native of Papua New Guinea who helped the Australians, especially the wounded, during World War II.

gab *v.* **1.** to talk idly; chatter. —*n.* **2.** idle talk; chatter. —*phr.* **3. the gift of the gab,** glib speech. [variation of *gob* mouth from Gaelic or Irish]

Gabba, the *n.* Woolloongabba Cricket Ground, Brisbane.

gabby *adj.* garrulous; loquacious.

gabfest *n.* **1.** a prolonged session of conversation or speeches; conference. **2.** an informal opportunity to chatter or gossip. [See **gab** + *-fest* a suffix indicating a period of festive or enthusiastic activity]

gadabout *n.* a restless person, especially one who leads an active social life. Also, **gad about.**

gadfly *n.* a persistently annoying or irritating person. [from *gadfly* a bloodsucking fly which goads or stings domestic animals]

gaff *phr.*

1. blow the gaff, to disclose a secret; reveal the truth, often unintentionally.

2. stand the gaff, *Chiefly U.S.* to cope with difficulties or ridicule.

gaffer *n.* **1.** *British.* father. **2.** *Chiefly British.* an owner, senior partner, or the like. **3.** *Shearing.* foreman: *The gaffer rings the bell.* [variant of late Middle English *godfar* (contracted form of *godfather*)]

gag *v.* **1.** to deceive; hoax. **2.** to make jokes. —*n.* **3.** a joke. **4.** any contrived piece of word play or horseplay.

gaga *adj.* **1.** senile; stupid. **2.** mad; fatuously eccentric. **3.** besotted: *He is gaga about his new car.* Also, **gah gah.** [French: senile, a senile person (probably imitative)]

gage *n.* marijuana or a marijuana cigarette.

gal[1] *n.* *Chiefly U.S.* a girl.

gal[2] *n.* See **galvo.**

galah *n.* **1.** a fool; simpleton. **2.** a show-off. —*phr.* **3. mad as a gumtree full of galahs,** quite stupid. [from Aboriginal *galah* a noisy pink and grey cockatoo]

galah session *n.* a time set aside for the people of isolated outback areas to converse with one another by radio. [from the noisy sounds made by large groups of galahs]

gallery *phr.* **play to the gallery,** to seek applause by playing to popular taste rather than considered judgment.

gallop *phr.* **the gallops,** turf horseracing (opposed to *the trots,* and *the jumps*).

Galloping Greens *pl. n.* **1.** the GPS team in the Queensland Rugby Football Union. **2.** the Randwick team in the N.S.W. Rugby Football Union.

gallows bird *n.* one who deserves to be hanged.

galoot *n.* an awkward, silly fellow: *silly galoot.* Also, **galloot.**

galumph *v.* **1.** to bound exultantly. **2.** to move clumsily, as through excess of enthusiasm. [from *gallop* and *triumph* coined by Lewis Carroll]

galvo *n.* Also, **galv, gal iron.** galvanised iron or steel sheeting. [from a type of iron formed into corrugated sheets and used especially in many rural buildings or outbuildings]

game *n.* **1.** business or profession: *Andrew is in the building game now.* —*adj.* **2.** willing to undertake something hazardous or challenging: *I'm game to go bushwalking.* —*phr.*

3. game as Ned Kelly, with fighting spirit; plucky; resolute. [from the bushranger]

4. game, set and match, a convincing victory; complete triumph. [a conclusive tennis score]

5. give the game away, a. to reject or abandon. **b.** to reveal some strategy or secret.

6. have the game sewn up, to be master of the situation.

7. make a game of, to ridicule.

8. play the game, to act fairly or justly, or in accordance with recognised rules.

9. the game, a. sexual intercourse. **b.** prostitution.

gammon *n.* **1.** deceitful nonsense; bosh. —*v.* **2.** to make pretence. **3.** to deceive with nonsense.

gamp *n. British.* an umbrella. [said to be from the umbrella of Mrs Sarah *Gamp* in Dicken's *Martin Chuzzlewit]*

gander *n.* **1.** a look at something: *Have (or take) a gander at this.* **2.** a fool.

gang *v.* **1.** to form or act as a gang. —*phr.* **2. gang up on,** to attack in a gang; combine against.

gang bang *n.* an occasion on which a number of males have sexual intercourse with one female. Also, **gangie, gang slash, gang splash.**

gangster *n. N.Z.* a tough but likeable character.

gangway *interj.* clear the way!

ganja *n.* Indian hemp; marijuana. Also, **ganga.**

gaol-bait *n.* a girl below the legal age of consent. Also, **jail-bait.**

gaolbird *n.* one who is, or has been, confined in gaol. Also, **jailbird.**

gaolie *n.* gaolbird. Also, **jailie.**

gap *phr.* **bridge/close/stop/fill the gap,** to make up for a lack of something or a deficiency.

Gap, the *phr.* **jump off the Gap, a.** to commit suicide. **b.** to give up; give in. Also, **go over the gap.** [from the name of a cliff at Watsons Bay, Sydney, where many suicides have occurred]

garbage *phr.*
1. a load of garbage, a lot of nonsense.
2. garbage in, garbage out, See **GIGO.**

garbage guts *n.* a person who eats to excess.

garbo *n.* **1.** one employed to collect garbage; garbageman. **2.** a garbage container.

Garbo *phr.* **do a Garbo,** to avoid publicity; remain in seclusion. Also, **do a Greta Garbo.** [from the film actress Greta *Garbo* who shunned the limelight.]

garden *phr.* **lead up the garden path,** to mislead, hoax, or delude.

gardening *n. Cricket.* (of batsmen) the use of the bat to prod, sweep, etc., a turf pitch between deliveries, usually as a delaying tactic.

gargle *n.* a drink, usually alcoholic.

garn *interj.* an expression of incredulity. [contraction of *go on*]

garryowen *n. Rugby League.* a short, high punt kick, aimed at enabling the kicker and his team-mates to rush forward and regain possession of the ball; bomb. [from *Garryowen* Rugby Club in Ireland, famous for its use of this kick]

gas[1] *n.* **1.** empty talk. **2.** something great, wonderful. —*adj.* **3.** great, wonderful: *a gas idea.* —*v.* **4.** to talk nonsense or speak boastfully to. **5.** to indulge idly in empty talk.

gas[2] *n. Chiefly U.S.* **1.** petrol. —*phr.* **2. step on the gas,** to hurry. [shortened form of *gasoline*]

gasbag *n.* **1.** an empty, voluble talker; a windbag. —*v.* **2.** to talk volubly; chatter.

gash *n.* **1.** a second helping of food; anything superfluous or extra. **2.** the vagina. **3.** a woman as a sex object.

gasket *phr.* **blow a gasket,** to lose one's temper.

gasper *n. British.* a cigarette.

gasser *n.* something great, wonderful.

gassy *adj.* given to frivolous or empty-headed chatter.

gastro *n.* upset stomach: *He had a day off work with gastro.* [shortened form of *gastroenteritis*]

gate money *n.* money given a prisoner on release from prison.

gatepost *phr.* **between you, me and the gatepost,** in confidence.

gather *phr.* **gather up the threads,** See **threads.**

gawk *v.* **1.** to stare stupidly. —*phr.* **2. gawk at,** to stare. [apparently representing Old English word meaning fool, from *gagol* foolish]

gay *adj.* **1.** camp; homosexual (a term used especially by homosexuals to describe themselves). —*n.* **2.** a homosexual. [origin uncertain]

gay and hearty *n.* a party. [rhyming slang]

gay lib *n.* a movement which seeks to end discrimination against homosexuals. Also, **Gay Lib.**

gazob *n.* a stupid person; a blunderer.

gazunder *n.* **1.** a chamber-pot. **2.** *Cricket.* a low ball. Also, **gezunder, gozunder.** [contraction of *goes under*]

g'day *interj.* See **good day.**

gear *n.* **1.** clothes, especially those bought or worn by young people. **2. a.** marijuana. **b.** drugs generally. **3.** the apparatus used to prepare and inject drugs, especially heroin. **4.** external genitalia. *—adj.* **5.** *British.* fashionable, delightful, or excellent. *—phr.*
6. drop one's gear, to get undressed.
7. gear up, to make ready; prepare.
8. get your gear on, to get dressed.

gee[1] *v.* to tease or provoke. [?from *gee up* a command to horses]

gee[2] *interj.* a mild exclamation of surprise or delight. Also, **gee whizz.**

geebung *n.* a native-born Australian. See **stringy-bark.** [from the *geebung* a native Australian tree]

gee-gee *n.* **1.** (*in children's speech*) a horse. *—phr.* **2. the gee-gees,** the horseraces.

geek *n.* **1.** a look: *Have a geek at this.* Also, **geez, gig, gink. 2.** a sideshow performer who bites the heads off or eats live animals.

gee up *n.* a lift (as of spirits, enthusiasm, etc.): *The big crowd gave the players a gee up.*

gee whiz *interj.* **1.** an exclamation expressing surprise, admiration, etc. Also, **gee whizz.** *—adj.* **2.** (of a device, etc.) technologically sophisticated: *a gee whiz computer game.* [a euphemistic variation of *Jesus*]

geez *interj.* gee; gee whiz; Jesus.

geezer *n.* an odd character. Also, **geeser.** [?from dialect pronunciation of *guise* to go in disguise]

gelt *n.* money. Also, **gilt.** [?Dutch *geld* money]

gen *n.* **1.** general information. **2.** all the necessary information about a subject: *give me the gen. —phr.* **3. gen up,** to become informed, to learn or read up (about).

gendarme *n.* any policeman. [French, formed as a singular from *gens d'armes* men at arms]

gendarmerie *n.* a police station. Also, **gendarmery.**

gentle Annie *n. N.Z.* a gentle sloping hill. [originally a coachman's term]

gents *n.* a public lavatory for men. Also, **gentlemen's, gents'.**

Geordie *n.* **1.** a Scotsman. **2.** *British.* a miner, especially one from the region around the river Tyne. **3.** a person who works or lives near the river Tyne. [diminutive of *George* a proper name]

geri *n.* an old person. Also, **gerry; jerry.** [short for *geriatric*]

get *v.* **1. a.** to be under an obligation to; be obliged to: *You have got to go.* **b.** to be sure: *You have got to be joking.* **2.** to hit: *The bullet got him in the leg.* **3.** to make a physical assault on, especially in vengeance: *I'll get you for that.* **4.** to become: *It got quite humid this morning.* **5.** Also, **get it.** to grasp or understand the meaning or intention of (a person): *Now I get what you mean.* **6.** to have an unspecified effect upon, as irritation,

anger, amusement: *Her giggle really gets to me.* **7.** to kill. *—phr.*
8. do a get, to escape, run away.
9. get across, to communicate successfully.
10. get a load of, See **load.**
11. get along, a. to go; go off. **b.** nonsense! (an exclamation of disbelief). **c.** to agree or be friendly.
12. get a move on, See **move.**
13. get any, See **any.**
14. get at, a. to hint at or imply: *What's she getting at?* **b.** to tamper with, as by corruption or bribery.
15. get away, See **away.**
16. get away from it all, to leave business, work, worries, etc., for a holiday.
17. get away (with you), See **away.**
18. get back on, to take vengeance on.
19. get by, to manage; carry on in spite of difficulties.
20. get cracking, to begin vigorously; hurry.
21. get down on, to steal.
22. get even with, to square accounts with; take vengeance on.
23. get his, hers, etc. a. to get a just reward. **b.** to be killed.
24. get in good with, to ingratiate oneself with.
25. get in on (the act), to take part in, especially a restricted activity.
26. get inside, to achieve deep understanding of.
27. get into, to become involved or immersed in (an activity or situation): *We are getting into growing our own vegies.*
28. get it in the neck, See **neck.**
29. get it together, to become organised, coordinated or systematised.
30. get lost, to go away; desist: *Get lost, will you!*
31. get off, a. to escape; evade consequences. **b.** to cease to interfere.
32. get off one's bike, See **bike.**
33. get off with, a. to behave flirtatiously with. **b.** to have sexual intercourse with.
34. get on, a. to agree or be friendly. **b.** to successfully place a wager.
35. get one's act together, See **act.**
36. get (on) one's goat/nerves/works/tit/quince, etc., to annoy or irritate one.
37. get one's end in, See **end.**
38. get one's own back, to be revenged.
39. get on to, to discover.
40. get out, a. (of information) to become publicly known. **b.** to succeed in solving.
41. get out from under, to escape from a difficult or threatening situation; abandon one's responsibilities.
42. get outside (of), See **outside.**
43. get round, a. to outwit. **b.** to cajole or ingratiate oneself with (someone). **c.** Also, **get over.** to overcome (difficulties, etc.).
44. get set, to place a bet (in two-up, horseracing, etc.).
45. get (stuck) into, a. to set about a task vigorously. **b.** to eat hungrily.
46. get the axe/boot/chop/run/spear, etc., to be dismissed from a job.
47. get to (someone), a. to arouse deep feeling in (someone). **b.** to annoy or irritate (someone).
48. get up, to dress elaborately.
49. get up to, to be involved in (especially mischief, etc.).
50. get with, to have intercourse with.

get-at-able *adj.* that may be reached or attained; accessible.

getaway *n.* **1.** a getting away; an escape. **2.** the start of a race. *—adj.* **3.** pertaining to a getaway: *The getaway car was found.* *—phr.* **4. do a getaway,** to escape; leave quickly.

get-out *n.* **1.** a style of clothing, especially one which appears artificial. *—phr.* **2. as all get-out,** in the extreme: *as silly as all get-out.*

get-rich-quick *adj.* of or relating to a scheme, business, enterprise, etc., designed to make a lot of money in a short time, sometimes by shady means.

get-together *n.* **1.** a meeting. **2.** a small and informal social gathering.

get-up *n.* **1.** style of production; appearance: *the get-up of a book.* **2.** style of dress; costume.

get-up-and-go *n.* enthusiasm; energy; enterprise.

gewgaw *n.* **1.** a bit of gaudy or useless finery. *—adj.* **2.** showy, but paltry.

Ghan, the *n.* name of a train on the Adelaide/Alice Springs route. [the railway follows the route taken by (Af)*ghan* camel trains which took supplies to Alice Springs]

ghost *n.* **1.** a person to whom one owes money. *—phr.*

2. give up the ghost, a. to die. **b.** to despair. **c.** (of a piece of machinery) to break down completely.

3. lay a ghost, to resolve or put to rest old fears, painful memories, etc.

4. not a ghost of a chance, no hope at all.

G.I.[1] *n.* **1.** *U.S. Army.* government issue. *—adj.* **2.** of or standardised by the army: *G.I. shoes.* [abbreviation originally of *galvanised iron* used in U.S. Army book-keeping entries of articles made of it; then, by association with *government issue,* of the full range of articles issued, and, finally, of the soldiers themselves]

G.I.[2] *adj.* (of a person or place) inconveniently located; inaccessible. [*G(eographically) I(mpossible)*]

Giants, the *pl. n.* the Adelaide City team in the National Soccer League of Australia.

gibber gunyah *n.* a rock cave. [Aboriginal *gibber* stone, boulder + Aboriginal *gunyah* shelter, dwelling]

giddy goat *n.* tote; totalisator. [rhyming slang]

gift *n.* **1.** anything very easily obtained or understood. *—phr.* **2. gift of the gab,** See **gab.**

gift-horse *phr.* **look a gift-horse in the mouth,** to criticise a gift; accept a gift ungratefully.

gig[1] *v.* **1.** to taunt; provoke. *—n.* **2.** a fool; meddler. [British dialect]

gig[2] *n.* **1.** a booking for a jazz or pop musician to perform at a concert. **2.** the concert itself. **3.** any job or occupation. *—v.* **4.** to attend a gig.

gig[3] *v.* **1.** to watch; stare. *—n.* **2.** an observer; eye witness. **3.** a stickybeak. **4.** a look; geek.

giggle *n.* an amusing occasion: *a bit of a giggle.* [apparently back formation from Obsolete. *giglet* giddy, laughing girl]

giggle factory *n.* a lunatic asylum. Also, **gigglehouse, giggle house.**

giggle hat *n. Military.* a soft hat worn with fatigue dress.

gigglesuit *n.* **1.** *Military.* clothing worn to do domestic chores, usually as punishment; fatigue dress. **2.** *Prison. Obsolete.* prison clothes.

GIGO *n. Computers.* a situation in which a computer has, inadvertently, been fed misinformation, resulting in an incorrect output. [*G(arbage) I(n) G(arbage) O(ut)*]

gilgie *n. W.A.* a freshwater crayfish; yabby. [Aboriginal]

gill *phr.*

1. fed up to the gills, a. thoroughly exasperated. **b.** Also, **full to the gills.** completely surfeited.

2. white or **green at the gills,** white-faced through fear, exhaustion, nausea, etc. [from *gill* the respiratory organ of fishes, etc.]

gilt *phr.* **take the gilt off the gingerbread,** to make a proposition less attractive or desirable.

gimme *interj.* **1.** give me. *—adj.* **2.** *Golf.* of or pertaining to a putt in which the ball lies so close to the hole that one's opponent concedes that the putt need not be made. *—n.* **3.** *Golf.* a gimme putt. *—phr.* **4. gimme girl,** a girl or woman who goes out with or marries a man purely for mercenary motives.

gimmick *n.* **1.** a pronounced eccentricity of dress, manner, voice, etc., or an eccentric action or device, especially one exploited to gain publicity. **2.** any tricky device or means. **3.** *U.S.* a device by which a magician works a trick. [?blend of *gimmer* trick finger-ring and *magic*]

gin *n.* an Aboriginal woman. [Aboriginal]

gin burglar *n.* a white man who has sex with Aboriginal women on a casual basis. Also, **gin shepherd.**

ging *n.* a child's catapult.

ginger *n.* **1.** piquancy; animation. **2.** Also, **ginge.** a nickname for any red-haired person. *—v.* **3.** to impart spiciness or piquancy to; make lively. **4.** to steal, especially of a prostitute stealing from a client's clothing. [from *ginger* a cooking spice]

gin jockey *n.* a white man who seeks sexual partners among Aboriginal women.

gink[1] *n.* a look: *Have a gink at this.* [variant of *geek*]

gink[2] *n.* a fellow: *a silly gink.* [British dialect *geek, geke* a fool, simpleton]

gin palace *n.* (formerly) a low public house which sold cheap spirits, as gin.

gin's piss *n.* **1.** weak beer. **2.** weak tea.

gip[1] *v.* to swindle, cheat.

gip[2] *n.* **1.** severe pain: *My leg is giving me gip. —phr.* **2. give gip,** to admonish; upbraid. Also, **gyp, jip.**

gippo *n.* **1.** an Egyptian. **2.** a gipsy. *—adj.* **3.** Egyptian. **2.** gipsy.

girl *n.* **1.** a sweetheart. **2.** a woman. **3.** a male homosexual.

girl Friday *n.* a female secretary and general assistant in an office. [?from Man *Friday* in Daniel Defoe's *Robinson Crusoe* who was able to help his master survive under difficult conditions]

girlie *n.* **1.** a girl. *—adj.* **2.** illustrating or featuring nude or nearly nude women: *a girlie magazine.*

girls *phr.*
1. girls' night out, an evening on which a group of women have a night out together.
2. the girls, a group of female friends, usually of similar ages.

girl's week *n.* the menstrual period.

girly-girly *adj.* effusively effeminate in the manner of a small girl.

gismo *n.* a gadget. Also, **gizmo.**

git *n. British.* **1.** a fool: *Silly old git!* **2.** a bastard.

give *v.* **1.** *Originally. U.S.* tell; offer as explanation: *Don't give me that.* **2.** to happen; occur: *What gives in Perth?* —*phr.*
3. give away, a. to let (a secret) be known. **b.** to betray (a person). **c.** to abandon; give up: *Times were hard so I gave farming away.*
4. give (oneself) away, to reveal something about (oneself) accidentally.
5. give (it) away, to cease to do (something), usually in exasperation.
6. give (someone) a go, See **go.**
7. give out, a. to announce publicly. **b.** *Jazz.* to perform as well as one can.
8. give over, to desist.
9. give someone the drum, See **drum.**
10. give the game or **show away,** See **game.**
11. give what for, to berate; chastise.

give-away *n.* **1.** Also, **dead give-away.** a betrayal, usually unintentional. **2.** a premium given with various articles to promote sales, etc. —*adj.* **3.** (of a television program, etc.) characterised by the awarding of prizes, money, etc., to recipients chosen, usually through a question-and-answer contest.

give-up *n. Prison.* an informer.

gizzard *n.* the stomach.

glad[1] *phr.*
1. the glad eye, an inviting or flirtatious look: *She gave him the glad eye.*
2. the glad hand, (*usually ironic*) an effusive welcome, often public: *to give someone the glad hand.*

glad[2] *n.* gladiolus, a type of flower. Also, **gladdie.**

Gladiators, the *pl. n.* the Brisbane City team in the National Soccer League of Australia.

glad rags *pl. n.* best clothes donned for special occasions: *Get on your glad rags, we're going to town.*

glam *adj.* **1.** glamorous. —*n.* **2.** glamour. [shortened form]

glass can *n.* a small squat beer bottle; a stubby.

glasshouse *n. British. Military.* a military prison.

glass jaw *n. Boxing.* a jaw which is more than commonly susceptible to injury.

glassy *n.* **1.** a much-prized, clear-glass marble. —*phr.* **2. the glassy eye,** a glance of cold disdain.

glitch *n.* **1.** an extraneous electric current or signal, especially one that interferes in some way with the functioning of a system. **2.** a hitch; snag; malfunction.

glitterati *pl. n.* well-known personalities who attend glamorous social events, such as first nights, etc. [from *glitter* + *(liter)ati*]

glitzy *adj.* gaudy. [?*glit(ter)* + *(rit)zy*]

glob *n.* a rounded lump of some soft but pliable substance: *a glob of cream.* [?blend of *globe* and *blob*]

glory box *n.* a chest in which young women store clothes, linen, etc., in expectation of being married; bottom drawer; hope chest.

glue-sniffer *n.* one who inhales the fumes from plastic cement or glue, for their narcotic effect.

gluggy *adj.* sticky.

glumbum *n.* a pessimistic person.

go *v.* **1.** to compare; to be normally: *She's quite young as grandmothers go.* **2.** to copulate. *—n.* **3.** an attempt: *to give it a go.* **4.** energy, spirit, or animation: *to be full of go.* **5.** a plan or purpose. **6.** something that goes well; a success: *to make a go of something. —phr.*
7. all the go, in the current fashion.
8. be going places, to be likely to achieve notable success.
9. fair go, adequate opportunity: *fair go, spinner!*
10. from go to whoa, from beginning to end.
11. from the word go, from the beginning.
12. give (someone) a (fair) go, to give (someone) a (fair) chance.
13. go all the way, to have sexual intercourse.
14. go a meal, drink, etc., to need a meal, drink, etc.
15. go around with, to keep the company of (a member of the opposite sex).
16. go back on, to fail to keep (one's word, promise, etc.).
17. go bush, See **bush.**
18. go down, to be sent to prison.
19. go down on, to perform fellatio or cunnilingus on.
20. go for the doctor, See **doctor.**
21. go for broke, See **broke.**
22. go in for, to make (a thing) one's particular interest.
23. go native, See **native.**
24. go off, a. to come to dislike. **b.** Also, **go off pop,** to reprimand; to scold, become angry with. **c.** to get married. **d.** to have an orgasm.
25. go on, a. to behave; act. **b.** to chatter continually. **c.** nonsense! (an exclamation of disbelief).
26. go out, to lose consciousness: *He went out like a light.*
27. go soft, See **soft.**
28. go the whole hog, See **hog.**
29. go to it, to undertake any activity with gusto.
30. go under, to be overwhelmed; be ruined.
31. go with, to frequent the society of.
32. have (give it) a go, a. to make an attempt; try: *to have a go at swimming.* **b.** *Cricket.* to hit out recklessly.
33. it's a go, all's clear.
34. let oneself go, to become uninhibited.
35. open go, a. situation in which fair play prevails and no unfair restraints or limiting conditions apply: *The election was an open go.* **b.** a situation in which normal restraints do not apply: *It was open go at the bar that night.*
36. that's the way it or **she goes,** that's how things are.

goalie *n.* a goalkeeper.

goal sneak *n. Australian Rules.* **1.** a player who catches the opposition unawares and scores a goal. **2.** a full-forward.

goanna *n.* a piano. [rhyming slang]

goat *n.* **1.** a scapegoat; one who is the butt of a joke. **2.** a fool: *to make a goat of oneself.* **3.** a lecher; a licentious man. *—phr.*

4. act the goat, See **act.**

5. a hairy goat, a racehorse which does not perform well.

6. get (on) one's goat, to annoy; enrage; infuriate.

7. run like a hairy goat, a. to run very slowly. **b.** to run very fast.

8. separate the sheep from the goats, divide the good from the bad.

gob[1] *n.* a sailor.

gob[2] *n.* **1.** the mouth. **2.** a sticky mass of slimy substance. *—v.* **3.** to spit or expectorate. *—phr.* **4. shut your gob,** command to keep quiet. [Gaelic or Irish]

gobble *v.* to seize upon greedily or eagerly.

gobbledegook *n.* language characterised by circumlocution and jargon: *The gobbledegook of government reports.* Also, **gobbledy-gook.** [grotesque coinage modelled on *hobbledehoy.* Final element *gook* may be slang word for tramp, variant of dialect *gowk* simpleton]

go-by *phr.* **give (someone) the go-by,** to intentionally pass by; snob.

God *interj.* **1.** an oath or exclamation used to express weariness, annoyance, disappointment, etc. *—phr.* **2. God's own country, a.** the United States of America. **b.** Also, **Godzone.** (*usually ironic*) one's own country viewed as the ideal.

God almighty *interj.* an expression of surprise or anger.

God-almighty *adv.* extremely; with absurd exaggeration: *Don't be so God-almighty pleased with yourself.*

God-awful *adj.* terrible.

goddamn *U.S. interj.* **1.** an oath expressing irritation, fury, etc. *—adj., adv.* **2.** Also, **goddamned.** damned.

godfather *n.* one who exercises power and influence through unofficial channels.

god squad *n.* a company of convinced Christians: *He was a member of the god squad.*

Godzone *n.* one's own country, especially Australia or New Zealand: *Why would you ever want to leave Godzone?*

goer *n.* **1.** one who or that which moves fast. **2.** any activity, project, etc., having evident prospects of success.

go-getter *n.* an enterprising, aggressive person, especially one who will stop at nothing to get what he wants.

goggle box *n. British.* a television set.

go-in *n.* a fight: *They had a bit of a go-in behind the pub.*

going *phr.*

1. going on, nearly: *It is going on four o'clock.*

2. the going thing, the current fad.

going-over *n.* **1.** a thorough examination. **2.** a severe beating or thrashing.

goings-on *pl. n.* **1.** actions; conduct; behaviour (used chiefly in a disapproving sense): *We were shocked by the goings-on at the office party.* **2.** current events: *She only kept in touch with the goings-on at home through newspapers.*

gold-digger *n.* a woman who exploits personal attractiveness for financial gain.

golden *adj.* gifted; fortunate and destined for success: *the golden boy of swimming.*

golden handshake *n.* a gratuity or benefit, given to employees as a recognition of their services on the occasion of their retirement or resignation, or as a sop when they are dismissed.

Golden Mile *n.* any especially wealthy area of a city. [from the *Golden Mile* in Kalgoorlie, an exceptionally rich goldbearing reef]

golden oldies *pl. n. Radio.* songs which have maintained their popularity over many years. Also, **gold.**

goldfish *phr.* **throw the goldfish another cat,** a humorous invitation to be extravagant.

goldfish-bowl *n.* a state of helpless exposure to public curiosity; lack of privacy.

goldmine *n.* a source of anything required: *a goldmine of information.*

golliwog *n. Prison.* an informer. [rhyming slang *golliwog* dog. See **dog**]

gollop *v.* to eat quickly and noisily. [blend of *gulp* and *gallop*]

golly *v.* **1.** to spit. —*n.* **2.** spittle.

gone *adj.* **1.** exhilarated; in a state of excitement (as by the influence of drugs, jazz, etc.). —*phr.*
2. far gone, a. much advanced; deeply involved. **b.** extremely mad. **c.** extremely drunk. **d.** almost exhausted. **e.** dying.
3. gone a million, See **million.**
4. gone on, infatuated with.

goner *n.* a person or thing that is dead, lost, or past recovery or rescue.

gong *n.* a medal.

Gong, the *n.* Wollongong.

gonzo *adj.* eccentric; crazy. [Italian *gonzo* simpleton]

gonzo journalism *n.* a style of journalism which specialises in the bizarre.

goo *n.* sticky matter.

good *phr.*
1. a good question, a difficult or demanding question.
2. a good way, a considerable extent.
3. all to the good, generally advantageous, often used to justify an unpleasant event.
4. as good as, in effect; practically: *He as good as said it.*
5. as good as they come, the best.
6. be up to no good, to be doing wrong; breaking the law in some undisclosed way; behaving in a suspicious manner.
7. for good or **for good and all,** finally and permanently; for ever: *to leave a place for good (and all).*
8. good and, used to add emphasis to a statement: *You can wait until we are good and ready.*
9. good grief, See **grief.**
10. good iron, a. an exclamation of approval. **b.** a likeable person.
11. good on you, an expression of approval, encouragement, etc. Also, **good for him, good on you,** *Chiefly British.* **good for you.**
12. good one, an exclamation of delight, approval, etc.
13. make good, a. to make recompense for; pay for. **b.** to keep to an agreement; fulfil. **c.** to be successful. **d.** to prove the truth of; substantiate.

14. **no good to gundy,** See **gundy.**

15. **that's a good one,** an ironic expression of disbelief.

good day *interj.* a conventional expression used at a meeting or a parting during the day. Also, **g'day.**

good doer *n.* **1.** a person who eats well (also applied to animals). **2.** a person who knows the ropes. Also, **hard doer.**

good egg *n.* **1.** a pleasant, agreeable, trustworthy person. —*interj.* **2.** an exclamation of pleasurable surprise.

good guts *n.* **1.** accurate information. **2.** such information designed to be used to bring about someone's downfall.

good guy *n.* the hero.

goodie *n.* a good person, especially a hero in a story, play or film. Also, **goody.**

goodies *pl. n.* **1.** sweets or cakes. **2.** attractive foodstuffs, clothes, possessions, etc.

good lady *n.* a wife or de facto wife.

good life *phr.* **the good life,** a life filled with material luxuries and comfort.

good-looker *n.* a person who has good looks.

good nick *adj.* healthy; in good condition.

good night *interj.* **1.** used to express surprise, anger, dismay, disgust, etc., as when something is finished, ruined etc. —*phr.* **2. good night nurse,** *Originally World War I Military.* Also, *N.Z.* **good night McGuinness.** an expression of resigned incredulity, displeasure, or mild dismay.

good-oh *interj.* an exclamation indicating pleasure, agreement, etc. Also, **goodo.**

good oil *n.* correct (and usually profitable) information, often to be used in confidence; the drum. Also, **dinkum oil.** See **oil.**

goods *pl. n.* **1.** the genuine article. **2.** evidence of guilt, as stolen articles: *caught with the goods.* —*phr.*

3. deliver the goods, to do what is expected, promised or required.

4. have the goods on (someone), have evidence enough to incriminate (someone).

good sort *n.* See **sort.**

goodtime *phr.* **goodtime girl,** a young woman, especially one of easy virtue.

goody *interj.* **1.** wonderful! how nice! —*n.* **2.** goodie.

goody-goody *n.* **1.** a sentimentally or priggishly good person. —*adj.* **2.** affecting goodness.

gooey *adj.* **1.** like goo; sticky; viscoid. **2.** overemotional; sentimental.

goof *n.* **1.** a foolish or stupid person. —*v.* **2.** to blunder; slip up. **3.** to swallow (a drug, as amphetamine). —*phr.*

4. goof around, to play the fool to entertain others.

5. goof off, to daydream; fritter away time.

6. goof up, to bungle; botch.

[apparently variant of Obsolete. *goff* dolt]

goof ball *n.* an amphetamine pill.

goofy-foot *n.* a surfboard rider who surfs with his right foot as the lead foot. Also, **goofy-footer.**

goog *n.* **1.** an egg. —*phr.* **2. full as a goog, a.** extremely drunk. **b.** well-fed.

googy-egg *n.* (*in children's speech*) an egg.

gook *n. Chiefly U.S. Military.* (*offensive*) a foreigner; an Asian, especially a national of a country in which Western soldiers are fighting, as a Japanese, a Korean or a Vietnamese.

goolie *n.* **1.** a stone. **2.** a testicle. **3.** a glob of phlegm.

gooligum *n.* an indefinite name for a thing which a speaker cannot or does not designate more precisely.

goom *n.* methylated spirits.

goomie *n.* **1.** (*offensive*) a person addicted to drinking methylated spirits. **2.** a homeless Aborigine.

goon *n.* **1.** a stupid person. **2.** a hired thug: *goon squad.* **3.** a hooligan or tough.

goondie *n.* an Aboriginal hut or gunyah. [Aboriginal]

goony bird *n.* **1.** *U.S.* an albatross. **2.** a Dakota aeroplane; DC3. Also, **gooney bird.**

goop *n.* **1.** *U.S.* bad-mannered or rude person. **2.** a silly or foolish person.

goori *n.* **1.** *N.Z.* a Maori-bred dog; a mongrel. **2.** (*offensive*) a Maori. Also, **goorie, goory.**

goose[1] *n.* **1.** a silly or foolish person; a simpleton. —*phr.*

2. a wigwam or **whim wham for a goose's bridle,** a reply to any unwanted question; a put-down; a snub.

3. cook someone's goose, to frustrate or ruin a person's hopes or plans.

goose[2] *v.* **1.** to poke someone between the buttocks, usually in fun and unexpectedly. —*n.* **2.** an unexpected poke between the buttocks.

gooseberry *phr.* **play gooseberry,** to embarrass or restrict two people, who might like to be alone, by accompanying them.

goose's neck *n.* a cheque. [rhyming slang]

gopher *n.* a person employed to run errands, give general assistance, etc. Also, **gofer.** [from *go for* by association with *gopher*]

gorblimey *interj.* **1.** an expression of surprise or amazement. —*adj.* **2.** vulgar; of or pertaining to the poor or working classes: *He wears gorblimey trousers.* Also, **gorblimy.** [alteration of *God blind me*]

gorgeous *adj.* very good, pleasing, or enjoyable: *I had a gorgeous weekend.*

gorgeous gussies *pl. n.* frilly panties. [named after *Gussy* Moran, U.S. tennis player]

gorilla *n.* **1.** an ugly, brutal fellow. **2.** a rolled cigarette.

gormless *adj.* (of a person) dull; stupid; senseless. [variant of dialect *gaumless* from *gaum* attention, heed]

gory *adj.* distasteful or unpleasant: *He read the gory details of the accident.*

Gosford dog *n.* See **Dapto dog.** [from *Gosford* a town north of Sydney]

go-slow *n.* **1.** a deliberate curtailment of output by workers as an industrial sanction; work-to-rule. **2.** *Prison.* a punishment cell.

got *phr.*

1. have got 'em bad, a. to be in a nervous condition. **b.** to be suffering from withdrawal symptoms, especially from alcohol.

2. have got into (someone), to be causing (someone) to display anger, irritation, etc.: *What has got into that man?*

3. have got it bad for, to be infatuated with.

gotcher *interj.* **1.** an exclamation accompanying the capture of a person. **2.** an exclamation indicating comprehension and agreement. Also, **gotcha.** [contraction of *(I have) got you*]

Gov *n. Chiefly British.* **1.** a form of address to a male member of a superior social class. **2.** a boss; employer. Also, **gov, Guv, guv.** [abbreviation of *Governor*]

Government house *n.* the main homestead on a sheep or cattle station.

government man *n.* (formerly) a euphemistic term for a convict.

government stroke *n.* the easy pace at which work is done supposed to be typical of those working for the government, originally used specifically of convict road labourers.

governor *n.* **1.** one's employer. **2.** any person of superior status. **3.** one's father.

Gowings *phr.* **gone to Gowings, a.** deteriorating financially. **b.** ill, especially with a hangover. **c.** failing dismally, as of a horse in a race, a football team, etc. [from *Gowings* Bros. Ltd., a Sydney retail firm]

goy *n.* a non-Jew; gentile. [Yiddish, from Hebrew *goi* people]

G.P. *Prison. —adj.* **1.** Governor's Pleasure; as of an indefinite sentence, when a person is found not guilty by reason of insanity. *—n.* **2.** a prisoner serving such a sentence.

grab *v.* **1.** to affect; impress: *How does that grab you? —phr.*

2. grab by the balls, to impress very favourably.

3. up for grabs, open to anyone to do something about it.

grace *phr.* **in someone's good graces,** looked upon favourably.

gracious *interj.* an exclamation of surprise, etc.

grade *phr.*

1. grade-A prick or **twerp, etc.,** (*offensive*) a highly unpleasant person.

2. make the grade, to reach a desired minimum level of achievement or qualification.

graduate *n. Prison.* a criminal who has been in child welfare institutions.

graft *n.* **1.** work, especially manual labour. *—v.* **2.** to work; to work hard. *—phr.* **3. hard graft,** work. [?from *grave* to dig]

grafter *n.* **1.** a very hard worker; an industrious person. **2.** *Chiefly U.S.* an extortionist.

grain *phr.*

1. go against the grain, contrary to one's natural inclinations.

2. with a grain of salt, with some reserve; without wholly believing.

grampers *n.* grandfather. Also, **gramps, grandpa, granddad.**

gran *n.* grandmother. Also, **grandma, granny.**

grand *adj.* **1.** first-rate; very good; splendid: *to have a grand time, grand weather.* *—n.* **2.** a thousand dollars.

grandmother *phr.*

1. teach your grandmother to suck eggs, See **egg.**

2. tell that to your grandmother! an expression of disbelief.

granny *n.* **1.** a grandmother. **2.** an old woman. **3.** a fussy person. **4.** a granny knot. **5.** the Sydney Morning Herald. Also, **grannie.**

granny flat *n.* a self-contained extension to a house, designed either to accommodate an elderly relative, as a grandmother, or to rent.

Granny Smith *n.* a variety of apple with a green skin and crisp juicy flesh, suitable for eating raw or cooking. Also, **grannie, granny.** [from Marie Ann *Smith* died 1870, who first produced them at Eastwood, Sydney]

grape *phr.*

1. a grape on the business, an interloper; an unwelcome person.

2. in the grip of the grape, addicted to drink, esp. wine.

3. the grape, wine.

grapes *pl. n.* haemorrhoids.

grapevine *n.* the means by which any form of information is passed, especially word of mouth.

grappler *n.* a professional wrestler.

grass *n.* **1.** marijuana. *—v.* **2.** to inform (on). *—phr.*

3. at grass, *Mining.* (of ore) stacked ready for shipment.

4. let the grass grow under one's feet, to be lax in one's efforts; miss an opportunity.

5. grass a catch, *Cricket.* to drop a catch.

6. on the grass, a. *Prison.* to be free, as a prisoner released from gaol. **b.** on strike.

7. put out to grass, a. to withdraw (a racehorse) from racing, etc., due to old age. **b.** to retire (a person.)

8. the grass, *Prison.* freedom.

grasser *n.* an informer.

grasshopper *phr.* **knee-high to a grasshopper,** small; young.

grassroots *Originally U.S.* *—pl. n.* **1.** the basic essentials or foundation. *—adj.* **2.** pertaining to, close to, or emerging spontaneously from the people.

grass widow *n.* **1.** a woman whose husband is temporarily absent. **2.** a woman neglected by her husband.

grave *phr.*

1. dig one's own grave, to cause one's own downfall or ruin.

2. have one foot in the grave, to be infirm, old, or near death.

3. to turn in one's grave, (of a dead person) to be thought likely to have been offended or horrified by a recent event or events.

grave-digger *n.* **1.** *British. Military.* strong drink. **2.** *Cricket.* the last batsman in the batting order of a team.

gravel rash *phr.* **get gravel rash,** to act in a sycophantic fashion, as if by crawling.

graveyard bark *n.* a loud and distressing cough. Also, **graveyard cough.**

graveyard chompers *pl. n.* false teeth.

graveyard shift *n.* the night shift; dogwatch.

gravy *n.* any bonus or perk; money easily acquired.

gravy train *n.* any course of action which results in the receipt of a perk or bonus: *on the parliamentary gravy train.*

grease *phr.* **grease (someone's) palm,** to bribe (a person).

greased lightning *phr.* **like greased lightning,** very quickly.

grease monkey *n.* a mechanic.

greaser *n.* **1.** (*offensive*) a member of a Latin-American race, especially a Mexican. **2.** a toadying, sycophantic person. Also, **greasepot.**

greasespot *phr.* **feel like a greasespot,** to be in a state of sweatiness; to experience lassitude induced by great heat.

greasies *pl. n.* **1.** fish and chips. **2.** greasy hair.

greasy *n.* **1.** one who, in a camp, attends to the chores of cooking, cleaning, etc. **2.** a shearer. **3.** (*offensive*) a Greek or Italian immigrant.

greasy spoon *n.* a cheap restaurant or cafe, very plain, often unclean, and providing cheap food.

great *n.* **1.** a great person; a person who has accomplished great achievements: *a cricket great.* —*phr.*

2. a great one for, enthusiastic about: *He's a great one for reading.*

3. go great guns, to be successful.

4. great at, skilful or expert in.

greatest *phr.* **the greatest, a.** exceptional in one's field of expertise. **b.** appealing; very nice.

Great Scott *interj.* a euphemistic variant of *Great God.* Also, **Great Scot.**

greedy-guts *n.* a glutton; one who is prepared to take more than his share of a dish.

Greek *phr.* **Greek love,** anal intercourse. Also **go Greek.**

greenback *n.* **1.** *U.S.* a U.S. legal-tender note, usually printed in green on the back, originally issued against the credit of the country and not against gold or silver on deposit. **2.** *Surfing.* an unbroken wave.

green ban *n.* a refusal by employees to work or allow work to proceed on a building site, when such work would necessitate destroying something of natural, historical or social significance.

green cart *n.* a van which, in popular fancy, takes people to a lunatic asylum. Also, **rubber cart.**

greengrocer *n.* **1.** a drug pusher specialising in marijuana. **2.** a large, bright green cicada, common to southern and eastern Australia. Also, **green grocer, green Monday, yellow Monday.**

greenhorn *n.* **1.** a raw, inexperienced person. **2.** a person easily imposed upon or deceived. [originally applied to an ox with green or young horns]

greenie *n.* **1.** a conservationist. **2.** *Surfing.* a large green unbroken wave. —*adj.* **3.** sympathetic with moves to conserve the environment, produce whole foods organically, and live more simply: *Tinned food is not for greenies.*

green light *n.* permission; authorisation: *We have the green light from the council on our building extensions.*

Green Rats, the *pl. n.* the Warringah team in the N.S.W. Rugby Football Union.

Greens, the *pl. n.* the Randwick team in the N.S.W. Rugby Football Union.

green thumb *n.* skill in gardening and plant-growing. Also, **green fingers.**

Gregory Peck *n.* **1.** neck. **2.** cheque. [rhyming slang from *Gregory Peck* U.S. film actor]

gremmie *n.* a surfboard rider whose behaviour in the surf is objectionable. Also, **gremmy.**

grey *n.* **1.** a penny which has two tails. **2.** a double-headed penny. [from English thieves' slang]

grey death *n.* prison stew.

grey ghost *n. N.S.W.* parking meter inspector. Also, *Vic.* **grey meanie.** [from the grey uniform]

greyhound *n.* **1.** a swift ship, especially a fast ocean liner. **2.** a thinly rolled cigarette. Also, **racehorse.** [from *greyhound* a tall slender dog noted for fleetness]

grey matter *n.* brains or intellect. [from the nervous tissue of the brain and spinal cord, of a dark reddish grey colour]

grey nurse *n.* purse. [rhyming slang]

grey power *n.* the influence, especially political, exerted by the elderly.

grid *n.* a bicycle. [back formation of *gridiron*]

grief *phr.* **good grief,** an expression of surprise, vexation, etc.

griffin *n. N.Z.* genuine information; the facts. Also, **griff.**

grill *v.* to subject to severe and persistent cross-examination or questioning.

grin *phr.* **grin and bear it,** to endure without complaint.

grind *v.* **1.** to work or study laboriously. **2.** to rotate the pelvis during or as during sexual intercourse or erotic dancing. *—n.* **3.** laborious or monotonous work; close or laborious study. **4.** a diligent or laborious student. **5.** act of sexual intercourse: *the old bump and grind.*

grip *n.* **1.** occupation; regular employment. *—phr.* **2. come** or **get to grips with,** to deal with, tackle (an enemy, a problem, etc.).

gripe *v.* to complain constantly; grumble.

grist *phr.* **it's all grist to the mill,** everything available can be used.

grit *n.* firmness of character; indomitable spirit; pluck.

grizzleguts *n.* a person given to complaining.

grocessor *n.* a turkey or other poultry producer who both grows the birds and processes them for market. [*grow* + *processor*]

grog *n.* **1.** alcohol, particularly cheap alcohol. *—phr.* **2. grog on,** to indulge in a long session of drinking. [from *grogram* the material of the cloak of Admiral Vernon ('Old Grog') who in 1740 ordered water to be issued with sailors' pure spirits]

grog artist *n.* a heavy drinker.

groggery *n.* a saloon.

groggy *adj.* **1.** staggering, as from exhaustion or blows. **2.** drunk; intoxicated.

grog-on *n.* a drinking party.

grog shop *n.* **1.** a shop selling alcohol. **2.** (formerly) a cheap tavern.

grommet *n. Surfing.* a young surfer, usually in his or her early teens.

groove *phr.*
1. groove on, a. to be in a state of euphoria. **b.** to be delighted or pleased with. **2. in the groove,** in an excited or satisfied emotional state, as through listening to jazz.

groovy *adj.* **1.** exciting, satisfying, or pleasurable. **2.** appreciative: *a groovy audience.*

grope *v.* **1.** to fondle, embrace clumsily and with sexual intent. *—phr.* **2. go the grope,** to fondle with a sexual purpose.

groper *n.* a West Australian. Also, **sandgroper.**

Groperland *n.* Western Australia.

gross *adj.* repulsive; disagreeable; objectionable.

grot *n.* **1.** filth. **2.** a filthy person. [back formation from *grotty*]

grotty *adj.* **1.** dirty; filthy. **2.** useless; rubbishy. [alteration of *grotesque*]

grouch *v.* **1.** to be sulky or morose; show discontent; complain. *—n.* **2.** a sulky or morose person or mood.

grouchy *adj.* sullenly discontented; sulky; morose; ill-tempered.

ground *phr.*
1. go to ground, to withdraw from public attention and live quietly.
2. get off the ground, See **off.**

groundbreaking *adj.* innovative.

ground floor *n.* the most advantageous position or relationship in a business matter or deal: *to get in on the ground floor.*

grouper *n.* any member of a right-wing faction in the Australian Labor Party, or a member of the Democratic Labour Party.

group grope *n.* sexual intercourse mutually undertaken at the same time by three or more people.

groupie *n.* **1.** a girl who travels with and makes herself available sexually to the male members of a pop or rock group. **2.** an admirer of a well-known personality.

grouse *adj.* **1.** very good. *—phr.* **2. extra grouse,** excellent.

grouter *n.* **1.** an unfair advantage. *—phr.* **2. come in** or **be on the grouter, a.** to take an unfair advantage of a situation. **b.** *Two-up.* to bet on a change in the fall of the coins.

grow *phr.*
1. grow like Topsy, to grow in an unplanned, random way.
2. grow on, to win the affection or admiration of by degrees.

growl *n.* female pudendum.

growly *adj.* good; excellent.

grub *n.* **1.** food or victuals. **2.** a dull plodding person; drudge. *—v.* **3.** to supply with food. **4.** to lead a laborious or grovelling life; drudge.

grubber *n.* **1.** *Cricket.* Also, **mullygrubber.** a ball delivered in such a manner that on contact with the ground it does not bounce. **2.** Also, **grubber kick.** *Rugby Football.* a kick which sends the ball along the ground. **3.** a hospital.

grubstake *n.* provisions, outfit, etc., furnished to a prospector or the like, on condition of participating in the profits of his discoveries.

gruff nut *n.* a small ball of hardened faeces which has adhered to an anal pubic hair. Also, **weenie.**

grundies *pl. n.* underwear. [See **reginalds**]

gub *n.* (*offensive, used by Aborigines*) a white person. Also, **gubber, gubba, gubbah.** [Aboriginal: white demon]

guck *n.* slimy, objectionable matter: *guck and goo.* [?blend of *goo* and *muck*]

guernsey *phr.* **get a guernsey,** to succeed, win approval (originally to be selected for a football team). [from *guernsey* a close-fitting knitted jumper, worn by seamen, footballers, etc., from the Isle of *Guernsey* in the English Channel]

guess *phr.*

1. anyone's guess, difficult to know what is correct or what action should be taken.

2. guess and by God, a matter of guesswork when normal methods seem ineffective.

3. keep someone guessing, not to let someone know what is happening.

guesstimate *n.* **1.** an estimate made chiefly by guessing. —*v.* **2.** to estimate in this way.

guff *n.* empty or foolish talk; humbug; nonsense.

guineapig *n.* a person used as the subject of any sort of experiment. [from the *guineapig* a short-eared, short-tailed rodent much used in scientific experiments]

guiver *n.* See **guyver.**

Gulf, the *n.* the Gulf of Carpentaria or the country near it.

gull plugger *n.* one who shoots indiscriminately at any bird, for sport.

gully-rake *v.* to steal stock.

gully-raker *n.* **1.** a person who steals stock. **2.** a stock-whip.

gum *phr.* **gum up the works,** to interfere with or spoil something.

gum-digger *n.* a dentist. Also, **gum-puncher.**

gumleaf mafia *n.* any group of expatriate Australians, as show business personalities living in the Unites States, especially Los Angeles.

gummy *n.* an old toothless sheep.

gummys *pl. n.* gumboots.

gumption *n.* **1.** initiative; resourcefulness. **2.** shrewd, practical commonsense. [originally Scottish]

gumshoe *n.* **1.** *U.S.* **a.** one who goes about softly, as if wearing rubber shoes. **b.** a policeman or detective. —*v.* **2.** *U.S.* to go softly as if wearing rubber shoes; move or act stealthily.

gumsucker *n.* a native of or resident in Victoria. [? from the habit of some colonial youths of chewing or sucking the transparent lumps of gum from the silver wattle]

gum tree *phr.*

1. have gum trees growing out of one's ears, to be an awkward, clumsy, country yokel.

2. have seen one's last gum tree, about to die.

3. up a gum tree, a. in difficulties; in a predicament. **b.** completely baffled.

gun *n.* **1.** a champion, especially in shearing. **2.** a large surfboat for riding big waves. —*v.* **3.** *Aeronautics.* to cause to increase in speed very quickly. —*adj.* **4.** of or pertaining to one who is expert, especially in shearing: *a gun shearer.* —*phr.*

5. beat or **jump the gun,** to begin prematurely; be overeager. [from racing sense, to begin before the starting gun]

6. carry or **hold (big) guns,** to be in a powerful or strong position.

7. go great guns, to have a period of success at something.

8. gun for, to seek (a person) with the intention to harm or kill.

9. have the guns for, to have the ability to do something.

10. stick to one's guns, to maintain one's position in an argument, etc., against opposition.

gundy *phr.* **gone to gundy** and **no good to gundy,** unsatisfactory; broken; beyond repair; ruined; worthless. Also, **Gundy.** [?from Aboriginal *gundy, goondie, gunyah* an Aborigine's hut made of boughs and barks]

Gunga Din *phr.* **bung it in, Gunga Din,** See **bung.**

gunk *n.* **1.** a food judged to be bad or inappropriate, especially oversweet and cloying. **2.** medicated ointment or the like. **3.** rubbish. **4.** nonsense. Also, **guk.**

Gunners, the *pl. n.* the South Melbourne team in the National Soccer League of Australia.

gunnie[1] *n. Prison.* an armed criminal.

gunnie[2] *n.* a girl who is free with sexual favours. [short for *gun moll*]

gun shearer *n.* an expert shearer, often the best in the shed.

gunyah *n.* **1.** an Aborigine's hut made of boughs and bark; humpy; mia mia; wurley. **2.** a small, rough hut or shelter in the bush. Also, **gunya.** [Aboriginal]

gup *n.* (usually a derogatory term used by Aborigines to refer to whites) fool; idiot.

gurgler *n.* **1.** a plughole. —*phr.* **2. down the gurgler,** ruined; irretrievably lost or destroyed.

guru *n.* an influential teacher or mentor. [Hindi: a preceptor and spiritual guide in Hinduism]

gush *v.* to express oneself extravagantly or emotionally; talk effusively.

gusher *n.* **1.** a person who gushes. **2.** a flowing oilwell, usually of large capacity.

gushy *adj.* given to or marked by gushing or effusiveness.

gussie *n.* an effeminate man. [diminutive of *Augustus*]

gussies *pl. n.* frilly pants for women. Also, **gorgeous gussies.** [from *Gussy* Moran, American tennis player]

gussy *phr.* **all gussied up,** smartly dressed. [?variant of *gussie*]

gut *adj.* **1.** relating to feelings, emotions, intuition: *a gut response.* **2.** relating to that which may engender feelings, emotions, etc.: *a gut issue.*

gut-buster *n.* an activity requiring great effort, especially great physical effort.

gutful *n.* more than enough: *I've had a gutful of this.* Also, (*especially N.Z.*) **gutsful.**

gutless *adj.* **1.** cowardly. **2.** lacking in power, especially of a car, money, etc. *—phr.* **3. a gutless wonder,** a person or thing whose performance does not live up to expectation.

gutrot *n.* unhealthy food or drink, especially any cheap alcoholic drink.

guts *pl. n.* **1.** the stomach or abdomen. **2.** one greedy for food; a glutton. **3.** courage; stamina; endurance: *to have guts.* **4.** essential information: *the guts of the matter.* **5.** *Two-up.* wagered money in the centre of the ring. **6.** the essential parts or contents: *Let me get to the guts of the motor.* *—v.* **7.** to cram (oneself) with food. *—phr.*

8. good guts, correct information; the news.

9. hate (someone's) guts, to loathe or detest (someone).

10. have (someone's) guts for garters, to extract revenge on (someone).

11. hold one's guts, to remain silent.

12. in the guts, *Two-up.* in the ring, as bets placed.

13. rough as guts, See **rough.**

14. spill one's guts, to reveal information.

15. work one's guts out, to work excessively hard.

gutser *n.* **1.** a person who eats too much. Also, **gutzer.** *—phr.* **2. come a gutser, a.** to fall over. **b.** to fail as a result of an error of judgment.

gutsy *adj.* **1.** full of courage; full of guts. **2.** warmly wholehearted; unreserved. **3.** strong, full-bodied: *a gutsy wine.*

guy *n.* **1.** a fellow or man: *guys and dolls.* **2.** (*usually pl.*) a person of either sex. **3.** a boyfriend. *—v.* **4.** to jeer at or make fun of; ridicule. *—phr.*

5. the good guys, the heroes.

6. the bad guys, the villains.

[from *Guy* Fawkes, the leader of the Gunpowder Plot to assassinate the King in Parliament, London, 5 November, 1605]

guyver *n.* affected talk and manner; foolish talk or nonsense; ingratiating behaviour. Also, **gyver, guiver, givor, givo, gyvo, guivo.**

n. a school-teacher. [from the tables learnt by rote, 'one *guzinter* two, two *guzinter* four', etc.]

guzzle *v.* to drink (or sometimes eat) frequently and greedily: *They sat there all evening guzzling beer.*

guzzle-guts *n.* a person who eats and drinks to excess, or in a disgusting and noisy manner.

gyp *v.* **1.** to swindle; cheat; defraud or rob by some sharp practice. **2.** to obtain by swindling or cheating; steal. *—n.* **3.** a swindle. **4.** a swindler or cheat. Also, **gip.**

gyppo *n., adj.* an Egyptian. Also, **gippo.**

gypster *n.* a con man. Also, **gipster.**

hack *n.* **1.** a taxi. **2.** Also **party hack.** See **party.** —*v.* **3.** to put up with; endure: *I can't hack it.*

hacker *n.* a hard worker.

Hackers, the *pl. n.* the Port Hacking Team in the N.S.W. Rugby Football Union.

hackie *n.* a taxidriver. Also, **hack pusher.** [from *hack* a horse kept for common hire]

had *phr.*

1. be had, to be cheated or duped.

2. have had, to be utterly exasperated with: *I have had this government.*

3. have had it, a. to be utterly exasperated. **b.** to be exhausted.

hades *n.* a euphemistic variant of hell. [Greek *Haidēs* in Greek mythology, the subterranean abode of departed spirits or shades]

ha-ha *phr.* **have the ha-has,** to be insane.

ha-ha duck *n.* a kookaburra. Also, **ha-ha pigeon.**

hair *phr.*

1. get in someone's hair, to irritate or annoy someone.
2. hair of the dog (that bit you), an alcoholic drink taken to relieve a hangover.
3. have by the short hairs or **by the short and curlies,** to have a person in one's power.
4. keep your hair on, keep calm; do not get angry.
5. let one's hair down, to behave in an informal, relaxed, or uninhibited manner; abandon oneself to pleasure.
6. make one's hair stand on end, to fill with terror; terrify.
7. (put) hair on (someone's) chest, (usually of a man) something, as a strong drink, etc., which will make one feel fitter, more virile, etc.
8. tear one's hair out, to show extreme emotion, as anger, anxiety, etc.

hair pie *n.* cunnilingus.

hair-raiser *n.* anything, as a story, that arouses fear or terror.

hairy *adj.* **1.** difficult: *That's a hairy problem.* **2.** frightening: *a hairy drive.* **3.** *N.Z.* rough; ramshackle; dilapidated. **4.** angry; excited. **5.** exciting; stimulating: *a hairy party.* *—phr.*

6. hairy goat, See **goat.**

7. run like a hairy goat, See **goat.**

hairy-legs *n.* **1.** a person responsible for maintaining the main running track of a railway. *—phr.* **3. rack off, hairy-legs!** an exclamation of dismissal, contempt, etc.

hairy Mary *n.* a species of beachworm found in northern N.S.W. and Queensland.

half *n.* **1.** a half-pint, especially of beer. *—interj.* **2. not half!** certainly; indeed. *—phr.*

3. and a half, of an exceptional nature: *He's a man and a half.*

4. by half, by a great deal; by too much: *too clever by half.*

5. half a mo, See **mo.**

6. half your luck, See **luck.**

7. not half, a. not really; not at all: *His first poems were not half bad.* **b.** very; surprisingly: *His paintings are not half good.*

8. the half of it, a more significant part of something: *You think we're in trouble, but that's not the half of it.*

half-axe *n.* one who is foolish or eccentric.

half-baked *adj.* **1.** not completed: *a half-baked scheme.* **2.** lacking mature judgment or experience: *half-baked theorists.*

half-dead *adj.* very tired; exhausted.

half-inch *v.* **1.** to steal. **2.** to arrest. [rhyming slang *half-inch* pinch. See **pinch**]

half-mast *phr.* **at half-mast,** (of long trousers) too short, not extending to the ankles.

half-pie *adj.* half-hearted; mediocre.

half-pint *n.* **1.** a small person, especially a small woman. *—adj.* **2.** of or pertaining to a small person.

half-seas-over *adj.* intoxicated.

half spot *n.* fifty dollars.

ham *n.* **1. a.** an actor who overacts. **b.** overacting. **2.** an amateur: *a radio ham.* *—v.* **3.** to act with exaggerated expression of emotion; overact: *to ham it up.*

hambone *n.* a male striptease. Also, **bone.**

hammer *n.* **1.** back. *—phr.* **2. be on (someone's) hammer,** to watch (someone) closely; badger. [rhyming slang, *hammer and tack* back]

hammer and tongs *adv.* with great noise, vigour, or violence.

hammering *n.* **1.** a beating; hiding. *—phr.* **2. take a hammering, a.** to be beaten up. **b.** to be subjected to intense cross-examination or criticism.

hand *phr.*

1. hand in one's dinner plate, to prepare to die.

2. hand it to, to give due credit to.

3. hand off, *Rugby Football.* to thrust off an opponent who is tackling.

4. hand over fist, a. easily. **b.** in large quantities: *to make money hand over fist.*

5. hands and heels, *Horseracing.* without the use of a whip.

6. have one's hands full, to be fully occupied.
7. hold the hand out, a. to exploit the benefits given out by the government and other welfare organisations. **b.** to demand bribe money.
8. lay one's hands on, to obtain.

handful *n.* a thing or a person that is as much as one can manage.

handicap event *n.* a social function, etc., to which one is expected to bring one's spouse.

handle *n.* **1.** a title in front of a name. **2.** a person's name. **3. a.** a beer glass with a handle. **b.** the contents of such a glass. **4.** a confidence trick. *—v.* **5.** to con; deceive. *—phr.*
6. get a handle on, a. to be able to utilise something. **b.** to understand.
7. fly off the handle, to lose one's temper, especially unexpectedly.

hand-me-down *n.* an article of clothing handed down or acquired at second hand. Also, **reach-me-down.**

handout *phr.* **live on handouts,** to subsist on the social benefits offered by charity, private and public. Also, **hand-out.**

hang *v.* **1.** used to add emphasis to a statement: *I'll be hanged if I do.* *—n.* **2.** the precise manner of doing, using, etc., something: *to get the hang of a tool.* **3.** meaning or force: *to get the hang of a subject.* *—phr.*
4. hang about or **around,** to loiter.
5. hang (around) with (someone), to spend time in (someone's) company.
6. hang, draw and quarter, to punish severely.
7. hang five, to ride a surfboard standing on the nose of the board with the toes of one foot over the edge.
8. hang in, a. (of horses) to veer away from the most direct course, toward the fence. **b.** (in surfboard riding) to ride close to the breaking part of the wave. **c.** to persevere.
9. hang it out, to go all out: *She really hung it out in the finals.*
10. hang loose, to relax; fill in time.
11. hang of a, Also, **hanguva.** in an exceptionally great (hurry, difficulty, etc.)
12. hang on, a. to wait: *Hang on! I'm not quite ready.* **b.** *Boxing.* to clinch. **c.** *Australian Rules.* to hold an opponent when he does not have the ball in his possession.
13. hang one on (someone), to punch (someone).
14. hang out, a. to live at or frequent a particular place. **b.** *Horseracing.* (of horses) to veer away from the most direct course, away from the fence.
15. hang out for, a. to remain adamant in expectation of (a goal, reward, etc.): *I'll hang out for a higher price before I'll sell.* **b.** to be in need of; crave: *He's hanging out for some dope.*
16. hang ten, *Surfing.* to ride a surfboard while standing on the nose of the board with all one's toes over the edge.
17. hang up, a. to tether (a horse). **b.** *Shearing.* to stop work, as by hanging up shears.
18. let it all hang out, a. to allow oneself to speak one's mind or show emotion freely. **b.** to be uninhibited in manner, dress, etc.

hanger-on *n.* one who clings to a person, place, etc.; follower.

hangman *n. N.Z.* a character; an eccentric person.

hang-out *n.* a place where one lives or frequently visits.

hangover *n.* **1.** something remaining behind from a former period or state of affairs. **2.** the after-effects of excessive indulgence in alcoholic drink.

hang-up *n.* something which occasions unease, inhibition, or conflict in an individual.

hanky-panky *n.* **1.** trickery; subterfuge or the like. **2.** sexual play.

happy *adj.* **1.** showing an excessive liking for, or quick to use an item indicated (used in combination): *trigger-happy.* *—phr.*
2. happy as a bastard on Father's Day, See **bastard.**
3. happy as Larry, very happy.
4. happy days! (*sometimes ironic*) have a good time!

happy hour *n.* the time in a hotel, club, etc., when drinks are either free or sold at a reduced price.

happy hunting ground *n.* a suitable place for an activity: *Abandoned gold-rush towns are happy hunting grounds for amateur prospectors.* [from North American Indian mythology, the place inhabited by souls after death.]

happy pill *n.* a tranquillising pill.

hard *n.* **1.** an erect penis. *—phr.*
2. get a hard on, to have an erection.
3. hard act to follow, See **act.**
4. hard cheese/cheddar/luck, a. bad luck. **b.** an off-hand expression of sympathy. **c.** a rebuff to an appeal for sympathy.
5. hard up, urgently in need of something, especially money.
6. put the hard word on, a. to ask a favour of. **b.** to ask another for sexual intercourse.

hard case *n.* **1.** a tough, cynical person. **2.** a witty, consistently amusing person. Also, **hard doer.** **3.** an incorrigible person. **4.** a person suffering from drug addiction, especially to alcohol.

hard-core *adj.* **1.** explicit, blunt, unequivocal: *a hard-core film.* **2.** physically addictive: *hard-core drugs.*

hard-done-by *adj.* unfairly treated: *Don't feel so hard-done-by.*

hard-earned *n.* money.

hard hat *n.* a construction worker, working in an area in which safety helmets must be worn.

hard-knocker *n.* a black bowler hat.

hard knocks *pl. n.* difficult experiences; misfortunes: *He came up through the school of hard knocks.*

hardliner *n.* a person, especially a politician, who takes a tough, stubborn view on an issue: *He is an abortion hardliner.*

hard lines *pl. n.* bad luck; unfair treatment.

hard luck story *n.* an account of personal misfortune, intended to arouse the sympathy of others.

hardnose *n.* **1.** a person who repeatedly or habitually relapses into crime. **2.** an uncompromising person.

hard-nosed *adj.* ruthless, especially in business.

hard stuff *n.* **1.** strong alcoholic liquor; spirits. **2.** hard drugs, as heroin, etc.

harpoon *n.* **1.** an additional period taught by a teacher because of the absence of a colleague. —*v.* **2.** to require (a teacher) to take an additional period.

has-been *n.* a person or thing that is no longer effective, successful, popular, etc.

hash *n.* flowering tops, leaves, etc., of Indian hemp, smoked, chewed, or otherwise used as a narcotic or intoxicant; hashish.

hash house *n.* a cheap restaurant.

hassle *n.* **1.** a quarrel, squabble. **2.** a struggle; period of unease: *Today was a real hassle.* —*v.* **3.** worry; harrass: *Don't hassle me!*

hat *n.* **1.** a rank of office among many: *Which hat is he wearing now?* —*phr.*

2. at the drop of a hat, on the spur of the moment; without preliminaries.

3. bad hat, a bad or immoral person.

4. eat one's hat, to be very surprised if a certain event happens: *If he wins this game I'll eat my hat.*

5. hats off to, an expression of approval for (something).

6. old hat, (of ideas, etc.) old fashioned; out of date.

7. talk through one's hat, to talk nonsense; speak without knowledge of the true facts.

8. throw one's hat in first or **the door,** to test the warmth of one's reception in company, as when arriving late.

9. throw one's hat into the ring, to join in a competition or contest.

10. under one's hat, secret, confidential: *Keep this information under your hat.*

11. wear two hats, to hold two official appointments at the same time.

hatch *phr.* **down the hatch,** drink up!

hatchet *phr.*

1. bury the hatchet, See **bury.**

2. dig up or **take up the hatchet,** to prepare for war.

hatchet face *n.* a sharp, narrow face.

hatchet man *n.* **1.** a man employed or delegated to perform unpleasant tasks, such as cutting costs, firing personnel, etc., for an employer. **2.** a hired assassin.

hatful *n.* a considerable quantity or number: *a hatful of runs.*

hatrack *n.* **1.** a thin or scrawny animal, as a horse, cow, etc. **2.** a very thin person.

hatter *n.* **1.** a miner who works alone. **2.** a lonely and eccentric bush dweller.

haul *n.* **1.** the taking or acquisition of anything, or that which is taken. —*phr.*

2. haul over the coals, to rebuke; scold.

3. haul up, to bring up, as before a superior, for reprimand; call to account.

have *v.* **1.** to hold at a disadvantage: *He has you there.* **2.** to outwit, deceive, or cheat: *She is a person not easily had.* **3.** to possess sexually; to copulate with. —*n.* **4.** a delusion; trick: *It was a bit of a have.* —*phr.*

5. have had it, a. to be fated beyond hope of recovery, to die, be defeated, etc. **b.** to have failed to take advantage of a last chance. **c.** to become out of fashion or no longer popular.

6. **have had (someone** or **something),** to be annoyed or exasperated with: *I've had you; I've had this job.*

7. **have it away,** *British.* to have sexual intercourse.

8. **have it coming to one,** to deserve an unpleasant fate.

9. **have it in for,** to hold a grudge against.

10. **have it made,** See **made.**

11. **have it off,** to have sexual intercourse.

12. **have it out,** to have a candid argument; to discuss a matter extensively.

13. **have oneself on,** to delude oneself, especially as a result of pandering to one's own ego.

14. **have someone on, a.** to tease or deceive a person. **b.** to accept a fight or competition with a person.

15. **have up,** to bring before the authorities, especially in court: *He was had up for theft.*

16. **let someone have it,** to launch a strong attack upon someone.

haves *phr.* **haves and have-nots,** rich and poor (of countries and people).

haw-haw *adj.* affectedly superior in enunciation.

hawk *phr.* **1. hawk about,** to spread, especially news and the like.

2. hawk the fork/dot/hips, to be a prostitute.

[back formation from *hawker*]

Hawkesbury clock *n.* a kookaburra. [from the *Hawkesbury* river, north of Sydney]

Hawkesbury rivers *pl. n.* alcoholic shakes. [rhyming slang *Hawkesbury rivers* shivers]

Hawks, the *pl. n.* **1.** the Hawthorn team in the Victorian Football League. **2.** the West Adelaide team in the National Soccer League.

hay *n.* **1.** money. *—phr.*

2. hit the hay, to go to bed.

3. make hay, to scatter everything in disorder.

4. roll in the hay, to sport sexually or copulate.

Hay *phr.* **Hay, Hell and Booligal,** hot and uncomfortable places: places to be avoided. [popularised by the poem *Hay, Hell and Booligal* by A.B. (Banjo) Paterson; Hay and Booligal are towns in inland N.S.W.]

hayburner *n.* a horse.

hayseed *n.* a countryman or rustic. [from the grass seed shaken out of the hay which is a symbol of a simple country person]

haystack *n.* *Canoeing.* a stationary white-headed wave caused by fast-flowing water moving into slow water.

hazed-off *adj.* sun-dried, as land.

head *n.* **1.** a person who uses drugs regularly, especially marijuana and LSD. *—phr.*

2. get a big or **swelled head,** to become conceited.

3. have heads over, to punish the people reponsible for (a blunder).

4. have one's head screwed on the right way, to have a lot of commonsense.

5. have (someone's) head, to punish (someone) severely.

6. head over heels/tail/turkey, a. upside-down; headlong as after a somersault. **b.** completely, utterly.

7. **head them,** (in two-up) to make the coins land with heads upwards.
8. **keep one's head above water,** to remain in control of a difficult situation, especially a financial one.
9. **lose one's head, a.** to panic, become flustered, especially in an emergency. **b.** to behave irrationally or out of character.
10. **make head or tail of,** to understand; work out: *I can't make head or tail of this question.*
11. **make heads roll,** to demote or dismiss people as a punishment.
12. **need one's head read,** (*humorous*) to be insane.
13. **off one's head,** mad; very excited; delirious.
14. **off the top of one's head,** impromptu; without prior thought.
15. **one's head off,** to an extreme; excessively: *talk one's head off.*
16. **out of one's head,** out of one's mind; demented; delirious.
17. **pull one's head in,** to mind one's own business.
18. **win against the head,** *Rugby Football.* to hook the ball despite the opposing team's advantage in having a loose head.

headache *n.* a troublesome or worrying problem.

head'em *v.* to play two-up.

head-hunting *n.* **1.** the eliminating of political enemies. **2.** (in a business or other organisation) the seeking out of a scapegoat for a misfortune or setback. **3.** the search for new executives, usually senior, through personal contacts rather than advertisements. **4.** *Rugby Football.* illegal tackling about the head, often in the hope of injuring and incapacitating players of the opposing team.

headless chook *phr.* **run around like a headless chook,** to aggravate a state of confusion by frantic and disorganised activity.

head-puller *n.* a person who directs a shop assistant's attention elsewhere while an accomplice steals.

head serang *n.* the person in charge; boss. Also, **head sherang, head shebang.** [*head* + Persian *serang* boatswain]

headshrinker *n.* Also, **shrink.** a psychiatrist.

health kick *n.* an obsession with maintaining a healthy lifestyle.

healthy *adj.* powerful: *a few healthy whacks in the ribs.*

healy *n.* **1.** a confidence trick. **2.** an illegal means of earning a living: *What's your latest healy?* —*v.* **3.** to catch on; to understand.

heap *n.* **1.** a great quantity or number; a multitude: *a heap of trouble.* **2.** something very old and dilapidated, especially a motor car. —*phr.*
3. **for heaps,** for a long time.
4. **give it heaps,** to treat with firmness, in order to exact good performance.
5. **give (someone) heaps, a.** express strong displeasure with (someone); criticise severely. **b.** to tease; provoke to anger, annoyance, etc., by banter or mockery.
6. **strike all of a heap,** to dumbfound; amaze; overwhelm.

hear *phr.* **hear things,** to imagine noises; hallucinate.

heart *phr.*
1. **be all heart,** (*usually ironic*) to be full of consideration and kindness.

2. eat one's heart out, See **eat.**
3. have one's heart in one's mouth, to be very frightened.

hearts-and-flowers *n. Chiefly U.S.* excessive sentimentality.

heart-starter *n.* **1.** an alcoholic drink, especially one taken early in the day, often as a remedy for a hangover. **2.** any drink, as strong coffee, etc., taken before one begins the day's activities.

heart-stopper *n.* an event or scene full of suspense.

heart-throb *n.* the object of an infatuation, as a pop singer, film star, or the like.

heat *n.* **1.** pressure of police, prison or other investigation or activity: *Lie low while the heat's on.* —*phr.*
2. put the heat on, to put pressure on.

heater *n. U.S.* a pistol or revolver.

heavens *interj.* an expression of surprise, etc.

heavy *adj.* **1.** coercive; threatening: *The cops were really heavy.* **2.** serious; intense: *It was really heavy, really terrible.* —*n.* **3.** Also **heavie.** a person who is eminent and influential in the sphere of his activities, as a senior student, an important business man, etc. **b.** a person of some status who unpleasantly exercises his authority or seeks to intimidate. **c.** a man who attempts to intimidate a woman into sexual submission. **4.** a detective. —*v.* **5.** to confront, put pressure on. —*phr.* **6. do a heavy,** to exert authority; intimidate.

heavyweight *n.* a person of considerable power, influence, or forcefulness in a certain field, as a writer, philosopher, or statesman.

heck *n., interj.* a euphemism for hell: *What the heck! Get the heck out of here.*

heebie-jeebies *pl. n.* **1.** a condition of nervousness or revulsion. **2.** a violent restlessness due to excessive indulgence in alcohol, characterised by trembling, terrifying visual hallucinations, etc.; delirium tremens. [coined by W. De Beck, 1890-1942, U.S. cartoonist]

heel *n.* **1.** a despicable person; cad. —*phr.*
2. cool or **kick one's heels,** to be kept waiting, especially as deliberate policy.
3. dig one's heels in, to maintain an immovable position in debate, etc.; be stubborn.
4. down at heel, in straitened circumstances; impoverished.
5. kick up one's heels, to enjoy oneself in an exuberant manner.

heifer *n.* **1.** a girl or woman. —*phr.* **2. legs like a Mullingar/Mungindie/Menindee heifer,** physically unattractive. [from *heifer,* a young cow]

heifer dust *n.* nonsense; rubbish. See also **bulldust.**

heifer paddock *n.* a girl's school.

Heinz *n.* an animal or person of mixed stock. [from advertisement for *Heinz* soup, emphasising many varieties]

hell *interj.* **1.** an exclamation of annoyance, disgust, etc. —*phr.*
2. beat hell out of, to physically assault in a vindictive manner.
3. blast hell out of, to verbally assault; to severely reprimand.
4. for the hell of it, for no specific reason; for its own sake.
5. get the hell out of, into, etc., move rapidly or energetically.

6. give (someone) hell, to make things unpleasant for (someone).

7. hell for leather, at top speed; recklessly fast.

8. hell of a. Also, **helluva. a.** appallingly difficult, unpleasant, etc. **b.** notable; remarkable.

9. hell's bells, a mild oath.

10. hell's teeth, an exclamation of astonishment, indignation, etc.

11. hell to pay, serious unwanted consequences.

12. like hell, a. very much (used as general intensive). **b.** not at all; definitely not.

13. merry hell, an upheaval; a severe reaction; severe pain.

14. not a hope in hell, not the slightest possibility.

15. play hell with, a. to cause considerable damage, injury or harm to. **b.** to reprimand severely.

16. raise hell, cause enormous trouble.

17. the hell with it, an expression of disgust or rejection.

18. what the hell, an exclamation of contempt, dismissal, or the like.

hellhole *n.* a highly unpleasant place.

hell's angels *n.* a group of leather-jacketed motor-cyclists. Also, **angels.** [originally a group of lawless motor-cyclists known for their disturbance of civil order in the United States, especially California]

helluva *adj.* **1.** very good: *He's a helluva fellow.* **2.** very bad: *It was a helluva shock.* **3.** (used to add emphasis to a statement): *We had a helluva good time.*

he-man *n.* a tough or aggressively masculine man.

hen *n.* a woman, especially a fussy or foolish woman.

hen-cackle *n. N.Z.* a mountain considered easy to climb.

hen fruit *n.* hen's eggs.

henhouse *n.* (*humorous*) a place inhabited mainly by women.

hen party *n.* (*humorous*) a party exclusively for women.

henpeck *v.* (of a wife) to domineer over (her husband).

henry *n.* a signature. [short for U.S. colloquialism John *Henry,* an autographed signature]

Henry the Third *n.* a turd. [rhyming slang]

hens' night *adj.* an evening on which a group of women have a night out together; girls' night out. See also **stag party.**

hen's teeth *phr.* **scarce** or **rare as hen's teeth,** extremely rare; non-existent.

hep[1] *adj. Chiefly U.S.* See **hip.**

hep[2] *n.* hepatitis. [shortened form]

hepcat *n.* (in the 1950s) an expert performer, or a knowing admirer, of jazz.

herb *n.* **1.** (*pl.*) power, especially of cars: *this car has plenty of herbs.* —*v.* **2.** (of a motor vehicle) to travel at speed. **3.** to convey at speed. —*phr.*

4. give it herbs, to accelerate a motor vehicle.

5. herb over, to pass: *Herb over that bottle, mate.*

6. the herb, marijuana.

herd *n.* **1.** a large company of people. —*phr.* **2. the herd,** the common people; the rabble.

here *phr.*
1. here goes! an exclamation to show one's resolution on beginning some bold or unpleasant act.
2. here today and gone tomorrow, (of someone or something) staying in one place for only a short time.
3. here we go again! an exclamation indicating exasperation or resignation at a course of action about to occur yet again.

herk *v.* to vomit.

hetero *adj.* **1.** showing or relating to sexual feeling for a person (or persons) of the opposite sex. —*n.* **2.** a heterosexual person.

het-up *adj.* anxious; worried. Also, **het up.** [alteration of *heated-up*]

Hexham grey *n.* an extremely large and voracious variety of mosquito found in the locality of Hexham, near Newcastle, NSW.

hey-diddle-diddle *n.* **1.** the middle. **2.** urination; a piddle. —*phr.* **3. through the hey-diddle-diddle,** *Football, etc.* through the middle; a goal. Also, **hi-diddle-diddle.** [rhyming slang]

hiccups *pl. n.* temporary difficulties at an early stage: *After a few hiccups the system worked well.*

hick[1] *n.* **1.** an unsophisticated person. **2.** a farmer. —*adj.* **3.** relating to or characteristic of hicks. [familiar form of *Richard* man's name]

hick[2] *n.* *Originally U.S.* a pimple. Also, **hickey.**

hide[1] *phr.* **hide one's head,** to be ashamed.

hide[2] *n.* **1.** the human skin. **2.** impudence: *He's got a hide!* —*phr.*
3. a thick hide, insensitivity to criticism.
4. no hide no Christmas box, no reward is to be had without impudent initiative.

hideaway *n.* a place of concealment; a refuge.

hi-diddle-diddle *n.* See **hey-diddle-diddle.**

higgledy-piggledy *adv.* **1.** in a jumbled confusion. —*adj.* **2.** confused; jumbled.

high *adj.* **1. a.** intoxicated or under the influence of drugs. **b.** elated, as from the effects of drugs or alcohol. —*n.* **2.** euphoric state induced by drugs. —*phr.*
3. high as a kite, a. under the influence of drugs or alcohol. **b.** in exuberant spirits.
4. high and dry, abandoned; stranded; deserted.
5. on a high, experiencing a euphoric state, especially as one induced by drugs.

highbrow *n.* **1.** a person who has pretensions to superior taste in artistic matters. —*adj.* **2.** relating to that which highbrows approve of: *highbrow music.* See also **lowbrow.**

high-camp *adj.* affected; ostentatious, as typical of certain homosexuals.

higher-up *n.* one occupying a superior position.

highfalutin *adj.* pompous; haughty; pretentious. Also, **hifalutin, highfaluting.**

high-flier *n.* one who is extravagant or goes to extremes in aims, pretensions, opinions, etc.

high-hat *v.* **1.** to snub or treat condescendingly. —*adj.* **2.** snobbish; affectedly superior.

high horse *phr.*

1. get off one's high horse, to stop being arrogant, self-righteous, etc.

2. on one's high horse, assuming an arrogant or superior air.

high jinks *phr.* See **jinks.**

high jump *n.* **1.** *Prison.* a criminal court. **2.** execution by hanging. —*phr.* **2. for the high jump(s), a.** about to face an unpleasant experience, especially a punishment or reprimand. **b.** *Prison.* up for trial.

Highlanders, the *pl. n.* the Gordon team in the N.S.W. Rugby Football Union.

high-stepping *adj.* **1.** (of a person) fashionably dressed; having fashionable pretensions. **2.** dedicated to the pursuit of pleasure; leading a hectic life.

high strikes *n.* outlandish or eccentric behaviour. [corruption of *hysterics*]

hightail *phr.* **hightail it,** to move away quickly: *He hightailed it out of town.*

highway robbery *n.* a shameless attempt to rob, overcharge or cheat someone. Also, **daylight robbery.**

high words *pl. n.* a heated argument; row.

hike *phr.* **take a hike,** (*often offensive*) an order to leave (a place, etc.).

hill *phr.*

1. as old as the hills, very old.

2. over the hill, past prime efficiency; past the peak of physical or other condition, etc.

Hill, the *n.* the grassed spectator area in front of the new scoreboard at the Sydney Cricket Ground.

himself *n. Chiefly British.* a man who is dominant in a house, office, etc., especially the master of the house: *Is himself at home today?*

hip[1] *phr.* **have (someone) on** or **upon the hip,** to have (someone) at a disadvantage.

hip[2] *adj. Chiefly U.S.* having inside knowledge, or being informed of current styles, especially in jazz: *to be hip to swing music. Also,* **hep.**

hipped *phr.* **hipped on,** *U.S.* greatly interested in; having an obsession: *He's hipped on playing the tuba.*

hippie *n.* one who rejects conventional social values in favour of new standards of awareness (sometimes drug-induced), universal love or union with nature, etc. Also, **hippy, hipster.**

hippie trail *n.* **1.** a travel route which takes in those places in the world where hippies congregate, as India, Bali, etc. **2.** a similar route within Australia, especially along the north-eastern coast.

hippo *n.* a hippopotamus.

hip-pocket *adj.* concerned with money: *a hip-pocket issue.*

hip-pocket nerve *n.* an imaginary nerve which is sensitive to demands for one's money, especially through government action to increase taxation, etc.

history *phr.*

1. be ancient history, to be finished or gone irrevocably.

2. be history, a. to be dead. **b.** to be ruined or incapacitated. **c.** to be broken beyond repair.

hit *v.* **1.** to inject any form of drugs. —*n.* **2.** a shot of heroin or any drug; a fix. —*phr.*
3. hit for six, to confuse or disturb greatly: *The bad news hit him for six.*
4. hit home, to make an impact or impression upon.
5. hit if off, to get on well together; agree.
6. hit off, to make a beginning; commence.
7. hit (someone) hard, to have a severe and distressing effect upon (someone).
8. hit the anchors, to apply the brakes of a motor vehicle.
9. hit the bottle or **booze,** See **bottle.**
10. hit the ceiling or **roof,** to display extreme anger or astonishment.
11. hit the deck, See **deck.**
12. hit the headlines, to gain publicity; to achieve notoriety.
13. hit the lip, to ride a surfboard off the extremity of a wave.
14. hit the nail on the head, a. to sum up with clarity and incisiveness. **b.** to give perfect satisfaction. **c.** to say or do exactly the right thing.
15. hit the road/bitumen/toe, to set out; depart.
16. hit the sack or **hay,** to go to bed.
17. hit the sky, to buy a round of drinks. [from rhyming slang *skyrocket* pocket]
18. hit the spot; to fulfill a need; satisfy.
19. not to know what hit one, to be taken unawares; be thrown into confusion or dismay.
20. hit up, to take a drug, as heroin, usually by injecting it into the bloodstream.

hit and miss *n.* **1.** the act of urinating. —*v.* **2.** to urinate. [rhyming slang *hit and miss* piss]

hitch *v.* **1.** to obtain or seek to obtain a ride from a passing vehicle. —*n.* **2.** *U.S.* a period of military service. **3.** a ride from a passing vehicle.

hitched *adj.* married.

hitchhike *v.* to travel by obtaining rides in passing vehicles.

hit man *n.* a hired assassin.

hive *phr.* **hive off,** to depart; to break off from a group: *to hive off by oneself.*

hock[1] *phr.* **in hock,** pawned.

hock[2] *n.* an active male homosexual.

hockshop *n.* *U.S.* a pawnshop.

hodad *n.* a swimmer who annoys or impedes surfboard riders. Also, **ho-dad.**

hoddie *n.* a bricklayer's assistant; a hodman.

hoe *phr.*
1. hoe in, a. to commence to eat heartily. **b.** to begin something energetically.
2. hoe into, a. to eat (food) heartily. **b.** to attack (a person) vigorously, usually verbally. **c.** to undertake (a job) with vigour.

hog *n.* **1.** a selfish, gluttonous, or filthy person. —*v.* **2.** to appropriate selfishly; take more than one's share of. —*phr.*
3. go the whole hog, to do completely and thoroughly; to commit oneself unreservedly to a course of action.
4. live high on the hog, to live luxuriously, extravagantly.

hoick *v.* to clear the throat and spit. [imitative]

hoist *v.* **1.** to steal, especially to shoplift. **2.** to throw. *—n.* **3.** a theft; housebreaking.

hoister *n. Prison.* a shoplifter.

hokonui *n. N.Z.* any illicitly-distilled spirits. [from *Hokonui* district of Southland Province, New Zealand]

hokum *n. U.S.* nonsense; bunk. [blend of *hocus-pocus* and *bunkum*]

hold *v.* **1.** to have adequate money or assets: *How are you holding? —phr.*

2. get a hold on oneself, to get control over oneself.

3. hold it, to stop; wait.

4. hold out, *Originally U.S.* to keep back something expected or due.

5. hold (someone's) hand, to provide with moral support; encourage.

hole *n.* **1.** an embarrassing position or predicament: *to find oneself in a hole.* **2.** any of certain apertures of the body, as the mouth, anus, or female genitals. *—phr.*

3. hole up, to hide (often from the police).

4. score a hole in one, (of a man) to have a woman yield to sexual intercourse on their first night out.

hole-and-corner *adj.* furtive; secretive; underhand.

Hole and Corner Man *n.* (formerly) a person who advocated severity in dealing with transported convicts and was opposed to the emancipists.

hollies *pl. n.* holidays.

hollow *adv.* **1.** utterly: *to beat someone hollow. —phr.* **2. have hollow legs,** to have a large appetite.

hollow log *n.* **1.** a new Australian; wog. **2.** a dog. [rhyming slang]

hollywood *phr.* **do a hollywood,** to gain attention by one's dramatic behaviour. [from *Hollywood* centre of the film industry in the United States]

hols *pl. n.* holidays.

holts *phr.* **in holts (with),** fighting or in dispute (with).

holus-bolus *adv.* **1.** all at once. **2.** in its entirety. [mock Latin]

holy *phr.* **holy terror,** a person difficult to deal with; an alarming or frightening person.

Holy Jesus *interj.* an exclamation indicating surprise, indignation, etc. Also, **Holy Christ.**

holy Joe *n.* **1.** a member of a religious order. **2.** one of unpleasantly rigid religious principle. **3.** a Roman Catholic.

home *phr.*

1. home and hosed/dried/with a rug on, finished successfully. [from horseracing]

2. home on the pig's back, certain to succeed.

3. nothing to write home about, not remarkable; unexciting; inferior.

4. who is he when he's at home? an exclamation, usually scornful, indicating that the person referred to has an undeservedly high opinion of himself.

home stretch *phr.* **on the home stretch,** in the last stages of any project or undertaking.

homework *phr.* **do one's homework,** to undertake preparatory work for a specific activity, meeting, interview, competition, etc.

homey *adj.* homelike; comfortable; friendly.

hominy *n. Prison.* breakfast; any porridge-like food.

hominy gazette *n. Prison.* news or rumours that circulate within a gaol when the cells are opened and breakfast is distributed.

homo *n., adj.* homosexual.

honest *phr.*

1. **honest to dinkum,** true, vouched for.

2. **make an honest woman of,** to marry.

honey *n.* **1.** a person or thing which inspires affectionate admiration: *That machine is a honey.* **2.** sweet one; darling (a term of endearment).

honey cart *n.* a sanitary cart.

honey pot *n.* **1.** sweet one; darling; honey. **2.** action of dive-bombing. **3.** a toilet, especially a sanitary can. **4.** the female pudendum. *—v.* **5.** to jump into water, as a swimming pool, creek, etc., with the knees tucked under the chin.

honk *v.* **1.** to emit an offensive smell. *—n.* **2.** an offensive smell. *—phr.* **3. on the honk,** emitting an offensive smell.

honky *n. U.S.* (*offensive*) a white man.

honky-tonk[1] *n.* **1.** *Chiefly U.S.* a cheap, sordid nightclub, dance hall, etc. *—adj.* **2.** *U.S.* of or pertaining to a honky-tonk.

honky-tonk[2] *n.* wine. [rhyming slang *honky tonk* plonk]

honours *phr.* **do the honours,** to preside over an event, as by pouring the drinks, etc.

hooch *n. U.S.* **1.** alcoholic beverages. **2.** alcoholic liquor illicitly distilled and distributed.

hood *n.* a hoodlum.

hoodoo *n.* **1.** a person or thing that brings bad luck. *—v.* **2.** to bring or cause bad luck to.

hooey *Originally U.S. —interj.* **1.** an exclamation of disapproval. *—n.* **2.** silly or worthless stuff; nonsense.

hoof *n.* **1.** (*humorous*) the human foot. *—v.* **2.** to dance. *—phr.* **3. hoof it,** to walk.

hoofer *n. U.S.* one who makes dancing an occupation, as a chorus girl.

hoo-ha *n.* a fuss; turmoil; argument.

hook *v.* **1.** to seize by stealth, pilfer, or steal. **2.** to marry: *She's managed to hook a rich man.* **3.** to depart; clear off. *—phr.*

4. **hook into,** to become an integral part of.

5. **hook it,** to depart; clear off.

6. **hook, line, and sinker,** completely.

7. **off the hook,** (of a garment) ready-made; off the peg.

8. **on one's own hook,** on one's own responsibility.

9. **on the hook, a.** waiting; being delayed. **b.** in a difficult predicament.

10. **put the hooks into,** to borrow from; cadge.

11. **sling one's hook,** to depart.

12. **the hooks,** the fingers.

13. **your hook!** a demand that one buy the next round of drinks.

hooked *adj.* **1.** married. *—phr.* **2. hooked on,** addicted; obsessed.

hooker[1] *n.* a prostitute.

hooker[2] *n. Chiefly U.S., Canada.* a glass of spirits, especially whisky.

hooks *pl. n.* the chevrons on an army uniform.

hooky *phr.* **play hooky,** to be unjustifiably absent from school, work, etc.Also, **hookey.**

hooley *n.* a wild party.

hooligan *n.* a hoodlum; young street rough.

hoon *n.* **1.** a loutish, aggressive, or surly youth. **2.** a foolish or silly person, especially one who is a show-off. **3.** one who lives off the proceeds of prostitution.

hoop *n.* **1.** a jockey. *—phr.*

2. go through the hoop, go through a bad time; undergo an ordeal.

3. jump through hoops, to obey without question, in the manner of a trained dog.

4. put through (the) hoops, subjected to a series of often unreasonable tests or trials.

hooroo *interj.* goodbye. Also, **hooray, ooray, ooroo.**

hoosegow *n. U.S.* a gaol. Also, **hoosgow.**

hoot[1] *n.* **1.** a thing of no value: *I don't give a hoot.* **2.** an amusing or funny thing.

hoot[2] *n. Chiefly N.Z.* money, especially money paid as recompense. Also, **hootoo, hout, hutu.** [Maori *utu*]

hooter *n.* the nose.

hop *v.* **1.** to dance. **2.** to go, come, move, etc.: *Hop into the car and we'll go.* *—n.* **3.** a flight of an aeroplane. **4.** a dance, or dancing party. *—phr.*

5. hop into, a. to set about something energetically: *He hopped into the job at once.* **b.** to put (clothes) on briskly: *He hopped into his cossie.*

6. hop into bed, to have casual sex.

7. hop it, to go away; leave.

8. hop to, to come or act quickly: *Hop to it.*

9. hop up and down, to express agitation.

10. on the hop, a. unprepared. **b.** busy, moving.

hope *phr.* **great white hope,** a person from whom or a thing from which exceptionally great successes or benefits are expected.

hophead *n.* **1.** one who drinks alcoholic beverages to excess. **2.** one addicted to drugs.

hori *n. N.Z. (offensive)* a Maori.

horn *n.* an erect penis.

horny *adj.* randy; sexually excited.

horror *n.* something considered atrocious or bad: *That hat is a horror.*

horse *n.* **1.** heroin. *—phr.*

2. back the wrong horse, to support the wrong or losing contender.

3. eat like a horse, to have a prodigious appetite.

4. horse about or **around,** to act or play roughly or boisterously.

5. hungry enough to eat a horse and chase the rider or **eat a horse if you took its shoes off,** very hungry.

6. jump a horse over the bar; eat or **drink a horse,** (formerly) to sell or exchange a horse for liquor.

7. willing horse, a willing worker.

horse and cart *n.* fart. [rhyming slang]

horse marines *phr.* **tell that** or **it to the horse marines,** See **marines.**

horse sense *n.* plain, practical, common sense.

horses for courses *n.* the notion that a person should be matched with a position, task, etc., suited to his or her particular talents. [from the theory that a horse which races well on one track should not run on a different track to which it is not suited]

horseshit *n.* nonsense; rubbish; bullshit.

horse's hoof *n.* a homosexual. [rhyming slang, *horse's hoof* poof, homosexual]

horse-trading *n.* shrewd and close bargaining.

hostie *n.* an air hostess.

hostile *phr.* **go hostile,** *N.Z.* become angry.

hot *adj.* **1.** fashionable and exciting. **2.** recently stolen or otherwise illegally obtained. **3.** (of a person) wanted by the police. *—phr.*
4. a bit hot, unfair; dishonest; high-priced.
5. blow hot and cold, change attitudes frequently; vacillate.
6. go hot and cold all over or **go all hot and cold,** to experience, or exhibit signs of, shock or embarrassment.
7. hot as Hades, very hot.
8. hot as Hay, Hell and Booligal, very hot. See **Hay.**
9. hot under the collar, angry; annoyed.
10. hot up, a. to escalate: *He hotted up his attack.* **b.** to stir up: *to hot things up a bit.* **c.** to tune or modify (a motor vehicle) for high speeds.
11. in hot water, in trouble.
12. like a cat on a hot tin-roof or **on hot bricks,** in a state of extreme agitation.
13. make it hot for, to make life unpleasant for.
14. not so or **too hot, a.** not very good; disappointing. **b.** unwell.
15. sell or **go like hot cakes,** to sell or be removed quickly, especially in large quantities.
16. the hots, a strong sexual attraction: *to have the hots for Sadie.*

hot air *n.* empty, pretentious talk or writing.

hot dog *n.* **1.** a short surfboard designed to turn quickly back and forth across the waves. *—interj.* **2.** *U.S.* an exclamation indicating enthusiasm, admiration, surprise, etc. Also, **hot-dog.**

hot line *n.* **1.** an especially important telephone connection. **2.** any instant communication channel, with restricted access. [from the *hot line* the direct telephone connection open to immediate communication in an emergency, as between the heads of state of the Soviet Union and the United States]

hotpants *n.* **a.** strong sexual desires. **b.** one who has strong sexual desires: *She's a real little hotpants.*

hotpoint *v. Prison.* to cheat; deceive. Also, **point.**

hot potato *n.* a risky situation, difficult person, or any other thing which needs careful handling.

hot property *n.* a person or thing highly valued for its commercial potential.

hot rod *n.* a car (usually an old one) whose engine has been altered for increased speed.

hot seat *n.* a position involving difficulties or danger.

hot-shot *adj.* **1.** exceptionally proficient. —*n.* **2.** one who is exceptionally proficient, often ostentatiously so.

hot stuff *n.* **1.** a woman or girl who is sexually exciting. **2.** something or someone of great excellence or interest.

hot-water boat *n.* (*humorous*) a small powered pleasure boat as a runabout (opposed to *sailing boat*).

hound *n.* **1.** a mean, despicable fellow. **2.** *U.S.* an addict.

house *phr.*

1. like a house on fire, very well; with great rapidity.

2. on the house, free; as a gift from the management.

3. safe as houses, completely safe.

4. the house, the main homestead on a sheep or cattle station.

5. the little house, an outside toilet.

how *n.* **1.** way or manner of doing: *to consider the hows of a problem.* —*phr.*

2. and how, very much indeed; certainly.

3. how about that! an exclamation of surprise (sometimes ironic) or of triumph.

4. how come? how did this happen; why?

5. how's that? a. what is the explanation of that? **b.** See **howzat?**

6. how's things or **tricks,** a form of greeting.

Howe, Jacky *n.* See **Jacky Howe.**

howlie *n.* a U-turn.

howling *adj.* enormous; very great: *His play was a howling success.*

howling jackass *n.* a kookaburra.

howzat *interj. Cricket.* an appeal by the fielding side to the umpire to declare a batsman out.

hoy *v.* **1.** to throw. —*n.* **2.** bingo.

Hoyts *phr.*

1. dressed up like the man outside Hoyts, to be elaborately overdressed.

2. the man outside Hoyts, a. the imaginary source of all unverified rumours. **b.** a mythical figure of authority. [from the elaborately dressed commissionaire outside *Hoyts* Theatre, Melbourne, in the early part of this century]

hubby *n.* husband.

hubcap *n. Sport.* a kneecap.

huddle *n.* a conference held in secret.

huff *phr.* **huff and puff,** to bluster. [from the children's story 'The Three Little Pigs']

hug *phr.* **hug oneself,** congratulate oneself; be self-satisfied.

Hughie *n.* a jocular name for the powers above used when encouraging a heavy rainfall or a good surf: *send her down, Hughie! whip 'em up, Hughie!*

hui *n.* a boisterous party or gathering. [Maori]

hum[1] *v.* **1.** to be in a state of busy activity: *to make things hum.* **2.** to smell strongly, especially disagreeably.

hum[2] *n.* **1.** a person who persistently cadges or scrounges. —*v.* **2.** to cadge. [short for *humbug*]

humdinger *n.* a person or thing remarkable of its kind.

humming *adj.* extraordinarily active, intense, great, or big.

hump *n.* **1.** a good surfing wave. —*v.* **2.** (of men) to have sexual intercourse with a woman. —*phr.*
3. hump the bluey, or **humping bluey,** See **bluey.**
4. over the hump, over the worst part or period of a difficult, dangerous, etc., time.
5. the hump, a fit of bad humour: *to get the hump.*

humpy *n.* any rough or temporary dwelling; bush hut. [Aboriginal]

hundred *phr.* **a hundred to one,** of remote possibility.

hung *phr.* **hung, drawn and quartered,** See **hang.**

hung-over *adj.* suffering the after-effects of drinking too much alcohol. [back formation from *hangover*]

hungry *phr.* **the hungry mile,** (formerly) a strip of Sussex Street, Sydney, near the waterfront, along which wharf labourers tramped to and from work.

hung-up *adj.* (of a person) displaying emotional stress.

hunk *n.* a sexually attractive male.

hunky *n. U.S.* (*offensive*) an unskilled or semiskilled workman of foreign birth, especially a Hungarian; bohunk.

hunky-dory *adj.* perfectly all right, satisfactory. Also, **hunky.**

hunt *phr.* **hunt and peck,** to type with two fingers.

hunting *phr.* **good hunting,** an expression wishing someone luck or success.

hurdy-gurdy *phr.* **on a hurdy-gurdy,** involved in a mindless and seemingly never-ending round of activities.

hurl *v.* **1.** to vomit. —*n.* **2.** the act of vomiting.

hurry-up *phr.* **give (someone) a hurry-up,** to remind or prompt (someone) to action.

husband-beater *n.* a very long, narrow, loaf of bread. Also, **wife-beater.**

hush-hush *adj.* highly confidential.

hush money *n.* a bribe to keep silent about something.

hustle *v.* **1.** to solicit for or as a prostitute. **2.** *U.S.* **a.** to obtain (money) by questionable methods. **b.** to pursue sales with aggressive energy.

hustler *n.* an energetic person who sweeps aside all obstacles in his or her path.

hype[1] *n.* **1.** fraud; racket; swindle. **2.** hypocrisy; pretentiousness. —*v.* **3.** to bluff; con, as by false publicity.

hype[2] *n.* **1.** a hypodermic needle. **2.** a drug addict. —*v.* **3.** to persuade, exhort to greater achievement, as a coach convincing a team that success can be theirs. —*phr.* **4. hype up, a.** to stimulate; make excited. **b.** to increase the power, speed, etc., of a car engine, etc.: *He hyped up his F.J.* [shortened form of *hypodermic*]

hyper *adj.* nervous; on edge; overstimulated.

hypo *n.* a hypodermic needle or injection.

ice *n.* **1.** a diamond or diamonds. —*phr.* **2. cut no ice,** to make no impression; be unconvincing: *His excuses cut no ice with me.*

iceberg *n.* **1.** a regular winter swimmer. **2.** a cold, reserved person. —*phr.* **3. tip of the iceberg,** a small part of a larger whole, usually a problem: *The leaking roof was only the tip of the iceberg.*

icebreaker *n.* anything which breaks down reserve or reticence.

ice-cold *n.* a cold can of beer.

ickle *adj.* little; small.

icky *adj.* **1.** sticky; gooey. **2.** difficult to deal with; disagreeable; troublesome. —*n.* **3.** a difficult situation.

idea-monger *n.* an inventive, creative person.

identity *n.* **1.** an odd, interesting, famous or infamous, person: *a well-known Sydney racing identity.* —*phr.* **2. old identity,** a person who has lived in a town, suburb, etc., for a very long time, and is interesting because of that.

idiot box *n.* a television set.

iffy *adj.* dubious; odd.

Ikey Mo *n.* **1.** a Jew. **2.** a moneylender. **3.** a tipster; bookmaker. —*adj.* **4.** cunning; nifty. **5.** mean; parsimonious. Also, **ikeymo.** [from familiar abbreviations of typical Jewish names *Isaac* and *Moses*]

illywhacker *n.* a confidence man; trickster.

imbo *n.* **1.** a simpleton; fool. **2.** *Prison.* a criminal's victim, especially the victim of a confidence trick.

immense *adj.* very good or fine.

import *n.* a person who has been brought into a business, team, etc., from outside.

improve *phr.* **on the improve,** getting better.

in *adj.* **1.** in favour; on friendly terms: *He's in with the managing director.* **2.** in fashion. *—n.* **3.** influence, pull; connection: *She has an in with the management — she married a director.* *—phr.*
4. in for it, about to be reprimanded or punished.
5. in like Flynn, See **Flynn.**
6. in on, having a share in or a part of, especially something secret, or known to just a few people.
7. nothing in it, (in a competitive situation) no difference in performance, abilities, etc., between the contestants.
8. well in, *Horseracing.* (of a horse) given a light handicap.

include *phr.* **include (someone** or **something) out,** to count as not participating: *You can include me out in this match.*

incog *adj., adv., n.* incognito.

indaba *n.* concern; affair. [Zulu: affair, business]

Indian *n.* a worker. [from the saying *too many chiefs and not enough Indians*]

Indian giver *n.* a person who gives something as a gift to another and later takes or demands it back.

indulge *v.* to drink alcohol or take drugs in excessive amounts.

influence *phr.* **under the influence,** drunk.

info *n.* information.

infra dig *adj.* beneath one's dignity.

Injun *n.* a North American Indian.

ink *n.* cheap wine. [?rhyming slang. See **pen and ink**]

inked *adj.* drunk; intoxicated.

inner man *n.* the stomach or appetite.

innings *phr.* **have had a good innings,** to have had a long life or long and successful career.

Innisfail *phr.* **in gaol at Innisfail,** to be in a difficult situation. See also, **Tallarook.**

inquest *n.* an inquiry into the reasons for the failure of a project, etc.

insane *adj.* fantastic; wonderful.

insect *n.* **1.** a contemptible person. *—adj.* **2.** contemptible.

inside *adv.* **1.** to or in prison. *—phr.* **2. get inside,** See **get.**

inside job *n.* a crime committed by or with the assistance of someone who is a member of the group or organisation against which the crime is directed.

insider *n.* **1.** one who is within a limited circle of persons who understand the actual facts of a case. **2.** one who has some special advantage.

insides *pl. n.* the internal parts of the body, especially the stomach and intestines.

instance *phr.* **give (someone) a for instance,** to give (someone) an example.

insurance *n.* **1.** an alternative to fall back on if one's main objective is lost: *She already has a boyfriend, so this bloke is just insurance.* **2.** protection money. *—phr.* **3. buy insurance,** to protect oneself against a possible future setback.

interesting *phr.* **interesting condition.** pregnancy.

interfere *phr.* **interefere with, a.** to molest sexually. **b.** to add alcoholic liquor to (coffee, tea, etc.).

into *adj.* **1.** devoted to the use or practice of; having an enthusiasm for: *I am into health foods.* *—phr.* **2. get into,** See **get.**

intro *n.* an introduction.

invite *n.* an invitation.

I.P. *n.* (usually among school teachers) an irate parent.

Irish *adj.* **1.** containing an inherent contradiction. *—phr.* **2. get one's Irish up,** to become angry.

Irish curtains *pl. n.* cobwebs.

Irishman's hurricane *n. Nautical.* a dead calm.

iron *n.* **1.** *British.* a pistol.*—v.* **2.** to assault; flatten. *—phr.*
3. iron out, a. to smooth and remove (problems and difficulties, etc.). **b.** to flatten (someone); knock down.
4. iron (oneself) out, to get drunk.

iron balls *phr.* **have iron balls, a.** to be very tough. **b.** to be impervious to illness, etc.

iron maiden *n.* a particularly difficult and cantankerous woman.

iron man *n.* **1.** a man of exceptional physical strength. **2.** Also, **ironman.** a participant in the Iron Man surfing contest. [from the *Iron Man* event at surf carnivals, a race consisting of surf ski, board, swim and sprint sections] **3.** (formerly) a pound note.

iron pot *phr.* **talk the leg off an iron pot,** to be very talkative.

Isa, the *n.* Mount Isa, Queensland.

it *n.* **1.** sex appeal. *—phr.* **2. with it, a.** in accordance with current trends and fashions; fashionable. **b.** well-informed and quick-witted.

itchy *phr.* **have itchy feet,** to have the desire to travel.

Itie *n.* an Italian. Also, **eyetie, eytie.**

itsy-bitsy *adj.* small; insubstantial. Also, **itty-bitty.**

ivory *n.* **1.** tooth, or the teeth. *—phr.* **2. the ivories, a.** the keys of a piano, accordion, etc. **b.** dice.
3. tickle the ivories, to play the piano.

J *n.* a marijuana cigarette. [See **jay**]

jab *n.* an injection with a hypodermic needle. Also, **job.**

jabberwocky *n.* nonsense. [coined by Lewis Carroll in 'Through the Looking Glass' (1871)]

jacaranda juice *n.* Grafton beer. Also, **jacka.** [Grafton, a town in northern N.S.W., is famed for its jacaranda trees]

jack *n.* **1.** a policeman, especially a detective. **2.** a double-headed coin. **3.** venereal disease. **4.** the anus. —*v.* **5.** *N.Z.* to arrange, organise, prepare. —*phr.*
6. jack of, fed up with.
7. jack off, to masturbate.
8. jack up, a. to refuse; be obstinate; resist. **b.** to prop up. **c.** to raise (prices, wages, etc.) **d.** *N.Z.* to fix up, renovate. *phr.*
1. I'm all right Jack, an expression of selfish complacency on the part of the speaker.
2. the house that Jack built, a venereal disease clinic.

jack-a-dandy *n.* See **dandy.**

jackass *n.* **1.** a very stupid or foolish person. **2.** Also, **laughing jackass,** a kookaburra.

jacked-up *adj.* infected with venereal disease.

jackeroo *n.* **1.** an apprentice station hand on a sheep or cattle station. —*v.* **2.** to work as a trainee on such a station: *He's jackerooing in Queensland this year.* Also, **jackaroo.**

Jack-'n'-Jill *n.* **1.** the bill. **2.** a dill; fool. **3.** a cash register; till. [rhyming slang]

Jacko *n.* the kookaburra. Also **jack, jacky.**

jackpot *phr.* **hit the jackpot, a.** to win chief prize on a gambling machine. **b.** achieve great success; be very lucky.

Jack Robinson *phr.* **before one can say Jack Robinson,** very quickly.

jackshay *n.* a tin quart pot used by bushmen for boiling water. Also, **jackshea.**

jacksy *n. N.Z.* the posterior; the buttocks.

Jack the Painter *n.* a strong, green, tea, used in the bush, which discolours utensils.

Jacky *n.* **1.** a male Aborigine. *—phr.* **2. sit (up) like Jacky,** to behave with full confidence; to be on one's best behaviour.
3. work like Jacky, to work very hard. Also, **Jackey.**

Jacky Howe *n.* **1.** a navy or black woollen singlet worn by labourers, bushmen, etc. **2.** any similar singlet. Also, **Jimmy Howe.** [?from name of Australian world champion shearer of 1892]

jade *n.* (*offensive*) a woman.

jag *n.* **1.** a drinking bout. **2.** any sustained single activity, often carried to excess: *an eating jag, a fishing jag.* [?from U.S. *jag,* a load carried on the back; thus as much drink as one can carry]

jagged *adj.* intoxicated, drunk.

jagging *phr.* **go jagging, a.** to gatecrash. **b.** to make unannounced social visits.

jail-bait *n.* See **gaol-bait.**

jailie *n.* See **gaolie.**

jake *adj.* all right: *She'll be jake, mate.* Also, **jakerloo.**

jalopy *n.* an old, decrepit, or unpretentious motor car.

jam[1] *n.* **1.** a difficult or awkward situation; a fix. *—phr.*
2. not all jam, not always pleasant and enjoyable.
3. put on jam, to affect a self-important manner.

jam[2] *n. U.S.* (homosexual use) a heterosexual.

jamas *pl. n.* pyjamas. Also, **jamies.**

jamboree *n.* a carousal; any noisy merrymaking. [apparently a blend of *jabber* and French *soirée* with *-m-* from *jam* crowd]

jammy *adj.* **1.** easy; requiring no effort. **2.** lucky.

jam-packed *adj.* crowded.

jam tart[1] *n.* a female; tart. Also, **J.T.**

jam tart[2] *n.* a fart. [rhyming slang]

jandals *pl. n. N.Z.* thongs.

jane *n.* a woman.

Jap *adj., n.* Japanese.

Japanese *phr.* **have a Japanese bladder,** to suffer from frequent urination.

jar *n.* a glass of beer.

jargon *n.* any talk or writing which one does not understand.

jarrah-jerker *n.* a timber-getter in Western Australia; axeman. [from *jarrah* a timber tree of Western Australia]

J. Arthur *n.* an act of masturbation. [from rhyming slang *J. Arthur Rank* wank]

jaunty *n.* Also, **jaundy, jonty.** *Naval.* the master-at-arms, the officer responsible for the maintenance of disciplinary regulations on board ship.

jaw *n.* **1.** talkativeness; continual talk. **2.** moralising or reproving talk. *—v.* **3.** to talk at length; gossip. **4.** to talk reprovingly; lecture; admonish.

jawbone *v.* to talk, especially at length, as in expounding an idea, presenting an argument, etc.

jaw-breaker *n.* **1.** a word hard to pronounce. **2.** a large, hard or sticky sweet.

jay *n.* a marijuana cigarette. [?from *j(oint)* or from *(Mary) J(ane)*]

jaybird *v.* **1.** to do housework in the nude. —*n.* **2.** a person who does this.

jazz *n.* **1.** liveliness; noisiness; spirit. **2.** pretentious or insincere talk. —*v.* **3.** to play or perform jazz music. **4.** to act or proceed with great energy or liveliness. —*phr.* **5. and all that jazz,** and all that sort of thing; et cetera. **6. jazz up,** to put vigour or liveliness into.

jazzy *adj.* wildly active or lively.

jeepers creepers *interj. Originally U.S.* an expression of surprise. [euphemism for *Jesus Christ*]

jell *v.* to take shape; crystalise; become definite.

jelly *n.* **1.** gelignite. —*phr.* **2. turn to jelly,** to become weak with fear.

jellyfish *v.* to act without courage, moral force, etc.; to be spineless.

jelly-leg *n.* one who yields readily to opposition.

jerk *n.* **1.** a stupid or naive person. —*phr.* **2. jerk off,** to masturbate.

jerk-off *n.* **1.** an act of male masturbation. **2.** a male masturbator. **3.** a very foolish person; an idiot.

jerry[1] *n.* **1.** a chamber-pot. **2.** lavatory. —*phr.* **3. full as a family jerry, a.** full (of food). **b.** drunk. [apparently from *jeroboam,* a very large wine bottle]

jerry[2] *phr.* **jerry to,** to understand, realise: *He jerries to what's going on.*

jerry[3] *n.* See **geri.**

Jerry *n.* **1.** a German, especially a German soldier. **2.** (collectively) Germans.

Jessie *phr.* **more hide/arse/cheek than Jessie,** brash; bold; impudent. [from *Jessie* the elephant at Taronga Park Zoo, who died 1939]

Jesus *interj.* an exclamation indicating surprise, indignation, etc.: *Jesus! Who was that?* Also, **sweet Jesus, jumping Jesus.**

Jesus-freak *n.* one, often outside the established churches, whose involvement in his religious experiences is the dominating factor in his personality.

Jets, the *pl. n.* **1.** the Newtown team in the N.S.W. Rugby Football League. Also, **the Blues. 2.** the Ipswich team in the Queensland Rugby Football League.

Jew *n.* **1.** (*offensive*) a usurer; miser; one who drives a hard bargain. —*adj.* **2.** (*offensive*) of Jews; Jewish.

jeweller's shop *n. Mining.* a rich find of gold, opal, etc.

jewish lightning *n.* the putative cause of fire when a building has been burned down for insurance money.

jewish stocktake *n.* the practice of setting fire to one's own business in order to claim insurance money.

jib *phr.* **cut of one's jib,** one's general appearance.

jiffy *n.* a very short time: *Hang on, I'll do it in a jiffy.* Also, **jiff.**

jig[1] *phr.* **the jig is up,** the game is up; there is no further chance.

jig[2] *v.* **1.** to play truant. *—phr.* **2. jig it,** to play truant. **3. jig school,** to stay away from school without permission.

jigger *n.* a name for any mechanical device, the correct name of which one does not know.

jiggered *adj.* **1.** a word used as a vague substitute for a taboo word: *I'm jiggered if I know; the machine is jiggered.* *—phr.* **2. jiggered up, a.** exhausted; tired. **b.** broken; destroyed.

jiggery-pokery *n.* dishonest dealing; trickery.

jigjig *n.* sexual intercourse.

jim *n.* a £1 note, or that amount. [abbreviation of obsolete British rhyming slang *Jimmy* O'Goblin, a sovereign]

Jim Crow *n. U.S.* (*offensive*) Negro.

jiminy *interj.* a mild exclamation of surprise. Also, **jiminy cricket.**

jimjams *pl. n.* **1.** extreme nervousness. **2.** delirium tremens, characterised by trembling, terrifying visual hallucinations, etc., and caused by excessive indulgence in alcohol.

Jimmy Brits *pl. n.* the shits. Also, **Jimmie Britt.** [rhyming slang, with reference to a former lightweight boxing champion of the world, *Jimmy Britt,* who was on vaudeville tour in Australia during World War I]

Jimmy Burke *n.* an act of male masturbation. [from rhyming slang *Jimmy Burke* jerk. See **jerk-off**]

Jimmy Grant *n. Obsolete.* immigrant. [rhyming slang]

Jimmy Riddle *n.* a piddle. [rhyming slang]

Jimmy Woodser *n.* **1.** one who drinks alone in a bar. **2.** an alcoholic drink consumed alone. [from the name of a fictitious character, *Jimmy Woods,* who always drank alone]

jingaloes *interj.* an expression of surprise.

jingling Johnny *n.* **1.** a person who shears sheep by hand. **2.** (*pl.*) hand shears.

jingo *phr.* **by jingo, a.** an exclamation of surprise. **b.** an expression of earnest affirmation. [origin uncertain; first used in conjurer's jargon]

jinks *phr.* **high jinks,** romping games or play; boisterous, unrestrained merrymaking.

jinx *n.* a person, thing, or influence supposed to bring bad luck. [from *jynx* a bird used in witchcraft, hence, a spell]

jitney *n. Originally U.S.* **1.** a motor car or small bus which carries passengers, for a low fare, originally five cents each. **2.** *Obsolete.* a five-cent piece. Also, **jit.**

jitter *v.* **1.** to behave nervously. *—phr.* **2. the jitters,** a feeling of nervousness: *Spiders really give me the jitters.*

jittery *adj.* nervous; jumpy.

jive *v. Originally U.S. Negro.* to tease; joke with.

Joan of Arc *n.* a shark. [rhyming slang]

job[1] *n.* **1.** an affair, matter, occurrence, or state of affairs: *He made the best of a bad job.* **2.** a difficult task. **3.** a finished product, as a car.

4. a theft or robbery, or any criminal deed. *—phr.*

5. **a good job,** a lucky state of affairs.

6. **big jobs,** (*in children's speech*) defecation.

7. **do a job on,** to destroy (someone's) reputation, etc.

8. **give up as a bad job,** to abandon as unprofitable an undertaking already begun.

9. **jobs for the boys,** government appointments made as a reward for political support.

10. **just the job,** exactly what is required.

11. **make a good job of,** to complete satisfactorily.

12. **on the job,** busy; occupied.

13. **the devil's own job,** an extremely difficult or frustrating experience.

job² *v.* **1.** jab; hit; punch: *Shut up or I'll job you.* *—phr.* **2. do a job on (someone),** to bash (someone); assault.

jobbies *pl. n.* (*in children's speech*) defecation.

jock *n.* **1.** a jockstrap. **2.** a male athlete, especially one in college or university. [from *jockstrap*] **3.** a jockey.

jockey *n.* **1.** a person accompanying a taxidriver who, if a potential passenger gives a destination which does not suit the driver, pretends that he has already hired the driver to take him to another destination, thus giving the driver an excuse for not accepting the passenger. *—phr.* **2. jockey for position,** to attempt to gain an advantageous position (in a race, contest, etc.).

jockstrap *n.* a support for the genitals, usually of elastic cotton webbing, and worn by male athletes, dancers, etc.

joe¹ *n.* ewe: *a bare-bellied joe.*

joe² *v.* **1.** to jeer at; abuse. *—n.* **2.** a fool. *—phr.* **3. make a joe of oneself,** to behave in a foolish manner. [from the practice in the gold diggings of yelling *Joe* at the troopers. See **Joe¹**]

Joe¹ *n.* (formerly) a trooper; a military policeman. [from C. *Joseph* La Trobe, whose mining regulations the troopers were enforcing on the gold diggings]

Joe² *n. U.S.* **1.** a man; an average fellow: *He's a good Joe.* **2.** an enlisted man in the U.S. army: *a G.I. Joe.* Also, **joe.**

Joe Blake *n.* a snake. Also, **Joe.** [rhyming slang]

Joe Blakes *pl. n.* delirium tremens, characterised by trembling, terrifying visual hallucinations, etc., and caused by excessive indulgence in alcohol. Also, **Joes, joes.** [rhyming slang *Joe Blakes* shakes]

Joe Blow *n.* the man in the street; the average citizen. Also, **Joe Bloggs.** [from **Joe²** + *Blow* probably just for the rhyme]

Joe Gardiners *pl. n. N.S.W. Obsolete.* boots. [from *Joe Gardiner Ltd,* Sydney, former boot manufacturer]

Joe Mark *n.* a shark. [rhyming slang]

Joe Public *n.* See **John Citizen.**

joes¹ *pl. n.* Also, **Joes.** See **Joe Blakes.** [rhyming slang *Joe Blakes,* shakes]

joes² *pl. n.* a mood of depression; blues.

joey *n.* **1.** a baby kangaroo. **2.** See **Joe**[1]. **3.** a young child. *—phr.* **4. to carry a joey,** to be pregnant. **5. to slip a joey,** to have a miscarriage. [Aboriginal]

Joey *n.* See **Joe**[1]. Also, **joey.**

john[1] *n.* a policeman. Also, **John.** [?from *gendarme* and rhyming slang *John* Hop, cop]

john[2] *n.* a toilet.

john[3] *n.* a Chinese. [from popular name *John* Chinaman]

john[4] *n.* a prostitute's client. Also, **John.**

John *n.* See **John Thomas.**

John Barleycorn *n.* whisky; spirits. [from the grain used in the production of whisky]

John Citizen *n.* the man in the street.

John Henry *n. U.S.* a signature. Also, **John Hancock.**

John Hop *n.* a policeman. [?from *gendarme* and rhyming slang *cop*]

johnny *n.* fellow; man.

Johnny Bliss *n.* an act of urination. [rhyming slang *Johnny Bliss* piss]

Johnny-come-lately *n.* a late arrival. Also, **Jill-come-lately, Shirley-come-lately.**

Johnny Hopper *n.* a policeman. [rhyming slang *Johnny Hopper* copper]

Johnny Horner *n.* corner. [rhyming slang]

Johnny-on-the-spot *n.* **1.** a person who is present and thus able to turn to advantage any opportunities which might occur: *He was Johnny-on-the-spot when the lead singer fell ill.* **2.** a person who is present when needed.

Johnny Raw *n.* an inexperienced person; new chum. Also, **Jacky Raw.**

Johnny Woodser *n. N.Z.* See **Jimmy Woodser.**

John Thomas *n.* the penis. Also, **John.**

joint *n.* **1.** the house, unit, office, etc., regarded as in some sense one's own: *Come round to my joint.* **2.** a disreputable bar, restaurant, or nightclub; a dive. **3.** a concealable firearm. **4.** a marijuana cigarette.

joke *phr.* **the joke is on (someone),** said of a person who has become the object of laughter or ridicule, usually after a reversal of fortune.

joker[1] *n.* a fellow or bloke: *a funny sort of joker.*

joker[2] *n.* **1.** a hidden clause in any paper, document, etc., which largely changes its apparent nature. *—phr.* **2. the joker in the pack,** a person whose behaviour is unpredictable. [from *joker,* an extra playing card in a pack, used in some games, often counting as the highest card or to represent a card of any denomination or suit the holder wishes]

jolly *adj.* **1.** amusing; pleasant. *—n.* **2.** a bit of agreeable talk or action to put a person in good humour. *—phr.* **3. jolly along,** *Chiefly British.* to talk or act agreeably to (a person) in order to keep him in good humour, especially with the purpose of gaining something.

jonnop *n.* a policeman. [contraction of rhyming slang *John Hop* cop]

josh *v. U.S.* **1.** to chaff; banter in a teasing way: *Don't worry, he was just joshing.* *—n.* **2.** a chaffing remark; a piece of banter.

joss *n.* boss.

josser *n.* a parson. [from *Joss* a Chinese deity or idol, from Latin *deus,* god]

journo *n.* a journalist.

joy *phr.* **not to have any joy,** to be unsuccessful.

joy-ride *n.* **1.** a pleasure ride, as in a motor car, especially when the car is driven recklessly or used without the owner's permission. **2.** a trip by a politician or official body a public expense. —*v.* **3.** to take such a ride.

J.T. *n.* See **jam tart**[1]. [initials]

judy *n.* a woman, or girl.

jug *n.* **1.** prison or gaol. —*v.* **2.** to commit to gaol, or imprison.

juggins *n.* a simpleton; naive person; fool. [originally a surname]

juice *n.* **1.** electric power. **2.** petrol, oil, etc., used to run an engine. **3.** any alcoholic beverage.

juice-freak *n.* one who seeks stimulation from alcohol rather than drugs.

July fogs *n. Shearing.* the dead season when no shearing is done.

jumbo *n.* **1.** an elephant. **2.** anything bigger than usual. [named after *Jumbo,* an elephant at the London Zoo, subsequently sold to Phineas T. Barnum]

jumbuck *n.* a sheep. [?Aboriginal corruption of *jump up*]

jump *v.* **1.** (of a wound, etc.) to hurt; throb. **2.** to attack suddenly without warning. —*n.* **3.** a head start in time or space; advantageous beginning. **4.** an act of coitus. —*phr.*

5. for or **on the (high) jump(s),** See **high jump.**

6. get the jump on, take by surprise; get an advantage over.

7. go jump (in the lake), an expression of annoyance or dismissal.

8. jump bail, See **bail.**

9. jump down (someone's) throat, to speak suddenly and sharply to (someone).

10. jump on or **upon,** to scold; rebuke; reprimand.

11. jump out of one's skin, to be frightened suddenly.

12. jump the box, to give evidence in the witness box.

13. jump the counter, *Prison.* to commit an armed robbery.

14. jump the gun, to start prematurely; obtain an unfair advantage.

15. jump the queue, to obtain something out of one's proper turn.

16. jump the rattler, to ride illegally on a freight train.

17. jump to it, to move quickly; hurry.

18. jump up and down, to complain strongly.

19. one jump ahead, in a position of advantage.

20. take a (running) jump at yourself, a. an impolite dismissal indicating the speaker's wish to end the conversation. **b.** an impolite instruction to a person to reconsider his attitude or performance.

jumped-up *adj.* upstart; conceited.

jumper *n. Mining.* one who usurps another miner's claim.

jumpers *phr.* **go jumpers,** to abscond from a ship's service.

jumps *pl. n.* hurdle or steeplechase horseraces.

jump-up *n.* an escarpment.

jungle bunny *n.* (*offensive, used by some white people*) Aborigines, Melanesians, Negroes, etc.

jungle juice *n.* **1.** *World War II Military.* a rough alcoholic drink originally made by soldiers in New Guinea. **2.** any drink considered to be as rough.

junk *n.* **1.** anything that is regarded as worthless or mere trash. **2.** any narcotic drug. —*v.* **3.** to cast aside as junk, discard as no longer of use.

junkie *n.* an addict, especially a person addicted to hard drugs. Also, **junky.**

k *n.* kilometres per hour: *We were only doing 100 k, officer.* Also, **k's.**

kack *n., v.* See **cack.**

kadoova *phr.* **off one's kadoova,** deranged, insane.

kafuffle *n.* argument; commotion; rumpus. Also, **kafoofle, kerfuffle, kerfoofle.**

kai *n.* food; a meal. [Melanesian Pidgin *kaikai* food]

Kal *n.* Kalgoorlie, Western Australia.

kale *n. U.S.* money. Also, **kail, kale-seed.** [from *kale* a type of green-leaved cabbage, with reference to the colour green frequently used for banknotes. See **lettuce**]

kamikaze *n.* **1.** (in surfboard-riding) a deliberate wipe-out. *—adj.* **2.** dangerous; suicidal: *his kamikaze driving.* [from the Japanese World War II *kamikaze* suicide squad pilots]

Kanakalander *n. Obsolete.* a Queenslander. [from the *kanakas* Pacific Islanders brought to Queensland as labourers]

kanga[1] *n. Prison.* a warder. Also, **kangaroo.** [rhyming slang *kangaroo* screw]

kanga[2] *n.* wages; money. [rhyming slang *kangaroo* screw]

kanga[3] *n.* a jackhammer. [?brand name]

kangaroo *n.* **1.** *(pl.) British Stock Exchange.* Australian mining and other shares. *—v.* **2.** to release the clutch of a car unevenly so that (a car) moves forward in a series of jerks. **3.** to squat over a toilet seat, while avoiding contact with it. **4.** *Chiefly U.S.* to convict (someone) with false evidence. *—phr.* **5. have kangaroos in the top paddock,** to be insane.

kangaroo court *n. Chiefly British, U.S.* an unauthorised or irregular court conducted with disregard for legal procedure, as a mock court by prisoners in a gaol, or by trade unionists in judging workers who do not follow union decisions.

Kangaroos, the *pl. n.* **1.** the Rugby League Football team representing Australia internationally. **2.** the Moorabbin team in the Victorian Football Association. **3.** Also, **Kangas.** the North Melbourne team in the Victorian Football League. Also, **Roos.**

kapai *interj. N.Z.* **1.** an exclamation of pleasure, approval, etc. *—adj.* **2.** good, agreeable.

kaput *adj.* **1.** smashed; ruined. **2.** broken, not working. [German]

kark *v.* See **cark.**

karma *n.* an intuition or feeling about the quality, mood or atmosphere of a person or place. [See **bad karma**]

Kath *n.* a long period of time, especially a long gaol sentence. Also, **Kathleen Mavourneen.** [from the Irish song *Kathleen Mavourneen* the refrain of which is 'It may be for years, it may be forever']

kayo *n.* **1.** a knockout. *—v.* **2.** to knock (someone) out. [from the initials *K(nock) O(ut)*]

keelhaul *v.* to reprimand severely. [from the former nautical punishment *keelhauling* hauling (a person) under the keel of a vessel]

keep *phr.*
1. for keeps, a. for keeping as one's own permanently: *to play for keeps.* **b.** permanently; altogether.
2. keep at, to badger, hector or bully.
3. keep down, to retain or continue in, as a job.
4. keep in, to detain (a child) after school.
5. keep in with, to keep oneself in favour with.
6. keep nit, to keep watch (usually while an illegal activity is afoot).
7. keep on, to persist.
8. keep track of, or **tabs on,** to keep account of.
9. keep up with the Joneses, to compete with one's neighbours in the accumulation of material possessions, especially as status symbols.

kelly[1] *n.* **1.** a crow. **2.** a prostitute. **3.** an inspector, as a ticket-inspector, etc.

kelly[2] *n.* **1.** an axe. *—phr.* **2. swing kelly,** to work with an axe. [brandname]

Kelly, Ned *phr.* **game as Ned Kelly,** See **game.**

Kembla *n.* change, as a returned balance of money. [rhyming slang *Kembla Grange*]

kennel *n.* a wretched abode.

kerfuffle *n.* See **kafuffle.** Also, **kerfoofle.**

kero *n.* kerosene.

kerplunk *phr.* **go kerplunk, a.** to fall heavily into water. **b.** to be a failure.

kewpie *n.* a prostitute. [rhyming slang *kewpie doll* moll]

key *n.* **1.** a prison term of indefinite length, at the Governor's pleasure. [from the idea of *throwing away the key*] *—phr.* **2. given the key,** *Prison.* declared an habitual criminal.

keyman *n. Prison.* an habitual prisoner.

Khyber *n.* rump; buttocks. [rhyming slang *Khyber Pass* arse]

kibitzer *n. U.S.* **1.** a spectator at a card game who looks at the players' cards over their shoulders. **2.** a giver of unwanted advice.

kibosh *n.* See **kybosh.**

kick *v.* **1.** to resist, object, or complain. *—n.* **2.** an objection or complaint. **3.** any thrill or excitement that gives pleasure; any act, sensation, etc., that gives satisfaction. **4.** a stimulating or intoxicating quality in alcoholic drink. **5.** vigour, energy, or vim. **6.** a shout of drinks. **7.** the start of a race. **8. a.** trouser pocket. **b.** financial resources: *nothing in the kick.* **9.** *British.* sixpence. *—phr.*

10. a kick in the arse, a. a setback. **b.** retribution.

11. a kick in the pants, a sharp reprimand.

12. for kicks, for the sake of gaining some excitement or entertainment.

13. hit the kick, a. to pay up. **b.** to pay the bill.

14. kick about or **around, a.** to maltreat: *The way they kick that dog about is disgusting.* **b.** to discuss or consider at length or in some detail (an idea, proposal, or the like).

15. kick against the pricks, See **prick.**

16. kick in, to contribute, as to a collection for a presentation.

17. kick off, a. to start, commence. **b.** *U.S.* to die.

18. kick on, a. to carry on or continue, especially with just adequate resources: *We'll kick on until the fresh supplies get here.* **b.** to continue a party or other festivity: *We kicked on until the early hours.*

19. kick oneself, to reproach oneself.

20. kick out, to dismiss; get rid of.

21. kick the bucket, to die.

22. kick/flog/whip the cat, to give way to suppressed feelings of frustration by venting one's irritation on someone.

23. kick the habit, a. to give up cigarettes, alcohol, etc., to which one has become addicted. **b.** to forego any pleasure.

24. kick the tin, to give money; contribute. [?from rattling of collection box]

25. kick up, to stir up; to cause (disturbance, trouble, noise, etc.): *to kick up a fuss.*

26. kick up one's heels, See **heels.**

27. kick upstairs, to promote someone to a position which has status but no real power; to remove someone from a particular office by promotion.

kickapoo juice *n.* any illicit home-brewed liquor, as liquor made in gaol. Also, **kickapoo joy juice.**

kickback *n.* **1.** a response, usually vigorous. **2.** any sum paid for favours received or hoped for.

kicking *phr.*

1. be alive and kicking, to be in good health and spirits.

2. be kicking along, to be coping adequately with one's life.

3. be kicking around, a. to lie scattered around. **b.** to be freely available.

kick-off *n.* the beginning or initial stage of something. [from the *kick-off* which starts a football game]

kid[1] *n.* **1.** a child or young person. **2.** *Horseracing.* an apprentice jockey.

kid[2] *v.* **1.** to tease; banter; jest with. **2.** to humbug or fool.

kiddie *n.* a child. Also, **kiddy.**

kiddiewinks *pl. n.* very young children viewed affectionately.

kiddo *n.* a familiar form of address.

kidman's mixture *n.* **1.** treacle. **2.** golden syrup. Also, **kidman's delight.** [from Sir Sidney *Kidman* (1857-1935), a cattleman whose pastoral empire included properties in Victoria, N.S.W. and Queensland. Golden syrup was usually included in rations to station hands and drovers]

kids' beer *n.* low alcohol beer.

kid-stakes *pl. n.* **1.** nonsense; joking pretence: *The pre-season competition is just kid-stakes compared to the premiership.* **2.** a small amount, especially of money; small stakes.

Kilkenny cats *phr.* **fight like Kilkenny cats,** to fight ferociously. [from *Kilkenny,* a town and county in south-east Ireland]

kill *v.* **1.** to cancel (a word, paragraph, item, etc.). **2.** to defeat or veto (a legislative bill, etc.). **3.** to have an irresistible effect: *dressed to kill.*

killer *n.* **1.** something particularly effective: *That joke is a killer.* **2.** the final fact which makes a situation unbearable; the last straw.

killing *n.* a stroke of extraordinary success, as in a successful speculation on the Stock Exchange.

kinchella *phr.* **put kinchella on (shears),** to sharpen (hand shears). Also, **kinchela.** [from *Kinchella,* the author of a pamphlet on caring for and sharpening shears]

kind *phr.* **kind of,** after a fashion; to some extent; somewhat; rather: *The room was kind of dark.*

king *n.* one who displays the greatest expertise in some field: *king of the road; cattle king.*

King Billy *n.* **1.** *Obsolete.* a mythical Aboriginal patriarch. **2.** a male Aborigine.

king hit *n.* **1.** a knock-out blow. **2.** any sudden misfortune.

king-hit *v.* to punch forcibly and without warning. Also, **king.**

kingpin *n.* **1.** the principal person in a company, etc. **2.** the chief element of any system or the like.

king-size *adj.* larger than the usual size: *a king-size hangover.* Also, **king-sized.**

kink *n.* a deviation, especially sexual.

kinky *adj.* **1.** appealing in an individual way. **2.** having unusual tastes; perverted. **3.** eccentric; mad.

kip[1] *n.* **1.** a sleep. —*v.* **2.** to sleep. **3.** to stay (somewhere) on a temporary basis: *He's kipping at Tom's for a couple of days.* —*phr.* **4. kip down,** to go to bed; sleep. [?Old English *cip,* brothel]

kip[2] *n.* **1.** a small, thin piece of wood used for spinning coins in two-up; bat; kylie. —*phr.* **2. pass the kip,** See **flute.** [?variant of *chip*]

kipper *n.* an Englishman. [from *kippers* fish cured by splitting, salting, drying and smoking, a popular English breakfast]

kipsie *n.* a house; the home. [from *kip* a place to sleep]

kiss *phr.*

1. kiss the dust, a. to be killed. **b.** to be humiliated.

2. **kiss my arse,** an expression of contempt or incredulity.

3. **kiss my foot,** an expression of astonishment, disbelief, etc.

kiss-and-ride commuter *n.* a commuter who is driven to and from the point of contact with the public transport system, usually by the partner, spouse, etc.

kisser *n.* the mouth.

kiss of death *n.* any act, fact, influence, relationship, etc., which proves disastrous.

kiss of life *n.* **1.** artificial respiration performed by the mouth-to-mouth or mouth-to-nose method. **2.** anything which revives a project, association, etc.

kit *phr.* **the (whole) kit and caboodle, a.** the whole thing; an item with all its parts. **b.** the whole group.

kitchen *phr.*

1. **get out of the kitchen (if one can't stand the heat** or **fire),** to withdraw from a difficult situation with which one is unable to cope.

2. **the rounds of the kitchen,** a severe scolding.

kitchen sink *phr.* **everything but the kitchen sink,** a large number of miscellaneous items.

kite *n.* **1.** a spinnaker. **2.** an aeroplane. **3.** a hang-glider. **4.** a cheque, especially one forged or stolen. **5.** *Prison.* a newspaper. *—phr.*

6. **fly a kite, a.** to pass off a forged cheque. **b.** to test public opinion by spreading rumours, etc.

7. **go fly a kite,** a peremptory dismissal.

kitten *phr.* **have kittens,** to be anxious or alarmed.

Kiwi *n.* **1.** a New Zealand soldier or representative sportsman, especially a Rugby League representative. **2.** any New Zealander. [from the bird unique to New Zealand]

Kiwiland *n.* New Zealand.

Kiwis, the *pl. n.* the Rugby League Football team representing New Zealand internationally.

knackered *adj.* exhausted; worn out.

knackers *pl. n.* testicles.

knee-high *phr.* **knee-high to a grasshopper, a.** very small in height. **b.** very young.

knee-jerk reaction *n.* automatic aversion or predictable opposition to a person or idea.

knees-up *n.* a party.

kneetrembler *n.* the act of sexual intercourse when both parties are standing.

knickers *phr.* **get one's knickers in a twist** or **knot,** *Originally British.* to become agitated or flustered.

knife *phr.*

1. **get one's knife into,** to bear a grudge against; desire to hurt.

2. **put the knife into,** to destroy a reputation or person maliciously.

3. **knife in the back,** to betray, especially to destroy a person's reputation or career in his or her absence.

4. **the knives are out,** an expression indicating that hostilities have begun.

knob *n.* **1.** the penis. *—phr.* **2. with knobs** or **bells on,** an expression used to add emphasis to a statement: *Don't worry, I'll be there, and with knobs on!*

knock *v.* **1.** to criticise; find fault with: *People are always knocking the Government.* **2.** to copulate with. **3.** *Prison.* to kill. —*n.* **4.** adverse criticism. **5.** an act of intercourse. **6.** a promiscous woman. —*phr.*
7. knock about or **around, a.** to wander in an aimless way; lead an irregular existence. **b.** to treat roughly; maltreat.
8. knock around with, to keep company with.
9. knock back, a. to consume, especially rapidly: *He knocked back two cans of beer.* **b.** to refuse. **c.** to set back; impede.
10. knock down, a. to reduce the price of. **b.** to spend freely, especially on liquor: *He knocked down his cheque at the local pub.* **c.** *N.Z.* to swallow (a drink).
11. knock into a cocked hat, to defeat; get the better of.
12. knock it off, stop it (usually used to put an end to an argument, fight, criticism, etc.)
12. knock off, a. to stop an activity, especially work. **b.** to deduct. **c.** to steal. **d.** to compose (an article, poem, or the like) hurriedly. **e.** to defeat, put out of a competition. **f.** to kill. **g.** (of a man) to have sexual intercourse with. **h.** to eat up; consume. **i.** (of police) to arrest (a person) or raid (a place).
14. knock oneself out, to exhaust oneself by excessive mental or physical work.
15. knock one's eye out or **socks off,** to cause one to feel excessive admiration.
16. knock on the head, to put an end to.
17. knock out, a. to overwhelm with success or attractiveness. **b.** to destroy; damage severely. **c.** to earn: *He knocked out a living as a station hand.* **d.** to make; produce: *Pete knocked out a few pots.*
18. knock rotten or **silly,** to strike heavily.
19. knock spots off, to defeat; get the better of.
20. knock the bottom out of, to refute (an argument); render invalid.
21. knock together, to assemble (something) hastily; put together roughly.
22. knock up, a. to arouse; awaken. **b.** to construct (something) hastily or roughly. **c.** *Sport.* to score (runs, tries, etc.). **d.** *Tennis.* to practise. **e.** to exhaust; wear out. **f.** to become exhausted. **g.** to make pregnant.
23. take a knock, to suffer a reverse, especially a financial one.
24. take the knock, *Horseracing, etc.* (of a punter) to admit that one is unable to settle one's debts with one's bookmaker.
25. take the knock on (someone), to cheat (someone) of his share in part or whole; to welsh on.

knockabout *n.* a station hand; odd-job man.

knock-back *n.* a refusal; rejection, especially of sexual overtures.

knockdown *n.* a formal introduction.

knocker *n.* **1.** a persistently hostile critic or carping detractor. —*phr.*
2. on the knocker, a. at the right time, punctual: *He was there on the knocker.* **b.** immediately; on demand: *cash on the knocker.*

knockers *pl. n.* breasts.

knocking shop *n.* a brothel.

knock-off time *n.* the time at which work officially finishes.

knockout *n.* a person or thing of overwhelming success or attractiveness.

knockout drop *n.* a sedative pill, often added to a drink. Also, **knock-drop.**

knockover *n.* an easy success; pushover.

knocktaker *n.* something which is a certainty to win, as a racehorse.

knot *n.* **1.** a swag. *—phr.* **2. be on the knot** or **push the knot,** to carry a swag; hump the bluey; waltz matilda.
3. put or **tie a knot in one's bluey, a.** to make ready for travel, action, etc. **b.** to leave one's job.

knotted *phr.* **get knotted,** go away, leave me alone.

know *phr.*
1. in the know, having inside knowledge.
2. know chalk from cheese, to be able to note differences; distinguish the specific qualities of various articles, people, etc.
3. know the ropes, to understand the details or methods of any business or the like.
4. know which side one's bread is buttered, to know where the advantage lies.

know-all *n.* **1.** one who claims to know everything, or everything about a particular subject. **2.** one who appears to know everything.

knuckle *v.* **1.** to assault, with fists or knuckle-dusters. *—phr.*
2. fond of the knuckle, keen on fighting.
3. go the knuckle, to fight, especially with the fists.
4. knuckle down, to apply oneself vigorously or earnestly, as to a task.
5. knuckle down or **under,** to yield or submit. **6. near** or **close to the knuckle,** (of a remark, joke, etc.) near the limit of what is permitted or acceptable.

knucklehead *n.* a fool.

knuckle sandwich *n.* a punch in the mouth: *Are you looking for a knuckle sandwich, mate?*

K.O.J. *adj.* (of children) enthusiastic; over-earnest: *He's too K.O.J. for me.* [*K(een) O(n the) J(ob)*]

konk *n.* See **conk.**

kook *n.* **1.** a strange or eccentric person. **2.** *Surfing.* a beginner, especially one who imitates others badly.

kooka *n.* kookaburra.

kooky *adj.* eccentric; odd.

koori *n.* See **kuri**[2]

kosher *adj.* **1.** good. **2.** proper.

K.P. *n.* a prostitute. [?from police records *K(nown) P(rostitute)*]

kraut *n.* (*usually offensive*) a German.

kuri[1] *n.* **1.** *N.Z.* a Maori-bred dog; a mongrel. **2.** (*offensive*) a Maori. Also, **goori, goorie.** [Maori]

kuri[2] *n.* **1.** an Aborigine. *—adj.* **2.** Aboriginal. Also, **koori, koorie.**

kway *n.* a thief who does not specialise in any one line of activity; an all-rounder.

kybosh *n.* **1.** nonsense. *—v.* **2.** to put an end to; stop. Also, **kibosh.** *—phr.* **3. put the kybosh on,** to put a stop to.

kylie *n.* See **kip**[2]. [Aboriginal: a boomerang having one side flat and the other convex]

la *n.* a toilet. Also, **lala.**

label *n.* the trade name, especially of a gramophone record company: *an old group on a new label.*

Laborite *n.* a member or supporter of the Australian Labor Party.

lace *v.* **1.** *British.* to lash, beat, or thrash. —*phr.* **2. lace into,** to attack (someone) verbally or physically.

lacker band *n. Vic.* a rubber band.

lad *n.* **1.** (in familiar use) any male. **2.** a devil-may-care, dashing man: *Watch out for him, they reckon he's a bit of a lad!*

ladder *n. Football.* the ranking of teams according to the number of games won and points scored for and against; premiership table.

la-di-da *adj.* **1.** affectedly pretentious, especially in manner, speech, or bearing. —*n.* **2.** the toilet. Also, **lah-di-day.**

ladies *n.* a toilet for women. Also, **ladies'.**

lady *phr.* **lady of the night,** a female prostitute.

Lady Blamey *n.* a beer glass made by cutting the top off an empty beer bottle. [from *Lady Blamey* wife of General Blamey, 1884-1951]

lady-killer *n.* a man supposed to be dangerously fascinating to women.

Lady Muck *n.* a woman who affects the manner of a grand lady, usually in an overbearing and unconvincing manner.

lady's waist *n.* **1.** a small thin beer glass. **2.** the contents of such a glass. [from the shape of the glass]

lag *v.* **1.** to send to prison. **2.** to arrest. **3.** to report the misdemeanours of (someone); inform on. —*n.* **4.** a convict, especially an habitual criminal: *an old lag.* **5.** a term of penal servitude.

lagger *n.* a police informer.

lagging *n.* a term of imprisonment.

lair *n.* **1.** a flashily dressed young man of brash and vulgar behaviour. *—phr.*
2. lair it up, to behave in a brash and vulgar manner.
3. lair up, a. to dress up in flashy clothes. **b.** to renovate or dress up something in bad taste. Also, **lare, mug lair.**

lairise *v.* to behave like a lair; show off; indulge in brash, vulgar exhibitionism.

lairy *adj.* **1.** exhibitionistic; flashy. **2.** vulgar. Also, **leary, leery.**

lam[1] *v.* **1.** to beat; strike. *—phr.* **2. lam into,** to thrash; rain down a succession of blows on.

lam[2] *U.S.* *n.* **1.** precipitate escape. *—v.* **2.** to run quickly; run off or away. *—phr.*
3. on the lam, escaping or fleeing.
4. take it on the lam, to flee or escape in great haste.

lamb *phr.* **lamb down, a.** to induce (someone) to spend in a reckless fashion. **b.** to spend (money) in a reckless or lavish fashion. **c.** to swindle; cheat; fleece. [from to *lamb,* meaning to fleece or swindle]

lamber-down *n.* the proprietor of a shanty, a small, rough bush hotel, often unlicensed.

lamebrain *n.* *U.S.* a foolish, unintelligent person.

lame dog *n.* See **dog.**

lame duck *n.* See **duck.**

lamp *n.* **1.** (*pl.*) the eyes. *—v.* **2.** to observe; look at.

lance jack *n.* a lance corporal.

land[1] *v.* **1.** to secure; make certain of; gain or obtain: *to land a job.* *—phr.* **2. land on one's feet, a.** to have good luck. **b.** to emerge successfully from an adverse situation.

land[2] *phr.*
1. land of the long white cloud, New Zealand.
2. land of the wattle, Australia.

language *n.* **1.** strong, vulgar language: *His language made her blush.* *—phr.*
2. speak the same language, to be in sympathy with; have the same mode of thinking; share the same jargon.
3. speak a different language, to be out of sympathy or accord especially as a result of different background, education, etc.

La Perouse *n.* alcoholic beverage. Also, **larper.** [rhyming slang *La Perouse* (a promontory at Botany Bay, N.S.W.) booze]

lark *n.* **1.** a merry or hilarious adventure; prank. **2.** a frolic. *—v.* **3.** to play pranks; have fun. Also, **skylark.**

larrikin *n.* **1.** (formerly) a lout, a hoodlum: *larrikins of the push.* **2.** a mischievous young person. **3.** an independent or wild-spirited person, usually having little regard for authority, accepted values, etc. [?British (Warwickshire and Worcestershire) dialect *larrikan* mischievous youth]

larrup *v.* to beat; thrash.

Larry *phr.* **happy as Larry,** See **happy.**

Larry Dooley *phr.* **give (someone) Larry Dooley,** to give (someone) a beating; assault, especially with the fists. [?from Larry Foley, a well-known boxer, trainer and promoter of boxing in Australia in the late 19th century]

lash *n.* **1. a.** sexual intercourse; coitus. **b.** anything which thrills or pleases. *—v.* **2.** *Prison.* to welsh on a debt. *—phr.*

3. have a lash at, to attempt.

4. lash out, a. to spend money freely. **b.** to strike vigorously, as with a weapon, whip, or the like. **c.** to burst into violent action or speech.

lashing *phr.* **lashings of,** large quantities; plenty: *lashings of cream with fruit salad.*

last *n.* **1.** the final mention or appearance: *to see the last of that woman.* *—phr.* **2. on one's last legs,** on the verge of collapse.

last straw *n.* the culminating irritation, mishap, etc., which is followed by a strong outburst or reaction. [from the saying 'the (last) straw that broke the camel's back']

latch *phr.* **latch on to, a.** to fasten or attach (oneself) to. **b.** to understand; comprehend. [from Middle English *lacche,* to take hold of, catch, take]

later *phr.* **see you later,** an expression of farewell, not usually referring to a definite later meeting.

latest *phr.* **the latest,** the most recent disclosure, gossip, fashion, advance, development, etc: *Have you heard the latest?*

lather *v.* to beat or flog.

laugh *phr.*

1. don't make me laugh! an exclamation indicating disbelief.

2. have the (last) laugh, to prove ultimately successful; win after an earlier defeat.

3. laugh fit to kill, to laugh extremely heartily.

4. laugh like a drain, to laugh loudly.

5. laugh in or **up one's sleeve,** to laugh inwardly at something.

6. laugh off or **away,** to dismiss (a situation, criticism, or the like) by treating lightly or with ridicule.

7. laugh on the other or **wrong side of one's face** or **mouth,** to evince disappointment, chagrin, displeasure, etc.

laughing *phr.*

1. be laughing, to be in a fortunate position.

2. no laughing matter, a serious matter.

laughing jackass *n.* a kookaburra. Also, **laughing kookaburra, laughing jack, laughing john.**

lav *n.* a lavatory. Also, **lavvy.**

law *phr.* **lay down the law,** to tell people authoritatively what to do, or state one's opinions authoritatively.

lay *v.* **1.** to have sexual intercourse with. *—n.* **2. a.** a person considered as a sex object: *a good lay, an easy lay.* **b.** the sexual act. *—phr.*

3. lay a ghost (to rest), See **ghost.**

4. lay an egg, a. to drop a bomb. **b.** to defecate. **c.** to blunder; flop; fail to elicit the desired result.

5. lay in (there), to maintain a course of action despite opposition, setbacks, etc.

6. lay into, a. to beat. **b.** to criticise severely.

7. lay it on, a. to exaggerate. **b.** to chastise severely.

8. **lay off, a.** to desist. **b.** to cease to annoy (someone). **c.** to protect a bet or speculation by taking some off-setting risk.
9. **lay out, a.** to exert (oneself) for some purpose, effect, etc. **b.** to strike down, especially to knock unconscious.
10. **lay up,** to cause to remain in bed or indoors through illness.

layabout *n.* one who does not work; a loafer; an idler.

lazybones *n.* a lazy person.

lazy ration *n.* (formerly) a subsistence ration supplied to a recalcitrant convict.

lead[1] *phr.*
1. **lead (someone) a chase** or **dance,** to cause (someone) unnecessary difficulty or trouble.
2. **lead (someone) by the nose,** to enforce one's will on (someone), especially unpleasantly.
3. **lead (someone) on,** to induce or encourage (someone) to a detrimental or undesirable course of action.

lead[2] *phr.*
1. **go down like a lead balloon,** to fail dismally; fail to elicit the desired response.
2. **put lead in one's pencil,** (of a male) to increase sexual capacity.
3. **swing the lead,** to be idle when there is work to be done.

leading light *n.* a person outstanding in a particular sphere.

leak *n.* **1.** an act of passing water; urination. —*v.* **2.** to pass water; urinate. **3.** to disclose (information, especially of a confidential nature) especially to the media.

leaky *adj.* allowing confidential information to be made public: *That department is very leaky.*

lean *phr.*
1. **lean on,** to intimidate; apply pressure on.
2. **lean over backwards,** to go to a great deal of trouble.

leap *phr.* **leap at,** to accept eagerly: *He leapt at the offer to go.*

leary *adj.* See **lairy.**

leather *n* **1.** a wallet. —*v.* **2.** to beat with a leather strap.

leatherie *n.* (formerly) a leather-jacketed motorbike rider, during the bodgie and widgie era.

leatherjacket *n.* a mixture of flour and water, fried and cooked on red-hot embers.

leatherneck *n.* **1.** a handyman on a station; rouseabout. **2.** *U.S.* a person serving on shipboard as a soldier.

leave *phr.*
1. **leave cold,** to make little or no impression on.
2. **leave (someone) for dead,** to outclass or outstrip (someone) in a competition.
3. **leave it at that,** to go no further; do or say nothing more.
4. **leave off,** to cease doing (something): *Leave off crying now.*

leery[1] *adj.* **1.** doubtful; suspicious. **2.** knowing; sly. —*phr.* **3. leery of the brush,** (of a man) nervous about getting married.

leery[2] *adj.* See **lairy.**

left *phr.* **have two left feet,** to be clumsy.

left foot *phr.* **kick with the left foot,** to be a Roman Catholic.

left-footer *n.* a Roman Catholic. [British]

leftie *n.* having socialist or radical political ideas; leftist. Also, **lefty.**

leg *phr.*
1. get a leg in, to make a start.
2. leg it, to walk or run.
3. not have a leg to stand on, not to have any good reason at all.
4. pull (someone's) leg, to make fun of; to tease.
5. run out of legs, to be physically exhausted, applied especially to football players, etc., at the end of a gruelling game.
6. shake a leg, to hurry up.
7. show a leg, to make an appearance.

legal eagle *n.* a lawyer.

leggie *n.* **1.** (in cricket) a ball bowled so as to change direction from leg to off when it pitches; leg break. **2.** a bowler who specialises in leg breaks; leg spinner.

legit *adj.* truthful. [shortened form of *legitimate*]

legitimate *n.* (formerly) a convict.

leg opener *n.* an acoholic drink calculated to facilitate the seduction of a woman.

leg-rope *v.* to delay; halt: *measures designed to leg-rope inflation.*

leisure *phr.* **lady of leisure,** (often ironic) a woman who has no regular employment, especially a married woman.

lemon *n.* **1.** something distasteful or unpleasant. **2.** something unsatisfactory, inferior or worthless: *The car was a lemon, nothing but trouble from the day it was bought.* **3.** a foolish, sour, or ugly person. **4.** (*pl.*) **a.** *Australian Rules.* three-quarter time. **b.** *Rugby, Soccer.* half-time.

lemon avenue *n.* a female wowser; a woman who does not drink alcohol.

lemonhead *n.* a surfie with bleached hair.

lemon squeezer *n. N.Z.* (formerly) a brimmed and peaked infantry dress hat of similar shape to a lemon sqeezer.

lemon time *n. Australian Rules.* the three-quarter time break.

lemony *adj.* angry; irritable.

length *phr.* **slip (someone) a length,** (of a man) to have sexual intercourse with (someone).

Leopards, the *pl. n.* the Marconi team in the National Soccer League.

leper *phr.* **social leper,** a person ostracised by society. [from the isolation in the past of *lepers,* i.e. those suffering from leprosy]

leprechaun *n.* an Irishman. [in Irish folklore, a little spirit or goblin]

leso *n., adj.* lesbian. Also, **les.**

let *phr.*
1. let alone, not even; certainly not: *I didn't get a cup of tea, let alone a meal.*
2. let (someone) down, to disappoint (someone); fail.
3. let fly, to express one's anger without restraint.
4. let go, a. to break wind; fart. **b.** to express one's anger without restraint.
5. let in for, to oblige to do something without prior consent or knowledge: *What have I let myself in for?*
6. let in on, to share secret information with.
7. let it be, to allow a situation to take its own course.
8. let off, a. to excuse; to exempt from (something arduous, as a punishment, or the like). **b.** to break wind.

9. let off steam, See **steam.**

10. let on, a. to divulge information, especially indiscreetly. **b.** to pretend: *He let on that he was a detective.*

11. let oneself go, to neglect oneself.

12. let one's hair down, See **hair.**

13. let out, to free from imputation of guilt: *That lets him out.*

14. let up, to slacken or stop.

let-out *n.* an escape from an obligation.

lettuce *n.* money in the form of banknotes; paper money. [?from the colour green frequently used for banknotes]

let-up *n.* a slackening; cessation; pause.

level *phr.*

1. one's level best, one's very best; one's utmost.

2. on the level, sincere; honest.

level-pegging *adj.* on equal terms. [from cribbage, in which the score is kept with rows of pegs]

lezzy *n.* a lesbian. Also, **lezz, lez.**

liar *phr.* **make a liar of,** to prove wrong: *He won the race, making liars of all the knockers.*

Lib *n.* a member or supporter of the Australian Liberal Party.

libber *n.* a liberationist, especially one involved in women's liberation.

liberal *phr.* **small-l liberal,** a person with conservative tendencies, but who prides himself on an independent, progressive point of view.

liberate *v.* to shoplift; steal.

lice ladders *pl. n.* short whiskers extending from the hairline to below the ears and worn with an unbearded chin; sidelevers.

lick *v.* **1.** to overcome in a fight, etc.; defeat. **2.** to outdo; surpass. *—phr.*

3. for the lick of one's life, at great speed.

4. lick and a promise, a. a quick perfunctory wash with a promise to be more thorough next time. **b.** a feeble, perfunctory or superficial attempt at doing something. **c.** a hasty tidy-up.

5. lick into shape, to bring to a state of completion or perfection; make efficient.

6. lick (someone's) boots, to act in a subservient manner; fawn upon.

7. lick or **bite the dust, a.** to be killed or wounded. **b.** to grovel; humble oneself abjectly.

licking *n.* **a.** a beating or thrashing. **b.** a defeat; setback.

lid *n.* **1.** a hat. *—phr.*

2. dip one's lid , a. to lift one's hat as a mark of respect. **b.** to respect someone.

3. lift the lid on, to reveal or expose (secret dealings, etc.).

4. put the lid on, a. to clamp down on or put an end to: *a law to put the lid on prostitution.* **b.** to remove as a possibility: *That puts the lid on our holiday.*

lie[1] *phr.* **lie like a pig in mud** or **shit,** to tell untruths convincingly, unabashedly or with practiced skill.

lie[2] *phr.*

1. let sleeping dogs lie, to avoid any disturbance, or a controversial topic or action.

2. lie down under, to accept (abuse, etc.) without protest.

3. **take it lying down,** to submit without resistance or protest.

life *n.* **1.** *Cricket.* a reprieve, as when a batsman survives a dropped catch, etc. *—phr.*
2. as large as life, (and twice as ugly), actually; in person.
3. for dear life, urgently; desperately.
4. for the life of one, with the greatest effort: *For the life of me I can't understand him.*
5. go for your life, an expression of encouragement or consent.
6. life of Riley, a life of ease and luxury: *After I won the pools it was just the life of Riley.*
7. not on your life, absolutely not.
8. such is life! an exclamation indicating resignation or tolerance. [said to be the last words of Ned Kelly]
9. take one's life in one's hands, to risk death.
10. that's life, an exclamation indicating resignation or tolerance.

lifer *n.* one sentenced to imprisonment for life.

lifesaver *n.* **1.** any person who restores another to good spirits with comfort, help, etc. **2.** anything restorative or beneficially rectifying: *That bank loan was a real lifesaver.*

lift *v.* to steal or plagiarise.

light[1] *phr.*
1. come to light, See **come.**
2. out like a light, unconscious, especially after being struck, or receiving an anaesthetic.
3. see the light, a. to be converted, especially to Christianity. **b.** to suddenly understand something.

light[2] *adj.* **1.** relating to a state of temporary financial embarrassment: *I'll pay you later, I'm a bit light today. —phr.* **2. light on,** in short supply; scarce.

light[3] *phr.* **light out,** to depart hastily.

light fingers *n.* a petty thief; pickpocket.

lightweight *adj.* **1.** unimportant, not serious; trivial. *—n.* **2.** a person of little mental force or of slight influence or importance.

like *adv.* **1.** as it were: *It's a bit tough, like. —conj.* **2.** just as, or as: *He did it like he wanted.* **3.** as if: *He acted like he was afraid.*

likely *phr.* **a likely story,** (with emphasis) indicating a lack of truth.

lily *n.* **1.** (*offensive*) a man who does not conform to some conventional notion of masculinity, as an artist, a homosexual, etc. *—phr.* **2. lily on a dustbin, a.** a person or thing rejected or neglected. **b.** something incongruous, as to be inappropriately or overdressed at an informal gathering.

lilywhite *adj.* beyond reproach; innocent.

lime-juicer *n.* See **limey.**

limelight *n.* **1.** the glare of public interest or notoriety. *—phr.* **2. steal the limelight,** to make oneself the centre of attention.

limey *n.* **1.** a British sailor or ship. **2.** (formerly) an Englishman, especially a recent immigrant. *—adj.* **3.** British. Also, **lime-juicer.** [from the prescribed use of lime juice against scurvey in British ships in the 18th century]

limit *phr.* **the (dizzy) limit,** someone or something that exasperates to an intolerable degree.

limmo *n.* a limousine.

line[1] *n.* **1.** a bush road. **2.** (*pl.*) *British.* a certificate of marriage. *—phr.*
3. do a line for, to flirt with.
4. do a line with, to enter into an amorous relationship with.
5. get a line on, to obtain information about.
6. get one's lines crossed, to misunderstand.
7. hard lines, bad luck.
8. lay it on the line, to state the case openly and honestly.
9. line up, to get hold of; make available: *We must line up a chairman for the conference.*
10. on the line, in jeopardy.
11. pay on the line, to pay promptly.
12. shoot a line, to boast.
13. the line, the equator.
14. toe the line, to conform; obey.

line[2] *phr.* **line one's pockets,** to furnish with money; improve one's financial situation.

linen *phr.* **wash one's dirty linen in public,** to discuss disagreeable personal affairs in public.

lingo *n.* **1.** language, especially peculiar or unintelligible language. **2.** language or terminology peculiar to a particular field, group, etc.; jargon.

lion *phr.* **the lion's share,** the largest portion of anything.

Lions, the *pl. n.* **1.** the Brisbane team in the National Soccer League. **2.** the Coburg team in the Victorian Football Association. **3.** the Fitzroy team in the Victorian Football League. **4.** the Subiaco team in the West Australian Football League.

lip *n.* **1.** impudent talk. *—phr.*
2. bite one's lip, a. to show vexation. **b.** to stifle one's feelings, expecially anger or irritability.
3. button the lip, to be silent.
4. give (someone) a fat lip, to punch (someone) in the mouth.
5. give lip, to talk, especially to a superior, in a cheeky or insolent manner.
6. keep a stiff upper lip, to face misfortune bravely, especially without outward show of perturbation.

lippie *n.* lipstick. Also, **lippy.**

liquid amber *n.* beer. Also, **amber fluid, amber liquid.**

liquid laugh *n.* vomit.

liquid lunch *n.* alcoholic drink, usually beer, consumed instead of food at the normal lunchtime.

liquor *phr.* **to liquor up,** to furnish with liquor or drink.

lit *phr.* **lit up,** drunk; intoxicated.

little *adj.* **1.** denoting approval: *little beauty, little ripper.* **2.** denoting disapproval: *little fool, little turd.* *—phr.* **3. little green man,** an imaginary person from outer space.

little Aussie battler *n.* See **battler.**

little house *n.* a toilet.

little woman *phr.* **the little woman,** one's wife.

live[1] *phr.*
1. live dangerously, to take risks; live with little regard to one's personal safety.
2. live high, to live at a high standard; live luxuriously.
3. live it up, to live wildly and exuberantly; go on a spree.

4. **live together,** to dwell together as lovers; cohabit.

5. **live with, a.** to dwell together with, as a husband or wife or lover; to cohabit with. **b.** to be obliged to put up with the consequences of (an act, decision, policy, etc.).: *I suppose we'll just have to live with it.*

live[2] *adj.* of present interest, as a question or issue: *The environment is a live issue at present.*

lived-in *adj.* untidy or homely.

lizard *n.* **1.** an idler or lounger in places of social enjoyment, public resort, etc., especially one who associates with women. See **lounge lizard. 2. a.** a sheep musterer. **b.** a property employee who maintains boundary fences [?from rests taken lying in the sun, like a lizard] **3.** a shearing handpiece. *—phr.*

4. **flat out like a lizard drinking,** extremely busy.

5. **starve** or **stiffen the lizards,** an exclamation of great surprise, protest or disbelief: *Well starve the lizards, he's won the lottery!*

load *n.* **1.** (*pl.*) a great quantity or number: *loads of people.* **2.** an infection of venereal disease, usually gonorrhoea. **3.** a sufficient quantity of liquor drunk to intoxicate. *—phr.*

4. **cop a load,** See **cop.**

5. **get a load of, a.** to look at; observe. **b.** to listen; to hear.

6. **load the dice,** to place in an especially favourable or unfavouable position.

loaded *adj.* **1.** Also, **loaded up.** unjustly incriminated; framed. **2.** very wealthy. **3.** incapacitated through excess alcohol or drugs.

loaf *n.* **1.** head; intelligence; brains. *—phr.* **2. use one's loaf,** to think; apply one's intelligence. [probably rhyming slang *loaf of bread* head]

loan *phr.* **have a loan** or **lend of (someone),** to practice a deception on or tease (someone).

loan shark *n.* one who lends money at an excessive rate of interest.

lob *v.* **1.** to move or go: *Where are you lobbing?* **2.** to arrive. *—phr.*

3. **lob in,** to arrive unannounced or unexpectedly.

4. **lob off,** to depart; go away.

5. **lob onto,** to find by chance.

lobby *n. Qld.* a fresh water crayfish; a yabby. [shortened form of *lobster* + *-y*]

lobster *n. History.* a trooper; redcoat.

local *phr.* **the local,** the closest or preferred hotel in the neighbourhood.

lock *phr.*

1. **lock into,** to become a permanent part of, to become fixed in: *She wants a career and not to be locked into marriage and motherhood.*

2. **lock on with,** to fight.

3. **locked away,** institutionalised, as in a gaol or mental asylum.

loco *n.* **1.** a mad or crazy person. *—adj.* **2.** insane; crazy.

lofty *n.* a term of address for a tall person.

log *n.* **1.** *Nautical.* a slow boat. **2.** a marijuana cigarette. **3.** Also, **log of wood. a.** a fool. **b.** a lazy person. **4.** (*pl.*) *Convict Obsolete.* a gaol; prison. [from the *logs* used in construction of early gaols] *—v.* **5.** to be put in gaol.

loghead *n.* a thick-headed or stupid person; a blockhead.

logie *n.* **1.** something extreme: *a logie of a cold.* **2.** a fictitious prize: *He got the logie for bad temper.* [from the *Logie* Awards for the Australian television industry]

log jam *n. Chiefly U.S.* a delay; blockage.

l.o.l. *n.* little old lady. [initials]

lollipop person *n.* a person employed to supervise a pedestrian crossing outside a school.

lolly *n.* **1. a.** the head. **b.** the temper. **2.** money. **3.** anything easy to understand. **4.** a timid person. **5.** the victim of a confidence trickster. *—phr.* **6. do one's lolly,** to lose one's temper.

lolly legs *n.* **1.** a tall, ungainly person with skinny legs. *—interj.* **2.** *Football.* a cry of derision shouted by supporters of the opposing team in an attempt to disconcert a player about to kick for a goal.

lolly water *n.* a sweet soft drink.

London *phr.* **bet London to a brick,** See **brick.**

long *phr.*

1. a long chance, not likely.

2. long in the tooth, elderly.

3. put the long ones in, *Prison.* to run; escape.

4. the long and the short of, the kernel; substance of; gist.

long-hair *n.* **1.** a long-haired man. **2.** an intellectual. **3.** a hippie.

long-haired *adj.* (*sometimes offensive*) highbrow; intellectual.

long john *n. N.Z.* an oblong loaf of bread.

long johns *pl. n.* a pair of long warm underpants.

long jump *n.* a hanging. Also, **high jump.**

longshanks *n.* a tall man.

long shot *n.* **1.** an attempt which has little hope of success, but which if successful may offer great rewards: *It's a long shot, but it's our only chance.* **2.** an outsider in a horserace.

long-sleever *n.* **1.** a tall glass. **2.** a long drink. **3.** a pint glass of beer. Also, **sleever.**

longtail *n.* treacle.

long'uns *pl. n.* long trousers.

long white cloud *n.* See **land.**

loo *n.* a toilet. [mincing variant of *lavatory*]

look *phr.*

1. for the look of the thing, for the sake of appearances.

2. looking at, liable for: *You'd be looking at $100,000 for a house in that area.*

3. look down one's nose at, to regard with barely concealed contempt.

4. look on the bright (the worst) side, to consider something with optimism (with pessimism).

5. not to know which way to look, to feel embarrassed.

6. look daggers at, See **daggers.**

looker *n.* Also, **good looker.** an unusually good-looking person.

look-in *n.* a chance of participating: *She was such a good player that he didn't get a look-in.*

lookout *n.* the proper object of one's watchful care or concern: *That's his lookout.*

looksee *n.* a visual search or examination.

loony *adj.* **1.** lunatic; crazy. **2.** extremely or senselessly foolish. —*n.* **3.** an insane person. Also, **looney, luny.** [variant of *luny* familiar shortening of *lunatic*]

loony bin *n.* a lunatic asylum.

loop-the-loop *n.* soup. [rhyming slang]

loopy *adj.* slightly mad or eccentric.

loose *phr.*

1. at a loose end, a. in an unsettled or disorderly condition. **b.** unoccupied; having nothing to do.

2. cut loose, See **cut.**

3. on the loose, a. free from restraint. **b.** on a spree. **c.** (of women) living by prostitution.

loot *n.* money.

Loo, the *n.* Woolloomooloo. [an inner-city dockside suburb of Sydney]

loppy *n.* a handyman on a station; rouseabout; leatherneck.

lord *phr.* **lord it over (someone),** to play the lord; behave in a lordly manner; domineer.

Lord Muck *n.* a man who affects the manner of a grand gentleman, usually in an overbearing and unconvincing manner.

lordship *phr.* **his lordship, a.** an arrogant, overbearing man. **b.** a humorous term for one's husband.

lose *phr.*

1. lose one's head, See **head.**

2. lose out on, to fail to achieve (a goal, etc.): *I lost out on that deal.*

3. lose out to, to be defeated or bettered: *I lost out to my rival.*

4. lose sleep over, to worry about excessively.

loser *phr.*

1. a born loser, an unsuccessful person; someone who seems destined to misfortune and failure.

2. a good (bad) loser, someone who loses with good (bad) grace.

loss *phr.*

1. a dead loss, a. a complete failure. **b.** an utterly useless person or thing.

2. cut one's losses, See **cut.**

lost *phr.* **get lost,** (*often offensive*) to go away; depart.

lost weekend *n.* **1.** a weekend spent secretly with one's lover. **2.** a weekend spent in drunken oblivion. **3.** a weekend spent in prison as part of a sentence.

lot *n.* **1.** a person of a specified sort: *a bad lot.* —*phr.*

2. have a lot going for it or **one,** to have many favourable qualities.

3. the lot, *Prison.* a life sentence.

lotus land *phr.* **live in lotus land,** to live without giving thought to the future.

lounge lizard *n.* **1.** (formerly) a ladies' man who frequents the lounge bar of hotels, where women congregate. **2.** *U.S.* a man who visits girls or women for amorous play, but does not entertain them away from their own homes.

louse *n.* **1.** a despicable person. —*phr.* **2. louse up,** to spoil.

lousy *adj.* **1.** mean, contemptible or unpleasant. **2.** inferior, no good. **3.** well supplied: *He's lousy with money.* **4.** trifling; mere: *He was fined a lousy $100.* **5.** unwell. **6.** relating to a bad mood or fit of depression.

lout *n.* **1.** a rough, uncouth and sometimes violent young man. **2.** an awkward, stupid person; boor.

[perhaps from obsolete *lout* stoop, bow low]

lovebird *n.* a man or a woman acting lovingly towards a member of the opposite sex.

love-juice *n.* **1.** an aphrodisiac. **2.** a sexual secretion.

love-nest *n.* a retreat for lovers, especially illicit.

lovey-dovey *adj.* affectionate, often noticeably in public.

lowbrow *n.* a person of low intellectual calibre or culture.

low-down *adj.* low; dishonourable; mean: *It was a really low-down thing to do.*

lowdown *n.* the actual unadorned facts or truth on some subject.

lowheel *n.* a prostitute.

low-life *n.* the vulgar or more seamy elements of society.

l.s.d. *n.* (formerly) money. [abbreviation for *pounds, shillings and pence* from Latin *librae, solidi, denarii* Roman coinage]

L.S.D. *n.* lysergic acid diethylamide, a hallucinogen, characteristically self-administered, especially in the late 1960s, to produce temporary hallucinations in the user; acid.

lube *n.* a lubrication, especially of a motor vehicle.

lubra *n.* (*sometimes offensive*) an Aboriginal woman. [Aboriginal]

luck *phr.*
1. down on one's luck, in poor or unfortunate circumstances.
2. half your luck, an expression indicating envy at someone else's good luck.
3. here's luck! an expression of goodwill, especially as a toast.
4. just one's luck, typical of one's luck, regarded as invariably bad.
5. no such luck, (*usually ironic*) unfortunately not.
6. one's luck is in, one is experiencing a continued run of good fortune.
7. push one's luck, to try to stretch one's luck too far.
8. strike me lucky, See **strike.**
9. the luck of the draw, the outcome of chance.
10. the luck of the Irish, good luck.

Lucky Country *n.* Australia (especially in the 1960s). [from the title of the book by Donald Horne, Australian writer]

lucky dip *n.* a chance; an undertaking of uncertain outcome.

lucky shop *n. Vic.* a government-run betting shop; TAB.

lucre *n.* money.

lug *n.* **1.** an ear. *—v.* **2.** to play music by ear. *—phr.* **3. blow down someone's lug,** to harangue someone. [?special use of *lug* pull]

lugger *n.* a person who subjects others to long, unwelcome monologues; an earbasher.

lulu *n.* **1.** an amazing person, event, or thing. **2.** a tall story. [origin uncertain]

lumber *v.* **1.** to foist off on or leave with, as with something or someone unwelcome or unpleasant: *to lumber with the bill.* **2.** to place under arrest.

lummox *n.* a clumsy, stupid person, usually overweight.

lump¹ *n.* **1.** a stupid, clumsy person. *—phr.* **2. have a lump in the throat,** to feel very emotional.

lump² *v.* **1.** to endure or put up with (a disagreeable necessity): *If you don't like it, you can lump it.* **2.** to carry (usually something heavy or cumbersome).

lumper *n. Originally British.* a wharf labourer.

lunatic soup *n.* any alcoholic drink, especially a strong drink, as brandy.

lunch *v.* **1.** Also, **drop one's lunch.** to fart. *—phr.* **2. cut (someone's) lunch,** to cuckold (someone).

lungs *pl. n.* female breasts, especially large ones.

lunkhead *n. U.S.* a stupid fellow; blockhead. Also, **lunk.**

lurgy *n.* **1.** a fictitious, very infectious disease. **2.** any illness. Also, **lurgi.**

lurk *n.* **1.** a place of resort; hide-out. **2.** a dodge; a slightly underhand scheme. **3.** a convenient, often unethical, method of performing a task, earning a living, etc. **4.** a job.

lurk man *n.* **1.** one who lives by his wits, as a confidence man, or an inventive layabout. **2.** one who is adept at exploiting an institutional system for his own benefit.

lush[1] *adj.* **1.** characterised by luxury and comfort. **2.** sexually attractive.

lush[2] *n.* **1.** intoxicating liquor. **2.** a drunken person. **3.** a drinking bout. **4.** one who takes alcoholic drinks, especially regularly. *—v.* **5.** to drink liquor. *—adj.* **6.** drunk.

lushington *n.* a drunkard. Also, **Alderman Lushington.**

lushy *adj.* drunk; tipsy.

luv *n.* a term of endearment [variant of *love*]

ma *n.* mamma; mother.

mac *n.* a mackintosh.

macaroni *n.* nonsense. Also, **maccaroni.**

machine *phr.* **the machine,** *N.Z.* the totalisator.

Mack *n.* *Chiefly U.S.* an informal term of address to a man. Also, **Mac.**

mad *adj.* **1.** moved by anger. *—phr.* **2. like mad, a.** in the manner of a madman. **b.** with great haste, impetuosity, or enthusiasm.
3. mad as a cut snake, insane; crazy.
4. mad as a gumtree full of galahs, quite mad; crazy.
5. mad as a meat axe, a. extremely annoyed. **b.** eccentric; behaving erratically.

mad dog *n.* an unpaid bill. Also, **a dog tied up.**

made *phr.* **have (got) it made,** to be assured of success.

madhouse *n.* a place of commotion and confusion.

mad mick *n.* **1.** a pick. *—phr.* **2. mad mick and banjo,** a pick and shovel. [rhyming slang]

mad mike *n.* *Qld.* a push bike. [rhyming slang]

mad mullah *n.* one whose behaviour is wild and unrestrained. [from Turkish, Persian and Hindustani *mulla* patron, lord]

madwoman *n.* **1.** a woman whose behaviour is considered to be outrageous or eccentric. *—phr.* **2. be all over the place** or **look like a madwoman's breakfast/ washing / custard / knitting/ lunch box/piss,** to be in complete confusion and disarray.

mag *v.* to chatter; to talk rapidly and to little purpose.

maggie *n.* a magpie.

maggot *n.* **1.** a term of abuse. *—phr.* **2. mad as a maggot,** very angry.

maggoty *adj.* angry.

magic sponge *n. Football.* the sponge or towel applied by trainers to slightly injured players.

maginnis *n.* **1.** Also, **crooked maginnis.** a wrestling hold from which there is no escape. *—phr.* **2. put the maginnis on (someone),** to have (someone) completely in one's power. [?from a wrestler named *McGinnis*]

magpie *n.* **1.** a chattering person. **2.** Also, **bower bird,** one who collects useless objects. **3.** a South Australian.

magpie clothing *n.* (formerly) black and yellow convict's clothing. Also, **magpie dress.**

Magpies, the *pl. n.* **1.** the Brunswick team in the Victorian Football Association. **2.** the Collingwood team in the Victorian Football League. **3.** the Glenorchy team in the Tasmanian Australian National Football League. **4.** the North Darwin team in the Northern Territory Football League. **5.** the Port Adelaide team in the South Australian National Football League. **6.** the Southern Suburbs team in the Queensland Rugby Football League. **7.** the Southern Districts team in the Queensland Rugby Football Union. **8.** the Western Suburbs team in the N.S.W. Rugby Football League. [from the team colours, black and white, the same as the markings of a magpie]

magsman *n.* a person who tells stories; raconteur.

mag wheel *n.* a magnesium alloy wheel used for its lightness on some motor cars. Also, **mag.**

mahatma *adj.* handy. [rhyming slang *Mahatma* Gandhi]

mail *n.* information.

mainbrace *phr.* **splice the mainbrace,** to invite an assembly to have a drink. [from the order to issue a tot of rum to a crew; originated as a rare or unlikely order, the mainbrace being replaced rather than repaired by splicing]

mainland *phr.* **the mainland,** *Tas.* continental Australia.

mainlanders *n. Tas.* residents of mainland states of Australia. [Tasmanians may be referred to as islanders or Taswegians. Tasmanians may retaliate by calling the residents of other states the mainlanders]

mainline *v.* to inject a narcotic drug directly into the vein.

mainliner *n.* one who uses addictive drugs injected directly into the vein.

major *n. Australian Rules.* a goal.

make *v.* **1.** to seduce or have sexual intercourse with. *—phr.*
2. make a play (for), to flirt with.
3. make a splash, to make a big impression socially, as by lavish entertainment, publicity stunts, etc.
4. make eyes at, to flirt with.
5. make heavy weather (of), to have difficulty; progress laboriously (with): *to make heavy weather of a simple calculation.*
6. make it, a. to achieve one's object. **b.** to arrive successfully. **c.** to purchase drugs.
7. make it (with), to have intercourse with.
8. make off with, to steal; take.
9. make one's alley good (with), to conciliate; placate.

10. make out, to have sexual intercourse: *I made out last night.*
11. make tracks, to depart.
12. make up to, to try to be on friendly terms with; fawn on.
13. make water, to urinate.
14. make waves, to deliberately create a disturbance; upset existing standards or notions.
15. on the make, a. intent on gain or one's own advantage. **b.** looking for a sexual partner.

makings *phr.* **the makings,** the tobacco and paper used to hand roll a cigarette.

Malabar rat *n.* a bandicoot.

male chauvinist *phr.* **male chauvinist pig,** a male who discriminates against women by applying to them stereotyped ideas of female incompetence, inferiority, female roles, etc., especially in an extreme way.

mallee *phr.*
1. strong or **fit as a mallee bull,** very strong or fit.
2. take to the mallee, go bush. **3. the mallee,** any remote, isolated or unsettled area. [Aboriginal: a low growing species of eucalypt, predominant in unsettled outback areas]

mallee root *n.* a prostitute. [rhyming slang]

maluka *n.* **1.** the boss. **2.** (*cap.*) a form of address to one's superior. [Aboriginal, made familiar by its use in the book 'We of the Never Never' (1908) by Mrs Aeneas Gunn]

mamma *n.* mother. Also, **mammy.**

man *n.* **1.** a term of address to a man or woman. *—phr.*
2. man outside Hoyts, See **Hoyts.**
3. the man on the land, farmers in general.

mandied *adj.* under the influence of Mandrax, a sedative.

mandy *n.* a sedative tablet; Mandrax.

man-eater *n.* a woman who habitually dominates lovers and then discards them.

mangle *n.* **1.** a bicycle. *—phr.* **2. put through the mangle,** exhausted, especially emotionally.

mangy *adj.* mean: *He's a mangy bastard.*

manic *adj.* extremely excited, energetic or agitated. [Greek *manikōs* insane]

mantrap *n.* a seductive woman.

Maori *adj. N.Z.* rough; uncivilised.

Maoriland *n.* New Zealand. [from *Maori* as the original inhabitants]

Maori P.T. *n. N.Z.* a rest; spell. [*Maori + P(hysical) T(raining)*]

map *phr.*
1. off the map, a. out of existence, into oblivion: *Whole cities were wiped off the map.* **b.** crazy.
2. put on the map, to make widely known; make famous.

marble *phr.*
1. lose one's marbles, to act irrationally; go mad.
2. make one's marble good (with), ingratiate oneself (with).
3. pass or **throw in one's marbles,** to die.

Marble Bar *phr.* **till it rains in Marble Bar,** never. [from *Marble Bar* a town in north-west W.A., one of the hottest and driest areas in Australia]

marble orchard *n.* a cemetery.

marching orders *n.* orders to leave; dismissal (from a job, etc.).

marguerite *phr.* **chase marguerites,** to clean fly-struck sheep. [French: daisy, pearl, from Latin *margarita* pearl]

marine *phr.*
1. dead marine, an empty and discarded beer, wine or spirits bottle.
2. tell it or **that to the (horse) marines!** *Originally U.S.* an expression of disbelief, especially at an unlikely story.

mark *phr.*
1. be quick off the mark, to be prompt in recognising and acting upon the possibilities of a situation.
2. be slow off the mark, to be sluggish or slow to start something.
3. give (someone) a good (bad) mark, to approve (disapprove) of (someone).
4. give full or **top marks (to),** to approve warmly.
5. off the mark, inaccurate; irrelevant.
6. overshoot the mark, to err by overestimating the requirements of a situation.
7. overstep the mark, to go beyond the bounds of convention or accepted standards.

marker *n.* an IOU.

market *phr.*
1. in the market, of a racehorse, etc., considered to have a chance of winning, consequently causing the betting odds to be short.
2. go to market, to become angry, excited, unmanageable.

Marlins, the *pl. n.* the North Queensland team in the Queensland Rugby Football League.

Maroons, the *pl. n.* the Queensland State representative sporting team. [from the team colours]

marrowbones *pl. n.* the knees.

martie *n.* a tomato. Also, **marty.**

mary *n.* **1.** (*offensive*) a woman, especially a black woman. **2.** a homosexual male.

Mary Jane *n.* marijuana. Also, **Mary Jay.**

Mary Lou *phr.* **on the Mary Lou,** on credit, as in making a purchase or placing a bet. [rhyming slang *Mary Lou* blue, credit (opposed to red, debit)]

Mary Pickford *phr.* **a Mary Pickford in three acts,** a quick wash; perfunctory wash of the face, hands and crotch. [*Mary Pickford* was an early U.S. film actress]

massive *adj.* tremendous; extraordinary; unusual.

mat *phr.* **(put) on the mat,** to reprimand.

mate *n.* **1. a.** an habitual associate; comrade; friend; equal; intimate: *They've been good mates from way back.* **b.** a form of address (usually amongst men): *How are you going, mate?* —*phr.* **2. be mates with,** to be good friends with.

matey *adj.* **1.** comradely; friendly. —*n.* **2.** (a form of address) comrade; chum.

mateyness *n.* hearty, good-natured friendship.

matilda *n.* **1.** a swag. —*phr.* **2. waltz Matilda,** to carry one's swag. [from German *Mathilde* female travelling companion, bed-roll, from the girl's name; ?taken to the goldfields by German speakers from South Australia]

matinee *n.* sexual intercourse in the afternoon.

mausoleum *n.* a large, old, gloomy building.

mayhem *phr.* **in mayhem,** in disorder or confusion.

mazuma *n.* money. [Yiddish]

McCoy *phr.* **the real McCoy,** the genuine article. [?nickname of U.S. boxer Kid *McCoy* (1873-1940) to distinguish him from another boxer with the same surname]

MCP *n.* male chauvinist pig. [from the initials]

m.d.o. *n. N.Z.* a sickie. [*M(aori) d(ay) o(ff)*]

meadow cake *n.* cow dung; cow pat.

meal ticket *n.* any means or source of financial support, as a pimp's prostitute, a spouse, a university degree, etc.

mean *adj.* **1.** small, humiliated, or ashamed: *to feel mean over some ungenerous action.* **2.** (of a person involved in a competitive activity, as sport, business, warfare, etc.) sufficiently accomplished and determined to make success very difficult for an opponent: *a mean bowler.* **3.** powerful, effective, having a vicious energy: *a big mean motor.*

measly *adj.* wretchedly poor or unsatisfactory; very small.

measure *phr.*

1. get the measure of (someone), to achieve equality with (someone), especially a competitor.

2. measure one's length, to fall flat on one's face.

meat *n.* **1.** male or female genitalia. *—phr.*

2. meat in the sandwich, See **sandwich.**

3. strong meat, books, films, etc., which would shock anyone with a nervous or squeamish disposition, as those depicting violence.

meataxe *phr.* **mad as a meataxe,** See **mad.**

meat pie *phr.* **as Australian as a meat pie,** unmistakeably Australian. [from the prominence of the meat pie in the Australian diet]

meat-pie western *n.* a film made in Australia in the genre of the U.S. western.

meat tag *n.* an identity disc, especially those worn by members of the armed forces. Also, **meat ticket.**

mega- a prefix denoting to a very great degree: *a megatrendy.*

megabuck *n.* a large amount of money.

megastar *n.* a singer, actor or showbusiness personality of immense fame and popularity.

Melba *phr.* **to do a Melba,** to make a habit of returning from retirement, in a number of 'farewell' performances. [from Dame Nellie *Melba* 1861-1931 who had several 'farewells']

Melbourne Cup *n. N.Z.* a chamberpot. [from *Melbourne Cup* a major Australian horserace, first run in 1861]

melon *n.* **1.** a head. **2.** a stupid person; fool. *—phr.* **3. cut a melon,** *U.S.* to declare a large extra dividend to shareholders.

melonhead *n.* a fool; idiot.

mend *phr.* **on the mend, a.** recovering from sickness. **b.** improving in state of affairs.

mental *adj.* **1.** foolish or mad. —*phr.* **2. chuck** or **crack a mental,** to lose one's temper in a violent manner.
3. drive (someone) mental, to annoy or irritate (someone).

merchant *n.* (*usually preceded by a defining term*) a person noted or notorious for the specified aspect of his behaviour: *panic merchant, standover merchant.*

merino *phr.* **pure merino,** (formerly) a free settler (opposed to *legitimate*).

mermaid *n.* an inspector at a weighbridge who is in charge of the scales.

merry *adj.* slightly intoxicated.

mesc *n.* mescaline, a hallucinatory drug.

mess *n.* **1.** a person whose life is confused or without coherent purpose, often due to psychological difficulties. —*phr.*
2. mess around or **about, a.** to waste time. **b.** to play the fool.
3. mess around with, to associate, especially for immoral or illegal purposes.
4. mess in, to meddle officiously.
5. mess someone around or **about,** to cause inconvenience to (a person).
6. mess with, to associate with; have dealing with: *Don't mess with him, he's trouble.*

message *phr.*
1. do the messages, to do errands.
2. get the message, to understand.

meter maid *n.* **1.** an attractive young woman employed to ensure that customers' cars at a shopping centre, visitors' cars at a holiday resort, etc., are not booked for being parked at an expired meter. **2.** *British.* a female traffic warden.

meth *n.* methedrine.

metho *n.* **1.** methylated spirits. **2.** one addicted to drinking methylated spirits. Also, **meths.**

Metho *n.* a Methodist.

metho artist *n.* a drinker of methylated spirits.

method *phr.* **method in one's madness,** reason or sense underlying one's apparent stupidity.

Mexican *n.* *N.S.W.* a Victorian (one south of the border).

Mexican stand-off *n.* a situation in which two opponents threaten each other loudly but neither makes any attempt to resolve the conflict.

michael *n.* the female pudendum. Also, **mick, mickey.**

mick[1] *n.* **1.** the obverse side of a coin; head. **2.** the reverse side of a coin; tail. —*v.* **3.** to throw (the coins) in two-up so that both tails are facing up.

mick[2] *n.* the female pudendum. Also, **michael, mickey.**

Mick *n.* **1.** a Roman Catholic (especially of Irish extraction). **2.** an Irishman.

mickery *n.* behaviour allegedly typical of Roman Catholics; Roman Catholic influence: *There's a lot of mickery in that Party.* Also, **Mickery.**

Mickey *n.* a policeman.

Mickey Doolan *n. N.Z.* a Roman Catholic (especially of Irish extraction).

mickey finn[1] *n.* a drink, usually alcoholic, which has been surreptitiously laced so as to cause to fall asleep, to discomfort or in some way to incapacitate the person who drinks it. Also, **Mickey Finn, mickey.**

mickey finn[2] *n.* the sum of five pounds in money (£5). [rhyming slang *mickey finn* spin, a colloquialism for five pounds]

mickey mouse[1] *adj.* **1.** cheap and not always reliable, as a piece of machinery, watch, etc. **2.** (of crossbred dairy and beef cattle) part Friesian with white face and mainly black body. [from *Mickey Mouse* a cartoon character created by Walt Disney, 1901- 1966, U.S. animator and film producer]

mickey mouse[2] *adj.* splendid, excellent. [rhyming slang *Mickey Mouse* grouse]

Mickhead *n.* a Roman Catholic.

micky[1] *n.* a young, wild bull. Also, **mickey.** [?from *Mick* an Irishman (by association with a young bull)]

micky[2] *phr.*

1. chuck a micky, to throw a fit, panic.

2. take the micky out of, to make seem foolish; tease.

micky quick *v.* to depart in a hurry.

microdot *n.* LSD in tablet form.

middlebrow *n.* **1.** a person of mediocre or limited intellectual calibre or culture. **2.** a person of middle-class or bourgeois taste. *—adj.* **3.** mediocre; bourgeois.

middling *adj.* **1.** in fairly good health. **2.** mediocre; second-rate. *—adv.* **3.** moderately; fairly. *—phr.* **4. fair to middling,** in average health or spirits; so-so.

middy *n.* **1.** a medium size glass, primarily used for serving beer. **2.** a measure of beer.

midge *n.* a small or diminutive person.

midnight *phr.* **burn the midnight oil,** See **burn.**

midsummer madness *n.* a temporary lapse into folly or foolishness, especially during the summer.

miff *n.* **1.** petulant displeasure; a petty quarrel. *—v.* **2.** to give offence to; offend.

miffed *adj.* annoyed; displeased.

miffy *adj.* extremely sensitive; easily offended.

mighty *adj.* **1.** great in amount, extent, degree, or importance. *—adv.* **2.** *U.S.* very: *to be mighty pleased.*

mike *n.* a microgram, especially of LSD.

mileage *phr.* **get mileage out of,** to gain advantage from. Also, **milage.**

milk *phr.* **milk the till,** to steal money from a cash register.

milk bar *phr.* **milk bar cowboy,** a bikie who frequents a milk bar meeting place.

milkie *n. N.Z.* a milkman. Also, **milko.**

milk run *n.* **1.** a routine or uneventful flight in an aeroplane, especially if regular. **2.** a routine trip taken in turns by each one in a group of mothers to take their children to and from school.

million *phr.*
1. gone a million, ruined; lost; done for.
2. one in a million, someone or something of great rarity or worth.

mincemeat *phr.* **make mincemeat of, a.** to assault and do harm to. **b.** to make a successful verbal attack on; berate.

mind *phr.*
1. mind one's own fowlhouse or **duckhouse,** mind one's own business.
2. pissed/smacked/freaked out of one's mind, to be in a state of diminished control of one's actions as the result of alcohol or drugs.

mind-blowing *adj.* exceptionally exciting, stimulating, euphoric: *a mind-blowing experience.* Also, **mind-bending.**

mind-boggling *adj.* overwhelming; stupendous.

minder *n. Originally British.* a bodyguard.

mingy *adj.* mean and stingy. [blend *m(ean)* + *(st)ingy*]

mini *n.* something small in size or dimension, as a skirt or motor vehicle.

mint *v.* **1.** to make or gain money rapidly. *—phr.* **2. mint condition,** brand new; perfect.

minter *n.* a woman who grants sexual favours to a man at the conclusion of, and as if in payment for, an expensive evening out. Also, **after-dinner mint.**

minus *adj.* lacking: *The profits were minus.*

minute man *n.* a man who, in intercourse, ejaculates after a very short time.

mirror *phr.* **done with mirrors,** done by sleight of hand or subterfuge.

miserable *adj.* stingy; mean.

miseries *phr.* **in the miseries,** unhappy; unwell; depressed.

miseryguts *n.* a person who is always whinging or complaining.

miss *v.* **1.** to fail to menstruate at the usual time. *—n.* **2.** an omission or neglect, usually deliberate: *Give it a miss. —phr.*
3. miss out on, to fail to be present, as at a function, or to fail to receive, especially something desired.
4. miss the boat or **bus,** to be too late; fail to grasp an opportunity.
5. not or **never miss a trick,** never to fail to exploit an opportunity, press an advantage, etc.

missus *n.* **1.** a man's wife; the woman with whom a man cohabits. **2.** the mistress of a household. Also, **missis.** [spoken form of Mrs]

missy *n.* **1.** a young girl. **2.** a forward young girl; a disagreeably pert girl or woman.

mistake *n.* an unplanned or unwanted pregnancy.

mister *n.* a term of address for a man: *Hey mister, you forgot your keys.*

mitt *n.* a hand. [variant of *mitten*]

mittens *pl. n.* boxing gloves.

mix *v.* Also, **mix it.** to fight vigorously, as with the fists.

mixed *phr.* **mixed up,** mentally confused.

mix-up *n.* a fight.

mo[1] *n.* **1.** a moment. *—phr.* **2. half a mo, a.** just a moment. **b.** (humorous) a toothbrush moustache. See mo².

mo² *n.* **1.** a moustache. *—phr.* **2. curl the** or **a mo!** an exclamation of admiration, delight, etc.

moan *n.* **1.** a grumble. *—v.* **2.** to grumble.

mob *n.* **1.** any assemblage or aggregation of persons, animals, or things; a crowd. **2.** a group of people, as friends, not necessarily large: *We'll invite the mob over for Saturday night.* *—phr.* **3. big mobs of,** a large number or amount: *I'll have big mobs of mashed potato, please.*

mobile wrestling *n.* the game of Rugby League. [a reference to the rough, bruising, body contact nature of the game]

mocker¹ *phr.* **put the mocker(s) on,** to bring bad luck to; jinx. Also, **put the mock(s) on.**

mocker² *n.* **1.** clothing. *—phr.* **2. mocker up,** to dress up: *He's all mockered up.*

mocker³ *n.* *N.Z.* a bellbird. [Maori *makomako*]

mod *n.* **1.** a young person of the early 1960s, especially in Britain, who was reasonably conventionally dressed, had ungreased hair, and rode a motor scooter in preference to a motor bike. **2.** a young person in the early 1980s espousing similar dress, music, etc. Also, **Mod.**

mod cons *pl. n.* the modern conveniences.

mod squad *n.* *Prison.* plain-clothes warders who handle sudden transfers of prisoners to other gaols.

moggy *n.* a cat. Also, **mog, moggie.**

moke *n.* **1.** a donkey. **2.** an inferior horse. **3.** a type of open-roofed motor vehicle. **4.** *U.S. Offensive.* a Negro.

mole *n.* a spy.

moll *n.* **1.** the girlfriend or mistress of a gangster, thief, etc. **2.** a girl or woman, especially of low character; a tart. **3.** a prostitute. **4.** the girlfriend of a surfie, bikie, etc. [short for *Molly* variant of Mary]

mollydooker *n.* a left-handed person. Also, **mollydook.**

molo *adj.* drunk.

mom *n.* *U.S.* a mother.

Mondayise *v.* *N.Z.* to shift (a public holiday) falling during a weekend to the following Monday.

Mondayitis *n.* lassitude and general reluctance to work as is often experienced on Mondays.

money *phr.*

1. for one's money, as far as one's own choice or preference is concerned; in one's own opinion.

2. give (someone) a run for their money, to put (someone) under the pressure of strong competition.

3. in the money, rich.

4. smart/wise/clever money, a. *Horseracing, etc.* bets placed by those with especially good or inside knowledge. **b.** *Stock Market, etc.* clever, sophisticated investors.

moneybags *n.* a wealthy person.

money-grubber *n.* an avaricious person; one devoted entirely to the making of money.

money-spinner *n.* a business enterprise or property which is very profitable.

mong *n.* a mongrel dog.

mongrel *n.* something difficult: *It's a mongrel of a job.*

moniker *n.* a person's name; a nickname. Also, **monicker, monniker.** [blend of *monogram* and *marker*]

monkey *n.* **1.** a sheep. **2.** the female pudendum. **3. a.** (formerly) the sum of fifty pounds in money (£50). **b.** *British.* the sum of five hundred pounds in money (£500). **c.** the sum of $500. **4.** *N.Z.* a mortgage. —*phr.*

5. a monkey on one's back, any obsession, compulsion, or addiction, seen as a burden, as a compulsion to work or an addiction to drugs.

6. brass monkey, See **brass monkey.**

7. get one's monkey up, to become angry or enraged.

8. I'll be a monkey's uncle, an exclamation of disbelief, surprise, etc.

9. make a monkey of, to make a fool of.

10. monkey (about) with, to play or trifle idly; fool.

11. monkey business, trickery; underhand dealing.

12. monkey tricks, mischief.

monkey bars *pl. n.* a three-dimensional structure of vertical and horizontal bars for children to play on; jungle gym.

monkey-dodger *n.* a shepherd; drover.

monkeyshine *n. U.S.* a mischievous or clownish trick or prank.

monkey suit *n.* a dinner suit.

monster *v.* to harass.

monte *n.* **1.** a certainty: *a monte to win.* —*phr.* **2. for a monte,** for certain. Also, **monty.**

month *phr.* **month of Sundays,** a very long time.

monthly *n.* a menstrual period.

moo *n. British.* a stupid person, especially a woman: *silly old moo.*

mooch *v.* **1.** to skulk or sneak. **2.** to hang or loiter about. **3.** to slouch or saunter along. **4.** to steal. **5.** to get without paying or at another's expense; cadge. —*phr.* **6. mooch off,** to depart. Also, **mouch.**

moolah *n.* money. Also, **moola.**

moon *v.* **1.** to wander (about) or gaze idly, dreamily, or listlessly. —*phr.* **2. once in a blue moon,** See **blue.**

Moonie *n.* a follower of the Unification Church founded by the Rev. Sun Myung Moon.

moonlighter *n.* one who does a job in addition to regular employment.

moonlight flit *n.* **1.** a departure by night with one's possessions in order to avoid payment of rent. **2.** any sudden departure, especially to avoid a disagreeable circumstance.

moonlighting *v.* working at a job in addition to one's regular, full-time employment.

moonshine *n.* smuggled or illicitly distilled liquor.

moonshiner *n.* **1.** an illicit distiller. **2.** one who pursues an illegal trade at night.

moony *adj.* dreamy, listless, or silly.

moosh *n.* **1.** the mouth; the face. **2.** prison food, especially porridge. Also, **mush.**

mopoke *n.* a slow, stupid or miserable-looking person. Also, **morepork.** [from *mopoke* an owl found in Australia and New Zealand]

moppet *n.* a doll.

moral *n.* a certainty: *It's a moral to win.*

morgue *n. Journalism.* **a.** the reference library of clippings, mats, books, etc., kept by a newspaper, etc. **b.** the room for it.

morning after *n.* the morning after a drinking bout.

morning-after *adj.* pertaining to a hangover: *that morning-after feeling.*

morning glory *n.* **1.** *Horseracing.* a horse which performs well in morning track work, but not in races. **2.** sexual intercourse had upon awakening. [from *morning glory* a climbing plant with delicate flowers, which are at their peak in the morning]

moron *n.* a stupid person.

morph *n.* morphine.

mortal *adj.* **1.** extreme; very great: *in a mortal hurry.* **2.** possible or conceivable: *of no mortal use.*

mortal sins *phr.* **ugly as a plateful of mortal sins,** very ugly.

mortgage *phr.* **have a mortgage on,** to be certain of (a win, etc.).

Moscow *n.* **1.** a pawnshop. —*v.* **2.** to pawn. —*phr.* **3. gone to** or **in Moscow,** in pawn.

mosey *phr.* **mosey along** or **off,** *U.S.* to stroll; saunter.

mossback *n. U.S.* **1.** a person attached to antiquated notions. **2.** an extreme conservative.

moss-grown *adj.* old-fashioned.

mother-fucker *n.* a person or thing which arouses exasperation, irritation, contempt, etc. Also, **mother.**

motherless *adv.* absolutely; completely: *stone motherless broke.*

mother's ruin *n.* gin. Also, **aunty's downfall.**

motorbike *n.* a motorcycle.

motser *n.* **1.** a large amount of money, especially a gambling win. —*v.* **2.** to bet a large amount of money: *to motser a horse.* Also, **motza.**

mouch *v.* See **mooch.**

mountain dew *n.* **1.** Scotch whisky. **2.** any whisky, especially when illicitly distilled.

mountain oysters *pl. n. N.Z.* the testicles of a lamb.

mouse *n.* **1.** a black eye. **2.** a person who is very quiet and shy, especially a girl or woman. —*phr.* **3. tight as a mouse's ear,** mean with money.

mouse cheese *n.* any type of cheddar cheese.

mouse-trap *n.* Also, **mouse-trap cheese.** an inexpensive, often tasteless, type of cheese.

mouth *phr.*

1. a shut mouth catches no flies, a piece of general advice against gossiping.

2. down in the mouth, unhappy; depressed.

3. shut one's mouth, to be quiet.

4. shut your mouth there is a bus coming, (to a garrulous person) a request for silence.

mouthman *n. Prison.* a con man.

mouthpiece *n.* a legal representative or adviser; an official spokesperson, usually for clients operating outside the law.

move *phr.*
1. get a move on, hurry up.
2. if it moves, shoot it, if it doesn't, chop it down, a humorous summing up of the Australian attitute to clearing and settling the land.
3. move with the times, to alter one's own attitudes or ideas in conjunction with changes in society.

moxie *n. U.S.* courage; audacity; liveliness. [from *Moxie* the name of a U.S. soft drink]

mozz *n.* **1.** a hex. *—v.* **2.** to hex. **3.** to inconvenience; hinder. *—phr.* **4. put the mozz on (someone),** to hex (someone). [short for *mozzle*]

mozzie *n.* a mosquito.

mozzle *n.* luck, especially bad luck. [Hebrew *mazzal* luck]

much *phr.* **much of a muchness,** (of two or more objects, concepts, etc.) very similar; having little to choose between them.

muck *n.* **1.** something of no value; trash. *—v.* **2.** to spoil; make a mess of. *—phr.*
3. make a muck of, to spoil; impair; disrupt.
4. muck about or **around,** to idle; potter; fool about.
5. muck in, a. to share, especially living accommodation. **b.** join in.
6. muck on the pluck, out of sorts, miserable.
7. muck up, a. to spoil. **b.** to misbehave.

muckrake *v.* to expose, especially in print, political or other corruption, real or alleged.

muckraker *n.* one who attempts to discover and disseminate information unfavourable to an associate, enemy, etc.

muckshifter *n. Building Trades.* an excavation worker.

muck-up *n.* **1.** a fiasco; muddle. **2.** (*in children's speech*) one who misbehaves, especially in school.

mud *phr.*
1. one's name is mud, one is in disgrace.
2. stick-in-the-mud, See **stick**[1]**.**
3. throw or **sling mud at,** speak ill of; abuse; vilify.

mudhook *n.* an anchor.

mudlark *n.* **1.** a street urchin. **2.** *Horseracing.* a horse that performs well on wet tracks.

mudpie *phr.* **throw mudpies,** to smear or discredit.

mudrunner *n. Horseracing.* a mudlark.

mud-slinger *n.* a person who sets out to discredit an opponent, especially one in public office, by hurling abusive accusations at him.

muff *n.* **1.** the female pudendum. **2.** any failure. *—v.* **3.** to perform clumsily, or bungle.

muff-diver *n.* a person who practises cunnilingus.

mug *n.* **1.** the face. **2.** *British.* the mouth. **3.** *British.* a grimace. **4.** a fool; one who is easily duped. *—v.* **5.** to assault by hitting in the face. **6.** to assault and rob. **7.** *U.S.* to take a photograph of (a person), especially in compliance with an official or legal requirement. **8.** *British.* to grimace. *—adj.* **9.** stupid: *a mug copper, a mug alec.* *—phr.* **10. mug up,** to study hard.

mug alec *n.* a foolish man, often loud-mouthed.

muggins *n.* a fool, often used comically by a speaker to refer to himself: *Who has to finish the job? Muggins!*

mug lair *n.* (a term of general contempt) a flashily dressed young man of especially brash and vulgar behaviour.

mug's game *n.* any activity which is unrewarding, not worth doing: *Door-to-door selling is a mug's game.*

mule *n.* **1.** a stupid or stubborn person. **2.** *Prison.* a courier of contraband, especially of drugs.

mulga *n.* **1.** a falsehood. *—phr.*
2. go mulga, to go bush.
3. up or **in the mulga,** in the bush. [from *mulga* a species of the genus *Acacia* found in the drier inland parts of Australia]

mulga wire *n.* an unofficial communication, as by word of mouth; bush wire; bush telegram.

mullet *phr.* **stunned mullet.** See **stunned.**

mullock *n.* **1.** anything valueless. *—v.* **2.** to work in a slipshod way. *—phr.* **3. poke mullock at,** to ridicule; make fun of. [from British dialect *mull* rubbish]

mullygrubber *n.* *Cricket.* a ball delivered in such a manner that on contact with the ground it does not bounce.

mum *n.* mother.

mumbo jumbo *n.* **1.** unintelligible speech or writing, often intended to be impressive; gibberish. **2.** superstition; witchcraft. [from the name of a deity formerly worshipped by certain West African tribes]

mummy *n.* mother.

munchies *pl. n.* **1.** anything to eat, especially snacks between meals. *—phr.* **2. have the munchies,** to experience a craving for food, especially one resulting from smoking marijuana.

munga *n.* *Originally military.* food. Also, **mungaree.** [?French *manger* to eat, adopted by Australian soldiers stationed in Egypt during World War I and World War II]

murder *n.* **1.** an uncommonly laborious or difficult task: *Gardening in the heat is murder. —phr.*
2. get away with murder, to avoid blame or punishment for (something).
3. scream/yell/cry blue murder, See **blue.**

murphy *n.* a potato. [special use of *Murphy* Irish surname (potatoes being seen as a staple in Ireland)]

Murray cod *phr.*
1. bet on the Murray cod, to bet on credit.
2. give (someone) the Murray cod, to signal (someone) to proceed. [rhyming slang *Murray cod* nod. See **nod**]

Murrumbidgee jam *n.* brown sugar added to a little cold tea to form a thick syrup to be spread on damper.

Murrumbidgee oyster *n.* a drink made from a raw egg; prairie oyster.

Murrumbidgee whaler *n.* a swagman frequenting the Australian inland rivers, who sustains himself by begging and fishing. [from *Murrumbidgee* a river in N.S.W.

and *whaler* a person who fishes for Murray cod]

muscle *v.* **1.** to make or shove one's way by sheer brawn or force. —*phr.*
2. muscle in (on), to force one's way in(to), especially by violent means, trickery, or in the face of hostility, in order to obtain a share of something.
3. put the muscle in, to throw one's weight around.

muscle man *n.* **1.** a very strong man; a man of unusually impressive and powerful physique. **2.** a man who regularly uses violence, or the threat of violence, to further the interests of his employer or himself.

museum piece *n.* anything old-fashioned or which has outlived its usefulness.

mush[1] *n.* weak or maudlin sentiment or sentimental language.

mush[2] *n.* See **moosh.**

mushroom *n.* a person who is deliberately kept ignorant and misinformed. [from the practice of growing mushrooms in the dark and feeding them with manure]

mushy *adj.* weakly sentimental: *a mushy valentine.*

music *phr.* **face the music,** to face the consequences, usually unpleasant, of one's actions; accept responsibility for what one has done.

musical milk *n.* methylated spirits.

muso *n.* a musician.

Mustangs, the *pl. n.* the Box Hill team in the Victorian Football Association.

mustard *phr.*
1. be unable to cut the mustard, to fail; to not come up to standard.
2. cut the mustard, to achieve an adequate level of performance; be capable.
3. keen as mustard, extremely keen or eager.

muster *n. Prison.* a roll call in gaol.

mute *n. Banking.* a mutilated banknote.

mutt *n.* **1.** a dog, especially a mongrel. **2.** a simpleton; a stupid person.

mutton *n.* **1.** the penis. —*phr.*
2. dead as a mutton chop, obviously dead.
3. dead as mutton, dull; boring.

mutton-bird *n.* Also, **mutton-bird eater.** an inhabitant of northern Tasmania. Also, **muttonbird.**

mutton-head *n.* a stupid or dull person.

muzzy *adj.* confused; dazed; tipsy.

mystery bag *n.* **1.** Also, **bag of mystery.** a sausage. **2.** a meat pie.

myxo *n.* a highly infectious viral disease of rabbits, artificially introduced into Britain and Australia to reduce the rabbit population; myxomatosis.

nab *v.* **1.** to catch or seize, especially suddenly. **2.** to capture or arrest.

nag *n.* a horse.

nail *v.* **1.** to secure by prompt action; catch or seize. **2.** to catch (a person) in some difficulty, a lie, etc. **3.** to detect and expose (a lie, etc.). —*phr.*

4. hard as nails, (of a person) stern; tough.

5. hit the nail on the head, See **hit.**

6. on the nail, on the spot, or at once: *cash on the nail.*

nail-biter *n.* an exciting game or match with a close finish: *the grand final was a real nail-biter.*

name *phr.* **the name of the game,** the central issue or the essential part of an operation, business, etc.

name-dropper *n.* one who introduces casually into a conversation names of prominent people as though they are personal friends, in order to impress.

nana *n.* **1.** the head. **2.** a fool; one easily duped. —*phr.*

3. do one's nana, to lose one's temper.

4. off one's nana, mad; deranged. [shortened form of *banana*]

nancy boy *n. Chiefly British.* **1.** an effeminate man. **2.** a homosexual.

nanna *n.* a grandmother. Also, **nana.**

nanny goat *n.* tote. [rhyming slang]

nap[1] *n.* blankets. [from the *nap* or pile of the fabric]

nap[2] *phr.* **not to go nap on,** to be unenthusiastic about. [from the bid in the card game *nap* or *napoleon* to win all five tricks of a hand]

nap[3] *Horseracing. n.* **1.** a good tip. —*v.* **2.** to name (a certain horse) as the winner of a race. [special use of *Napoleon* the card game]

napoo *adj.* **1.** finished; used up. **2.** doomed; done for. [alteration of French *il n'y a plus* there is no more]

narc *n.* a member of the narcotics squad in the police force. [shortened form of *narcotics;* ?pun on *nark*]

nark *n.* **1.** an informer; a spy, especially for the police. **2.** a scolding, complaining person; one who is always interfering and spoiling the pleasure of others. —*v.* **3.** to nag; irritate; annoy. **4.** to act as an informer. [Gipsy *nāk* nose]

narked *adj.* upset; irritated; angry.

narky *adj.* irritable; bad-tempered.

nasho *n.* **1.** national service. **2.** one who has been called up for national service. [abbreviation of *national service* compulsory service in the armed forces]

nasties *n. British.* the genital organs.

nasturtiums *phr.* **cast nasturtiums,** a humorous term for casting aspersions.

nasty *n.* sexual intercourse.

natch *adv.* naturally. [shortened form]

national *phr.* **the national game,** two-up.

native *phr.* **go native,** to turn one's back on the comforts of civilisation and adopt a native way of life.

natter *v.* **1.** to chatter; gossip. —*n.* **2.** a chat.

natural *n.* a thing or a person that is by nature satisfactory or successful: *a natural for the part.*

natural foot *n.* a surfer who rides with the left foot in front of the right.

nature *phr.*

1. call of nature, the need to urinate or defecate.

2. nature's need, the need to defecate.

3. one of nature's own (gentlemen), a solid, honest, reliable person.

naughty *n.* sexual intercourse: *a nightly naughty.*

naughty forty-eight *n.* a dirty weekend.

navel *phr.* **contemplate one's navel,** See **contemplate.**

nbg *adj.* no bloody good. [from the initials]

near thing *n.* a close call.

neat *adj.* fine, pleasing, excellent.

nebbie *n.* a nembutal, a form of sleeping pill.

necessary *phr.* **the necessary,** money.

neck *v.* **1.** to play amorously. **2.** *Prison.* to commit suicide by hanging. —*phr.*

3. get it in the neck, to be reprimanded or punished severely.

4. go under (someone's) neck, to act without regard for (someone's) authority or responsibility.

5. neck of the woods, a specific area, particular place: *We don't often see you in this neck of the woods.*

6. neck oneself, to hang oneself; be the cause of one's own misfortune or downfall.

7. pain in the neck, See **pain.**

8. pull one's neck or **head in,** to withdraw; to mind one's own business.

9. stick one's neck out, to act, express an opinion, etc., so as to expose oneself to criticism, hostility, danger, etc.

10. up to one's neck, a. to the limit: *I've had it up to my neck.* **b.** overwhelmed: *He's up to his neck with work.*

neckful *n.* an extreme amount: *I've had a neckful of my mother-in-law.*

necking *n.* the act of playing amorously.

necktie *n. U.S.* a hangman's rope.

ned *phr.* **ned them** (in two-up) to throw two heads; head 'em. [from the Edward VII (*Ned*) pennies traditionally used for two-up]

neddy *n.* a horse. Also, **ned.**

Ned Kelly *n.* **1.** a person or business seen as unprincipled and extorting; bushranger. *—phr.* **2. as game as Ned Kelly,** See **game.**

need *phr.* **need like a hole in the head,** to be severely inconvenienced or distressed by.

needle *n.* **1. a.** tension. **b.** aggression; unpleasantness. *—adj.* **2.** involving unpleasantness or aggression: *a needle game.* *—v.* **3.** to tease or heckle.

neg driving *n.* the offence of negligent driving.

Nellie Bly *n.* **1.** a pie. **2.** a lie. **3.** a tie. Also, **Nelly Bligh.** [rhyming slang]

nelly *n.* **1.** Also, **Nelly's death.** cheap wine, especially red wine. *—phr.* **2. not on your nelly!** absolutely not! [from *Nelly* a girl's name]

nerd *n.* an idiot; fool. Also, **nurd.**

nerve *n.* impertinent assurance.

nerves *pl. n.* **1.** nervousness: *a fit of nerves.* **2.** a common term for a variety of medical or psychiatric disorders associated with tension, stress or nervous collapse: *Mum saw the doctor about her nerves.* *—phr.* **3. get on one's nerves,** to irritate.

nettle *phr.* **grasp the nettle,** to approach an unpleasant task with courage and resolution.

never-never *n.* **1.** *Originally British.* the hire-purchase system: *on the never-never.* *—adj.* **2.** imaginary: *never-never land.* *—phr.* **3. the Never-Never,** sparsely inhabited desert country; a remote and isolated region.

New Australian *n.* an immigrant, especially one whose native tongue is not English; migrant.

new chum *n.* **1.** a novice; one inexperienced in some field: *a new chum on the job.* **2.** (formerly) **a.** a newly transported convict. **b.** a newly arrived British immigrant.

new-laid *adj.* inexperienced; immature in judgment; green.

news *phr.* **bad news,** someone or something from whom or which nothing good is to be expected: *Those kids are bad news.*

newshawk *n.* a newspaper reporter, especially one with a keen eye for news. Also, **newshound.**

newsy *adj.* full of news.

next-door *phr.* **the boy (girl) next-door, a.** a boy (girl), often associated with one's family circle, whose virtues or attractions tend to be overlooked. **b.** an unsophisticated, wholesome boy (girl).

Niagara Falls *pl. n.* the testicles. [rhyming slang *Niagara Falls* balls]

nibs *phr.* **his nibs,** an arrogant or self-important man.

nick[1] *n.* **1.** prison. **2.** the act of stopping work and leaving the job, especially one finished ahead of schedule: *the nick's on, the 4 o'clock nick.* —*v.* **3.** to capture or arrest. **4.** to steal. —*phr.*

5. do a nick, to run away; escape.

6. in good nick, in good physical condition.

7. nick off, to leave, disappear.

8. nick out, to go out for a short period.

nick[2] *phr.* **in the nick,** in the nude; naked.

nicked *phr.* **get nicked, a.** to go away. **b.** to be caught.

nicker *n.* (formerly) the sum of one pound, especially a £1 note; quid.

nicky *phr.* **nicky up,** to improve, especially superficially, the appearance of (a car, house, etc.).

niff *n.* an unpleasant smell.

nifty *adj.* **1.** smart; stylish; fine: *a nifty little car.* —*n.* **2.** *Chiefly U.S.* something nifty, as a smart or clever remark. [originally theatrical slang]

nig *n.* (*offensive*) a Negro. [short for *nigger*]

nigger *n.* **1.** (*offensive*) a Negro. **2.** (*offensive*) a member of any dark-skinned race. —*adj.* **3.** (*offensive*) denoting or pertaining to a Negro or dark-skinned person. —*phr.* **4. nigger in the woodpile,** a hidden snag.

night-bird *n.* one who is habitually up or prowling at night; a nighthawk.

nightcap *n.* an alcoholic or other drink, especially a hot one, taken before going to bed.

night cart *n.* a sanitary cart.

nighthawk *n.* one who is habitually up or prowling about at night.

night owl *n.* a person who often stays up late.

nightwatchman *n. Cricket.* a low order batsman, who is sent in to bat late in the afternoon when the batting-side captain wishes to preserve his better batsmen for the next day's play.

nig-nog *n. Offensive.* **1.** a simpleton. **2.** *British.* a black person.

nine *phr.* **dressed (up) to the nines,** smartly dressed or overdressed.

niner *n.* a nine gallon keg of beer. Also, **nine.**

nineteen *phr.* **talk nineteen to the dozen,** to talk very quickly or excitedly.

nineteenth hole *n.* the bar in a golf clubhouse.

ning-nong *n.* a simpleton.

nip *v.* **1.** to steal. —*phr.*

2. nip away or **off,** to move or go suddenly or quickly.

3. nip away or **up,** to snatch or take suddenly or quickly.

4. put the nips in, to extract money, etc., by putting pressure on someone.

Nip *n.* (*offensive*) a Japanese, especially a Japanese soldier. [short for *Nipponese*]

nipper *n.* **1.** a small boy or younger brother. **2.** a child. **3.** a junior life-saver. **4.** *Building Trades.* one employed on a construction site to do small odd jobs, as make tea, buy lunch, etc. **5.** (*pl.*) handcuffs.

nippy *adj.* nimble; active.

nit[1] *n.* a foolish or stupid person.

nit[2] *n.* watch; guard: *to keep nit.*

nite *n.* night.

nit-keeper *n.* a person who keeps watch; a lookout.

nitpick *v.* to be unduly critical; to be concerned with insignificant details.

nitty *adj.* foolish.

nitty-gritty *n.* the hard core of a matter: *Let's get down to the nitty-gritty.*

nitwit *n.* a slow-witted or foolish person.

nix *n.* **1.** nothing. *—adv.* **2.** no. *—interj.* **3.** *U.S.* a signal warning of someone's approach: *Nix, the cops!* [German, variant of *nichts* nothing]

Noah *phr.* **when Noah was a boy,** a long time ago.

Noahs *n.* a shark. Also **Noah's ark.** [rhyming slang *Noah's Ark* shark]

nob[1] *n.* the head.

nob[2] *n.* a member of a social elite.

nobble *v.* **1.** to disable (a horse), as by drugging it. **2.** to win (a person, etc.) over by underhand means. **3.** to swindle. **4.** to add water to a drink of spirits. **5.** to catch or seize. *—phr.* **6. nobble with,** to give someone a task, etc., which they may be unwilling to perform: *The headmaster nobbled him with the task of reorganisation.*

nobbler *n.* **1.** any small glass of spirits. **2.** a glass of spirits with water added.

nobby[1] *adj.* **1.** smart; elegant. **2.** first-rate.

nobby[2] *n.* an opal. Also, **knobby.**

nobody *phr.* **like nobody's business,** energetically; intensively: *He worked like nobody's business.*

nod *phr.*

1. get the nod, a. to gain approval or permission. **b.** to get unofficial assurance of a job, position, etc.

2. give the nod, a. to permit. **b.** to make a signal to.

3. go to the land of nod, to fall asleep.

4. nod off, to go to sleep.

5. nod one's head, *Prison.* **a.** to plead guilty. **b.** to accept blame or responsibility for something.

6. on the nod, on credit: *to bet on the nod.*

noddle *n.* the head.

nog *n.* (*offensive*) a coloured person, especially a Vietnamese. Also, **noggy.**

noggin *n.* the head.

no-good *adj.* worthless.

no-hoper *n.* **1.** one who displays marked incompetence: *He is a real no-hoper at tennis.* **2.** a social misfit. **3.** a social outcast, vagrant. **4.** an unpromising animal, as a second-rate racehorse, greyhound, etc.

noise *phr.* **1. a big noise,** an important person: *He's a big noise around this town.*

2. make a big noise, to skite about one's abilities, contacts, plans, etc.

3. make a noise, to pay up, as for a round of drinks, etc.

nonch *n.* *Prison.* **1.** a method of shop-lifting in which the goods stolen are concealed in a coat or jacket not worn but nonchalantly draped over one shoulder, ostensibly because of the heat. *— v.* **2.** to steal in such a manner.

non-com *n.* a non-commissioned officer.

non compos *adj.* **1.** in a vague or dazed state of mind. **2.** unconscious, as from alcoholic drink. **3.** not of sound mind; mentally incapable. [short for Latin *non compos mentis*]

nong *n.* a fool; an idiot. Also, **ning nong.**

no-no *n.* **1.** something not to be done on any account: *In this house smoking is a no-no.* **2.** an impossible undertaking. **3.** a complete failure.

no-nonsense *adj.* **1.** practical. **2.** strict. **3.** unpretentious.

non-pro *n., adj.* non-professional.

non-starter *n.* something which has no chance of success, as an idea that is discounted as inherently impracticable.

non-U *adj.* not appropriate to or characteristic of the upper class. [*non + U(pper class)*]

noodle *n.* **1.** the head. *—phr.* **2. use one's noodle,** to think for oneself. [?variant of *noddle* (with *oo* from *fool*)]

nooky *n.* sexual intercourse. Also, **nookie.**

noongar *n.* *W.A.* an Aborigine. [Aboriginal]

nope *adv.* an emphatic form of *no.*

norks *pl. n.* breasts. [?from *Norco* butter, the wrapping of which at one time featured a cow's udder]

north and south *n.* the mouth. [rhyming slang]

nose *phr.*

1. by a nose, by a very narrow margin.

2. get up (someone's) nose, to irritate or annoy (someone).

3. keep one's nose clean, to follow rules and regulations meticulously so as to avoid any blame.

4. look down one's nose at, to despise; disdain.

5. on the nose, a. smelly, objectionable, decayed, stinking (especially of rotten organic matter, as food). **b.** unpleasant, distasteful. **c.** (in gambling) for a win only; straight out.

6. pay through the nose, to pay an excessive amount.

7. put (someone's) nose out of joint, to thwart or upset (someone).

8. turn one's nose up, to be contemptuous or ungrateful.

9. under one's nose, in an obvious place.

10. up your nose with a rubber hose, See **up.**

nose rag *n.* a handkerchief.

nosh *v.* **1.** to eat; have a snack or a meal. *—n.* **2.** anything eaten, especially a snack. **3.** a titbit eaten between meals.

noshery *n.* **1.** a cheap eating house or cafe. **2.** *Joc. or ironic* a good restaurant: *a big bash in a fashionable Melbourne noshery.*

nosh-up *n.* a meal.

nosy *adj.* prying; inquisitive. Also, **nosey.**

nosy parker *n.* a person who continually pries; a meddler; a stickybeak. Also, **nosey parker.**

note *n.* (formerly) a one pound (£1) note.

nothing *phr.*

1. in nothing flat, in no time at all; very quickly.

2. nothing doing, definitely no or not.

3. nothing to write home about, not worthy of special mention; ordinary.

nous *n.* common sense.

novel *phr.* **the great Australian novel,** the novel which will say it all, but which is yet to be written.

now *adj.* very fashionable, trendy: *a now happening.*

nowhere *phr.* **get nowhere,** to achieve nothing.

nowler *n. Shearing.* a sheep that is difficult to shear because of dirt, burrs, etc., in the wool.

nozzle *n.* the nose.

nth *phr.* **the nth degree** or **power,** the utmost extent.

nub *n.* the point or gist of anything: *the nub of the matter.*

nubbies *pl. n.* women's breasts.

nuddy *phr.* **in the nuddy,** in the nude.

nudge *phr.*

1. nudge the bottle, to drink alcoholic liquor to excess.

2. give it a nudge, to make an attempt; have a try at something.

nugget *n.* **1.** a short muscular man or animal. **2.** an unbranded calf. —*v.* **3.** to steal cattle, by branding unbranded calves; duff.

nuke *n.* a nuclear device.

number *n.* **1.** an article of merchandise: *Her new dress is a cute little number.* **2.** a marijuana joint. —*phr.*

3. do a number, a. to perform a specified piece, or routine. **b.** to lose one's temper.

4. one's number is up, a. one is in serious trouble. **b.** one is due to die.

number crunching *n.* long, laborious, recursive mathematical computation, especially as done by computers.

number one *n.* **1.** urination. **2.** *Navy.* the first officer. **3.** (*pl.*) *Navy.* dress uniform.

number two *n.* defecation.

numbskull *n.* a dull-witted person; a dunce; a dolt. Also, **numskull.**

numero uno *n.* **1.** oneself. **2.** the leader or most important person in any situation. [Italian: number one]

nummy *adj.* (of food) tasty; delectable.

nun *phr.* **dry as a nun's nasty,** See **dry.**

nunty *adj.* dowdy; out of fashion.

nurd *n.* See **nerd.**

nut *n.* **1.** the head. **2.** an enthusiast. **3.** a foolish or eccentric person. **4.** an insane person. **5.** a testicle. —*phr.*

6. do one's nut, to be very angry, anxious, or upset.

7. hard nut to crack, a. a difficult question, undertaking, or problem. **b.** a person who is difficult to convince, understand, or know.

8. nut out, to think out; solve (a problem, a plan of action, etc.).

9. off one's nut, a. mad; insane. **b.** crazy; foolish.

10. work one's nut, to scheme.

nut case *n.* a foolish or eccentric person.

nuthouse *n.* a mental hospital. Also, **nut factory.**

nut man *n.* a homosexual.

nuts *interj.* **1.** an expression of defiance, disgust, etc.: *Nuts to you!* —*adj.* **2.** crazy; insane. —*phr.* **3.**

do one's nuts over, to become infatuated with.
4. nuts on or **over,** overwhelmingly attracted to: *I'm nuts over her.*

nuts and bolts *pl. n.* the most basic components; the essential factors of a situation, policy, plan, etc.

nuts-and-bolts man *n.* **1.** a man whose interests and aptitudes are essentially practical. **2.** a person who puts emphasis on analysing the essential factors of a situation, policy, plan, etc.

nutshell *phr.* **in a nutshell,** in very brief form; in a few words: *Tell me the story in a nutshell.*

nutter *n.* a crazy or foolish person.

nutty *adj.* **1.** silly or stupid; crazy. *—phr.*
2. nutty as a fruitcake, completely mad.
3. nutty over, overwhelmingly attracted to.

nympho *n.* a nymphomaniac.

O *phr.* **the big O,** female orgasm.

Oaks, the *pl. n.* the Oakleigh team in the Victorian Football Association.

oat *phr.*
1. feel one's oats, a. to feel gay or lively. **b.** to be aware of and use one's importance and power.
2. sow (one's) wild oats, to indulge in the excesses or follies of youth, especially in sexual promiscuity.

oath *phr.* **blood oath, my (colonial) oath,** a statement of emphatic agreement.

obscenity *n.* an action or state of affairs that is degrading and offensive to one's moral or aesthetic sense: *War is an obscenity.*

ocker *n.* **1.** the archetypal uncultivated Australian working man. **2.** a boorish, uncouth, chauvinistic Australian. **3.** an Australian male displaying qualities considered to be typically Australian, as good humour, helpfulness, and resourcefulness. *—adj.* **4.** relating to an ocker. **5.** distinctively Australian: *an ocker sense of humour.* Also, **okker.** [?variant of *Oscar* the character in the television program by Ron Frazer]

ockerdom *n.* (*offensive*) the society of boorish, uncouth, chauvinistic Australians.

ockerina *n.* a humorous term for an ocker's female counterpart.

ockie strap *n.* a stretchable rope with hooks on either end used for securing luggage to roof-racks, etc.; octopus strap.

OD *n.* **1.** an overdose, especially of an injected addictive drug, as heroin. **2.** a surfeit; oversupply. *—phr.* **3. OD on, a.** to give oneself a drug overdose of. **b.** (*jocular*) to consume to excess; have a surfeit: *I OD'd on icecream.*

oddball *n.* **1.** one who is unusual or peculiar; an eccentric. *—adj.* **2.** unusual; eccentric.

odd bod *n.* an eccentric person, especially one with a particular fixation.

odds *phr.*
1. **odds and ends,** odd bits; scraps; remnants; fragments.
2. **odds and sods,** a miscellaneous collection of people or things.
3. **over the odds,** too much.
4. **what's the odds,** what difference does it make?

odds-on *adj.* (of a chance) better than even; that is more likely to win, succeed, etc.

off *prep.* **1.** refraining from (some food, activity, etc.): *to be off gambling.* —*adj.* **2.** Also, **off tap.** *Prison.* convicted. **3.** offensive: *an off joke.* **4.** unwell. **5.** rotten: *That meat is off.* —*phr.*
6. **get off the ground,** to begin, get under way; in progress: *The job has not got off the ground yet.*
7. **off like a bride's nightie** or **a robber's dog, a.** acting promptly. **b.** leaving rapidly.
8. **off the air,** See **air.**

off-beat *adj.* unusual; unconventional.

off-centre *adj.* (of a person) unstable.

off-colour *adj.* **1.** Also, **off.** unwell. **2.** of doubtful propriety or taste: *an off-colour story.*

office *phr.* **the office,** a hint or signal.

officer's pox *n. Military.* venereal disease. Also, **padre's pox.**

offie *n. Cricket.* **1.** a ball bowled so as to change direction from leg to off when it pitches; off break. **2.** a bowler who specialises in off-break deliveries.

offish *adj.* aloof.

off-putting *adj.* disconcerting; discouraging.

off-the-peg *adj.* relating to clothes which are sold ready-made. Also, **off-the-rack.**

oh boy *interj.* **1.** an exclamation indicating surprise, delight, etc. **2.** an exclamation indicating dismay, trepidation, etc.

oil *n.* **1.** flattery; bribery. —*phr.*
2. **the good** or **dinkum oil,** correct (and usually profitable) information, often to be used in confidence; the drum.
3. **the oil,** *N.Z.* an excellent person or thing.

oiled *adj.* drunk.

oil painting *phr.* **be no oil painting,** to lack good looks.

oil rag *phr.* **live on the smell of an oil rag,** to survive on the barest amount of food, money, etc.

okay *adj.* **1.** all right; correct. —*adv.* **2.** well; effectively; correctly; acceptably. —*v.* **3.** to put an 'okay' on (a proposal, etc.); endorse; approve; accept. —*n.* **4.** an approval, agreement or acceptance. Also, **ok, OK, O.K., o.k.** [origin much debated but probably abstracted from or popularised by the 'O.K. Club', formed in 1840 by partisans of Martin Van Buren, 1782-1862, President of the United States 1837-41, who allegedly named their organisation with the initials of 'Old Kinderhook' (Kinderhook, New York State, being Van Buren's birthplace)]

OK card *n.* a union card.

old *adj.* **1.** relating to a long acquaintance or friendly feeling: *good old Henry.* **2.** carried to great lengths; great: *a fine old spree.* —*n.* **3.** old beer. —*phr.*

4. old as the hills, See **hill.**

5. the olds, one's parents.

old boy *n.* **1.** *Chiefly British.* a husband; a father. **2.** the penis.

old chum *n.* **1.** one experienced in some field; old hand. **2.** (formerly) an experienced colonist; one who had spent some years in the colony, especially in the outback.

Old Country *n.* Britain, especially England.

old Dart *n.* Britain, especially England.

old Dutch *n.* **1.** a wife. **2.** a cleaning lady. [shortened form of *duchess*]

old fellow *n.* **1.** (as a term of address) friend, companion. **2.** the penis. Also, **old boy.**

old girl *n.* **1.** an old woman. **2.** *Chiefly British.* a wife; a mother.

old hand *n.* **1.** (formerly) a convict who had spent some years in the colony. **2.** an ex-prisoner. **3.** one with experience in some field.

oldie *n.* **1.** one regarded as old by the speaker. **2.** a parent.

old lady *n.* **1.** a mother, usually one's own. **2.** a wife, usually one's own.

old maid *n.* a person with the alleged characteristics of an old maid, such as primness, prudery, fastidiousness, etc.

old man *n.* **1.** a father, usually one's own. **2.** a husband, usually one's own. **3.** one in a position of authority, as an employer. **4.** the penis. **5.** a very large kangaroo. —*adj.* **6.** relating to anything of exceptionally large size, intensity, etc.: *an old man of a storm.*

olds *pl. n.* parents.

old stager *n.* a person of long experience; a veteran.

oldster *n.* an old or older person. [*old* + *ster* modelled on *youngster*]

old-timer *n.* **1.** one whose residence, membership, or experience dates from a long time ago. **2.** an old man.

old woman *n.* **1.** a wife, usually one's own. **2.** a mother, usually one's own. **3.** a fussy, silly person of any age or sex.

ology *n.* any science or branch of knowledge. [abstracted from words like *biology, geology*]

Olympians, the *pl. n.* the Sydney Olympic team in the National Soccer League.

on *prep.* **1.** indulging to excess: *on the bottle; on the turps.* —*phr.*

2. be on, a. to be willing or in agreement. **b.** to have a bet accepted.

3. be on about, be concerned about; to complain about.

4. be on at, to nag.

5. be on to a good thing, a. to have hit upon a successful, especially money-making, scheme, project, etc. **b.** (of a man) to be optimistic of having sexual intercourse with a woman.

6. go on at, to berate; scold.

7. have oneself on, to think oneself better, more skilled or more important, than one really is.

8. not on, not a possibility; not allowable: *To buy a car now is just not on.*

9. on for young and old, in a state of general disorder and lack of restraint.

10. on it, on a drinking bout.

11. on to, in a state of awareness; knowing or realising the true meaning, nature, etc.: *The police are already on to your little game.*

once-over *n.* **1.** a quick or superficial examination, inspection, treatment, etc., especially of a person viewed as a sexual object. **2.** a beating-up; act of physical violence.

oncus *adj.* See **onkus.**

one *phr.*

1. get/guess/have it in one, to hit on the correct answer, attain a goal, etc., at one's first attempt.

2. to one them, *Two-up.* to throw a head and a tail.

one-armed bandit *n.* a poker machine.

one-eyed trouser snake *n.* the penis. [phrase popularised by Australian humorist Barry Humphries]

one-horse *adj.* **1.** involving one person only: *a one-horse affair.* —*phr.*

2. one-horse race, any contest or activity where the winner or outcome is regarded as being obvious.

3. one-horse town, an insignificant or backward town.

one-man band *n.* **1.** a musician who alone, especially as a street entertainer, plays many instruments, held or strapped to his body. **2.** one who undertakes alone all the tasks presented by a situation.

one-man show *n.* an enterprise which is dominated by the personality and wishes of one man.

one-night stand *n.* **a.** a chance encounter with a person, involving sexual intercourse, but not developing into a steady relationship. **b.** a person with whom one has had such an encounter.

one o'clock *phr.* **get on like one o'clock,** to be compatible; to get on well. [from the position of the hands on the face of a clock at one o'clock]

one-out *adj.* alone; unaided: *The Soviet State stood one-out against the world.*

one-track *adj.* restricted; preoccupied with one idea: *a one-track mind.*

onion *n.* **1.** the head. **2.** the practice, especially amongst bikie gangs, of a number of men having sexual intercourse with one girl on the one occasion. **3.** a girl with whom a number of men have sexual intercourse on the one occasion. —*phr.*

4. know one's onions, to know one's job thoroughly; be experienced.

5. off one's onion, mad; insane.

onka *n.* a finger. [rhyming slang *Onkaparinga* (a Victorian town) finger]

onkus *adj.* **1.** bad; unacceptable. Also, **oncus. 2.** (of machinery) out of order.

on ya *interj.* good on you. See **good.**

oodles *pl. n.* a large quantity: *oodles of money.*

oomph *n.* **1.** vitality; energy. **2.** sex appeal.

OO7 *n. Prison.* a bond. [from the character James Bond, agent *OO7*]

oozle *v.* to steal.

O.P. *n.* (usually of a cigarette) other people's. Also, **o.p.**

open go *n.* See **go.**

open list *n.* the list of drugs which can be prescribed, as a pharmaceutical benefit, without restriction.

open slather *n.* a situation in which there are no restraints, often becoming chaotic or rowdy; free-for-all.

opera house *n.* a hotel which is an early opener. [from ironic reference to the Sydney Opera House which opened several years later than originally planned]

operator *n.* one who successfully manipulates people or situations: *He's a smooth operator.*

opt *phr.* **opt out,** to decide to take no part in the accepted social institutions and conventions.

optic *phr.* **have an optic at,** to look at.

optics *pl. n.* the eyes.

orange people *pl. n.* **1.** adherents of yoga who wear uniformly orange clothing. **2.** followers of Rajneesh who wear clothing coloured orange, red, pink, etc., the colours of the sunset.

orange-time *n.* a scheduled break in a game, as hockey, football, etc., during which oranges are eaten by the players as a refreshment.

order *phr.*

1. a tall order, a difficult task or requirement.

2. in short order, speedily; promptly.

order of the boot *n.* the sack; dismissal.

order of the day *n.* a plan for the day's activities, as a family picnic, journey, etc.

original *adj.* mentally ill; insane: *Poor Dick went a bit original after that.*

ornery *adj.* **1.** stubborn. **2.** low or vile. [contraction of *ordinary*]

orry-eyed *adj.* drunk.

ort *n.* the anus: *In your ort, sport.*

o.s. *adj., adv.* abroad; overseas.

oscar *n.* cash, money. [rhyming slang *Oscar* Asche, Australian actor, 1871-1936]

ossified *adj.* in a drunken stupor.

o.t. *adj., adv.* overseas. [*o(ver) t(here)*]

other *phr.*

1. the other half, a. either of the two classes into which society is divided, the rich or the poor (but especially the poor): *See how the other half lives.* **b.** See **better half.**

2. the other side, *N.Z.* Australia.

otherie *n.* an other. Also, **othery.**

othersider *n.* *W.A.* a person living on the other side of the Nullarbor plain; t'othersider; eastern stater.

out *phr.*

1. go all out, to extend oneself; pursue an interest, goal, etc., with the utmost energy.

2. out of it, a. incapacitated as a result of taking drugs or alcohol. **b.** in a dreamy or vague state of mind, as if under the influence of drugs or alcohol.

3. out to it, a. unconscious. **b.** asleep. **c.** dead drunk. **d.** completely exhausted.

outback *n.* **1.** Also, **Outback.** remote, sparsely inhabited country; the bush. *—adj.* **2.** related to, or located in the outback. *—adv.* **3.** in or to the outback: *to live outback.*

outdoors *phr.* **the great outdoors,** the natural environment, especially wilderness areas.

outer *n.* **1.** an open betting place near a racecourse. *—phr.*
2. on the outer, excluded from the group; mildly ostracised.
3. the outer, *Prison.* the world outside prison.

Outer Mongolia *n.* any remote, isolated and therefore culturally backward place.

outfit *n.* a syringe and needle used for taking drugs.

outlander *n.* an outsider.

outside *prep.* **1.** Also, **outside of.** with the exception of. *—n.* **2.** *Prison.* the world outside prison. *—phr.*
3. at the outside, at the utmost limit: *Not more than ten at the outside.* *—phr.* **4. get outside,** to eat or, especially, drink: *Get yourself outside this beer.*

over *phr.*
1. all over, characteristic of; typical of: *That's him all over.*
2. all over with, done with; finished.
3. be all over, to show great affection towards; be excessively attentive to: *She was all over him as soon as he entered the room.*
4. over the fence, unreasonable; unfair.
5. over the hill, See **hill.**
6. over the top, excessive.

overkill *n.* the use of more resources or energy than is necessary to achieve one's aim.

overland *v.* to drove sheep or cattle across country for long distances, especially through remote areas, as from the Northern Territory to Adelaide.

overlander *n.* a drover who takes stock long distances overland.

overplay *phr.* **overplay one's hand,** to overestimate one's chance of success.

owl *n.* **1.** a person of nocturnal habits. **2.** a person of solemn appearance.

owled *adj.* drunk. Also, **oiled.**

own *phr.*
1. come into one's own, a. to receive an inheritance. **b.** to be in a situation where one's particular skills or attributes are evident.
2. get one's own back, to have revenge.
3. on one's own, on one's own account, responsibility, resources, etc.

Oxford *n.* a dollar. Also, **Oxford scholar.** [rhyming slang]

oxster *n.* *Chiefly British.* an armpit.

oyster *n.* a close-mouthed person.

Oz *n.* **1.** Australia. *—adj.* **2.** Australian.

P *phr.*

1. mind one's p's and q's, to behave properly and carefully.

2. silent like the P in swimming or **the ocean/sea/surf etc.,** a pun on the word 'pee', used humorously to explain the pronunciation of words beginning with a silent 'p', such as 'psalm', 'psuedo', 'ptomaine', etc. See **pee.**

pa *n.* papa; father.

pace *phr.* **put through one's paces,** to cause to perform or show ability.

pacifier *n.* a police baton.

pack *v.* **1.** to be capable of (forceful blows, etc.): *He packs a mighty punch. —phr.*

2. go to the pack, a. to degenerate; collapse. **b.** to give up; admit defeat.

3. pack death/it/shit, to be afraid.

4. pack the game or **it in,** to give up; desist from.

5. pack up or **in,** to cease to function; to become useless.

packet *n.* **1.** a large sum of money: *to drop a packet at the races; earn a packet.* **2.** a heavy or forceful blow, injury, setback, or the like: *He's caught a packet.*

pad *n.* **1.** a dwelling, especially a single room. **2.** a bedroom. **3.** a bed. *—v.* **4.** Also, **pad down.** to live or sleep (somewhere).

paddle *v.* **1.** *U.S.* to beat with or as with a paddle; spank. *—phr.* **2. paddle one's own canoe,** See **canoe.**

paddlefeet *n.* an awkward person, especially a person with a clumsy walk.

paddock-basher *n.* an off-road vehicle.

paddy *n.* an intense anger; a rage. [from *Paddy,* an Irishman (the Irish being seen as quick-tempered)]

Paddy *n.* an Irishman. [familiar variant of Irish *Padraig* Patrick]

Paddy's lantern *n. N.Z.* the moon.

paddy-wagon *n.* a police van for transporting prisoners.

paddywhack[1] *n.* **1.** Also, **paddy.** a rage. **2.** a spanking.

paddywhack[2] *n.* the back. [rhyming slang]

padre's bike *phr.*
1. gone for a ride or **shot through on the padre's bike,** a humorous response to an enquiry about someone's whereabouts.
2. go through on the padre's bike, to travel at great speed.

padre's pox *n. Military.* a euphemism for venereal disease.

pain *phr.* **pain in the neck** or **arse,** an irritating, tedious, or unpleasant person or thing.

paint *n.* **1.** colour, as rouge, used on the face. **2.** wine. *—phr.*
3. on the paint, *Horseracing.* in the inside lane when rounding a bend, near the rails on the edge of the track.
4. paint the town red, to have a spree; celebrate.

painters *phr.* **have the painters in,** to be menstruating.

pakapoo ticket *n.* something that looks confusing or incomprehensible: *marked like a pakapoo ticket.* [from *pakapoo* a type of Chinese lottery using slips of paper with characters written on them]

pakaru *adj. N.Z.* **1.** ruined, broken. *—v.* **2.** to ruin, break. Also, **puckeroo.** [Maori]

pakeha Maori *n.* a European who adopts the Maori way of life.

Paki *n.* a Pakistani.

pal *n.* **1.** a comrade; a chum. **2.** an accomplice. *—v.* **3.** to associate as pals. *—phr.*
4. be a pal! be a friend and help me!
5. pal up with, to become associated or friendly with.

pally *adj.* friendly.

palm[1] *phr.* **cross/grease/oil someone's palm,** to bribe.

palm[2] *phr.* **take the palm,** to outdo; cheat.

palooka *n.* a stupid or clumsy boxer, etc.

pan[1] *n.* **1.** the face, as in *dead-pan.* *—v.* **2.** to criticise or reprimand severely. *—phr.* **3. pan out,** result; turn out.

pan[2] *n.* a large truck; transport. Also, **pantech.** [from *pantechnicon* a furniture-moving van]

Pancake Day *n.* a remote or improbable time: *She won't be back till Pancake Day.* [from *Pancake Day* or Shrove Tuesday on which pancakes are traditionally eaten as the last eggs to be consumed before Lent]

panhandle *v. U.S.* to beg (usually in the street).

panic *phr.* **be at panic stations,** to be in a situation requiring extreme measures; be chaotic.

panic button *n.* **1.** an imaginary device said to be pressed when one gets flustered by an emergency. *—phr.* **2. press** or **hit the panic button,** to react as to an emergency; overreact.

pannikin boss *n.* an overseer of a small business or a small gang of labourers; a person of minor authority; foreman. Also, **panno.**

pannikin snob *n.* a person of small means who affects a snobbish attitude.

pansy *n.* **1. a.** an effeminate man. **b.** a male homosexual. *—v.* **2.** to move in an effeminate way: *He was pansying along the street.* **3.** to treat something fussily or in an effeminate way: *The golfer was pansying the ball.* *—phr.* **4. pansy up,** to overdecorate; pretty up: *to pansy up a room.*

Panthers, the *pl. n.* **1.** the Penrith team in the New South Wales Rugby Football League. **2.** the South Adelaide team in the South Australian National Football League. **3.** the Waverly team in the Victorian Football Association. **4.** the Western Suburbs team in the Queensland Rugby Football League.

pants *phr.*

1. get into (someone's) pants, to have sexual intercourse with.

2. wear the pants, to be in charge; dominate.

pants man *n.* a woman-chaser.

paper *phr.* **paper over,** to try to hide something.

paperbag *phr.*

1. paperbag job, (usually of a woman) someone physically unattractive, but nevertheless viewed as a sexual object (with the aid of an imaginary paperbag to hide the head).

2. unable to fight or **blow one's way out of a paperbag,** ineffectual; lacking strength; lacking spirit.

paperhanger *n. Prison.* one who passes valueless cheques.

papoose *n.* a baby. Also, **pappoose.** [Algonquian (a North American Indian language) *papeisses* child]

parachute *n.* a fart.

paralytic *adj.* completely intoxicated with alcoholic drink; very drunk.

paranoid *adj.* emotionally hypersensitive.

paras *pl. n.* paratroops.

parcel post *phr.* **have come up by parcel post,** to be inexperienced, as of a new arrival.

parcel-post man *n.* an inexperienced man; new chum.

pardner *n. U.S.* partner; friend. Also, **pard.**

parish *phr.* **on the parish,** poor; needy. [from obsolete sense of being in receipt of poor relief from the parish]

parish pump *n.* **1.** a gathering place for gossip. *—adj.* **2.** parochial.

park *v.* **1.** to put or leave: *Park your bag in the corner.* *—phr.* **2. park oneself,** to be seated.

parliament house *n.* a urinal. [a humorous comparison with *Parliament House,* 'where all the big heads hang out']

parlour pink *n.* a person who expresses enthusiasm for the philosophies of the Communist Party without actively supporting it.

parrot *phr.* **pissed as a parrot,** very drunk.

part *phr.* **part up with,** to hand over; pay.

party *phr.* **party on,** to continue a party.

party hack *n.* a person who is said to have given lengthy, mainly menial service to a political party.

party man *n. Politics.* a person whose actions, words and associations are almost entirely directed or dictated by party considerations.

party pooper *n.* a person who has a discouraging or depressing effect, especially at a party.

pash *n.* **1.** a passion; infatuation: *to have a mad pash on someone.* **2.** kissing or cuddling. —*v.* **3.** Also, **pash off, pash on.** to kiss or cuddle.

pass *phr.*

1. make a pass (at), to make an amorous overture or gesture.

2. pass the buck, to avoid responsibility by passing it to another.

3. pass up, a. to refuse; reject. **b.** to fail to grasp.

passenger *n.* a member of a team, staff, etc., who does not perform his fair share of work.

passion pit *n.* **1.** any cinema in which petting takes place among the audience, especially a drive-in. **2.** a bedroom decorated in a vulgar and lascivious manner.

paste *v.* **1.** to strike with a smart blow, or beat soundly, as on the face or body. **2.** to berate.

pasting *n.* **1.** a beating or thrashing. **2.** a tirade of abuse.

pat[1] *phr.*

1. a pat on the back, a gesture or word of encouragement or congratulation.

2. pat (someone) on the back, to congratulate or encourage with praise.

pat[2] *phr.* **sit pat,** to stick to one's decision, policy, etc.

pat[3] *phr.* **on one's pat,** alone. Also, **Pat.** [rhyming slang *Pat Malone* alone]

Pat *n.* an Irishman. [shortened form of common Irish name *Patrick*]

patch *phr.*

1. hit a bad patch, to suffer a series of misfortunes, especially financial.

2. not a patch on, not comparable to; not nearly as good as.

3. purple patch, See **purple.**

pathetic *adj.* **1.** miserably inadequate: *Her vegetables made a pathetic showing at the annual produce fair.* **2.** inviting scorn or pity because of patent shortcomings, pettiness, greed, rudeness, etc.

Pat Malone *n.* See **pat**[3]**.**

patriarch *n.* an authoritarian man.

patsy *n.* **1.** a scapegoat. **2.** a person who is easily deceived, swindled, ridiculed, etc. [? Italian *pazzo* fool, lunatic]

Patti Duke *n. Australian Rules.* crepe paper streamers in team colours attached to a very short stick; flogger. [from the U.S. television series *The Patti Duke Show,* which featured cheerleaders using similar streamers]

pav *n.* a pavlova. [the pavlova was invented in 1935 by Herbert Sachse, 1898-1974, Australian chef, and named by Harry Nairn of the Esplanade Hotel, Perth, after Anna *Pavlova,* 1885-1931, Russian ballerina]

paw *n.* **1.** the human hand. —*v.* **2.** to handle clumsily, rudely, or overfamiliarly. **3.** to use the hands clumsily or rudely on something.

pay *v.* **1.** to admit the truth of; acknowledge that one has been out-

witted, especially in repartee or argument: *I'll pay that.* —*phr.*
2. give (someone) a pay, to castigate or rebuke (someone).
3. pay off, a. to retaliate upon or punish. **b.** to bribe.
4. pay out, a. to retaliate for an injury; punish in revenge. **b.** to protest volubly.

pay-dirt *phr.* **strike pay-dirt,** to be successful, especially in gaining money. [from *pay-dirt* the earth or rock which yields a profit to the miner]

pay-off *n.* **1.** the final settlement of a salary, bet, bribe, or debt. **2.** the climax, as of a joke or routine. **3.** a final, sometimes unexpected consequence.

PD *adj.* of or pertaining to a police vehicle, as a police car, motorbike, paddy wagon, etc. [from the onetime *P(olice) D(epartment)* numberplates of such vehicles]

pea *n.* **1.** a racehorse that seems likely to win. **2.** one who is predicted to win or succeed as a favoured applicant for a job.

peach[1] *n.* a person or thing especially admired or liked.

peach[2] *v. British.* to inform against an accomplice or associate.

peachy *adj.* excellent; wonderful.

peacock *v.* **1.** (formerly) to buy up the best portions of a piece of land, as the land around waterholes or creeks, so that the remaining land is of little value to any other buyer, thereby gaining use of that land as well. **2.** to pick the best portions out of (something).

peacocker *n.* **1.** (formerly) a squatter who selected choice pieces of land, as around waterholes, by using dummy buyers. **2.** one who chooses the best parts, pieces, etc., of anything.

pea eater *n.* a foolish person.

peanut *n.* **1.** (*pl.*) any small amount, especially of money. **2.** an insignificant person.

peanut-butter sandwich *n. Prison.* an honest job.

pearler *adj.* excellent; pleasing. Also, **purler.**

pearly *n.* **1.** *British.* a button, especially one of a large number sewn on to the clothes of costermongers for ornament. **2.** *British.* a costermonger whose clothes are so ornamented.

Pearly Gates *pl. n.* **1.** the entrance to heaven. **2.** (*without caps.*) the teeth.

peasant *n.* an unsophisticated person; one unable to appreciate that which is cultured and tasteful; a boor.

pea souper *n.* an extremely thick fog.

pebble *n.* **1.** a man or animal which cannot be cowed or beaten. —*phr.* **2. (not) the only pebble on the beach,** (not) the only person or thing to be considered; (not) the only possibility.

peck *n.* a hasty kiss.

pecker *n.* **1.** *U.S.* the penis. —*phr.* **2. keep one's pecker up,** to remain cheerful; maintain good spirits, courage, or resolution.

peckish *adj.* feeling rather hungry.

peddle *v.* to give (as information); tell.

pedestal *phr.* **set on a pedestal,** to idealise.

pee *v.* **1.** to urinate. —*n.* **2.** an act of urination. —*phr.* **3. have a pee,** to urinate.

peel *v.* **1.** to undress. —*phr.* **2. keep one's eyes peeled,** to keep a close watch.

peeler *n. Obsolete.* a policeman. [named after Sir Robert *Peel* 1788-1850, British Prime Minister, who, as Home Secretary, founded the London Metropolitan Police]

peepers *pl. n.* the eyes.

peeve *v.* **1.** to vex or make cross. —*n.* **2.** an annoyance: *my pet peeve.*

peg *n.* **1.** a leg, sometimes one of wood. **2.** an occasion; reason: *a peg to hang a grievance on.* **3.** a degree: *to come down a peg.* **4.** a drink, usually made of whisky or brandy and soda water. —*v.* **5.** to aim or throw. **6.** to observe; identify the true nature of: *to have someone pegged.* —*phr.*

7. off the peg, (of a garment) available for immediate use; ready-made.

8. peg out, to die.

9. take down a peg, to humble.

10. square peg in a round hole, a misfit.

peggie *n.* the man who acts as nipper for wharf labourers.

pelf *n.* money or riches.

pelican's breakfast *n.* See **breakfast.**

pelican shit *phr.* **a long streak of pelican shit,** a very tall and thin person.

pen *n.* prison.

pen and ink *n.* an alcoholic drink. [rhyming slang]

penciller *n. Horseracing.* a bookmaker's clerk, who fills in racing ledgers and betting sheets.

penguin *n.* a nun. [from the similarity of the markings of the bird to the black and white nun's habit]

penny *n.* **1.** an unspecified sum of money: *I haven't got a penny.* —*phr.*

2. a bad penny, a bad, or undesirable person or thing.

3. a pretty penny, a considerable amount of money.

4. spend a penny, to go to the toilet.

5. the penny drops, the explanation or remark is understood.

6. turn an honest penny, to earn an honest living; earn money honestly.

penny-dreadful *n.* a cheap, lurid story, as of crime or adventure. Also, **dreadful.**

penny pincher *n.* a mean, niggardly person.

pen-pusher *n.* one who works with his pen, especially a clerk engaged in work considered to be drudgery.

pep *n.* **1.** spirit or animation; vigour; energy. —*phr.* **2. pep up,** to give spirit or vigour to. [short for *pepper*]

peppy *adj.* energetic.

pep talk *n.* a vigorous talk to a person or group calculated to arouse support for a cause, increase determination to succeed, etc.

percentage *n.* gain; advantage.

Percy *n.* **1.** the penis. —*phr.* **2. point Percy at the porcelain,** (of a male) to urinate.

peril *n.* See **red peril, yellow peril.**

perish *phr.* **do a perish, a.** to die. **b.** to suffer greatly from cold, hunger, etc.

perished *adj.* weakened or exhausted by cold or hunger.

perisher *n.* **1.** a bitterly cold day. **2.** an annoying or mischievous child.

perishing *adj.* **1.** bitterly cold. **2.** unpleasant; objectionable. *—adv.* **3.** very; extremely.

perk[1] *v.* to percolate.

perk[2] *n.* a fringe benefit or bonus; something given in addition to one's normal salary. [from *perquisite*]

perk[3] *phr.* **perk up,** to vomit.

pernickety *adj.* **1.** fastidious; fussy. **2.** requiring painstaking care.

person *n.* one not entitled to social recognition or respect: *that person!*

personal *phr.*

1. be personal, to make disparaging remarks about a person rather than directing oneself to an argument.

2. get personal, to touch on intimate or private matters.

persuader *n. Horseracing.* a jockey's whip.

persuado *adj., n.* a humorous variant of 'pseudo'; sham.

perv *n.* **1.** a pervert. *—phr.* **2. have a perv,** to look at lustfully. Also, **perv on, perve.**

pesky *adj. Chiefly U.S.* troublesome; annoying.

pestiferous *adj.* mischievous, troublesome, or annoying.

pet *v.* to fondle or caress.

peter *n.* **1.** a till; cash register. **2.** a prison cell. *—phr.*

3. black peter, a punishment cell devoid of light or furniture.

4. tickle or **rat the peter,** to ring up false amounts on a cash register, so as to pocket the extra money.

peter thief *n. Prison.* a prisoner who steals from other prisoners' cells.

peth *n.* pethidine, an analgesic.

petticoat *n.* a woman or girl.

pew *n.* a chair; any place to sit down: *Take a pew.* [from *pew* a benchlike seat in a church]

phiz *n.* the face. [short for *physiognomy*]

phizzgig *n.* See **fizgig.**

phone *n.* **1.** a telephone. *—phr.* **2. the phone,** *Prison.* empty sewerage pipes which are used as a means of communication between cells.

phoney *adj.* **1.** not genuine; spurious, counterfeit, or bogus; fraudulent. *—n.* **2.** a counterfeit or fake. **3.** a faker. Also, **phony.** [variant of *fawney* ring (used in confidence trick) from Irish *fāinne*]

phooey *interj. Originally U.S.* an exclamation denoting contempt, disbelief, rejection, etc. [variant of *phew*]

phut *phr.* **go phut,** to collapse, become ruined. [from Hindi *phatnā* to explode]

physical jerks *pl. n.* physical exercises, usually performed without apparatus, to improve the health of the body. Also, **jerks.**

physio *n.* **1.** physiotherapy. **2.** a physiotherapist.

pi *adj. British.* pious, especially hypocritically or smugly so.

pianna *n.* **1.** a piano. *—phr.* **2. play the pianna,** to ring up false amounts on a cash register, so as to pocket the extra money.

Picadilly *adj.* **1.** silly. **2.** chilly. [rhyming slang]

picayune *n.* **1.** an insignificant person or thing. *—adj.* **2.** Also **picayunish.** of little value or account; small; petty. [from *picayune* a Spanish coin formerly used in some southern states of the USA]

Piccadilly bushman *n.* (formerly) one who owns a country property in Australia but who lives in London's West End.

pick[1] *n.* **1.** a hypodermic needle used for taking drugs. *—phr.*

2. pick at, a. to find fault with, in a petty way. **b.** to eat very little of.

3. pick holes in, to criticise; find fault with.

4. pick on, a. to annoy; tease; criticise or blame. **b.** to single out (a person) especially for an unpleasant task.

5. pick (someone's) brains, to find out as much as one can, from someone else's knowledge of a subject.

6. pick to pieces, to criticise, especially in petty detail.

7. pick up, a. to acquire casually. **b.** to become acquainted with informally or casually. **c.** to improve. **d.** to arrest.

8. pick up the threads, See **threads.**

pick[2] *n.* an anchor.

pickie *n.* **1.** picture, as a photograph, illustration, etc. **2.** a motion picture. **3.** (*pl.*) a cinema. Also, **piccie, pikkie.**

pickle *n.* **1.** a predicament. *—phr.*

2. a rod in pickle, *Horseracing.* a horse which is kept for a race it is thought certain to win.

3. have a rod in pickle, have a punishment ready.

pickled *adj.* drunk.

pick-me-up *n.* **1.** a stimulating or refreshing drink, especially alcoholic. **2.** any restorative, such as a meal or drink.

pick-up *n.* **1.** an informal or casual acquaintance, especially one made in the hope of sexual adventure. **2. a.** an improvement. **b.** a pick-me-up.

picky *adj.* petty; mean-minded; obsessed with detail.

picnic *n.* **1.** an enjoyable experience or time. **2.** an easy undertaking. **3.** an awkward situation; a hullaballoo. *—phr.* **4. be no picnic,** (of an event, chore, etc.) to be difficult or unpleasant: *Compiling a dictionary is no picnic.*

picture *phr.*

1. get the picture or **be in the picture,** to understand the situation.

2. put in the picture, to make fully cognisant; inform.

3. the pictures, a cinema.

piddle *v.* **1.** to urinate. **2.** to do anything in a trifling or ineffective way; dawdle.

pie[1] *phr.*

1. have a finger in every pie, to have an interest in or play a part in many affairs.

2. pie in the sky, the illusory prospect of future benefits.

pie[2] *N.Z.* *—adj.* **1.** good; straight. *—interj.* **2.** an exclamation indicating delight, approval, etc. *—phr.*

3. be pie on, to be good at or keen on. [Maori *(e) pai ana*]

piece *n.* **1.** (*offensive*) a woman: *She's a nice little piece.* **2.** *Prison.* a concealable firearm. *—phr.*

3. a piece of cake or **piss,** an easily achieved enterprise or undertaking.

4. a piece of one's mind, outspoken criticism or reproach.

5. piece of work, a person, considered as an example of a specified quality: *a nasty piece of work.*

6. take a piece out of, to reprimand severely.

pie-eater *n.* (*sometimes offensive*) an Australian. [from a supposed preference for meat pies, seen as a characteristic Australian dish]

pie-eyed *adj.* drunk.

piffle *n.* **1.** nonsense; idle talk. *—v.* **2.** to talk nonsense.

piffling *adj.* trivial; petty; nonsensical.

pig[1] *n.* **1.** a person or animal of piggish character or habit. **2.** (*offensive*) a policeman. **3.** *Rugby Union.* a forward. *—v.* **4.** Also, **pig it,** to live, lie, etc., as if in a pigsty; live in squalor. *—phr.*

5. buy a pig in a poke, to enter into something, as a purchase, job, etc., without knowing all the facts.

6. home on the pig's back, successfully by an easy margin.

7. make a pig of oneself, to over-indulge oneself, as by eating too much.

8. pigs! Also, **(in a) pig's arse/bum/ear,** an exclamation of contempt, derision, denial, etc.

pig[2] *n. N.Z.* a demijohn or flagon of beer.

pigeon *n.* **1.** responsibility; concern: *That's his pigeon.* **2.** a dupe.

pig island *n.* New Zealand. [from the wild pigs said to have been released there by James Cook, 1728-79, English navigator and explorer]

pig's ear *n.* **1.** beer. [rhyming slang] *—phr.* **2. in a pig's ear,** See **pig.**

pigskin *n.* a saddle.

pig-swill *n.* inferior or unpleasant food. Also, *N.Z.* pig-tucker.

pike[1] *phr.*

1. pike on, to let down; abandon: *Don't pike on me now.*

2. pike out, to go back on an arrangement; to opt out: *He piked out on the deal.*

pike[2] *v.* to go quickly.

piker[1] *n.* **1.** one who opts out of an arrangement or challenge or does not do his fair share. **2.** one who, from lack of courage or from diffidence, does anything in a contemptibly small or cheap way. **3.** one who gambles, speculates, etc., in a small, cautious way.

piker[2] *n.* a wild bullock.

pikestaff *phr.* **as plain as a pikestaff,** unmistakably clear.

pikkie *n.* See **pickie.**

pile *n.* **1.** a large number, quantity, or amount of anything: *A pile of things to do.* **2.** a large accumulation of money. *—phr.*

3. pile on the agony or **pile it on,** to exaggerate.

4. pile up, (of a vehicle, driver, etc.) to crash.

pile-driver *n.* a powerful punch, kick, stroke, etc.

pile-up *n.* **1.** a crash or collision, usually involving more than one vehicle. **2.** an accumulation; backlog.

pill *n.* **1.** a disagreeable, insipid person. **2.** *Sports.* a ball, especially in football, tennis, etc. *—phr.*

4. sugar the pill, to make bearable some unpleasant experience.

5. the pill, a form of oral contraceptive.

pillow biscuit *n.* See **squashed fly biscuit.**

pimp *n.* **1.** an informer; a tale-bearer. **2.** one who solicits for a prostitute. *—v.* **3.** to inform; tell tales. **4.** to solicit for a prostitute.

pin[1] *n.* a leg.

pin[2] *phr.* **pull the pin,** to cause trouble. [from the *pin* in a hand grenade]

pinch *v.* **1.** to steal. **2.** to arrest. *—n.* **3.** an arrest. **4.** a theft. *—phr.*

5. with a pinch of salt, with some reserve; without wholly believing.

6. at a pinch, in an emergency, crisis, etc.; if necessary.

pineapple *n.* **1.** *Military.* a bomb or hand granade, especially of the fragmentation type, resembling a pineapple in appearance. *—phr.* **2. rough end of the pineapple,** a raw deal; the worst part of a bargain.

pinhead *n.* a stupid person.

pink *v.* *Shearing.* to shear so closely that the skin of the sheep is exposed.

pink elephant *n.* a hallucination, especially as reputedly experienced by alcoholics.

pinkeye *n.* a holiday, celebration; festival. Also, **pink-hi.** [Aboriginal]

pinkie *n.* **1.** a communist sympathiser. **2.** *Vic.* a parking ticket. **3.** the little finger or toe. **4.** cheap red wine.

pinnie *n.* a pinball machine.

pint *n.* *British.* a pint of beer.

pintail *n.* a surfboard with a pointed tail.

pint-size *adj.* (of a person, etc.) small or insignificant.

pin-up *n.* **1.** a picture, typically pinned to the wall by a personally unknown admirer, of an attractive member of the opposite sex, especially a film star, or a nude or nearly nude girl. **2.** the girl or man depicted.

pip[1] *n.* *Military.* a badge of rank worn on the shoulders of certain commissioned officers.

pip[2] *n.* **1.** a humorous term for any minor ailment in a person. *—v.* **2.** to annoy. *—phr.* **3. give (someone) the pip,** to annoy; irritate, especially without intention: *His stupidity gives me the pip.*

pip[3] *v.* **1.** to beat in a race, etc., especially by a small margin: *The favourite was pipped at the post.* **2.** to hit with a missile, as by shooting.

pipe *phr.*

1. pipe down, to become or keep quiet.

2. pipe up, a. to begin to talk, especially unexpectedly. **b.** to make oneself heard. **c.** to speak up, as to assert oneself.

pipeline *phr.* **in the pipeline** or **chute,** on the way; in preparation.

pip-emma *n.* p.m. [used by radio operators. See **ack-emma**]

pipe-opener *n.* See **heart-starter.**

pipes *pl. n.* the respiratory passages.

pipsqueak *n.* a small or insignificant person or thing.

piss *v.* **1.** to urinate. **2.** to rain heavily: *It was pissing down.* —*n.* **3.** urine. **4.** an act of passing water; urination. **5.** alcoholic drink, especially beer. —*adj.* **6.** very: *piss-awful, piss-easy.* —*phr.*
7. all piss 'n' wind, loquacious, but insincere.
8. on the piss, on a drinking spree.
9. piece of piss, See **piece.**
10. piss about or **around,** to mess about.
11. piss (something) away, to waste; squander.
12. piss in (someone's) pocket, to behave obsequiously towards (someone).
13. piss into the wind, to embark on a futile course of action.
14. piss off, (*sometimes offensive*) to go away.
15. piss (someone) off, a. to send (someone) away. **b.** to annoy (someone) intensely.
16. piss on, to drink considerable quantities of liquor, especially beer.
17. piss (all) over, to beat or confound utterly.
18. (one) wouldn't piss on (something or someone), to hold (something or someone) in utter contempt: *I wouldn't piss on him if he was on fire.*

pissant *phr.*
1. drunk as a pissant, very drunk.
2. game as a pissant, very brave.

pissaphone *n. Military.* (in World War II, especially in tropical areas) a large funnel resembling the mouthpiece of a wall telephone, set into the ground for use as a urinal.

pissed *adj.* **1.** drunk. —*phr.* **2. pissed as a parrot; pissed to the eyeballs; pissed from arsehole to breakfast time,** very drunk.

pissed-off *adj.* disgruntled; fed up; thoroughly discontent.

piss-elegant *adj.* with pretensions to elegance.

pisser *n.* **1.** a pub. **2.** a urinal.

piss-fart *v.* to waste time.

pisspot *n.* a drunkard. Also, **piss-head.**

piss-up *n.* an occasion on which a large quantity of alcohol is consumed by a group of people, as at a party, etc.

piss-weak *adj.* **1.** mean; despicable; shabby: *a piss-weak thing to do.* **2.** inadequate; disappointing; not up to standard. **3.** of weak character; cowardly; irresolute. Also, **piss-poor.**

pissy *adj.* **1.** unpleasant. **2.** mildly drunk.

pit *n.* **1.** a bed. —*phr.* **2. the pits,** the most unpleasant or most obnoxious (place, circumstance, condition, etc.).

pitch *phr.*
1. pitch a line, to attempt to impress by boastful and sometimes untruthful speech, often as a means of winning sexual favours from a woman.
2. pitch a tale/yarn/etc., to tell a story, especially one that is exaggerated or untrue.
3. pitch in, a. to contribute or join in. **b.** to begin vigorously, as eating, etc.

4. pitch into, a. to attack verbally or physically. **b.** to begin to do or work on (something).
5. queer someone's pitch, to upset someone's plans.

pitched battle *n.* any violent fight involving many people.

pit stop *n.* any stop made on a journey in a motor car, for refuelling, refreshment, etc.

Pitt Street *phr.*
1. not know whether it's Pitt Street or Christmas, to be in a state of confusion.
2. Pitt Street farmer, (in New South Wales) one who owns a country property, often for purposes of tax avoidance, but who lives and works in Sydney. See also, **Collins Street cocky, Queen Street bushie.** [from *Pitt Street* a major street in the city of Sydney]

pixie *phr.* **away** or **off with the pixies** or **fairies, a.** daydreaming. **b.** mentally unsound. **c.** incapacitated by alcohol.

P.J.s *pl. n.* pyjamas.

plagon *n.* a flagon. [Aboriginal pidgin for *flagon*]

plagon wagon *n.* (*offensive*) a car full of Aborigines.

planet *phr.* **off the planet,** fantastic; wonderful.

plank *v.* **1.** to lay or put down: *Just plank it down somewhere.* —*n.* **2.** a type of surfboard.

plant *n.* **1. a.** something or someone intended to trap, decoy, or lure. **b.** a spy. **2.** something hidden, often illegally. **3.** a place where stolen goods are hidden. **4.** a scheme to trap, trick, swindle, or defraud. —*v.* **5.** to deliver (a blow, etc.). **6.** to hide or conceal, as stolen goods. **7.** to place (evidence) so that, when discovered, it will incriminate an innocent person. **8.** to put (gold dust, ore, etc.) in a mine or the like to create a false impression of the value of the property. —*phr.* **9. plant one's foot,** to quickly accelerate a car, etc.

plaster *v.* **1.** to hit hard and often. **2.** to bomb heavily. **3.** *Gambling.* to wager a large sum of money: *to plaster the favourite.*

plastered *adj.* drunk.

plastic *adj.* (of people, society, etc.) artificial; fake.

plate *n.* **1.** a contribution of food brought to a social gathering: *Bring a plate, but the grog is on the house.* —*phr.*
2. on a plate, or **platter,** (of something offered) capable of being taken or dealt with without effort.
3. on one's plate, waiting to be dealt with; pending.

plate face *n.* (*offensive*) an Asian.

plates of meat *pl. n.* feet. [rhyming slang]

play *phr.*
1. play around, a. to philander. **b.** to be sexually promiscuous.
2. play ball, to cooperate.
3. play cat and mouse, a. to delay the inevitable defeat of an opponent so as to enjoy observing his struggles and discomfiture. **b.** (in racing) to speed up and slow down as a tactic to gain an advantage over the rest of the field.
4. play for keeps, to expect a decision, result, etc., to be permanent.
5. play (it) cool, to act cautiously.
6. play silly buggers, to act the fool.

7. **play the field, a.** to have as many flirtations as possible. **b.** to keep oneself open to advantage from a number of sources.
8. **play the game,** to play in accordance with the rules.
9. **play up, a.** to behave naughtily or annoyingly. **b.** to philander.
10. **play up to,** to attempt to get into the favour of.

played-out *adj.* exhausted; used up.

pleb *n.* **1.** a commonplace or vulgar person. *—adj.* **2.** vulgar, commonplace. [shortened form of *plebeian* from Latin *plebs* the commons of ancient Rome as contrasted with the patricians]

plenty *adv.* fully: *plenty good enough.*

plod *n.* **1.** a time sheet. **2.** a long, rambling narrative.

plonk[1] *v.* **1.** put: *Plonk it here.* *—adv.* **2.** exactly: *plonk in the middle.*

plonk[2] *n.* **1.** any alcoholic liquor, especially cheap wine. *—phr.* **2. plonk artist/fiend/freak/merchant,** an addict of cheap wine.

plonko *n.* a wine addict. Also, **plonkie.**

plop *n.* (*in children's speech*) faeces. Also, **plop plop.**

plough *n.* **1.** *British.* an examination failure. *—v.* **2.** *British.* **a.** to fail (someone) in an examination. **b.** to fail an examination. *—phr.*
3. plough back, to reinvest (profits of a business) in that business.
4. plough into, to attack energetically; to throw oneself into. Also, *Chiefly U.S.,* **plow.**

pluck *v.* **1.** to rob, plunder, or fleece. **2.** *British.* to reject, as after an examination.

plug *n.* **1.** the favourable mention of a product or the like on radio, television, etc.; an advertisement, especially unsolicited. **2.** *U.S.* a worn-out or unsaleable article. **3.** a punch. **4.** a worn-out or inferior horse. **5.** *U.S.* a man's tall silk hat. *—v.* **6.** to mention (a publication, product or the like) favourably and, often, repetitively as in a lecture, radio show, etc. **7.** to punch. **8.** to shoot. **9.** to work steadily or doggedly.

plughole *phr.* **down the plughole,** wasted (as effort, money, etc.); ruined.

plug-ugly *U.S.* *n.* **1.** a ruffian; a rowdy; a tough. *—adj.* **2.** characteristic of ruffians or the like.

plumb *adj.* **1.** downright or absolute. *—adv.* **2.** completely or absolutely. Also, **plum.**

plummy *adj.* choice, good, or desirable.

plunge *v.* **1.** to bet or speculate recklessly. *—phr.* **2. take the plunge,** to resolve to do something (usually unpleasant) and to act straightaway.

plunger *n.* a reckless punter or speculator.

plunk *v.* **1.** to shoot at. *—n.* **2.** *U.S.* a dollar.

pocket *phr.*
1. in one's pocket, under one's control.
2. line one's pockets, to gain, especially financially, at the expense of others.

pod *phr.* **in pod,** pregnant.

poddy calf *n.* half. [rhyming slang]

poddy-dodger *n.* one who steals unbranded calves.

podge *n.* a podgy person; one who is short and fat.

poet's day *n.* Friday, the day on which people often leave work a little early. [*p(iss) o(ff) e(arly) t(omorrow's) s(aturday)*]

po-faced *adj.* expressionless.

point *phr.* **point the bone,** See **bone.**

poisoner *n.* a cook, especially a shearers' cook, etc.

poke[1] *v.* **1.** to have sexual intercourse with. *—n.* **2.** a blow with the fist. **3.** the act of sexual intercourse. *—phr.*

4. better than a poke in the eye with a burnt stick, See **burnt.**

5. more than one can poke a stick at, a lot of; many; much.

6. poke fun/mullock/borak at, to ridicule or mock, especially covertly or slyly.

7. poke one's nose in, to interfere; pry; show too much curiosity.

8. take a poke at, to aim a blow at.

poke[2] *phr.* **buy a pig in a poke,** See **pig.**

poker face *n.* an expressionless face. [from the card game *poker* in which the players must not reveal by facial expressions the value of the cards in their hands]

poker machine *n.* a coin-operated gambling machine, usually operated by pulling a handle, the score being shown in the form of replicas of three or more playing cards, pictures of fruit, etc. Also **fruit machine.**

pokey *n.* gaol.

pokie *n.* a poker machine.

Polack *n. Chiefly U.S.* (*offensive*) a person of Polish descent.

pole[1] *phr.*

1. up the pole, a. in a predicament. **b.** slightly mad. **c.** completely wrong.

2. pole on, to impose on by loafing or cadging (originally of a horse, bullock, etc., which had the position of poler).

3. would not touch (something) with a forty foot or **barge pole,** to completely reject (something).

pole[2] *phr.* **poles apart,** having completely opposite or widely divergent views, interests, etc.

poler *n.* a lazy person; loafer. [from *poler* one of two bullocks or horses harnessed to the pole of a wagon; the polers do not take as much weight as the leaders, but are important in steering the wagon]

poley *n.* **1.** a dehorned or hornless animal; a polled beast. **2.** a saddle without kneepads. *—adj.* **3.** (of a cup) having lost its handle: *a poley cup.* **4.** (of an animal) dehorned or hornless. **5.** (of a saddle) without kneepads.

polish *v.* **1.** to praise. *—phr.*

2. polish off, to finish, or dispose of quickly: *to polish off an opponent.*

3. polish up, to improve.

political *adj.* interested in politics: *Sheila is not political.*

pollie *n.* a politician.

pollyanna *n.* a girl of unreasonable optimism, cheerfulness and goodwill. [after the chief character in *Pollyanna* (1913), a novel by Eleanor Porter, 1868-1920, American writer]

polony *n. W.A.* bologna sausage or devon.

pommy *n.* **1.** an Englishman. *—adj.* **2.** English. Also, **pom.** [origin uncertain]

pommyland *n.* England.

pommy's towel *phr.* **dry as a pommy's towel,** See **dry.**

pommy wash *n.* (*humorous*) a bathe with a wet sponge, rather than immersing in water.

ponce *n.* **1.** a pimp. **2.** a dandy, often effeminate. *—v.* **3.** to act as a pimp. *—phr.*
4. all ponced up, spruced up.
5. ponce about, to flounce; behave in a foolishly effeminate fashion.

pong *n.* **1.** a stink; unpleasant smell. *—v.* **2.** to stink.

Pong *n.* a Chinese.

pongo[1] *adj.* smelly.

pongo[2] *N.Z. n.* **1.** a British soldier. **2.** an Englishman. *—adj.* **3.** English.

Pongolia *n. N.Z.* Britain. [from *pongo* modelled on *Mongolia.* See **pongo**]

pongy *adj.* smelly.

pony *n.* **1. a.** (formerly) the sum of £25. **b.** the sum of $25. **2.** a small glass for beer or spirits. **3.** *U.S.* a crib for study purposes. *—v.* **4.** *U.S.* to prepare (lessons) by means of a crib. *—phr.* **5. pony up,** to pay (money), as in settling an account.

poo *n.* **1.** a euphemistic term for faeces. *—phr.* **2. in the poo,** in trouble or bad favour. Also, **pooh.**

pooch *n.* a dog.

poofter *n.* **1.** a male homosexual. **2.** (*offensive*) a person, especially one who is weak or cowardly. Also, **poof.**

poofter-basher *n.* one who participates in poofter-bashing.

poofter-bashing *n.* **1.** assault on a male homosexual. **2.** verbal attacks on men in public life reputed to be homosexuals.

poofter-rorter *n.* **1.** a procurer of male homosexuals. **2.** one who assaults and robs homosexuals.

pooh *n.* See **poo.**

Pooh-Bah *n.* a pompous person, especially an official. [from *Pooh-Bah* the Lord-High-Everything-Else in Gilbert and Sullivan's opera 'The Mikado']

poohey *adj.* disagreeable; unpleasant.

pool *v.* **1.** to inform on (someone). *—phr.* **2. in the pool,** in trouble; informed on.

poon *n.* (*offensive*) a stupid, useless person; an idiot; fool.

poonce *n.* (*offensive*) a male homosexual. [variant of *ponce*]

pooncey *adj.* effeminate.

poop[1] *v.* to tire or exhaust.

poop[2] *n.* excrement.

poopcatchers *pl. n.* loosely fitting short breeches, gathered at the knee; plus-fours; applecatchers.

pooped *adj.* exhausted.

poor white trash *n.* (*offensive*) poor whites collectively.

poove *n.* a poofter. Also, **pouf, pouffe.**

pop[1] *v.* **1.** to fire (a gun, etc.). **2.** to pawn. *—n.* **3.** each: *They cost five dollars a pop.* *—phr.*
4. pop off, a. to depart, especially abruptly. **b.** to die, especially suddenly.

5. pop the question, to propose marriage.

pop² *adj.* popular.

pop³ *n.* father, or grandfather.

pop-out *n.* a mass-produced surfboard.

poppa *n.* father.

popper *n.* a press-stud.

poppy *phr.* **tall poppy,** a person who is pre-eminent in a particular field; a person with outstanding ability, wealth, status. [in 1931 J.T. Lang, Premier of New South Wales, applied the term *tall poppy* to all those on government salaries above £10 per week]

poppycock *n.* nonsense; bosh.

poppy show *n.* an indecorous showing of a girl's or woman's upper legs and underwear.

pop-shop *n.* pawnshop.

popsy *n.* a girl, especially a sexually attractive one.

p.o.q. *v.* to depart in a hurry. Also, **P.O.Q.** [*p(iss) o(ff) q(uick)*]

porangi *adj. N.Z.* mad or crazy.

pork *n. U.S.* appropriations, appointments, etc., by the government for political reasons rather than for public necessity.

pork and bean *n.* a homosexual. Also, **pork'n.** [rhyming slang *pork and bean* queen. See **queen**]

pork barrel *n.* a government appropriation, bill, or policy which supplies funds for local improvements designed to ingratiate legislators with their constituents.

pork chop *phr.*
1. like a pork chop at a bar mitzvah, out of place; ill at ease.
2. off like a pork chop in the sun, departing hurriedly.

porky *adj.* fat.

porno *n.* **1.** Also, **porn.** pornography. **2.** one who delights in pornography. *—adj.* **3.** pornographic: *a porno film.*

porn shop *n.* a sex shop.

porridge *phr.*
1. do porridge, *British.* to do time in gaol.
2. save one's breath to cool one's porridge, See **breath.**
3. stir the porridge, to take one's turn relatively late in a pack-rape.

port *n.* **1.** a suitcase. **2.** *Qld.* a shopping bag. [shortened form of *portmanteau*]

Port *n.* a local name for various towns, suburbs or areas, such as Port Macquarie, Port Douglas, etc.

posh *adj.* **1.** elegant; luxurious; smart; first-class. *—phr.* **2. posh up,** to make smart or elegant. [origin uncertain; a popularly-held belief is that it comes from the initials *p(ort) o(ut) s(tarboard) h(ome),* with reference to the better (i.e. cooler) accommodation on vessels sailing from Britain to India, Australia, etc.]

positive *adj.* downright; out-and-out. Also, **positively.**

possie *n.* a place; position. Also, **pozzy.**

possum *phr.*
1. play possum, to dissemble; feign illness or death.
2. stir the possum, See **stir.**

post and rail *n.* a lie; a tall story. [rhyming slang *post and rail* tale]

post-and-rail tea *n.* strong bush tea.

poster *n. Australian Rules.* a kick which hits one of the goalposts, scoring a point.

postholes *phr.* **a load of postholes,** (used by road hauliers, etc.) an empty load.

postie *n.* a postman.

pot *n.* **1.** a large sum of money. **2.** *Horseracing.* a heavily backed horse; favourite. **3.** a medium sized beer glass; middy. **4.** the contents of such a glass. **5.** an important person: *a big pot.* **6.** a trophy or prize in a contest, especially a silver cup. **7.** (*pl.*) a large quantity. **8.** marijuana. —*v.* **9.** to take a pot shot; shoot. **10.** to capture, secure, or win. **11.** to put (a young child) on a potty. **12.** to make pregnant. —*phr.*
13. go to pot, to deteriorate.
14. put (someone's) pot on, to inform against.

potato *phr.* **not the clean potato, a.** (of a person) of questionable background, reputation, etc., as (formerly) an exconvict. **b.** (of an action, etc.) of dubious legality.

potatoes *pl. n.* **1.** See **small potatoes.** —*phr.* **2. strain the potatoes,** to urinate.

potato peeler *n.* a girl or woman; girlfriend. [rhyming slang *potato peeler* sheila. See **sheila**]

potboiler *n.* an inferior work of literature, piece of music, film, etc., produced merely for financial gain.

pothead *n.* a person who smokes marijuana frequently.

potluck *n.* a random or haphazard choice. [from *pot luck* whatever food happens to be at hand without special preparation or buying]

potted *adj.* **1.** abridged, summarised, or condensed: *a potted version.* **2.** *U.S.* drunk.

potty[1] *adj.* **1.** foolish; crazy. **2.** *British.* paltry; petty.

potty[2] *n.* a chamber-pot, especially one for a child. Also, **pottie.**

pot-valiant *adj.* brave only when drunk.

pot-walloper *n.* a heavy drinker.

poultice *n.* **1.** a large amount: *He has a poultice of money.* **2.** a mortgage.

pound *n. Prison.* **1.** solitary confinement. —*phr.* **2. do pound,** to be locked on one's own in a cell as a punishment for an offence committed whilst in gaol.

powder *phr.*
1. powder one's nose, (of a female) to visit the ladies' room or toilet.
2. take a powder, to depart; disappear.

power *n.* **1.** a large number or amount. —*v.* **2.** to move with great force: *He powered his way through the crowd.*

powerful *adj.* great in number or amount: *a powerful lot of money.*

powerhouse *adj.* strong; powerful: *a powerhouse tackle.*

powwow *n.* **1.** any conference or meeting. —*v.* **2.** to confer. [from Algonquian *pow wah* or *po-wah* a North American Indian ceremony, performed for the cure of disease, success in a hunt, etc.]

pox *n.* any venereal disease.

pox doctor *phr.* **dressed up like a pox doctor's clerk,** dressed flashily, but in poor taste.

pozzie *n.* See **possie.**

P.P. *n.* (usually among school teachers) a proud parent.

prac *adj.* practice: *prac teaching.*

prad *n.* a horse.

prang *v.* **1.** to crash (a car or the like). —*n.* **2.** a crash, especially a minor one, in a motor vehicle or the like. [from British *prang* to crash-land an aircraft]

pratfall *n. U.S.* an undignified fall on the buttocks, especially on the stage.

prawn *n.* **1.** a weak, spiritless, insignificant person: *He's a bit of a prawn.* —*phr.* **2. come the raw prawn** or **uncooked crustacean,** to try to deceive; delude: *Don't come the raw prawn with me.*

prawnie *n.* a prawn fisherman.

prawn night *n.* a social function at a club at which prawns and beer are served.

preachify *v. Chiefly U.S.* (*usually offensive*) to preach in an obtrusive or tedious way.

precious *adv.* extremely; very: *precious little.*

preggers *adj.* pregnant. Also, **preggie, preggies, preggo, prego.**

prelim *n.* any event, as an examination or sporting contest, which is preliminary to the main event. [abbreviation of *preliminary*]

pre-loved *adj.* second-hand.

premmie *n.* **1.** an infant born prematurely. —*adj.* **2.** of or pertaining to such an infant. Also, **prem, premie.**

prep *adj.* **1.** preparatory: *a prep school.* —*n.* **2.** preparation of schoolwork. —*v.* **3.** to prepare (a patient) for an operation.

Presbo *n.* a Presbyterian.

press *phr.* **go to press,** *Prison.* to put a statement in writing incriminating oneself or others.

pressie *n.* a present.

pretty *adj.* **1.** considerable; fairly great. —*phr.*

2. pretty penny, a considerable sum of money.

3. sitting pretty, in a satisfactory and unchallenged position.

prezzie *n.* a gift. [shortened form of *present*]

Prezzie *n.* a Presbyterian.

price *v.* **1.** to ask the price of. —*phr.* **2. what price, a.** what is the chance of. **b.** what do you think of.

priceless *adj.* delightfully amusing; absurd.

pricey *adj.* expensive.

prick *n.* **1.** the penis. **2.** an unpleasant or despicable person. —*phr.* **3. kick against the pricks,** to hurt oneself by vain resistance.

pricker *phr.* **have the pricker,** to be in a temper; to be very angry.

prickteaser *n.* one who withholds sexual favours from a man after having encouraged expectation of them. Also, **cockteaser.**

prima donna *n.* a temperamental, petulant person.

primed *adj.* drunk.

prissy *adj.* precise; prim; affectedly nice.

private eye *n.* a private investigator.

pro[1] *n.* a professional.

pro[2] *n.* a prostitute.

Prod *n.* **1.** a Protestant. —*adj.* **2.** Protestant. Also, **Proddie.**

prodgie *n.* a produce store.

prof *n.* professor.

prole *n.* a member of the proletariat. Also, **prol.** [shortened form]

promise *phr.* **be on a promise,** (of a man) to be assured of a particular sexual partner: *He was on a promise in every second house.*

pronto *adv.* promptly; quickly.

prop[1] *v.* to stop suddenly.

prop[2] *n.* a propeller.

proper *adj.* complete or thorough: *a proper thrashing.*

properly *adv.* completely.

proposition *n.* **1.** a proposal for sexual intercourse. —*v.* **2.** to propose sexual intercourse to.

props *pl. n.* the legs.

prossie *n.* a prostitute.

protection *n.* **1.** immunity from prosecution or harassment obtained by a person involved in illegal activities by means of bribes to appropriate officials. **2.** money paid to criminals as a guarantee against threatened violence. **3.** *Prison.* segregation in solitary confinement or the like because of the likelihood of violence from other prisoners.

proud *phr.* **do proud,** to treat lavishly.

pseud *n. Chiefly British.* a person who pretends to be what he is not. [from *pseudo* false]

psych *v.* **1.** to persuade by the application of psychological knowledge and techniques rather than overtly by argument, especially when leading others to perform better in a race, competition, etc.: *The coach psyched them into a brilliant display of tennis.* —*phr.* **2.** **psych (oneself) up, a.** to persuade (oneself) by similar techniques to perform better: *He psyched himself up to go into the exam.* **b.** to gain an advantage over (an opponent) by employing similar techniques, as by making him nervous or unsure.

psychedelic *adj.* having bright colours and imaginative patterns, as materials.

psycho *n.* **1.** an insane person. **2.** a psychopath. —*adj.* **3.** insane or obsessional. **4.** psychopathic.

pub *n.* a hotel. [short for *public house*]

pub crawl *n.* drinking at a series of hotels in succession either alone or in company with others.

pubes *pl. n.* pubic hairs.

pucka *n.* See **pukka.**

puckeroo *n. N.Z.* See **pakaru.**

Puddies, the *pl. n.* the Brothers team in the Queensland Rugby Football Union.

pudding *n.* **1.** Also, **pud, pudden.** a small, fat person. **2.** Also, **pudding head.** a stupid person. —*phr.* **3.** **in the pudding club,** pregnant.

pudding face *n.* a round, fat, smooth face.

puff *phr.* **run out of puff,** to exhaust all resources.

puffed *adj.* out of breath.

puffed-up *adj.* self-important.

pug *n.* a pugilist.

puggim *adj.* Aboriginal pidgin for fucking.

puke *n.* to vomit.

pukka *adj.* **1.** of colonial, especially Anglo-Indian, origin, behaviour, etc. **2.** proper and correct; socially acceptable. Also, **pucka.**

pukunui *n. N.Z.* a big belly.

pull *v.* **1.** to draw out for use, as a knife or a pistol. **2.** *Horseracing.* to prevent (a horse) from running on its merits. **3.** to have sexual intercourse with. *—n.* **4.** an advantage over another or others. **5.** influence, as with persons able to grant favours. **6.** the ability to attract or draw audiences, followers, etc.: *An actor with box office pull. —phr.*

7. pull a fast one or **swiftie,** to deceive; to play a sly trick.

8. pull in, a. to arrest. **b.** to earn (as a wage or salary).

9. pull no punches, to act in a straightforward manner; be brutally honest.

10. pull off, to succeed in achieving or performing (something).

11. pull oneself off, to masturbate.

12. pull one's finger out, to attack a job, task, etc. with energy after a period of inertia or laziness.

13. pull one's head in, to withdraw; to mind one's own business.

14. pull one's punches, to act with more show than effect, as by failing to follow through an initial move.

15. pull one's weight, to make a full and fair contribution to a task or undertaking.

16. pull someone's leg, See **leg.**

17. pull strings, to seek the advancement of oneself or another by using social contacts and other means not directly connected with one's ability or appropriateness.

18. pull the other one, or **leg, (it's got bells on it** or **plays Waltzing Matilda** or **the Marseillaise),** an expression of disbelief, etc.

19. pull the plug on, to prevent (someone) from continuing their present activities as by making some damning revelation, issuing an order, etc.

20. pull the rug from under someone's feet, to place someone in a position of disadvantage.

21. pull the wool over someone's eyes, to deceive; hoodwink.

22. pull through, a. Also, **pull round.** to recover, as from an illness, adversity, or the like. **b.** to make one's way through, as by a pull or effort.

23. pull to pieces, to analyse critically in detail.

24. pull up, a. to finish an activity, etc.: *How did he pull up after the weekend?* **b.** *Horseracing* (of a jockey) to prevent (a horse) from running a race to the best of its ability.

pullie *n.* a pullover.

pulverise *v.* to defeat overwhelmingly, as a fighter.

pump *phr. Surfing.* **the sets are pumping,** the waves are good. See **set.**

punch *n.* **1.** a vigorous, telling effect or force. *—phr.*

2. pack a punch, to have an extreme effect upon someone.

3. pull no punches, See **pull.**

4. pull one's punches, See **pull.**

5. punch the bundy, to operate a time clock; to start or finish a shift.

6. punch one through, (of a man) to have sexual intercourse.

Punch *phr.* **pleased as Punch,** delighted; highly pleased. [from *Punch* the chief character in the puppet show called 'Punch and Judy']

punch-drunk *n.* dull-witted; stupid or dazed.

punch-up *n.* a fight.

punchy *adj.* **1.** punch-drunk. **2.** forceful; vigorously effective.

punish *v.* to make a heavy inroad on (a supply, etc.).

punk *n.* **1.** something or someone worthless, degraded, or bad. **2.** a petty criminal. **3.** a follower of punk rock. *—adj.* **4.** relating to punk rock and an associated style of dress, hairstyle and behaviour.

punk rock *n.* a type of rock music, usually with a fast, energetic beat reminiscent of early rock, which is associated with rebelliousness, aggressiveness, violence and sexuality.

punt[1] *phr.* **have a punt,** to make an attempt.

punt[2] *v.* **1.** to wager. *—phr.* **2. take a punt,** to take a chance.

pup *phr.*

1. be sold a pup, to be the victim of some deception.

2. the night is (still) a pup, it is (still) early in the night.

pure Merino *n.* **1.** (formerly) a free settler of substance who opposed the social advancement of the emancipists. **2.** a member of an old and established Australian family of free, not convict, descent.

purge *n. N.Z.* any alcoholic beverage.

purple *phr.*

1. purple passage, (of prose) full of elaborate literary devices and pretentious effects.

2. purple patch, a period of good fortune.

push *v.* **1.** to place excessive or dangerous strain on: *You're pushing your luck.* *—n.* **2.** perservering energy; enterprise. **3. a.** a group or set of people who have a common interest or background: *the Balmain push.* **b.** (formerly) a gang of vicious city hooligans: *the Rocks push.* **4.** influence; power. *—phr.*

5. pushed for, in difficulties with; lacking: *We're pushed for time.*

6. push it, a. to work harder than normal, as to meet a deadline. **b.** to be exorbitant in one's demands.

7. push off, to leave; go away.

8. push on, to continue; proceed.

9. push shit uphill, to attempt the impossible.

10. push (someone's) barrow, See **barrow.**

11. push up daisies, to be dead and buried.

12. push up zeds, See **zed.**

13. the push, dismissal; rejection; the sack: *She gave him the push.*

pushover *n.* **1.** anything done easily. **2.** an easily defeated person or team.

pushy *adj.* aggressive; presuming.

puss[1] *n.* a girl or woman.

puss[2] *n.* **1.** the face. **2.** the mouth.

pussy *n.* the vulva.

pussyfoot *phr.* **pussyfoot around,** to act cautiously or timidly, as if afraid to commit oneself on a point at issue.

pussy-footer *n.* a sneak thief.

put *v.* **1.** *U.S.* to make off: *to put for home.* *—phr.*

2. put away, a. institutionalised for reasons of mental illness: *He should be put away.* **b.** sent to gaol.

3. **put down** or **away,** to destroy (an animal) especially mercifully, as for reasons of old age, disease, etc.

4. **put in, a.** to betray or report (someone) as for a misdemeanour. **b.** to nominate (someone absent) for an unpleasant task.

5. **put in the boot, a.** to attack savagely by kicking. **b.** to attack without restraint. **c.** to take unfair advantage.

6. **put in the fangs/hooks/nips/screws,** to borrow.

7. **put it across (someone),** to deceive or outwit.

8. **put it on (someone),** to confront (someone) directly on an issue.

9. **put one over,** to deceive; outwit; defraud.

10. **put out, a.** to subject to inconvenience. **b.** to annoy, irritate, or vex. **c.** (of a woman) to be free with sexual favours.

11. **put paid to,** to destroy finally: *Bankruptcy put paid to his hopes of becoming a millionaire.*

12. **put (someone) down,** See **down.**

13. **put the acid on,** to put pressure on (someone) for a favour, especially a loan.

14. **put the hard word on,** See **hard.**

15. **put up to,** to persuade to do.

16. **put up or shut up,** to be prepared to support what one says or else remain silent.

17. **put up the shutters,** to fail in business.

18. **put up with,** to endure; tolerate; bear.

put-down *n.* an insult or humiliation.

put-on *n.* a bluff; pretence.

putty *phr.* **up to putty,** worthless. Also, **upter.**

put-up *adj.* planned beforehand in a secret or crafty manner: *a put-up job.*

pyalla *v.* **1.** to talk. —*n.* **2.** talk. [Aboriginal]

QFRTB *adj.* very full. [*Q(uite) F(ull) R(eady) T(o) B(urst)*]

q.t. *adj.* **1.** quiet. *—phr.* **2. on the q.t.,** secretly.

quack *n.* any medical practitioner.

quandong *n.* **1.** a woman who refuses to have sex after being wined and dined. **2.** a person who cadges or imposes upon another. [unexplained: the *quandong* is an Australian tree yielding a fruit with a seed which has an edible kernel]

quean *n. Obsolete.* a male homosexual.

queen *n.* **1.** a male homosexual. *—phr.* **2. queen (it) up,** to adopt effeminate dress or manner.

Queensland dust *n.* sugar.

Queen Street bushie *n.* (in Queensland) one who owns a country property, often for purposes of tax avoidance, but who lives and works in Brisbane. See also, **Pitt Street farmer, Collins Street cocky.** [from *Queen Street* a major street in the city of Brisbane]

queeny *adj.* effeminate.

queer *adj.* **1.** of questionable character; suspicious; shady. **2.** mentally unbalanced or deranged. **3.** bad, worthless, or counterfeit. **4.** homosexual. *—v.* **5.** to spoil; jeopardise; ruin: *queer one's pitch.* *—n.* **6.** a male homosexual.

queer street *n.* a state of financial embarrassment. Also, **Queer Street.**

quickie *n.* **1.** something produced in a short space of time on a low budget and therefore of inferior quality. **2.** anything taken or done quickly, as a drink, sexual intercourse, etc. **3.** *Cricket.* a fast bowler.

quid *n.* **1.** (formerly) a pound in money, especially £1 as a pound note. **2.** Also, **quids.** money, especially a large amount: *I'll bet that cost a quid or two.* *—phr.*

3. a quick quid, money earned with little effort, often by dishonest means.
4. earn or **make a quid,** to earn some money.
5. have a quid, to be wealthy.
6. not for quids, never; for no inducement at all.
7. not get the full quid, not to obtain the full value for one's money.
8. not the full quid or **dollar,** mentally retarded; dull-witted.
9. turn an honest quid, to earn money by honest means.

quiet *phr.* **on the quiet,** secretly.

quietly *phr.* **just quietly,** in confidence.

quill *phr.* **drive a quill,** to work in an office.

quim *n.* the female genitalia.

quince *n.* **1.** a homosexual who is both active and passive. *—phr.*
2. do one's quince, to lose one's temper.
3. get on one's quince, to annoy, irritate.

quits *phr.* **call it quits, a.** to abandon an activity, especially temporarily. **b.** to give up a quarrel, rivalry, etc.; agree to end a dispute, competition, etc.

quiver *phr.* **a full quiver,** a large family.

quod *n.* a prison.

quoit *n.* **1.** Also, **coit.** the anus. **2.** Also, **coit.** the buttocks.

rabbit[1] *n.* **1.** a bottle of beer. **2.** a fool. **3.** *Cricket.* a poor batsman.

rabbit[2] *phr.* **rabbit on,** talk nonsense, usually at length. [rhyming slang *rabbit and pork* talk]

rabbit ears *pl. n.* an indoor television antenna with two adjustable arms. Also, **rabbit's ears.**

rabbit-killer *n.* a short, sharp blow on the nape of the neck or lower part of the skull. Also, **rabbit-punch, rabbit-chop.**

rabbit-oh *n.* a street vendor selling rabbits. Also, **rabbito.** [from the vendor's cry *rabbit-oh*]

Rabbitohs, the *pl. n.* the South Sydney team in the N.S.W. Rugby Football League. [?from the selling and raffling of rabbits (in the 1930s) to raise funds for the club]

rabbit-proof fence *phr.* **beyond the rabbit-proof fence,** in the far outback. [from the fences constructed in the 19th century in an attempt to halt the spread of rabbits; the best known fences followed the borders between Victoria and South Australia, New South Wales and South Australia, and New South Wales and Queensland]

rabble *v.* **1.** to create an uproar: *The class is rabbling.* —*phr.* **2. the rabble,** (*offensive*) the lowest class of people. [?akin to British *rabble* utter in a rapid confused manner]

race *phr.*
1. not in the race or **hunt,** having no chance at all.
2. race off, to seduce.
3. race off with, to steal.

racehorse *n.* **1.** a thinly rolled cigarette; greyhound. **2.** a swag rolled in a long thin roll, like a cigarette.

rack *phr.* **rack off,** to leave; go: *He racked off ages ago. Rack off, hairy legs!*

racket *n.* **1.** an organised illegal activity such as the extortion of money by threat or violence from legitimate businessmen: *the protection racket.* **2.** a dishonest scheme, trick, etc. **3.** one's legitimate business or occupation: *He's in the advertising racket.*

raddle-marked *adj.* (of a shearer) due for early dismissal because his sheep were badly shorn and thus marked with raddle, a red colouring.

radical *adj.* wonderful; fantastic. Also, **rad.**

rafferty *adj. N.Z.* rough, ragged. [British dialect]

Rafferty's rules *pl. n.* no rules at all, as of a contest or organisation run in a slipshod fashion. Also, **Rafferty rules.** [British dialect *rafferty* irregular, linked by association with the Irish surname *Rafferty*]

rag[1] *n.* **1.** a newspaper or magazine, especially one considered as being of little value. **2.** a wretched or worthless person or thing. *—phr.*

3. chew the rag, a. to argue or grumble. **b.** to brood or grieve.

4. on the rags, menstruating.

5. glad rags, fine clothes.

6. sky the rag, to surrender; give in.

rag[2] *v.* **1.** to scold. **2.** to tease; torment. **3.** to play rough jokes on. *—n.* **4.** any disorderly or high-spirited conduct, especially by a group of young people. **5.** (in certain student communities) an organised display of grotesque or absurd behaviour publicising a collection of money for charity, or the like.

rage *n.* **1.** an exciting or entertaining event: *That party was a rage.* *—v.* **2.** to set about enjoying oneself: *Let's go raging.*

rag top *n.* (*offensive*) an Arab.

rag trade *n.* the clothes-manufacturing trade.

Raiders, the *pl. n.* **1.** the Canberra team in the N.S.W. Rugby Football League. **2.** the Newcastle K.B. United team in the National Soccer League.

rail *phr.*

1. off the rails, in an abnormal condition; insane; out of control.

2. on the rails, functioning in a normal manner.

railroad *v.* **1.** to send or push forward with great or undue speed: *to railroad a bill through parliament.* **2.** to convict (a person) unjustly or with undue haste.

rain *phr.*

1. keep out of the rain, to keep out of trouble.

2. rain cats and dogs, to rain heavily.

3. right as rain, perfectly all right; safe; in good health.

rainbow *n. Military.* (World War I and World War II) a soldier who went to the front after the dangerous action had finished. [from *rainbow* after the storm]

raincheck *n.* **1.** a postponement (as of an invitation, etc.). **2.** a ticket issued by a store to guarantee supply of sale-priced goods which are temporarily unavailable. *—phr.* **3. take a raincheck,** to accept an invitation or make an agreement for which the specific details, as time and place, have not yet been fixed. [from U.S. *raincheck* a ticket for future use given to spectators at a baseball game, sports meeting, etc., stopped by rain]

raincoat *n.* a contraceptive sheath.

raining *phr.* **if it was raining virgins I'd be locked in the dunny with a poof** or **if it was raining pea soup I'd be hit on the head by a fork,** an expression of gloom and continual bad luck.

rainmaker *n. Australian Rules.* a very high kick.

raise *phr.*
1. raise Cain or **hell,** to create a disturbance, nuisance, or trouble.
2. raise the roof, See **roof.**

rake *n.* a comb. Also. **bug rake.**

rake-off *n.* **1.** a share or portion, as of a sum involved or of profits. **2.** a share or amount taken or received illicitly.

rally up *n. Prison.* a demonstration inside a gaol.

ralph *phr.* **ralph up,** to vomit. [imitative]

ram *n.* a trickster or confidence man, especially one who sets up victims for another.

rambunctious *adj. U.S.* **1.** boisterous; noisy. **2.** obstreperous; perverse; unruly. [variant of *rumbustious*]

ramp *n.* **1.** *Prison.* a search, especially of a prisoner's cell. **2.** a swindle, especially one depending on a rise in prices.

Rams, the *pl. n.* the Preston team in the National Soccer League.

ran *v.* See **run.**

randy *adj.* **1.** having a free indulgence of lust; lecherous. **2.** sexually aroused.

rangatira *n. N.Z.* a chief; boss; superior of any kind. [from Maori *rangatira* a noble]

ran-tan *phr.* **on the ran-tan,** on a drinking bout.

rap[1] *v.* **1.** to accelerate (a motor vehicle). *—n.* **2.** punishment or blame, especially of one who accepts punishment for a crime he did not commit: *to take the rap.* **3.** a criminal charge: *a housebreaking rap.* *—phr.*
4. give (a motor vehicle) a rap, to accelerate (a motor vehicle) and travel at full speed for a short period.
5. rap over or **on the knuckles,** to reprimand sharply; reprove.
6. rap up, to praise extravagantly.

rap[2] *n.* **1.** a conversation. *—phr.* **2. rap on,** to talk discursively. [?*rap(port)*]

rapt *v.* **1.** delighted: *She was rapt about her results.* **2.** infatuated: *He is really rapt in her.*

rare *phr.*
1. rare as hen's teeth, See **hen's teeth.**
2. rare as rocking horse shit, See **rocking horse.**

rash *phr.* **to be all over (someone) like a rash,** to make a nuisance of oneself by excessive and unwelcome attention to (someone).

raspberry *n.* **1.** a sound expressing derision or contempt made with the tongue and lips. *—v.* **2.** to make such a sound. *—phr.* **3. raspberries to (someone),** an expression of dismissal, derision, contempt, etc. [rhyming slang *raspberry tart* fart]

rat *n.* **1.** one who abandons his friends or associates, especially in time of trouble. **2.** a person considered as wretched or despicable. *—v.* **3.** to desert one's party or associates, especially in time of trouble: *a man who would rat on* *his friends.* **4.** to continue at work during a strike; to work as a scab. **5.** *Mining.* **a.** to pilfer opal from a miner's hiding place. **b.** enter someone's mine and take out opal rock. —*interj.* **6.** (*pl.*) an exclamation of annoyance, incredulity, denial, or disappointment. —*phr.*

7. go the rat, to do something completely wholeheartedly.

8. like a rat up a rope or **drainpipe,** very quickly.

9. rat on, a. to inform (on); betray. **b.** to go back on (a statement, agreement, etc.)

10. rat through, to sort through in a careless or hasty manner.

11. smell a rat, to be suspicious.

ratbag *n.* **1.** a rascal; rogue. **2.** a person of eccentric or non-conforming ideas or behaviour.

rat factory *n. N.Z.* a psychiatric hospital.

rat-fink *n. Originally U.S.* a despicable person.

ratshit *adj.* **1.** useless; broken. **2.** depressed or unwell. **3.** no good: *That exam was ratshit.* Also, **R.S.**

ratter *n.* **1.** a deserter or betrayer. **2.** *Mining.* one who takes opal from someone else's mine.

rattle *v.* **1.** to disconcert or confuse (a person). —*phr.* **2. rattle the pan,** to die.

rattlebrain *n.* a giddy, empty-headed chatterer. Also, **rattlehead, rattlepate.**

rattler *n.* **1.** *Chiefly U.S.* a goods train. —*phr.* **2. jump** or **scale the rattler,** board a train illegally.

rattletrap *n.* **1.** a shaky, rattling object, as a rickety vehicle. **2.** a garrulous person. **3.** the mouth.

rat-trap *n.* **1.** a difficult and involved set of circumstances. **2.** a bicycle pedal having deep serrations and a toe-clip to prevent the foot slipping.

ratty *adj.* slightly eccentric.

rave *v.* **1.** to talk or write with extravagant enthusiasm. **2.** to act boisterously or enthusiastically. —*n.* **3.** a long and animated conversation. **4.** a wild or hectic party or the like. [Middle English, probably from Old French *raver* wander, be delirious]

raver *n.* an enthusiastic person, usually a young one, as a fan of a pop singer.

raw *adj.* **1.** harsh or unfair: *a raw deal.* —*phr.*

2. come the raw prawn, See **prawn.**

3. the raw, a. a crude, uncultured state: *The play portrayed life in the raw.* **b.** naked; nude: *She sunbakes in the raw.* **c.** a particularly sensitive place, point, topic or the like: *Her remark touched him on the raw.*

razoo *n.* **1.** a gambling chip. —*phr.* **2. not have a brass razoo,** to have no money at all. [origin uncertain]

razor gang *n.* a Parliamentary committee responsible for making recommendations for reductions of government expenditure.

razz *v.* **1.** to deride; make fun of; chiack. —*n.* **2.** severe criticism; derision. [short for *raspberry*]

razza *n.* a Returned Services League (RSL) club. Also, **rissole.**

razzamatazz *n.* **1.** noisy and showy activity. **2.** any traditional style of jazz. Also, **razzmatazz.**

razzle-dazzle *n.* **1.** noisy and showy activity; razzamatazz. **2.** a spinning frame providing thrilling rides in a carnival or playground.

RBT *n.* Random Breath Test. Also, **rbt.**

reach-me-down *n.* See **hand-me-down.**

read *phr.*
1. read between the lines, to perceive the truth of a situation, regardless of its appearance.
2. you wouldn't read about it! an exclamation of astonishment, sometimes ironic.

ready *n.* **1.** ready money. **2.** scheme; racket; lurk. *—phr.* **3. work a ready,** to adopt a scheme or racket, usually dishonest or illegal.

real *adv.* **1.** very. *—phr.*
2. for real, a. actual, definite: *That overseas trip is for real.* **b.** genuine, sincere: *He's for real.*
3. the real McCoy or **thing,** the genuine article.

ream *n.* (*usually pl.*) a large quantity: *to write reams.* [from *ream* 500 sheets of paper]

recap *v.* **1.** to review by way of an orderly summary, as at the end of a speech or discourse; recapitulate. *— n.* **2.** the act of recapitulating: *Her recap made it all very clear.* [shortened form]

recce *n.* **1.** a reconnaissance: *do a recce.* **2.** *Films, Television.* research work for location shots or background material. Also, **reccy.** [shortened form]

red *adj.* **1.** ultraradical politically; communist. *— n.* **2.** Also, **Red.** an ultraradical; a communist. *—phr.*
3. paint the town red, to celebrate, especially wildly and extravagantly.
4. see red, to become angry or infuriated.
5. the red, loss or deficit: *to be in or out of the red.*

red-arse *n.* *N.Z. Military.* raw recruit; rookie.

red-baiting *n.* the act of denouncing or deprecating political opponents who are radical or left-wing.

Red Demons, the *pl. n.* the Melbourne team in the Victorian Football League. Also, **Red Legs.**

red-eye *n.* a black cicada of eastern Australia having ruby red eyes.

redfed *n.* a socialist, leftist, unionist, etc. [from *red* a communist + *Fed(eration of Labour)* an early Australian trade union congress]

Redfern *phr.* **get off at Redfern,** to practise coitus interruptus. [from *Redfern* a railway station immediately before Central Railway Station, Sydney; possibly derived from British expression 'getting out at Gateshead']

red hat *n.* *Military.* an officer of the rank of colonel and above.

Red Heavies, the *pl. n.* the University team in the Queensland Rugby Football Union. Also, **Varsity.**

red hots *pl. n.* *Horseracing.* harness racing. [rhyming slang *red hots,* the trots]

Red Legs, the *pl. n.* **1.** the Dandenong team in the Victorian Football Association. **2.** the Norwood team in the South Australian National Football League.

red-light district *n.* a neighbourhood with many brothels. [from former times when a brothel was indicated by a red light]

redneck *n.* *U.S.* (*offensive*) a southern U.S. white farm labourer, especially one who is ill-educated or ignorant.

red ned *n.* cheap red wine.

red peril *n.* the threatened expansion of communism.

red rag *n.* something that excites a person's anger or passion: *like a red rag to a bull.*

red shirt *n.* (formerly) a flogged back: *to give a convict a red shirt.*

red steer *n.* a bushfire. Also, **red bull.**

Reds, the *pl. n.* the Drummoyne team in the N.S.W. Rugby Football Union. Also, **Dirty Reds.**

red tape *n.* excessive attention to formality and routine. [from the *red tape* much used for tying up official papers]

reef *phr.*
1. **reef off,** to steal.
2. **reef out,** to remove, usually by force.

reefer *n.* a marijuana cigarette. [from *reef* a part of a sail which is rolled up]

ref *n.* referee. [shortened form]

reffo *n.* a refugee, especially one arriving in Australia before World War II. Also, **refo.** [shortened form of *refugee*]

reginalds *pl. n.* underwear. Also, **reggies, grundies.** [variation of rhyming slang *Reg Grundies* undies]

rego *n.* registration. [shortened form]

regular *adj.* **1.** complete; thorough: *a regular rascal.* —*n.* **2.** a regular customer: *regulars only.*

relation *phr.* **have relations,** to have sexual intercourse.

rellie *n.* a relative. [shortened form]

remittance man *n.* (formerly) an Englishman (often of aristocratic background) living in Australia because his presence was no longer acceptable in England, but supported by a remittance from his family.

rep *n.* a representative: *union rep.* [shortened form]

repat *n., adj.* repatriation: *repat pension.* [shortened form]

repro *n.* a copy or duplicate, especially of a picture or the like made by photoengraving or some similar process. [shortened form]

Reps *n.* House of Representatives. [shortened form]

rest *n.* **1.** *Prison.* a period of imprisonment of one year. —*v.* **2.** *Theatre.* to be unemployed, as an actor.

retread *n.* a person who has come out of retirement to take up work again.

reviver *n.* an intoxicating drink; a stimulant. Also, **corpse reviver.**

rhubarb *n.* **1.** confused noise, argument, etc., as in a quarrel. **2.** a quarrel or squabble; commotion. [from the word 'rhubarb' supposedly spoken by actors to simulate noisy conversation in the background]

rib *v.* to tease; ridicule; make fun of. [apparently short for *rib-tickle*]

ribuck *adj.* *Obsolete.* very good, genuine: *a ribuck shearer.* Also, **ryebuck.**

rice and sago *n.* an Italian. [rhyming slang *rice and sago* dago]

rich *adj.* **1.** highly amusing. **2.** ridiculous, absurd, or preposterous.

Richard *phr.* **to have had the Richard, a.** to be ruined. **b.** to be worn out. [from Dick, diminutive of *Richard.* See **dick**]

ride *v.* **1.** to harass or torment. —*phr.*
2. ride for a fall, to act in a way which will inevitably bring disaster.
3. take for a ride, a. to kidnap and murder. **b.** to deceive and wilfully mislead.

ridges *phr.*
1. have been (a)round the ridges, to be expert or experienced.
2. round the ridges, in and about brothels, pubs, clubs, etc.

ridgy-didge[1] *adj.* true; correct; genuine. Also, **ridge.**

ridgy-didge[2] *n.* refrigerator. [rhyming slang *ridgy-didge* fridge]

rig *n.* **1.** a powerful truck, as used for long-distance haulage. **2.** Also, **rig-out.** costume or dress, especially when odd or conspicuous. **3.** a male sheep, horse, etc., not properly castrated. —*phr.* **4. rig out** or **up,** to fit or deck with clothes, etc.

right *adj.* **1.** unquestionable; unmistakeable; true: *He's a right idiot.* —*adv.* **2.** very; really; extraordinarily: *He's right stupid.* —*phr.*
3. dead or **bang to rights,** *Prison.* in the very act of a crime or other deed; red-handed: *They caught him dead to rights.*
4. right as rain, See **rain.**
5. right up there, *Originally U.S.* close to success.
6. she'll be right. Also, **she's right,** an expression of confidence that everything will go well.
7. too right, an emphatic expression of agreement.

right-footer *n.* a Protestant.

righto *interj.* an expression indicating agreement. Also, **rightio, rightoh, righty-ho, right on.**

right-to-lifer *n.* a person who supports anti-abortion policies.

ring[1] *n.* **1.** anus. —*phr.*
2. ring in the jack, *Two-up.* to use a double-headed coin.
3. ring the board or **shed,** to shear more sheep than anyone else in the shearing shed.
4. run rings round, to be markedly superior to; easily surpass.
5. toss one's hat into the ring, to enter a contest or competition.

ring[2] *phr.*
1. ring a bell, to arouse a memory; sound familiar.
2. ring in, to insert or substitute, often dishonestly.

ringbolt *v. N.Z.* to stow away on a ship with the connivance of the crew.

ringer[1] *n.* a stationhand, especially a stockman or drover.

ringer[2] *n.* **1.** an athlete, horse, etc., entered in a competition under false representations as to identity or ability. **2.** Also, **dead ringer, dead ring.** a person or thing that closely resembles another: *He was a dead ringer for the local policeman.*

ringer[3] *n.* **1.** the fastest shearer of a group. **2.** any person of outstanding competence. [from British dialect *ringer* anything superlatively good]

ringie *n. Two-up.* the person in charge of a two-up school.

ring-in *n.* **1.** a person or thing substituted for another at the last moment. **2.** one belonging to a group only in appearance.

riot *n.* **1.** one who or that which causes great amusement, enthusiasm, etc.: *She's a riot at a party.* *—phr.* **2. run riot, a.** to act without control or restraint; disregard all limits. **b.** to grow luxuriantly or wildly.

rip[1] *v.* **1.** to move along with violence or great speed. *—phr.*

2. let it rip, to allow an engine, etc., to go as fast as possible by ceasing to check or control its speed.

3. let rip, a. to give free rein to anger, passion, etc. **b.** to utter oaths; swear.

4. rip in(to), begin rapidly, eagerly: *Let's rip into the housework.*

5. rip off, to overcharge; swindle.

6. rip out, to utter angrily; shout.

7. wouldn't it rip you? an expression of annoyance, exasperation, etc.

rip[2] *n.* **1.** a dissolute or worthless person. **2.** a worthless or worn-out horse. **3.** anything of little or no value.

ripe *adj.* **1.** drunk. **2.** obscene or pertaining to obscenity.

rip-off *n.* an excessive charge or price; swindle.

ripped *adj.* heavily under the influence of drugs or alcohol; stoned.

ripper *n.* something or someone exciting extreme admiration: *You little ripper!*

rip-roaring *adj.* boisterous; riotous; wild and noisy: *to have a rip-roaring time.*

ripsnorter *n.* a person or thing made remarkable by some outstanding characteristic, as great strength, excellence, liveliness, beauty, etc.

rise *phr.*

1. get or **take a rise out of,** to provoke to anger, annoyance, etc., by banter, mockery, deception, etc.

2. give rise to, to cause or produce.

3. make a rise, to be successful; gain prosperity.

risk *phr.* **no risk!** an exclamation of reassurance or approval.

rissole *n.* **1.** a Returned Services League (RSL) Club. *—phr.* **2.** *Theatre.* **do the rissoles,** (of entertainers) to perform shows in Returned Services League (RSL) Clubs.

ritzy *adj.* luxurious, elegant. [after the *Ritz* Hotel, London, a luxurious hotel]

river *phr.*

1. sell down the river, to betray; deceive.

2. up the river, in gaol or a corrective institution.

roach *n.* the butt of a marijuana cigarette. [origin uncertain]

road *phr.*

1. one for the road, a final alcoholic drink consumed before setting out on a journey, returning home from a public house, etc.

2. on the road, a. on tour, as a theatrical company. **b.** (formerly of convicts) employed in road building. **c.** (formerly) unemployed and on the road, as a swagman, etc.

3. hit the road, to set out on a journey.

road-hog *n.* a motorist who drives without consideration for other road users.

roadie *n.* a person associated with a pop group who arranges road transportation, sets up equipment, etc. Also, **road manager.**

roar *phr.* **roar (someone) up,** to scold angrily or abuse.

roaring *adj.* brisk or highly successful, as trade.

roast *v.* **1.** to criticise, rebuke or ridicule severely. *—n.* **2.** a severe criticism or rebuke.

robber's dog *phr.* **off like a robber's dog,** See **off.**

rock[1] *n.* **1.** a jewel, especially a diamond. **2.** (*pl.*) the testicles. *—phr.* **3. get one's rocks off,** (of a male) to ejaculate.
4. on the rocks, a. into or in a state of disaster or ruin. **b.** (of drinks) with ice-cubes: *Scotch on the rocks.*
5. the Rock, Ayers Rock.
6. the Rocks, the western arm of Sydney Cove, once characterised by rocky outcrops of sandstone, and in the late ninteenth century haunt of the Rocks push.

rock[2] *phr.* **rock the boat,** to cause a disturbance; make difficulties; raise awkward questions; threaten the status quo.

rock and roll *n.* dole. [rhyming slang]

rock-ape *n.* (*offensive*) an oaf; idiot.

rock chopper *n.* a Roman Catholic. [?from the initials *R.C.*]

rock college *n.* prison.

rocker *n.* **1.** a young person of the early 1960s, characterised by rough, unruly behaviour, who usually wore leather clothing, had greased-back hair and rode a motorcycle. See, **mod.** *—phr.* **2. off one's rocker,** crazy; mad; demented: *You must be off your rocker to suggest such a thing.*

rocket *n.* **1.** a severe reprimand; reproof. *—phr.* **2. go like a rocket, a.** to move fast. **b.** (of a machine) to function well.

rock-hop *v.* **1.** (in a sailing race) to sail very close inshore, risking running aground, to gain advantage. **2.** to fish from coastal rocks.

rock-hound *n.* a geologist.

Rockie *n.* a member of the Royal Australian Naval Reserve.

rocking horse *phr.* **rare** or **scarce as rocking horse manure/shit/dirt,** very scarce; non-existent.

rock spider *n. Prison.* a child molester.

rod *n.* **1.** revolver; pistol. **2.** the erect penis. *—phr.* **3. (have a) rod in pickle,** See **pickle.**

rod-walloper *n.* a male who masturbates.

roger[1] *interj.* an expression of agreement, comprehension, etc. [*Roger* (personal name) used in telecommunications as a name for *r*, used as an abbreviation for *received*]

roger[2] *n.* **1.** the penis, especially when erect. *—v.* **2.** (of a man) to have sexual intercourse with. [?from *Roger* personal name]

rogues' gallery *n.* **1.** a collection of portraits of criminals, as at a police station. **2.** any collection of portraits resembling this.

roll *v.* **1.** to luxuriate or abound (in wealth, etc.). **2.** to cast dice. **3.** to rob a person, often with violence. **4.** to defeat. *—n.* **5.** a wad of paper currency. **6.** any amount of money. **7.** the sexual act. —*phr.*

8. roll in, a. to arrive. **b.** to retire to bed.

9. roll over, (of a politician) to resign gracefully.

10. roll up, a. to arrive. **b.** to gather round.

rollie *n.* a hand-rolled cigarette. Also, **roll-your-own.**

rolling *adj.* **1.** drunk. **2.** Also, **rolling in money, rolling in it.** very rich.

rolling stone *n.* a wanderer; an itinerant; a person without ties or fixed address.

roll-ons *n.* woman's elasticised foundation garment; step-ins. See *N.Z.* **easies.**

Ron *phr.* **one for Ron,** a cigarette borrowed for later on. [*Ron* contraction of 'later on' ('ron)]

roo *n.* kangaroo.

roo bar *n.* a metal grid placed in front of a car to prevent damage to the vehicle in case of collison, especially with kangaroos, stray cattle, etc., on outback roads.

roof *v.* **1.** to kick or punch in the mouth. —*phr.*

2. hit the roof, become very angry; lose one's temper.

3. raise the roof, a. to create a loud noise. **b.** to make loud protests or complaints.

rookie *n.* a raw recruit, originally in the army, and hence in any service, sporting team, etc. Also, **rooky.**

room *phr.*

1. no room to swing a cat, confined, cramped or cluttered.

2. room in, (of a mother in a maternity hospital), to have her baby sleeping in the room with her.

3. room to move, scope to manoeuvre; options or choices.

4. the smallest or **littlest room,** a toilet or bathroom.

roost *phr.*

1. come home to roost, to come back upon the originator; recoil.

2. rule the roost, to be in charge; dominate.

rooster *phr.* **a rooster one day and a feather duster the next,** an expression describing the uncertainty of continued popularity or success.

Roosters, the *pl. n.* **1.** the Eastern Suburbs team in the N.S.W. Rugby Football League. **2.** the Geelong West team in the Victorian Football Association. **3.** the North Adelaide team in the South Australian National Football League.

Roos, the *pl. n.* **1.** Also, **the Kangaroos.** the Australian representative Rugby League Football team. **2.** the Clarence team in the Tasmanian Australian National Football League. **3.** Also, **the Kangaroos.** the North Melbourne team in the Victorian Football League.

root[1] *n.* **1.** the act of sexual intercourse. **2.** a sexual partner: *a good root.* —*v.* **3.** to have sexual intercourse with. **4. a.** to exhaust. **b.** to break; ruin.

root[2] *v.* *U.S.* to give encouragement to, or applaud, a contestant, etc. [?variant of *rout* make a loud noise]

rooted *adj.* **1.** exhausted. **2.** frustrated; thwarted. **3.** broken; ruined. *—phr.* **4. get rooted,** (*offensive*) go away.

rope *phr.*
1. at the end of one's rope or **tether,** an the end of one's endurance.
2. give someone enough rope to hang himself or **herself,** to allow freedom to someone to prove his or her unworthiness.
3. know the ropes, See **know.**
4. on the ropes, in a hopeless position; near to failure.
5. rope in, to draw, entice, or inveigle into something.

ropeable *adj.* angry; bad-tempered. Also, **ropable.**

ropy *adj.* **1.** worn; deteriorated; below the desired standard. **2.** bad-tempered. Also, **ropey.**

rort *n.* **1.** a trick; lurk; scheme. **2.** a wild party. *—v.* **3.** to gain control over (an organisation, as a branch of a political party) especially by falsifying records. [origin uncertain]

rorter *n.* a con man operating on a small scale, as with worthless goods.

roscoe *n. Prison.* a gun.

Rosebuds, the *pl. n.* the Adamstown (Newcastle) team in the National Soccer League.

rosella *n.* **1.** a sheep that has lost wool, especially from disease and is showing patches of bare skin; barebelly. **2.** a shearer who is stripped to the waist while he is working. [from *Rose Hill* an early settlement in N.S.W.]

rosiner *n.* a strong alcoholic drink. Also, **rosner, roziner.** [from obsolete British *rosin* to make drunk]

rot *n.* **1.** nonsense. *—interj.* **2.** an exclamation of dissent, distaste, or disgust. *—phr.* **3. wouldn't it rot your socks,** an exclamation of annoyance, disgust, etc.

rotate *phr.* **wouldn't it rotate you,** an exclamation of annoyance, disgust, etc.

rotgut *n.* alcoholic liquor of inferior quality.

rotten *adj.* **1.** wretchedly bad, unsatisfactory, or unpleasant: *to feel rotten, rotten work.* **2.** extremely drunk.

rotter *n. Chiefly British.* a thoroughly bad, worthless, and objectionable person.

rough *adj.* **1.** severe, hard, or unpleasant: *to have a rough time of it.* *—n.* **2.** a rough person; rowdy. *—phr.*
3. a bit rough, unfair, unreasonable.
4. cut up rough, to behave angrily or violently; be upset.
5. give (someone) the rough edge of one's tongue, to give (someone) a berating.
6. in the rough, in a rough, crude, unwrought, or unfinished state.
7. rough as guts/bags/mullet/goats' knees/a sand bag/a pig's breakfast, a. very rough; crude; of inferior quality. **b.** (of a person) uncouth; bad mannered. **c.** Also, **rough as bags and twice as dirty** or **nasty.** (World War I Army applied especially by Australian soldiers to London prostitutes) rough, uncouth.
8. rough it, to live without even the ordinary comforts or conveniences: *We roughed it all month long.*

9. rough on, a. severe towards. **b.** unfortunate for.

10. rough trot, a spell of bad luck or misfortune.

rough-house *n.* **1.** noisy, disorderly behaviour or play; rowdy conduct; a brawl. *—v.* **2.** to disturb or harass by a rough-house.

roughie *n.* **1.** one who is rough or crude. **2.** a shrewd trick; a cunning act. **3.** *Horseracing.* a horse with little chance of ever winning a race.

roughneck *n.* **1.** a rough, coarse person. **2.** *Chiefly U.S.* a member of an oil-drilling crew.

rough-up *n.* a brawl, fight.

round *phr.*

1. go the rounds, a. (of people) to make a series of visits. **b.** (of gossip, information, etc.) to become generally known.

2. round the bend or **twist,** insane.

3. round up, to collect (cattle, people, etc.) in a particular place or for a particular purpose.

rouse *phr.* **rouse on,** to castigate; criticise severely.

rouseabout *n.* a handyman on a station, in a hotel, etc.; bluetongue. Also, **rouser, rousie.**

Roy *n.* the archetype of the young, status-conscious Australian (opposed to *Alf*).

royal *n.* a member of a royal family, especially the British royal family.

Royal Alberts *pl. n.* See **alberts.**

Royal Alfred *n.* a large and elaborate swag, containing clothes, cooking utensils and tent.

Royals, the *pl. n.* the East Perth team in the West Australian Football League.

rozzer *n.* *Originally British.* a policeman.

R.S. *adj.* ratshit.

rub *phr.*

1. rub it in, to remind someone repeatedly of his mistakes, failures or short comings.

2. rub out, to kill.

3. rub shoulders or **elbows,** to come into social contact.

4. rub (up) the right way, to please.

5. rub (up) the wrong way, to annoy.

rubber *n.* **1.** *U.S.* a contraceptive sheath. **2.** a thong. **3.** a tyre.

rubber bands *phr.* **betting with rubber bands,** *Horseracing.* said of punters at the end of a losing day when all that they have left is the rubber bands which earlier held wads of money.

rubber cart *n.* See **green cart.**

rubber cheque *n.* a cheque which is dishonoured and so bounces back from the bank.

rubber duckie *n.* a small inflated rubber boat usually with a powerful outboard motor and often used for surf rescue.

rubberneck *n.* **1.** an extremely or excessively curious person. **2.** a tourist. *—adj.* **3.** pertaining to or for such person. *—v.* **4.** to look at things in an excessively curious manner.

rubber stamp *n.* one who gives approval without consideration, or without demur.

rubber-stamp *v.* to give approval without consideration.

rubbery *adj.* unreliable: *The budget estimates are a bit rubbery.*

rubbidy *n.* a pub. Also, **rubbity, rubbidy-dub.** [rhyming slang *rub-a-dub* pub]

rube *n. U.S.* an unsophisticated countryman. [short for *Reuben* a man's name]

rub-off *n.* a bonus; beneficial side effect.

ruby-dazzler *n.* **1.** Also, **bobby-dazzler,** *N.Z.* **rube.** an excellent thing or person. *—adj.* **2.** excellent.

ruckus *n. Chiefly U.S.* **1.** a commotion; rumpus. **2.** a violent disagreement.

ruction *n.* a disturbance, quarrel or row.

ruddy *adv.* extremely: *I've a ruddy good mind to hit him.* [a substitution for *bloody*]

rude *adj.* inconsiderate; unfair; dismissive: *That's a bit rude.*

rug *phr.* **cut a rug,** *Chiefly U.S.* to dance, especially with verve, as to jazz.

rugged *adj.* uncomfortable; entailing hardship.

rugger-bugger *n.* (*offensive*) **1.** a fanatical Rugby Union supporter. **2.** an aggressively self-confident man, often of good social standing and reactionary views, given to loud displays and larrikin behaviour.

rule *phr.*

1. bend the rules, to make an exception to the rules.

2. rule out, to exclude, refuse to admit, declare (something) out of the question.

Rules *n.* Australian Rules Football. Also, **rules.**

rum *adj.* **1.** odd, strange, or queer. *—phr.* **2. rum go,** harsh or unfair treatment. Also, **rummy.**

rumble *v.* **1.** to take part in a fight, as between gangs. **2.** to detect or become suspicious of, as a fraud. *—n.* **3.** a fight, especially between teenage gangs. **4.** a prank, usually undertaken by a group, sometimes involving violence.

rumpus *n.* **1.** disturbing noise; uproar. **2.** a noisy or violent disturbance or commotion.

run *phr.*

1. cut and run, to disengage and take to flight.

2. on the run, escaped or hiding from pursuit, especially by the police.

3. run a book, to accept bets.

4. run after, to seek to attract.

5. run all over, to outclass; defeat.

6. run around, to behave promiscuously.

7. run (around) with, to keep company with.

8. run away with, a. to elope with. **b.** to steal. **c.** to win easily: *He ran away with the election.* **d.** to use up (money, etc.) quickly. **e.** to accept (an idea), especially erroneously or with insufficient justification: *Don't run away with the idea that you can go on behaving so badly.*

9. run dead, See **dead.**

10. run down, to denigrate; make adverse criticism of.

11. run in, arrest.

12. run hot, to be performing well.

13. run off, to write or otherwise create quickly.

14. run off with, a. to steal. **b.** to elope with.

15. run out, *U.S.* to drive out; expel.
16. run out on, to desert; abandon.
17. run rings (a)round (someone), to perform with far greater success.
18. run short, to become scarce or nearly used up.
19. run up, *Dressmaking.* to make quickly.
20. run up against, a. to meet unexpectedly. **b.** to be impeded by.
21. the runs, diarrhoea.

run-around *n.* **1.** equivocation; evasion. *—phr.* **2. give (someone) the run-around,** to fob (someone) off with evasions and subterfuges. Also, **runround.**

run-in *n.* disagreement; argument; quarrel.

rush *v.* to heap attentions on.

Russians *pl.n.* (formerly) wild cattle. Also, **Rooshians.**

rust bucket *n.* a badly rusted motor vehicle.

rustle *phr.* **rustle up,** to move, bring, get, etc., by energetic action: *rustle up breakfast.*

ruth *phr.* **cry Ruth,** to vomit. [imitative]

ryebuck *adj. Obsolete.* See **ribuck.**

sack *n.* **1.** dismissal or discharge, as from employment. **2.** a bed. *—v.* **3.** to dismiss or discharge, as from employment. **4.** *Football.* to tackle and put to the ground. **5.** to abandon; give up; forget: *You can sack that idea. —phr.* **6. hit the sack,** to go to bed.

sadie *n.* a cleaning woman. [from the song *Sadie, the cleaning lady*]

sad sack *n.* **1.** a habitually (and often demonstratively) sad person. **2.** an ineffective person who always blunders despite good intentions, especially a soldier.

Saigon rose *n.* a form of venereal disease, thought to have been introduced by returned soldiers from Vietnam. [from *Saigon* capital of South Vietnam and *rose* euphemism for a syphilitic sore]

sail *phr.* **sail in,** to go boldly into action.

sailer *n. N.Z.* a falling tree-branch.

Saints, the *pl. n.* **1.** the St Kilda team in the Victorian Football League. **2.** the St George team in the N.S.W. Rugby Football League. **3.** the St Marys Club in the Northern Territory Football League.

sale *phr.* **have** or **make a sale,** *N.Z.* to vomit.

Sallie *n.* **1.** a member of the Salvation Army. *—phr.* **2. the Sallies,** the Salvation Army. Also, **Sal, Sally, Salvo.**

salt *n.* **1.** a sailor, especially an experienced one: *old salt. —phr.* **2. go through like a dose** or **packet of salts, a.** to make a quick passage through (something), as does a purgative. **b.** to make a brief visit causing great disturbance.

saltcellar *n.* either one of the hollows above the collarbone of thin people.

salted *adj.* experienced in some occupation, etc. [?from *old salt* an experienced sailor]

salt mine *n.* (*usually plural*) (*humorous*) a place of habitual confinement and drudgery. [?from the use of forced labour in the salt mines of Siberia]

salute *n.* Also, **Australian salute, Barcoo salute.** (*humorous*) the movement of hand and arm to brush away flies from one's face.

Salvo *phr.* **the Salvos,** the Salvation Army.

sambo *n.* **1.** (*usually cap.*) (*offensive*) a Negro. **2.** a sandwich. Also, **sammie.**

sandgroper *n.* a West Australian.

sandie *n.* **1.** a sandfly. **2.** a sand crab. **3.** a girl who frequents the beach.

sandshoe *interj.* (*humorous*) thankyou. [altered form]

sandwich *phr.* **the meat in the sandwich,** a person caught between two opposing parties, with each of whom he has some connection, but over whose actions he has no control.

san fairy Ann *phr.* it doesn't matter. [French *ça ne fait rien*]

sanger *n.* a sandwich.

sanny cart *n.* a sanitary cart.

sanny man *n.* a sanitary man.

sap *n.* a fool or weak person. Also, **saphead.**

sapheaded *adj.* silly; foolish.

sappy *adj.* silly or foolish.

sarky *adj.* sarcastic.

sassy *adj. Originally U.S.* saucy.

satchel swinger *n.* a bookmaker.

sauce *n.* **1.** impertinence; impudence. —*v.* **2.** to speak impertinently to. —*phr.* **3 fair suck of the sauce bottle,** See **suck.**

sausage *phr.*

1. not a sausage, absolutely nothing. [?rhyming slang *sausage and mash* cash]

2. (silly) sausage (*used affectionately, especially to children*) silly person.

sausage dog *n.* a dachshund.

sausage roll *n. Australian Rules.* a goal. Also, **sausage, saus.** [rhyming slang]

sav *n.* **1.** a saveloy. —*phr.* **2. fair suck of the sav,** See **suck.**

saved *phr.* **saved by the bell,** rescued from a predicament at the last minute. [from the boxing sense, when a boxer is saved from being counted out by the bell marking the end of the round]

saver *n.* a bet laid to offset another; a covering bet.

savvy *v.* **1.** to know; understand. —*n.* **2.** understanding; intelligence; commonsense.

sawbones *n.* a surgeon.

say-so *n.* **1.** one's personal statement or assertion. **2.** final authority. **3.** a command.

scab *n.* a strike-breaker; blackleg.

scabby *adj.* mean or contemptible: *That was a scabby trick.*

scad *n.* (*usually plural*) a large quantity: *He has scads of money.*

scale *v.* **1.** to defraud or cheat; steal. **2.** to ride on a public conveyance, especially a tram, without paying a fare: *scale a rattler.*

scaler *n.* one who evades paying a fare on a bus, train, etc.

scales *phr.*
1. tip the scales, to influence favourably.
2. tip the scales at, to weigh.

scalp *n.* **1.** a small profit made in quick buying and selling. —*v.* **2.** to buy and sell so as to make small, quick profits, as stocks. **3.** to buy (tickets) cheap and sell at other than official rates.

scaly *adj.* shabby; despicable.

scam *n.* an illegal business operation; a racket.

scandal *n.* gossip in general.

scarce *phr.*
1. make oneself scarce, to make off; keep out of the way.
2. scarce as hen's teeth, See **hen's teeth.**
3. scarce as rocking horse shit, See **rocking horse.**

scare *phr.* **scare the living daylights out of,** to greatly frighten.

scared *phr.*
1. scared stiff, very frightened.
2. run scared, to panic; retreat in disarray.

scaredy-cat *n.* (*especially in children's speech*) a coward.

scarper *v.* to run away; depart suddenly, especially leaving behind debts or other commitments. Also, **do a scarper.**

scat *v.* go off hastily (usually as a command).

scatter *n.* a drinking spree; bender.

scatterbrain *n.* one incapable of serious, connected thought.

scene *n.* **1.** an area or sphere of interest or involvement: *Politics is his scene.* —*phr.*
2. a good (bad) scene, a place or situation which has a good (bad) ambience.
3. on the scene, in fashion.
4. the scene, the contemporary fashionable world.

schizo *n.* **1.** a schizophrenic. —*adj.* **2.** schizophrenic.

schlanter *n.* See **slanter.**

schlemiel *n.* *U.S.* an awkward and unlucky person for whom things never turn out right. Also, **schlemihl.**

schmalz *n.* excessive sentimentality, especially in the arts. Also, **schmaltz.**

schmuck *n.* a stupid person; idiot; fool.

schnozzle *n.* the nose. Also, **schnoz.**

school *n.* **1.** a group of people settled (either on one occasion or habitually) into a session of drinking or gambling. **2.** *Prison.* the gaol and its associated 'culture'. —*phr.*
3. *Prison.* **old school,** a prisoner who has served a lot of time in gaol.
4. tell tales out of school, See **tell.**

schoolie *n.* **1.** a schoolteacher. **2.** a school prawn.

schoolmarm *n.* **1.** a schoolmistress, especially of the old-fashioned type. **2.** any prim woman.

school sores *pl. n.* impetigo, an infectious skin disease of children.

sci-fi *n.* **1.** science fiction. —*adj.* **2.** science-fiction.

scoff *v.* **1.** to eat greedily and quickly. —*n.* **2.** food.

sconce *n.* **1.** the head or skull. **2.** sense or wit.

scone *n.* **1.** the head. —*phr.*
2. do one's scone, to lose one's temper.

3. off one's scone, mad; insane.

scoob *n.* a marijuana cigarette.

scoop *n.* a big haul, as of money.

scorcher *n.* **1.** a very hot day. **2.** *Sport.* an outstanding performance; a blinder. **3.** an excessively fast driver.

score *n.* **1.** latest news or state of progress: *What's the score on the new space rocket?* **2.** a prostitute's customer. *—v.* **3.** to be successful in having sexual intercourse. *—phr.*
4. know the score, be aware of what is required.
5. pay off or **settle a score, a.** to avenge a wrong. **b.** to fulfil an obligation.
6. score off, gain an advantage over.

Scottie *n.* **1.** a Scotsman. **2.** a Scottish terrier. Also, **Scotty.**

scotty *adj.* bad-tempered.

scout *n.* **1.** a fellow: *a good scout.* *—phr.* **2. scout out** or **up,** to seek; search for: *Try to scout out an entertainer for Saturday night.*

scrag *n.* **1.** the neck of a human being. *—v.* **2.** to wring the neck of; hang; garrotte. **3.** to seize roughly by the neck.

scrag end *n.* the remains of a meal; left-overs.

scram *v.* to get out quickly; go away.

scrambled egg *n. Military, Navy, etc.* the gold braid on an officer's uniform.

scran *n.* food.

scrap *n.* a fight or quarrel.

scrape *n.* **1.** an embarrassing situation. **2.** a fight; struggle; scrap. **3. a.** *Medical.* a curettage of the uterus. **b.** an abortion. *—phr.*
4. scrape the bottom of the barrel or **bucket,** to dredge up scurrilous or irrelevant information in criticism of another.
5. scrape through, to manage to get by with difficulty; succeed by a narrow margin: *It was difficult having no money, but somehow we managed to scrape through.*

scrap heap *phr.* **on the scrap heap,** useless; no longer employable: *There he was, on the scrap heap at 49.*

scrappy *adj.* **1.** given to fighting. **2.** careless, incomplete or untidy.

scratch *adj.* **1.** done by or dependent on chance: *a scratch shot.* **2.** gathered hastily and indiscriminately: *a scratch crew, a scratch meal.* *—v.* **3.** to manage with difficulty: *scratch along on very little money.* **4.** (formerly) to flog with a cat-o'-nine tails. *—n.* **5.** money. *—phr.*
6. from scratch, from the beginning or from nothing.
7. up to scratch. a. conforming to a certain standard; satisfactory. **b.** *Boxing.* (of a boxer) arriving at the fight by an agreed time.

scream *n.* **1.** someone or something that is very funny. *—phr.* **2. scream blue murder,** to complain vociferously.
3. scream for, to want desperately, be in great need of.

screamer *n.* **1.** someone or something causing screams of astonishment, mirth, etc. **2.** *Printing.* an exclamation mark. **3.** *Australian Rules.* a very high mark: *to pull down a screamer.*

screw *n.* **1.** a hard bargainer; a miser. **2.** a broken-down horse. **3.** wages; money. **4.** a prison warder. **5.** sexual intercourse. *—v.* **6.** to extract or extort. **7.** to have sexual intercourse with. *—phr.*

8. have a screw loose, to be slightly eccentric; have crazy ideas.

9. put the screws on, to apply pressure; intimidate: *to put the screws on a debtor.*

10. screw (someone) up, to cause (someone) to become mentally and emotionally disturbed.

11. screw (up), to make a mess of; impair; frustrate.

screwball *n.* **1.** an erratic, eccentric, or unconventional person. *—adj.* **2.** erratic, eccentric, or unconventional.

screwed *adj.* **1.** drunk; intoxicated. *—phr.* **2. screwed up, a.** mentally and emotionaly disturbed. **b.** broken; impaired.

screwy *adj.* **1.** eccentric; crazy. **2.** strange; peculiar.

scrounge *v.* **1.** to borrow, sponge, or pilfer. *—phr.* **2. scrounge around,** to gather, as by foraging; search out.

scrub[1] *v.* **1.** to cancel; get rid of. **2.** *Horseracing.* (of a jockey) to move the whip or arms rhythmically to and fro, to encourage the horse during the final stretch of the race.

scrub[2] *phr.* **the scrub,** country areas in general, as opposed to the city.

scrubber *n.* **1.** the grey kangaroo (applied especially by kangaroo shooters). **2.** *British.* (*offensive*) a promiscuous or mercenary girl; a girl of loose morals.

scrubbing-brush *n.* *Obsolete.* bread with more chaff and bran in it than flour.

scum *n.* *Chiefly U.S.* semen.

scummy *n.* *British.* worthless; despicable. Also, **scumbag.**

scunge *n.* **1.** an unkempt, slovenly person. **2.** dirt, mess, slime, etc. **3.** messy, untidy objects: *I'll clear the scunge off this desk.*

scungy *adj.* mean, dirty, miserable, unpleasant. Also, **skungy.**

scupper *v.* *British.* to deprive of any chance of success; to defeat.

scuttlebutt *n.* rumour; gossip. [from *scuttlebutt* (formerly) a cask of drinking water on a ship, and thus a place for gossiping]

sea *phr.* **half seas over,** drunk.

Sea Eagles, the *pl. n.* the Manly Warringah team in the N.S.W. Rugby Football League.

seagull *n.* **1.** a man who does the interior painting in a ship. **2.** *N.Z.* a casual wharf labourer who is not a member of a trade union.

Seagulls, the *pl. n.* **1.** the Sandy Bay team in the Tasmanian Australian National Football League. **2.** the Williamstown team in the Victorian Football Association. **3.** the Wynnum-Manly team in the Queensland Rugby Football League.

sea-legs *n.* **1.** the ability to walk with steadiness or ease on a rolling ship. **2.** the ability to resist seasickness.

seatman *n.* a homosexual who is sexually aggressive.

seaweed *n.* a surfie.

sec *n.* a second: *Wait just a sec, please.*

secco *n.* a sex pervert.

second best *phr.* **come off second best,** to be defeated in a contest.

second fiddle *phr.* **play second fiddle,** See **fiddle.**

second-hand Sue *n.* **1.** a prostitute of faded charms. **2.** a passive homosexual who is either old or unattractive.

seconds *n.* **1.** (at a meal) **a.** a second helping. **b.** a second course.

—phr. **2. take seconds,** to reconsider.

second wind *phr.* **get one's second wind,** to experience a revival of interest, enthusiasm, etc., in a task in hand.

section *phr.* **last section,** the end, as of a marriage, friendship, etc.

security blanket *n.* something which affords a sense of security, well-being, or comfort to the possessor. [from the comic strip 'Peanuts' by U.S. cartoonist Charles Schulz in which a character, Linus, carries round a blanket for security]

see *phr.*

1. go and see a man about a dog, to depart without saying why.

2. see about, to deal with or attend to.

3. see here, an expression used to attract attention, for emphasis or the like.

4. see off, to turn away, especially forcibly; cause to leave.

5. see out, a. to continue in (an undertaking) until it is finished. **b.** to live until the end of (a period) or outlive (a person).

6. see things, to hallucinate.

seed *phr.* **go** or **run to seed,** to approach the end of vigour, usefulness, prosperity, etc.

seedy *adj.* out of sorts physically.

segro *phr.* **on segro,** *Prison.* (of a prisoner) segregated from other prisoners; on protection.

seine *n.* **a.** (formerly) the sum of £1. **b.** the sum of $1.

sell *v.* **1.** to cheat or hoax. *—n.* **2.** an act of selling or salesmanship. **3.** a hoax or deception. **4.** a disappointment. *—phr.*

5. be sold on, to be very enthusiastic about.

6. sell down the river, Also, **sell out.** to betray.

sell-out *n.* **1.** a betrayal. **2.** a play, show, etc., for which all seats are sold.

semi *n.* **1.** a semidetached house. **2.** semitrailer.

senator *n.* a schooner of beer. [rhyming slang *Senator Spooner*]

send *v.* **1.** to excite or inspire (as a jazz musician, listener, or other person). *—phr.*

2. send off, a. to cause to depart. **b.** to be present at a departure, as of a friend.

3. send packing, to dismiss; send away.

4. send up, a. to mock or ridicule; satirise. **b.** to imprison.

5. send (someone) about his business, to send away or dismiss abruptly or rudely.

6. send (someone) up the wall, See **wall.**

send-off *n.* **1.** a party or other social arrangement in honour of a person who is setting out on a journey, career, etc. **2.** a start given to a person or thing.

send-up *n.* a satire or parody.

septic *n.* a Yank. [rhyming slang *septic tank*]

serene *phr.* **all serene.** See **all.**

serve *n.* a strong rebuke; a tongue lashing: *She gave him a real serve when he came home drunk.*

sessionman *n.* a studio musician.

set *adj.* **1.** marked for dislike, attack, or destruction: *She had him set.* —*n.* **2.** *Surfing.* a series of waves. —*phr.*
3. dead-set, true; certain: *It's a dead-set rip-off at that price.*
4. (dead) set against, to have an unfavourable or hostile attitude to (something): *She was dead set against housework.*
5. get set, *Horseracing, Gambling.* (of a punter) to place a bet on a horse.
6. set back, to cost: *It set him back $10.*
7. set eyes on, to see.
8. set on, determined: *She was set on going to England.*
9. set (someone) up, to arrange a situation in which (a person) appears in a bad light or is incriminated falsely.
10. set to, to start to fight.

settler's clock *n.* a kookaburra.

settler's matches *pl. n.* the long pieces of dead bark which trail loosely from certain gum trees.

set-to *n.* a fight; argument.

set-up *n.* **1.** the way anything is arranged or organised: *What is the set-up here?* **2.** contest or undertaking which presents no real challenge or problems, as a fixed boxing match. **3.** a racket; swindle. **4.** a trap; ambush.

seven *phr.* **throw** or **chuck a seven, a.** to fall unconscious; to die. **b.** to become very angry; lose one's temper.

seventeen-door sedan *n.* (formerly) a sanitary cart.

seven-year itch *n.* marital boredom or discontent considered to develop after about seven years of marriage.

sex *phr.* **to have sex,** to have sexual intercourse.

sex kitten *n.* a sexually attractive and provocative young woman.

sexploitation *n.* **1.** the use of sex to help sell a product, film, etc. **2.** the exploitation of sex in books, films, magazines, etc.

sexpot *n.* a blatantly sexually attractive female.

sex-sell *n.* the use of sexually stimulating copy and illustrations in advertising.

sex shop *n.* a shop in which articles associated with sexual behaviour are sold.

s.f.a. *n.* very little; next to nothing. [abbreviation for *S(weet) F(anny) A(dams)* or *S(weet) F(uck) A(ll)*]

shack *phr.* **shack up, a.** to live at a place; reside: *You can come and shack up with us till your house is ready.* **b.** to live in sexual intimacy.

shades *pl. n.* sunglasses.

shady *adj.* **1.** uncertain; questionable; of dubious character or reputation. —*phr.* **2. on the shady side of,** beyond in age: *on the shady side of forty.*

shag[1] *phr.* **like a shag on a rock,** alone; deserted; forlorn.

shag[2] *v.* **1.** to have sexual intercourse with. **2.** to chase; run after. **3.** to masturbate. —*n.* **4.** an act or instance of sexual intercourse, especially group sexual activity. —*phr.*

5. get shagged, an expression of contempt or rejection.

6. shag out, to tire, exhaust.

shagger's back *n.* any pain in a man's back, jocularly attributed to the strains of sexual intercourse.

shaggin' wagon *n.* a panel van or station wagon, often luxuriously appointed with carpet, curtains, etc., as a suitable place for sexual intercourse.

shaggy dog story *n.* a generally long and involved funny story whose humour lies in the pointlessness or irrelevance of its conclusion.

shake *v.* **1.** to steal. *—n.* **2.** an earthquake. **3.** (*pl.*) a state of trembling, especially that induced by alcoholism, drugs or nervous disorder. **4.** a moment, a short time: *Just a shake. —phr.*

5. brace of shakes, a very short time; an instant.

6. in two shakes of a dog's tail, in a very short time.

7. no great shakes, of no particular importance; unimpressive.

8. shake down, a. to settle in or retire to a bed, especially a makeshift or temporary one. **b.** to settle comfortably in or adapt oneself to new surroundings, etc. **c.** to condition: *to shake down a vessel by a first voyage.* **d.** to extort money from. **e.** to search (someone); frisk.

9. shake hands with the wife's best friend, (of a male) to urinate.

10. shake off, a. to get rid of; free oneself from. **b.** to get away from; elude.

11. shake up, to disturb or agitate mentally or physically.

shakedown *n.* **1.** extortion, especially by blackmail or threatened violence. **2.** *U.S.* a thorough search.

shake-up *n.* a thorough change in a business, department or the like, as by dismissals, demotions, etc.

Shaky Islander *n.* a New Zealander.

Shaky Isles *n.* New Zealand.

shandygaff *n.* any mixture, as of wet and dry sheep for shearing. Also, **shandigaff.** [from the mixture of alcoholic and non-alcoholic beverages in the drink *shandy,* formerly *shandygaff*]

shandy system *n. Real Estate.* a property rating system based on both unimproved capital value and the commercial value. [from the mixture of basic beverages in the drink *shandy*]

shanghai *v.* **1.** to involve someone in an activity, usually without their knowledge or against their wishes. **2.** to steal. **3.** *Prison.* to transfer without warning to another gaol. *—n.* **4.** *Prison.* an unexpected transfer to another gaol.

shanks's pony *phr.* **ride shanks's pony,** to walk. Also, **ride shanks's mare.**

shape *phr.* **shape up or ship out,** to perform as required or leave.

shark bait *n.* one who swims where there is danger of a shark attack.

shark shit *phr.* **lower than shark shit,** (of behaviour, etc.) low; mean; despicable.

Sharks, the *pl. n.* **1.** the Cronulla-Sutherland team in the N.S.W. Rugby Football League. **2.** the East Fremantle team in the West Australian Football League.

shat *phr.* **be shat off,** to have had enough of, be upset. [from the supposed past participle of *shit*]

shaver *n.* a youngster; fellow.

she *phr.* **she'll be right,** See **right.**

shebang *n.* **1.** thing; affair; business. **2.** a hut; shanty; shack. **3.** a disturbance; brawl.

sheeny *n. British.* (*offensive*) a Jew.

sheep dip *n.* cheap, inferior liquor.

sheepo *n.* a shearing-shed hand employed to keep shearers' pens full of sheep.

sheep's eyes *pl. n.* amorous or love-sick glances.

sheet *phr.*

1. sheet home, to attach (blame, responsibility, etc.).

2. short a sheet of bark, See **bark.**

3. three sheets in or **to the wind,** intoxicated; drunk.

sheeted *v. Prison.* (of a prisoner) to have a complaint sheet put in against one by a warder for an offence alleged to have been committed whilst in gaol.

sheikh *n.* (formerly) a man who is boldly amorous towards many women. Also, **sheik.**

sheila *n.* **1.** a girl or woman: *a beaut sheila.* **2.** a girlfriend. [probably from *Sheila* an Irish girl's name]

shekels *pl. n.* money.

shelf *n.* **1.** an informer. *—v.* **2.** to inform on (someone). *—phr.* **3. on the shelf,** (of a woman) unattached or unmarried, and without prospects of marriage.

shell *phr.* **shell out,** to hand over; pay up.

shellacking *n.* a beating; thrashing.

shell-back *n.* **1.** an experienced sailor, especially an old one. **2.** a sailor who has crossed the equator.

shell-like *n.* (*humorous*) an ear: *A word in your shell-like, please.*

shemozzle *n.* **1.** a confused state of affairs; muddle. **2.** an uproar; row. [Yiddish, from Hebrew: bad luck]

shenanigan *n.* nonsense; deceit; trickery. Also, **shenanigans.**

she-oak *n. Obsolete.* beer.

shepherd's clock *n.* a kookaburra.

sherbet *n.* beer.

shicer *n.* **1.** an unproductive gold mine. **2.** a swindler.

shicker *n.* **1.** alcoholic drunk. **2.** a drunkard. *—phr.* **3. on the shicker,** intoxicated; drunk.

shickered *adj.* drunk; intoxicated. Also, **shicker.**

shift *phr.* **make shift,** to manage with effort or difficulty.

shillelagh *n. Horseracing.* a whip. [from *shillelagh,* (in Ireland) a cudgel]

shilling *phr.* **wouldn't say shit for a shilling,** See **shit.**

shimmy *n.* a chemise.

shindig *n.* **1.** a dance, party, or other festivity, especially a noisy one. **2.** a disturbance; quarrel; row.

shine *n.* **1.** a liking; fancy: *to take a shine to.* **2.** *Obsolete.* a caper; prank. *—phr.* **3. take the shine out of, a.** to remove or spoil the lustre or brilliance of. **b.** to surpass; get the better of.

shiner *n.* a black eye.

Shiners, the *pl. n.* the Sunshine team in the Victorian Football Association.

shingle *n.* **1.** a small signboard, especially that of a professional man. *—phr.*
2. be a shingle short, to be eccentric; mentally disturbed.
3. hang out one's shingle, to set up in professional practice.

shiny-arse *n.* **1.** a soldier in a base unit, as opposed to one in the fighting line. **2.** a male office worker.

ship *phr.* **ship out,** to send away or get rid of.

shiralee *n.* **1.** a burden or bundle. **2.** a swag.

shirt *phr.*
1. keep one's shirt on, to refrain from losing one's temper or becoming impatient.
2. lose one's shirt, to lose everything.
3. put one's shirt on, to bet heavily or put all one's resources on (a horse, etc.).

shirt-front *n.* **1.** *Australian Rules.* Also, **shirt-fronter.** a head-on charge aimed at bumping an opponent to the ground. *—v.* **2.** to bump (an opponent) in such a manner.

shirt-lifter *n.* a male homosexual.

shirty *adj.* bad-tempered; annoyed.

shit *v.* **1.** to defecate. **2.** to anger or disgust. *—n.* **3.** faeces; dung; excrement. **4.** the act of defecating. **5.** a contemptible or despicable person. **6.** nonsense; rubbish; lies. **7.** marijuana or hashish. **8.** *Prison.* tobacco. *—interj.* **9.** an exclamation expressing anger, disgust, disappointment, disbelief, etc. *—phr.*
10. get the shits, to become exasperated or angry.
11. give (someone) the shits, to arouse dislike, resentment, annoyance in (someone).
12. have shit for brains, to be extremely stupid.
13. have the shits, a. to have diarrhoea. **b.** to feel fed up or weary.
14. have the shits (with), to feel fed up or angry (with).
15. in the shit, a. in trouble. **b.** in an angry or resentful mood.
16. not worth a pinch of shit, completely worthless.
17. pack shit, See **pack.**
18. put shit on, a. to denigrate; criticise. **b.** to hoodwink; deceive.
19. shit (it) in, win or succeed easily.
20. shit out of luck, *U.S.* very unlucky; in a position of little hope. Also, **SOL.**
21. the shit hits the fan, the trouble begins.
22. up shit creek (without a paddle), in trouble; in difficulties.
23. up to shit, worthless; useless.
24. wouldn't say shit for a shilling, (of a person) snobbish; affected; arrogant.

shitcan *v.* to denigrate unmercifully: *One by one the lawyers shitcanned him.*

shithead *n.* **1.** a mean contemptible person. **2.** a no-hoper; dullard. **3.** a person who smokes marijuana regularly.

shithouse *n.* **1.** a toilet. *—adj.* **2.** foul; wretchedly bad. Also, **shouse.**

shitkicker *n.* **1.** an assistant, especially one doing menial or repetitive jobs. **2.** a person of little consequence.

shitless *adv.* completely; utterly (used to add emphasis to a statement): *scared shitless; bored shitless.*

shitman *n.* a low-ranking official; underling.

shit on the liver *phr.* **to have shit on the liver,** to be in a bad temper. Also, **SOL.**

shits *pl. n.* diarrhoea.

shit-scared *adj.* very frightened.

shit-stirrer *n.* **1.** an activist, especially in a political context. **2.** a trouble-maker.

shitty *adj.* **1.** annoyed; bad tempered. **2.** unpleasant; disagreeable; of low quality. *—phr.* **3. pack** or **crack a shitty,** to sulk.

shitty-livered *adj.* ill-tempered; angry.

shiv *n. British.* a knife.

shivers *phr.*

1. cold shivers, a sensation of fear, anxiety, distaste.

2. the shivers, a fit or attack of shivering.

shivoo *n.* a party; celebration; spree. Also, **chavoo, shavoo.**

shlemiel *n. U.S.* an innocent fool.

shmo *n.* a foolish, boring or stupid person. Also, **schmo.**

shocker *n.* **1.** a sensational work of fiction. **2.** an unpleasant or disagreeable person or thing.

shoe *phr.*

1. in (someone's) shoes, in the position or situation of (another): *I shouldn't like to be in his shoes.*

2. know where the shoe pinches, to know the cause or real meaning of trouble, misfortune, sorrow, etc., especially from personal experience.

shoemaker's children *pl.n.* children of technically skilled parents who get no benefit from their parents' expertise in a particular field.

shoestring *phr.* **on a shoestring,** with a very small amount of money.

shonk *n.* dealer in unreliable, dubious quality, machinery. [backformation from *shonky*]

shonky *adj.* **1.** of dubious integrity or honesty. **2.** mechanically unreliable. Also, **shonkie.**

shooftee *n.* See **shufty.**

shook *phr.* **shook on, a.** in love with or infatuated by (a person). **b.** disposed favourably towards (a thing or course of action).

shoot *v.* **1.** to begin, especially to begin to talk. *—phr.*

2. shoot a fairy, See **fairy.**

3. shoot a line, See **line.**

4. shoot down in flames, to completely destroy (another's argument, aspiration, etc.).

5. shoot for a win, to play in expectation of winning.

6. shoot off, to go away quickly.

7. shoot off one's mouth, a. to talk indiscreetly, especially to reveal secrets, etc.; talk wildly or tactlessly. **b.** to exaggerate; boast.

8. shoot one's bolt, a. to do one's utmost. **b.** to ejaculate prematurely.

9. shoot the moon, to abscond.

10. shoot through, a. to go away, usually absenting oneself improperly: *Instead of going to the exam, he shot through.* **b.** to move away rapidly: *to shoot through like a Bondi tram.*

11. shoot up, a. to take drugs intravenously. **b.** to cause damage, confusion, etc., by reckless or haphazard shooting.

shooter *n.* **1.** something that shoots; a gun, pistol, or the like. **2.** a fast surfboard. **3.** a photograph.

shooting iron *n.* a firearm, especially a pistol or revolver.

shooting match *phr.* **the whole shooting match,** everything under consideration.

shop *v.* **1.** to inform against, betray to the police, etc. *—phr.*
2. all over the shop, all over the place; in confusion.
3. set up shop, to set oneself up in business.
4. shop around, to make general and wide-ranging inquiries.
5. talk shop, to discuss one's trade, profession, or business.

shoppy *n. Prison.* a shoplifter.

Shop, the *n.* the University of Melbourne, Victoria.

short *v.* **1.** to short-circuit. **2.** to sell short. *—phr.*
3. be caught short, a. to discover an inconvenient lack of something, as money. **b.** to have a sudden and urgent need to urinate or defecate.
4. make short work of, to finish or dispose of quickly.
5. sell oneself short, a. to underestimate one's abilities or achievements. **b.** to behave in a fashion considered to be unworthy of one.
6. short a sheet of bark, See **bark.**

short and curlies *pl. n.* See **hair.**

short arm *n.* a penis.

short-arm parade *n.* in an establishment of the armed forces, a parade called for the medical inspection of the men's genitalia. Also, **short-arm inspection.**

short arse *n.* a person of below average height.

short-change *v.* to cheat.

short-circuit *v.* to abbreviate procedures, especially in relation to paperwork; bypass procedural steps: *I can short-circuit the system.*

short hairs *pl. n.* See **hair.**

shortie *n.* a short person or thing.

short-story writer *n. Prison.* a forger, especially of cheques.

short straw *phr.* an unfavourable outcome: *Someone had to cop the short straw.* [from drawing lots, the loser being the one who draws the short straw]

short-timer *n. Prison.* a prisoner serving a short sentence (usually less than two years).

shot *n.* **1.** an injection of a drug, vaccine, etc. *—phr.*
2. big shot, an important person.
3. call the shots, to be in command.
4. have a shot at, a. to criticise; ridicule. **b.** to make an attempt at.
5. like a shot, instantly; very quickly.
6. shot in the arm, something that gives renewed confidence, vigour, etc.
7. shot in the dark, a wild or random guess.
8. that's the shot! an exclamation of approval.
9. to be (get) shot of, to be (get) rid of.

shoulder *phr.*
1. give the cold shoulder to, to treat coldly; ignore; snub.
2. have broad shoulders, to be able to accept responsibility.
3. rub shoulders with, to associate with; come into contact with.

shouse *adj.* very bad; shithouse. [contraction of *shithouse*]

shout *v.* **1. a.** to stand (the company) a round of drinks. **b.** to pay for something for another person; treat: *I'll shout you to the pictures; I'll shout you a new dress.* —*n.* **2.** the act of shouting, as by providing drinks. **3.** one's turn to shout. —*phr.* **4. shout down,** to drown (another's words) by shouting or talking loudly.

shove *phr.*
1. shove it! an expression of dismissal, contempt, etc.
2. shove off, to leave; depart.

shovel *n.* a room; living quarters. [rhyming slang, *shovel and broom* room]

show *v.* **1.** to give an exhibition, display, or performance. —*n.* **2.** a theatrical performance or company. **3.** a party. **4.** a chance: *Do you think he stands a show?* —*phr.*
5. give the show or **game away,** to reveal all the details of a plan, scheme, etc.
6. run the show, to control or manage a business, etc.
7. show a clean pair of heels, escape.
8. show a leg, See **leg.**
9. show off, a. to behave in an attention-seeking manner. **b.** to exhibit for approval or admiration.
10. show up, a. to expose (faults, etc.); reveal. **b.** to appear superior to (another); outdo. **c.** to turn up; appear at a certain place.
11. steal the show, to attract most attention; be the most popular person or item.

shower *n.* **1.** a dust storm: *Bedourie shower, Bourke shower, Cobar shower, Darling shower, Wilcannia shower, Wimmera shower, etc.* **2.** *British.* an unpleasant group or person. —*phr.* **3. (someone) didn't come down in the last shower,** (someone) is not naive or gullible.

show-off *n.* one given to pretentious display or exhibitionism.

shrapnel *n.* small change, especially silver. [from *shrapnel* small metal fragments]

shrewdie *n.* **1.** Also, **shrewd head.** a shrewd person. **2.** a shrewd trick.

shrimp *n.* a diminutive or insignificant person. [from *shrimp* a small marine crustacean]

shrink *n.* See, **headshrinker.**

shucks *n.* **1.** something useless: *not worth shucks.* —*interj.* **2.** *U.S.* an exclamation of disgust or regret. [from *shuck* a husk, shell or pod]

shuffle *phr.* **shuffle off,** to die.

shufty *n.* a look. Also, **shufti. [Arabic]**

shunt *phr.* **get the shunt,** to be dismissed from a job.

shut *phr.*
1. keep one's mouth shut, to keep a secret.
2. shut one's eyes to, to refuse to notice; ignore.
3. shut up, a. to stop talking; become silent. **b.** to stop (someone) from talking; silence.

shut-eye *n.* sleep.

shutterbug *n.* a keen photographer. [from *shutter* a camera mechanism + *bug* an obsession]

shy *adj.* **1.** short: *shy of funds.* **2.** failing to pay something due, as one's ante in poker. —*phr.* **3. fight shy of,** to avoid; keep away from.

shypoo *n.* inferior quality liquor.

shyster *n.* **1.** one who gets along by petty, sharp practices. **2.** a lawyer who uses unprofessional or questionable methods. [apparently alteration of *Scheuster,* an unscrupulous 19th century New York lawyer; perhaps influenced by German *Scheisser* from *scheissen* shit]

sick *adj.* **1.** disgusted; chagrined. —*phr.*

2. be sick of, to feel fed up with; have had enough of: *He was sick of his employer's complaints about his work.*

3. sick as a dog, very sick.

4. sick up, to vomit.

sickie *n.* a day taken off work with pay, because of genuine or feigned illness.

sick list *phr.* **on the sick list,** not well; ill: *He has been on the sick list for a week now.*

side *n.* **1.** a pretentious or superior manner: *to put on side; to bung on side.* —*phr.* **2. on the side,** as a sideline; secretly.

sidekick *n.* **1.** an assistant. **2.** a close friend.

sidewinder *n. U.S.* a disabling swinging blow from the side.

sieve *n.* **1.** one who cannot keep a secret. —*phr.* **2. have a head like a sieve,** to be very forgetful.

sight *n.* **1.** something that looks odd or unsightly: *She looks a sight in her new hat.* **2.** a great deal: *It's a sight better here than at the last hotel.* —*phr.* **3. not by a long sight** or **shot,** on no account; definitely not.

silent *phr.* **silent like the P in swimming,** See **P.**

silent cop *n.* a small raised circular device in a roadway, used as a guide to control turning traffic; traffic dome.

silk[1] *n.* a jockey.

silk[2] *n. U.S.* a white person. [?from the texture of the hair]

silk department *adj.* **1.** top quality: *That girl is silk department.* —*n.* **2.** an easy, pleasant job that is often well-paid. Also, **the silk.**

silly *adj.* **1.** stunned. —*n.* **2.** Also, **silly billy.** a silly person. —*phr.* **3. silly as a wet hen/two-bob watch/square wheel,** stupid; idiotic; erratic.

Silver City *n.* Broken Hill. [from rich silver deposits in the area; ?also from the widespread use of galvanised iron as a building material]

silvertail *n.* **1.** a member of a wealthy elite. **2.** a highly placed official. **3.** a social climber. **4.** a prisoner who helps warders, as by informing, and receives special privileges.

simmer *phr.* **simmer down,** to become calm or calmer.

sin-bin *n.* **1.** a panel van or station wagon, often luxuriously appointed, as a suitable place for sexual intercourse; shaggin' wagon. **2.** (in team sports) an area adjoining the playing field set aside for penalised players; penalty box.

sing *v.* **1.** to turn informer. **2.** (formerly) to scream while being flogged. —*phr.* **3. sing out,** to call out in a loud voice; shout.

single *n.* an unmarried person: *singles' night; singles' bar.*

singleton *n. Australian Rules.* a minor score, as a behind.

sink *v.* **1.** to drink: *Let's sink a middy. —phr.*

2. sink or swim, to make a desperate attempt to succeed, when the alternative is complete failure.

3. sink the slipper, a. to kick someone as in a streetfight or brawl. **b.** *Football.* to kick the ball.

sin-shifter *n.* a priest.

sissy *n.* **1.** an effeminate man or boy. **2.** a timid or cowardly person. Also, **cissy.** [from *sister*]

sit *phr.*

1. be sitting pretty, See **pretty.**

2. sit in, (of a performer) to play as a temporary member of a band.

3. sit on or **upon, a.** to check; rebuke; repress. **b.** to prevent (a document) from becoming public knowledge so as to avoid the action recommended in it: *The government is sitting on the report.*

4. sit out, a. to stay till the end of: *Though the film was boring we sat it out.* **b.** to outlast.

5. sit pat, See **pat**[2]**.**

6. sit tight, a. to take no action; bide one's time: *I'll sit tight till I know what the decision is.* **b.** to stay exactly where one is: *Sit tight while I get some help.*

sitcom *n.* a situation comedy. [shortened form]

sit-down money *n.* unemployment benefits.

sitter *n.* something easily accomplished, as a catch in cricket, a mark to be shot at, etc.

sitting duck *n.* **1.** one who is easily duped or defeated. **2.** something easily accomplished.

six *phr.*

1. go for six or **a sixer,** to suffer a major setback, as falling over heavily.

2. hit for six, to dispatch or destroy utterly.

sixer *n.* **1.** *Cricket.* a hit scoring six runs. **2.** a patrol leader of cubs in the Boy Scouts or brownies in the Girl Guides.

six o'clock swill *n.* See **swill.**

sixpence *n.* a five cent coin.

six-shooter *n.* a revolver with which six shots can be fired without reloading.

six-stitcher *n.* a cricket ball with a leather covering sewn together with six rows of stitches around the seam.

sixty-nine *n.* simultaneous fellatio and cunnilingus; soixante-neuf.

size *n.* **1.** actual condition, circumstances, etc.: *That's about the size of it. —phr.* **2. to size up, a.** to form an estimate of. **b.** to come up to a certain standard.

sizzle *v.* to be very hot.

skag *n.* **1.** a cigarette. **2.** heroin.

skate[1] *phr.* **skate on thin ice,** to place oneself in a delicate situation; touch on a contentious topic.

skate[2] *n. U.S.* a person, a fellow.

skedaddle *v.* **1.** *Originally U.S.* to run away; disperse in flight. *—n.* **2.** a hasty flight. Also, **skidaddle.**

skeeter *n.* **1.** a mosquito. **2.** a nickname for a man of small build.

skeleton *n.* **1.** a very lean person or animal. *—phr.* **2. skeleton in the cupboard,** some fact in the history or lives of a family which is kept secret as a cause of shame.

skerrick *n.* a very small amount: *not a skerrick left.* [British dialect]

skew-whiff *adv.* askew. Also, **skewiff.**

skid *phr.*
1. on the skids, deteriorating fast.
2. put the skids under, to place in a precarious position; to ensure the downfall of. [from *skid* a plank on which something heavy may be slid along]

skiddoo *v. U.S.* **See skedaddle.**

skidlid *n.* a motorcyclist's crash-helmet.

skidmark *n.* a mark or smudge of faeces on underwear, etc.

skid row *phr.* **on skid row,** destitute.

skin *n.* **1.** a contraceptive sheath. —*v.* **2.** to strip of money or belongings; fleece, as in gambling. **3.** *U.S.* to slip off hastily. —*phr.*
4. by the skin of one's teeth, scarcely; just; barely.
5. get under one's skin, a. to irritate one. **b.** to fascinate or attract one.
6. skin (someone) alive, berate; scold: *Mum will skin me alive if I don't get home soon.*
7. to give (someone) some skin, *U.S.* to shake hands with (someone).

skin and blister *n.* sister. Also, **blister.** [rhyming slang]

skin flick *n.* a pornographic movie. Also, **skin flic.**

skinful *n.* **1.** a large amount, especially of alcoholic drink. —*phr.* **2. have had a skinful, a.** to be drunk. **b.** to be fed up.

skin game *n.* **1.** a dishonest activity, business, etc. **2.** a swindle. [from *skin* to strip of money or belongings + *game*]

skinhead *n.* a member of any group of young men identified by close-cropped hair and sometimes indulging in aggressive activities.

skinned *phr.* **keep one's eyes skinned,** to be extremely vigilant.

skinner *n.* **1.** *Horseracing.* a horse which wins a race at very long odds. **2.** *Horseracing.* a betting coup. —*phr.* **3. a skinner,** *N.Z.* **a.** empty; finished: *The beer's a skinner.* **b.** broke: *I'm a skinner.*

skinny-dip *v.* **1.** to bathe in the nude. —*n.* **2.** a nude swim.

skinpop *n.* **1.** an injection of a drug into the skin. —*v.* **2.** to inject drugs into the skin.

skins *pl. n.* drums; bongos.

skint *adj.* completely without money; broke. [from *skinned,* stripped of money or belongings]

skip[1] *v.* to go away hastily; abscond.

skip[2] *n. Chiefly Sport.* a captain or leader, as of a team; skipper.

skip[3] *n.* an offensive term for an Australian. Also, **skippy.** [from *Skippy* the name of a kangaroo in children's television series]

skirt *n.* a woman or girl.

skite *v.* **1.** to boast; brag. —*n.* **2.** a boast; brag. **3.** Also, **skiter.** a boaster; braggart. [Scottish and northern dialect *skate*]

skitebook *n.* a scrapbook in which actors, etc., keep notices of their performances, photographs, etc., especially for purposes of promotion.

skol *v.* to consume (a drink) at one draught. Also, **skoal.** [Scandinavian]

skull *n.* the head as the seat of intelligence or knowledge: *Get this into your thick skull.*

skun *v.* to drink all the liquor supplies (of someone).

skunk *n.* **1.** a thoroughly contemptible person. —*v.* **2.** *U.S.* (in games) to beat so completely as to keep from scoring.

sky *v.* **1.** to raise aloft; strike (a ball) high into the air. —*phr.*
2. sky the towel, to give up, admit defeat.
3. the sky's the limit, there is no limit.

skyjack *v.* **1.** to hijack (an aeroplane). —*n.* **2.** the act of hijacking an aeroplane.

skyman *n.* an aviator.

sky pilot *n.* **1.** a clergyman. **2.** *U.S.* an aviator.

sky rocket *n.* pocket. Also, **sky.** [rhyming slang]

skywonkie *n.* a person who can predict the weather.

slab *n.* **1.** a mortuary table. **2.** a tall, awkward fellow.

slabby *n.* *N.Z.* a timber worker who handles slabs of timber.

slab-heap *phr.* **be on the slab-heap,** to be set aside as no longer useful or needed.

slacker *n.* one who avoids work, effort, etc.

slag *n.* **1.** spittle. —*v.* **2.** to spit. —*phr.* **3. slag off,** to bad-mouth; criticise severely.

slam *v.* **1.** to criticise severely. —*n.* **2.** a severe criticism.

slammer *n.* gaol.

slanter *n.* **1.** a trick. —*phr.* **2. work a slanter,** to trick or con someone. Also, **schlanter, slinter.** [from South African *schlenter*]

slant-eye *n.* (offensive) an Asian.

slap *adv.* **1.** directly; straight. —*phr.*
2. a slap in the face or **eye,** a setback or humiliation.
3. slap down, to rebuke or suppress the enthusiasm of.

slap-bang *adv.* **1.** Also, **slam-bang.** violently; suddently. **2.** exactly; precisely; just: *slap-bang in the middle.*

slaphappy *adj.* **1.** cheerful. **2.** irresponsible; lackadaisical.

slaps *pl. n.* thick, rectangular thongs made from bamboo, cloth, etc.

slap-up *adj.* first-rate; excellent: *I think we deserve a slap-up meal.*

slash *n.* the act of urinating.

slashing *adj.* very large or fine: *a slashing fortune.*

slate[1] *phr.*
1. clean slate, a good record.
2. put it on the slate, to record a debt, as on a slate; give credit.

slate[2] *v.* to criticise or review adversely.

slather *phr.* **open slather,** complete freedom; free rein.

slats *pl. n.* **1.** bottom; buttocks. **2.** ribs.

slaughter *v.* to defeat thoroughly.

slavedriver *n.* **1.** a hard taskmaster, as a teacher, boss, etc. **2.** a wife.

slave-labour *n.* work considered as very badly paid.

slay *v.* to amuse (someone) greatly: *That comedian really slays me.*

sleaze *n.* low life; squalor; etc.

sledging *n.* (in cricket) the practice among bowlers and fielders of heaping abuse and ridicule on the batsman.

sleep *n.* **1.** *Prison.* a prison sentence longer than seven days but less than three months. *—phr.*
2. lose sleep over, See **lose.**
3. sleep around, to be sexually promiscuous.
4. sleep under the house, to be in disgrace or disfavour.

sleeper *n.* **1.** a sedative drug. **2.** someone or something that unexpectedly achieves success or fame, such as an unadvertised television show. **3.** a book, item of manufacture, etc., which has slow but constant sales. **4.** an agent or spy who remains inactive for a number of years.

sleeping dogs *phr.* **let sleeping dogs lie,** See **lie.**

sleepyhead *n.* a sleepy or lazy person.

sleeve *v. U.S.* to arrest.

sleever *n.* See **long sleever.**

slew *v. Prison.* to spy.

slewed *adj.* **1.** lost; astray. **2.** intoxicated; drunk.

slice *n.* **1.** (formerly) the sum of £5. **2.** the sum of $5. *—phr.* **3. a slice of the cake,** a share in the profits.

sliced bread *phr.* **the best** or **greatest thing since sliced bread,** excellent; first-rate: *He thinks he is the greatest thing since sliced bread.*

slick[1] *n.* **1.** a tyre with a smooth tread used on racing cars. **2.** a very large tyre with a smooth tread used in underground mining operations. *—adj.* **3.** shrewdly adroit; glib: *a slick operator.*

slick[2] *phr.* **slick up,** *U.S.* to make smart or fine.

slicker *n.* **1.** See **city slicker. 2.** *Military.* a ground sheet, also worn as a cape against rain.

Slickers, the *pl. n.* the Sydney City team in the National Soccer League of Australia. Also, **Sydney Slickers.**

slime *n.* **1.** servility; quality of being ingratiating. **2.** an unpleasant person: *Some slime stole my bike. —v.* **3.** to flatter.

slimy *adj.* servile; unpleasantly ingratiating.

sling *n.* **1.** money given as a bribe; protection money. *—v.* **2.** *Prison.* to bribe. *—phr.*
3. sling it in, to abandon an occupation, situation, etc.
4. sling the hook, to pass on the responsibility for a task, etc.
5. sling off at, to speak disparagingly of (someone): *He slings off at his teachers.*

slinter *n.* a trick; slanter.

slip *v.* **1.** to become somewhat reduced in quantity or quality: *The stock market slipped today. —phr.*
2. be slipping, to be losing one's acuteness, abilities, or the like.
3. give (someone) the slip, to escape from.
4. slip a joey, See **joey.**
5. slip it to, (of a male) to have sexual intercourse with.
6. slip into, to attack or criticise.

slipper *v.* **1.** to kick (someone) as in a fight. *—phr.* **2. sink the slipper,** See **sink.**

slippy *adj.* nimble, quick, or sharp.

slipslop *n.* **1.** a sloppy food or drink. **2.** meaningless, loose, or trifling talk or writing.

slip-up *n.* a mistake or blunder: *Several minor slip-ups in spelling.*

slit *n.* the vagina.

slob *n.* a stupid, clumsy, uncouth, or slovenly person.

slog *v.* **1.** to hit hard, as in boxing, cricket, etc. **2.** to drive with blows. **3.** to walk steadily and firmly; plod heavily. **4.** to toil. —*n.* **5.** a strong blow with little finesse. **6.** a spell of hard work or walking.

slop *n.* **1.** a choppy sea. —*phr.* **2. slop over,** (of persons, etc.) to be unduly effusive; gush.

slope[1] *v.* **1.** to move or go: *just sloping along.* —*phr.* **2. slope off,** to go away, especially furtively.

slope[2] *n.* (*offensive*) an Asian. Also, **slopie, slopy.**

sloppy *adj.* **1.** weak, silly, or maudlin: *sloppy sentiment.* **2.** loose, careless, or slovenly: *to use sloppy English.* **3.** untidy, as dress.

slops *pl. n.* beer.

slopshop *n.* a cheap clothing shop.

slosh *n.* **1.** *British.* watery or weak drink. **2.** a heavy blow. —*phr.* **3. sloshing wine,** wine (especially in large containers) for quaffing.

sloshed *adj.* drunk.

slot *n.* **1.** *Prison.* a cell. —*v.* **2.** *Prison.* to lock up in a cell. —*phr.* **3. work one's slot out,** to work very hard.

slot machine *n.* See **one-armed bandit.**

slouch *phr.* **no slouch,** efficient; quick; expert: *He's no slouch at brick laying.*

slough *n.* **1.** *Prison.* a cell. —*phr.* **2. slough up,** *Prison.* to lock someone in a cell, often for solitary confinement.

slow *phr.* **slow on the draw,** (of a person) dull witted; slow to comprehend something.

slow burn *phr.* **do a slow burn,** to smoulder with anger.

slowcoach *n.* a slow or dull person. Also, **slowpoke.**

slug *v.* **1.** to exact heavy payment, either by price or tax: *He slugged you for this motor car.* —*n.* **2.** a high price, tax: *The income tax these days is a real slug.* **3.** a disadvantage. **4.** a bullet. **5.** a drink or swallow of something, especially spirits.

slugger *n.* **1.** one who strikes hard, especially with the fists or a baseball bat. **2.** a prize-fighter.

sluice *n.* a brief wash, especially in running water.

slum *phr.* **slum it,** to be living in circumstances below one's usual or expected standard of living.

slummock *n.* **1.** a careless, untidy person. —*v.* **2.** to act in a slovenly way. [British dialect *slammock*]

slush *n.* silly, sentimental, or weakly emotional writing, talk, etc.

slush box *n.* *Motor.* an automatic transmission.

slushy *n.* Also, **slusher.** a cook's assistant; kitchen worker.

sly boots *n.* a deceitful person.

sly grog *n.* illegally sold liquor.

sly grogger *n.* a person selling alcoholic drink illegally.

sly grog shop *n.* premises on which liquor is sold without the requisite licence.

smack *n.* **1.** heroin. —*adv.* **2.** with a smack; suddenly and sharply. **3.** directly; straight. —*phr.* **4. have a smack at,** to attempt.

5. smack or **slap in the eye, a.** a snub. **b.** a setback or disappointment.

smacked-out *adj.* heavily under the influence of heroin.

smacker[1] *n.* a dollar (formerly a pound).

smacker[2] *n.* **1.** a young boy. **2.** (*in familiar address*) any young male.

smacker[3] *n.* the mouth.

smackhead *n.* a person who takes or is addicted to taking heroin.

small-arm parade *n.* See **short-arm parade.** Also, **small-arm inspection.**

smallest room *n.* a toilet. Also, **smallest room in the house.**

small potatoes *pl. n.* matters or persons of little or no importance. Also, **small beer.**

smalls *pl. n. Chiefly British.* small items of personal laundry; underclothing.

small screen *n.* television (as opposed to the cinema screen).

smarm *v.* to fawn ingratiatingly; flatter; be servile. [British dialect]

smarmy *adj.* ingratiating, falsely charming, or flattering.

smart alec *n.* one who is ostentatious in the display of knowledge or skill, often despite basic ignorance or lack of ability. Also, **smart aleck.**

smart arse *n.* a smart alec; know-all.

smartie *n.* a smug know-all.

smart money *n.* See **money.**

smartypants *n.* a foolishly conceited person.

smashed *adj.* incapacitated as a result of taking drugs, alcohol, etc.

smasher *n.* an extremely attractive person.

smash-hit *n.* an immediately and extremely successful play, film, record, or the like.

smell *phr.*
1. hang around like a bad smell, to make a nuisance of oneself by lingering.
2. live on the smell of an oil rag, See **oil rag.**

smeller *n.* the nose.

smelly *n.* a fart.

smithereens *pl. n.* small fragments.

smitten *adj.* very much in love. [from *smitten*, struck or hit hard]

smoke *phr.*
1. big smoke, the city, or any built-up or closely settled area. [?Aboriginal pidgin]
2. go into smoke, to disappear.
3. go or **end up in smoke,** to have no solid result; end or disappear without coming to anything.
4. in smoke, in hiding.

smoko *n.* **1.** a rest from work; tea-break. **2.** any informal gathering. Also, **smoke-o, smoke-oh.**

smooch *v.* to kiss; cuddle; behave amorously.

smoodge *v.* **1.** to kiss; caress. **2.** to flatter; curry favour. Also, **smooge.**

snack *n.* anything easily done.

snaffle *v.* **1.** to steal. **2.** to take away quickly before anyone else: *Early shoppers snaffled up the sales bargains.* Also, **snavel.**

snafu *Originally U.S. Military n.* **1.** chaos; a muddled situation. —*adj.* **2.** in disorder; out of control; chaotic. —*v.* **3.** to throw into disorder; muddle. [from the initial letters of *s(ituation) n(ormal): a(ll) f(ouled) u(p)*]

snag *n.* a sausage.

snagger *n.* **1.** a shearer who does his job roughly or inexpertly. **2.** a person employed to remove snags from the river.

snaily-horn *adj.* **1.** of cattle, having slightly twisted horns. —*n.* **2.** such a beast.

snake *phr.*
1. have a snake in one's pocket, to be slow or reluctant to buy a round of drinks.
2. like a cut snake, in a frenzy of activity.
3. snake in the grass, a very deceitful or treacherous person; a hidden enemy.

snake-charmer *n.* a fettler.

Snake Gully *n.* a mythical remote place.

snake juice *n.* any alcoholic beverage.

snake oil merchant *n.* a quack or a person selling quack medicine.

snakes *phr.* **go for a snakes,** to urinate. [rhyming slang *snakes hiss* piss]

snake's belly *phr.* **lower than a snake's belly,** unpleasant; mean; despicable.

snaky *adj.* annoyed; angry or spiteful.

snap *n.* **1.** briskness, vigour, or energy, as of persons or actions. **2.** an easy and profitable or agreeable position, piece of work, or the like. —*phr.* **3. snap out of it,** to recover quickly from a mood, as anger, unhappiness, etc.
4. snap one's fingers at, to disregard; scorn.
5. snap one's twig, See **twig.**
6. snap (someone's) head off, to speak angrily and sharply to.

snapper *n.* a ticket inspector on the railways; ticket snapper.

snappy *adj.* **1.** crisp, smart, lively, brisk, etc. —*phr.* **2. make it snappy,** to hurry up.

snarler *n. N.Z.* a sausage.

snarl-up *n.* any upset to the accustomed or expected progression of events, as a traffic jam.

snatch *v.* **1.** *Building Trades.* to leave (a task or contract) uncompleted. —*n.* **2.** a robbery by a quick seizing of goods. **3. a.** the female pudenda. **b.** a woman as a sexual object. —*phr.* **4. snatch one's time,** Also, **snatch it.** to leave a job.

snazzy *adj.* **1.** very smart; strikingly fashionable; stylish. **2.** (of a person) very well-dressed. **3.** brightly patterned; having gay designs.

sneak *v.* **1.** to take surreptitiously, or steal. —*n.* **2.** an act of sneaking; a quiet departure. —*phr.* **3. sneak out** or **off,** to leave quickly and quietly.

sneaker *n.* a shoe with a rubber or other soft sole used especially in gymnasiums.

sneeze *phr.*
1. sneeze at, to show contempt for, or treat with contempt. **2. not to be sneezed at,** worth consideration.

sniffles *phr.* **the sniffles,** a cold, or other condition marked by sniffling.

sniffy *adj.* inclined to sniff, as in disdain; disdainful; supercilious.

snifter *n.* **1.** a small drink of an alcoholic beverage. —*adj.* **2.** excellent; very fine.

snip *v.* to borrow or get money from: *Can I snip you for $2?*

snippet *n.* a small or insignificant person.

snitch[1] *v.* to snatch or steal. [?variant of *snatch*]

snitch[2] *v.* **1.** to turn informer. —*n.* **2.** Also, **snitcher.** an informer. **3.** a feeling of ill-will: *She has a snitch against you.*

snitchy *adj.* bad-tempered.

snooker *n.* **1.** a hiding place. —*v.* **2.** to obstruct or hinder (someone), especially from reaching some object, aim, etc. **3.** to hide.

snookums *n.* a term of endearment. Also, **snooks.**

snoop *v.* **1.** to prowl or pry; go about in a sneaking, prying way; pry in a mean, sly manner. —*n.* **2.** an act or instance of snooping. **3.** one who snoops, as a detective. [Dutch *snoepen,* take and eat (food and drink) on the sly]

snoot *n.* **1.** the nose. **2.** a snob.

snooty *adj.* **1.** snobbish. **2.** haughty; supercilious.

snooze *v.* **1.** to sleep; slumber; doze; nap. —*n.* **2.** a rest; nap. **3.** *Prison.* a period of imprisonment of three months.

snoozer *n.* *Obsolete.* a person; bloke.

snorker *n.* a sausage.

snort *v.* **1.** to laugh outright or boisterously. **2.** to sniff (a powdered drug, as cocaine). —*n.* **3.** an alcoholic drink.

snorter *n.* **1.** anything unusually strong, large, difficult, dangerous, as a fast ball in cricket, a gale, etc. **2.** an alcoholic drink.

snot *n.* **1.** mucus from the nose. **2.** a snotnose. [Old English *gesnot*]

snotnose *n.* a snobbish or affected person. Also, **snotty nose.**

snotty *adj.* **1.** of or pertaining to snot. **2.** (especially of a child) dirty. **3.** snobbish; arrogant. **4.** ill-tempered; cranky.

snotty nose *n.* See **snotnose.**

snout *n.* **1.** a person's nose, especially when large or prominent. —*v.* **2.** to rebuff. —*phr.* **3. have a snout on,** to bear a grudge, ill-will, against.

snow *n.* **1.** cocaine or heroin. **2.** Also, **snowie.** a nickname for a fair-haired person, usually male. —*v.* **3.** to misrepresent or purposely confuse; cover up. **4.** to be under the influence of cocaine or heroin. —*phr.*
5. be snowed under, to be overcome by something, as work.
6. it's snowing down south, a euphemism used to point out that someone's underwear is showing.

snowball *phr.* **a snowball's chance (in hell),** virtually no chance at all.

snowbird *n.* *U.S.* a cocaine or heroin addict.

snow bunny *n.* a young woman in whose mind the social events at a ski resort figure rather more than skiing. Also, **ski bunny.**

snowdrop *v.* to steal laundry from clothes lines.

snow job *n.* **1.** an attempt to distract attention away from certain aspects of a situation by supplying an overwhelming amount of often extraneous information; a cover-up. **2.** a quick and superficial attempt to improve the state or appearance of (something).

snoz *n.* the nose.

snuff *phr.* **snuff it,** to die.

snuffle *phr.* **the snuffles,** a condition of the nose, as from a cold, causing snuffling.

snufflebuster *n.* a prudish teetotaller; wowser.

snug *phr.* **snug as a bug in a rug, a.** warm and comfortable. **b.** content.

so *interj.* **1.** an expression indicating rejection, lack of interest, refusal to understand: *So? —phr.* **2. so what,** what does that matter?

soak *n.* **1.** a heavy drinker. **2.** a prolonged drinking bout. *—phr.* **3. soak up,** to drink (alcohol) especially to excess.

so-and-so *n.* a very unpleasant or unkind person: *He really is a so-and-so.*

soap *n.* **1.** *U.S.* money used for bribery. *—v.* **2.** See **soft soap.** *—phr.* **3. not know someone from a bar of soap,** not know someone at all.

soapbox *n.* any place, means, or the like, used by a person to make a speech, voice opinions, etc.

soapie *n.* a radio or television play presented serially in short regular programs, dealing usually with domestic problems, especially in a highly emotional manner. Also, **soapy, soap opera.** [so called because originally sponsored on U.S. radio by soap manufacturers]

soapy *adj.* flattering; given to using smooth words.

S.O.B. *n.* *U.S.* (offensive) a despicable or contemptible person. Also, **s.o.b., sob.** [*s(on) o(f) (a) b(itch)*]

sobersides *n.* a serious person.

sob-sister *n.* **1.** a woman journalist who writes a newspaper or magazine feature in a sentimental style. **2.** a person who plays on one's emotions.

sob-story *n.* **1.** a story full of sentiment and pathos. **2.** an excuse: *He arrived very late and gave them a sob-story about a broken clock.*

sob-stuff *n.* sentimental matter, as in a sob-story, designed to arouse the emotions.

Socceroos, the *pl. n.* the Australian representative Soccer team. [*soccer* + *(r)oo*]

social disease *n.* any venereal disease.

sock[1] *phr.*

1. pull one's socks up, to make more effort.

2. put a sock in it, to be quiet.

3. sock away, a. to accumulate a store of (something), especially in secret (from the habit of storing money in an old sock). **b.** to consume large quantities of (something, especially alcohol).

sock[2] *v.* **1.** to strike or hit hard. *—n.* **2.** a hard blow. *—phr.* **3. sock it to (someone), a.** an expression of encouragement, exhorting someone to do their best. **b.** to attack, usually verbally.

sod[1] *n.* a damper which is doughy from being badly cooked.

sod[2] *n.* **1.** a disagreeable person. *—phr.*

2. sod all, nothing.

3. sod it, an exclamation of annoyance, disgust, etc. [shortened form of *sodomite,* one who practises sodomy]

soda *n.* **1.** something which can be done easily and successfully: *The exam was a soda.* *—phr.* **2. on a soda,** secure in an easy, comfortable way of life.

soft *adj.* **1.** not hard, trying, or severe; involving little effort: *a soft job.* **2.** easily influenced or swayed, as a person, the mind, etc.; easily imposed upon. **3.** foolish; feeble; weak. *—phr.*

4. a soft touch, one who yields easily to requests for money, etc.

5. be soft on, a. to be sentimentally inclined towards. **b.** to act towards (someone) in a less harsh manner than expected.

6. have a soft spot for someone or **something,** to be fond of someone or something.

7. soft in the head, stupid; insane.

softie *n.* **1.** a generous or soft-hearted person. **2.** one who is easily duped. **3.** one (especially a man) who is not as brave or hardy as others consider proper. Also, **softy.**

soft-pedal *v.* **1.** to make concessions or be conciliatory, as in an argument. **2.** to tone down; make less strong, uncompromising, noticeable, or the like. [from *soft pedal* a piano pedal for lessening the volume]

soft soap *n.* smooth words; flattery.

soft-soap *v.* to ply with smooth words; cajole; flatter.

soixante-neuf *n.* simultaneous fellatio and cunnilingus; sixty-nine. [French: sixty-nine]

SOL[1] *adj.* bad-tempered; irritable. Also, **S.O.L.** [*S(hit) O(n) L(iver)*]

SOL[2] *adj.* *U.S.* unlucky; in a hopeless position. Also **S.O.L.** [*S(hit) O(ut of) L(uck)*]

sold *v.* See **sell.**

solid *adj.* **1.** on a friendly favourable, or advantageous footing. *—phr.* **2. a bit solid,** unreasonable; harsh; severe.

solitary *n.* *Prison.* solitary confinement.

sollicker *n.* anything or anyone remarkably big, good, great, etc.

so long *interj.* goodbye.

some *adj.* **1.** of considerable account or consequence; notable of the kind: *That was some storm.* *—adv.* **2.** to some degree or extent; somewhat. **3.** *U.S.* to a great degree or extent; considerably: *That's going some!*

somersault *v.* *Prison.* to alter one's plea from guilty to not guilty.

song *phr.* **for a song,** at a very low price.

song and dance *n.* a fuss or commotion: *Bill kicked up a great song and dance about that.*

sonky *adj.* foolish; silly.

son of a bitch *n.* **1.** (*offensive*) a mean and contemptible man. **2.** an object as a car, etc. which has incurred one's wrath. **3.** a familiar greeting: *How are you going, you old son of a bitch?*

son of a gun *n.* **1.** *Chiefly U.S.* an approving term for a man, especially one who has demonstrated his masculinity in some way. *—phr.* **2. I'll be a son of a gun!,** an exclamation of surprise, admiration, etc.

sook *n.* **1.** a timid or cowardly person. **2.** a dim-witted person.

sool *v.* to incite or urge (someone).

sooler *n.* one who incites or urges others to a course of action.

sooner *n.* **1.** a lazy person. **2.** a horse which pulls back. [one who would *sooner* loaf than work, *sooner* go backwards than forwards, etc.]

sop *n.* a weak or cowardly person.

soppy *adj.* excessively sentimental; mawkish; silly.

sore *adj.* **1.** *Originally U.S.* irritated, offended, or feeling aggrieved: *What are you sore about? —phr.*

2. done up like a sore finger or **toe, a.** dressed up in one's best clothes. **b.** overdressed.

3. stand or **stick out like a sore finger** or **toe,** to catch the eye by virtue of a contrast with surrounding people or objects.

sorehead *n.* a disgruntled or vindictive person, especially an unsportsmanlike loser.

sort *n.* **1.** a woman or girl: *a good sort, a drack sort. —phr.*

2. a good sort, a. a sexually attractive woman or man. **b.** one who is likeable, trustworthy, reliable.

3. out of sorts, not in a normal condition of good health, spirits or temper.

4. sort of, to a certain extent; in some way; as it were.

sound *phr.* **sound off, a.** to speak or complain frankly. **b.** to speak angrily; lose one's temper. **c.** to boast; exaggerate. **d.** *U.S.* to call one's name, sequence, number, etc.

soup *n.* **1.** *Surfing.* foaming water, caused by the breaking of a wave. *—phr.*

2. in the soup, in trouble.

3. soup up, to modify (an engine, especially of a motor car) in order to increase its power

sourpuss *n.* one having a sour or gloomy disposition; an embittered person.

souse *v.* **1.** to intoxicate. **2.** to drink to intoxication. *— n.* **3.** a drunkard. [from *souse,* to steep in pickling liquid]

south *phr.* **down south,** (of money) not to be spent; to be saved; put in the bank.

southerly buster *n.* a violent, cold southerly wind blowing on the south-eastern coast of Australia, causing a sudden drop in the temperature, and often accompanied by dust squalls; brickfielder.

southpaw *n.* **1.** a person who is left-handed. **2.** *Boxing.* a boxer who stands with his right arm and right leg forward. *— adj.* **3.** left-handed.

souvenir *v.* to pilfer.

sozzled *adj.* drunk. [obsolete *sozzle* drunken stupor (akin to *souse*)]

spaced-out *adj.* in a euphoric or dreamy state, as if under the influence of a hallucinogen. Also, **spaced, spaced out.**

spacey *adj.* dreamy; hallucinatory.

spade *n.* (*offensive*) someone of very dark skin, as a Negro, Aborigine, etc. [from *spade* a black figure used on playing cards]

spadger *n.* a sparrow. Also, **spag, spoggy.** [British dialect]

spag *n.* **1.** spaghetti. **2.** (*offensive*) an Italian. **3.** (*offensive*) the Italian language.

spaghetti *n.* *Electronics.* plastic sleeving for wires.

Spanish dancer *n.* cancer. [rhyming slang]

spanking *adj.* unusually fine, great, large, etc.

spanner *phr.* **spanner in the works,** any cause of confusion or impediment.

spare tyre *n.* a roll of fat around a person's midriff.

spark *v.* **1.** to kindle or stimulate (interest, activity, etc.). *—phr.*
2. bright spark, a clever or lively person.
3. spark off, to bring about; cause; precipitate.

sparkler *n.* a diamond.

sparks *n.* a radio operator or electrician.

sparring partner *n.* one with whom one has happily shared experiences but with whom it has seldom been possible to agree. [from one who practises with a boxer in training]

sparrow fart *n.* dawn; very early morning.

sparrowgrass *n.* asparagus.

spastic *n.* **1.** a fool. **2.** a clumsy person.

speak *phr.* **speak for yourself,** an expression of disagreement.

spear *n.* **1.** dismissal, the sack. **2.** a surfboard. *—v.* **3.** to move rapidly, especially in a restricted passage: *a racehorse spearing down the rails.* *—phr.*
4. get the spear, to be dismissed, as from employment, etc.
6. spear on, to proceed rapidly.

spec *n.* **1.** speculation. *—adj.* **2.** speculative: *spec builder.* *—phr.* **3. on spec,** as a guess, risk, or gamble: *to buy shares on spec.*

special *phr.* **on special,** available at a bargain price.

specimen *n.* a person as a specified kind, or in some respect a peculiar kind, of human being: *He's a poor specimen.*

specs *pl. n.* spectacles; glasses.

spectator sport *n.* any activity in which one plays no active part, being present merely as an onlooker.

speed *n.* amphetamines; pep pills.

speed-cop *n.* a policeman, often a motorcyclist, who enforces the observation of speed-limits.

Speed Gordon *phr.*
1. go or **shoot through like Speed Gordon,** to move very quickly.
2. in more trouble than Speed Gordon, in great trouble; facing many difficulties. Also, **Flash Gordon.** [from the character in the space adventure comic strip]

speed-merchant *n.* one who drives a motor vehicle extremely fast.

speedo *n.* speedometer. [shortened form]

Speewa *n.* a mythical outback town.

spell *n.* **1.** an interval or space of time, usually indefinite or short. **2.** a rest from work.

spic *n.* **1.** (*offensive*) any person of European descent. **2.** *U.S.* a person from a Spanish-speaking community. Also, **spick, spik.** [?Spanish mispronunciation of *speak*]

spice *n.* (formerly) plausible patter used by showmen to draw the public.

spider *n.* **1.** (formerly) a drink of lemonade or ginger beer with brandy. **2.** an aerated soft drink to which ice-cream is added. **3.** *Trotting.* a pacing gig. **4.** See **rock spider.**

spiderman *n. British.* one who works on high buildings, especially as an erector of the steel framework.

spiel *n.* **1.** glib or plausible talk, especially for the purpose of persuasion, swindling, seduction, etc. **2.** any talk or speech. *—v.* **3.** to talk plausibly; deliver a patter or sales talk. **4.** to attempt to lure, persuade, or deceive (someone) by glib talk. [German: play]

spieler *n.* **1.** one who delivers or is proficient at, delivering a spiel; a glib talker. **2.** a swindler; cardsharper.

spiffy *adj. British.* spruce; smart; fine. [British dialect *spif* smartness]

spiflicate *v.* to destroy utterly; hurt, punish, or damage; destroy or kill. Also, **spifflicate.**

spiflicated *adj.* drunk. Also, **spifflicated, spiffed.**

spike *n.* **1.** a High-Churchman. *—v.* **2.** (of a newspaper editor) to reject (a story). **3. a.** to add alcoholic liquor to a usually non-alcoholic drink. **b.** to add a hard drug to a soft drug. **c.** to add anything to any drink or food. *—phr.* **4. spike someone's guns,** to frustrate (someone's) plans.

spiky *adj.* High-Church Anglican.

spill *v.* to divulge, disclose, or tell: *spill the beans, spill one's guts.*

spin *n.* **1.** a state of confusion or excitement. **2.** experience or chance generally: *a rough spin, a fair spin.* **3.** Also, **spinner, spinnaker. a.** (formerly) the sum of £5. **b.** the sum of $5. **4.** *Prison.* a prison sentence of five years' duration. *—phr.* **5. spin a yarn, a.** to tell a tale. **b.** to tell a false or improbable story or version of any event.

spindleshanks *pl. n.* **1.** long, thin legs. **2.** (*construed as singular*) a tall, thin person with such legs.

spinebash *v.* **1.** to rest; loaf. *—n.* **2.** a rest.

spinner *phr.* **come in spinner,** See **come.**

spit *n.* **1.** Also, **dead spit,** the image, likeness or counterpart of a person, etc. *—phr.*
2. on the spit, under pressure; beset by difficulties.
3. spit chips, to be very annoyed.
4. spit it out, speak up.
5. the big spit, vomit.

spitting distance *n.* a very short distance: *They lived within spitting distance of each other.*

spitting image *n.* the image, likeness or counterpart of a person, etc. Also, **spit, dead spit.**

spiv *n. Originally British.* one who lives by his wits, without working or by dubious business activity, and usually affecting ostentatious dress and tastes. [back-formation from British dialect *spiving* smart. See *spiffy*]

splash *v.* **1.** to display or print very noticeably, as in a newspaper. **2.** to spend (money) freely. *—phr.*
3. make a splash, to be noticed; make an impression on people.
4. splash the boots, to urinate.

splice *v.* to join in marriage. [from *splice,* to join together or unite]

spliced *adj.* drunk.

split *v.* **1.** to divide something with another or others. **2.** to commit a betrayal by divulging information.

3. to leave hurriedly. —*n.* **4.** something combining different elements, as a drink composed of half spirits, half soda-water. **5.** a dish made from sliced fruit (usually banana) and ice-cream, and covered with syrup and nuts. **6.** a drink containing only half the usual quantity. **7.** an act or arrangement of splitting, as of a sum of money. —*phr.*

8. split on, to betray, denounce, or divulge secrets concerning.

9. split one's sides, to laugh heartily.

10. split up, to part; leave each other; become separated.

splosh *n.* money.

spoggy *n.* a sparrow. Also, **spadger, sproggy.** [British dialect]

spoke *phr.* **put a spoke in one's wheel,** to interfere with one's plans.

spondulicks *n.* money. Also, **spondulix, spons, spon.**

sponge *v.* **1.** to live at the expense of others. —*phr.*

2. sponge on, to take advantage of the generosity of; live at the expense of.

3. throw in the sponge, to give up; abandon hope or one's efforts.

sponger *n.* a freeloader; a person who lives off another's generosity.

spoof *n.* semen.

spook *n.* **1.** a ghost; a spectre. **2.** an agent of an intelligence organisation; a spy.

spooked *adj.* frightened; on edge; nervous.

spooky *adj.* **1.** like or befitting a spook or ghost; suggestive of spooks; eerie. **2.** (of a surf) difficult; unpredictable. Also, **spookish.**

spoon *v.* **1.** to show affection towards, especially in an openly sentimental manner. —*phr.* **2. be born with a silver spoon in one's mouth,** to inherit social or financial advantages and privileges.

spoony *adj.* **1.** foolishly or sentimentally amorous. **2.** foolish; silly. —*n.* **3.** one who is foolishly or sentimentally amorous. **4.** a simple or foolish person. Also, **spooney.**

sport *n.* **1.** (a term of address, usually between males) any person approached as a friend: *Goodday, sport.* **2.** one who is interested in pursuits involving betting or gambling. —*v.* **3.** to display freely or with ostentation: *to sport a roll of money.* —*phr.*

4. a good sport, a. a person of sportsmanlike or admirable qualities; one who exhibits boldness or good humour in the face of risk or ridicule. **b.** a person who is easygoing, good-natured and agreeable.

5. be a sport, a. a request to play fair. **b.** to accede to a request; be agreeable.

sporty *adj.* **1.** flashy; vulgarly showy. **2.** stylish.

spot *n.* **1.** a small quantity of something: *a spot of tea.* **2.** an alcoholic drink: *He stopped at the pub for a spot.* **3.** a predicament: *He was in a bit of a spot when the crash came.* **4. a.** (formerly) the sum of £100. **b.** the sum of $100. —*v.* **5.** to detect or recognise. —*phr.*

6. change one's spots, to alter one's fundamental character.

7. hit the spot, to satisfy (as of food or drink).

8. **knock spots off,** to outdo without difficulty or by a large margin.

9. **on the spot,** in trouble, embarrassment, or danger.

spot kick *n. Soccer.* a penalty kick.

spot-on *adj.* **1.** absolutely right or accurate; excellent. *—interj.* **2.** an exclamation of approbation, etc.

spout *v.* **1.** to utter or declaim in an oratorical manner. *—phr.* **2. up the spout, a.** ruined; lost. **b.** pawned. **c.** pregnant.

spread *v.* **1.** to exert (oneself) to an unusual extent to produce a good effect or fine impression. *—n.* **2.** a meal set out, especially a feast. **3.** a pretentious display. **4.** *Chiefly U.S.* a large property or area of land. *—phr.*

5. **a good spread,** a lot of publicity, especially in the various channels of the media.

6. **spread it on thick,** to exaggerate.

spring *v.* **1.** to catch out; to come upon unexpectedly. **2.** to cause or enable (someone) to escape from prison.

spring chicken *n.* (*usually with a negative*) a very young person: *I'm no spring chicken.*

sprog *n.* **1.** a child or youngster. **2.** a new recruit, as in an airforce.

sproggy *n.* a sparrow. Also, **spadger, spoggy, spridgy, spudgy, spug, sproug.** [British dialect]

spruik *v.* **1.** to harangue or address a meeting. **2.** (of a showman) to harangue prospective customers to entice them into his tent, stripjoint, etc. [origin uncertain; ?from Dutch *spreken* speak]

spruiker *n.* a showman who harangues prospective customers.

spud *n.* **1.** a potato. **2.** an Irishman. *—phr.* **3. not the clean spud** or **potato,** See **potato.**

spud-bashing *n.* the act of peeling potatoes.

spudder *n.* an employee at an oil well. [from *spudder,* a rig or oil rig, especially one used to begin a well]

spunk *n.* **1.** pluck; spirit; mettle. **2.** semen. **3.** a good-looking person. [blend of *spark,* a fiery particle and obsolete *funk,* spark, touchwood]

spunky *adj.* **1.** plucky; spirited. **2.** good-looking; attractive.

spurs *phr.* **win/gain/get one's spurs,** to achieve one's first distinction or success.

square *n.* **1.** one who is ignorant or uninterested in the latest fads. *—v.* **2.** to bribe. *—adj.* **3.** substantial or satisfying: *a square meal. —adv.* **4.** solidly or directly: *to hit a nail square on the head. —phr.*

5. **on the square, a.** fair; fairly. **b.** abstaining from alcohol; teetotal.

6. **square off, a.** to apologise; make recompense. **b.** Also, **get square.** to get revenge; pay someone back.

7. **square the circle.** to attempt the impossible.

squared ring *n.* the boxing ring. Also, **squared circle.**

squarehead *n.* **1.** a conservative in manners, dress and behaviour. **2.** *Prison.* a law-abiding person.

square one *n.* the beginning; the point of departure: *We're back to square one.*

square-shooter *n.* an honest, fair person.

square wheel *phr.* **silly as a square wheel,** ridiculous; very silly.

squash *v.* **1.** to silence, as with a crushing retort. —*n.* **2.** a social gathering or party; social.

squat *v.* **1.** (formerly) to settle on land, often large tracts, without government permission. **2.** to occupy a building without title or right. —*n.* **3.** a building which is occupied without title or right: *He shares a squat in Glebe.*

squatter *n.* **1.** (formerly) one who settled on Crown land to run stock, especially sheep, initially without government permission, but later with a lease or licence. **2.** one of a group of rich and influential rural landowners. **3.** one who occupies a building without right or title.

squatter's daughter *n.* water. [rhyming slang]

squattocracy *n.* the long-established and wealthy landowners who regard themselves as an aristocracy. [*squat(ter)* + *(aris)tocracy*]

squawk *v.* **1.** to complain loudly and vehemently. —*n.* **2.** a loud, vehement complaint.

squeak *n.* **1.** a narrow escape. —*v.* **2.** to confess or turn informer.

squeaker *n.* **1.** the short-nosed rat-kangaroo. **2.** any of various cicadas which make a squeaking noise.

squeaky-clean *adj.* **1.** very clean. **2.** morally irreproachable.

squeal *n.* **1.** a protest or complaint. —*v.* **2.** to turn informer. **3.** to protest or complain. **4.** to disclose or reveal, as something secret.

squeeze *n.* **1.** a situation from which extrication is difficult: *in a tight squeeze.* **2.** the act of blackmailing. —*v.* **3.** to put pressure upon (a person or persons) to act in a given way, especially by blackmail.

squeezebox *n.* an accordion.

squelch *n.* **1.** a crushing argument or retort. —*v.* **2.** to put down or suppress completely; silence, as with a crushing retort.

squib *v.* **1.** to behave in a fearful or cowardly manner. —*n.* **2.** Also, **damp squib.** any plan, project, etc., which does not eventuate.

squiffy *adj.* **1.** slightly intoxicated; tipsy. **2.** crooked; askew.

squint *n.* a furtive glance.

squire *n.* (*humorous*) a form of address to a man.

squirrel *n.* a person who hoards objects of little value. [from the hoarding habit of the squirrel]

squirrel grip *n.* *Football.* an illegal tackle in which pressure is applied to the testicles of the tackled player; Christmas hold. [a handful of *nuts* testicles]

squirt *n.* **1.** an insignificant, self-assertive fellow. **2.** a short person.

squish *v.* **1.** to squeeze or squash. **2.** (of water, soft mud; etc.) to make a gushing sound. —*n.* **3.** a noise made by squishing.

squitters *n.* diarrhoea.

squiz *v.* **1.** to look at quickly but closely. —*n.* **2.** a quick but close look.

stab *n.* **1.** an attempt; try. —*phr.* **2.** **stab (someone) in the back,** to do harm to (somebody), especially somebody defenceless or unsuspecting, as by making a treacherous attack upon his reputation.

stack *n.* **1.** a great quantity or number. **2.** a combination of amplifiers and speaker boxes. **3.** (*pl.*) a great amount. *—phr.*

4. blow one's stack, See **blow.**

5. stack it on, to exaggerate one's concern, grief, anger, etc.

6. stack on a blue, a. to start a fight. **b.** to create a disturbance.

stacked *adj.* Also, **well stacked.** (of a woman) having big breasts; buxom.

stack-up *n.* an accident involving a number of vehicles, railway carriages, etc.

stag *n.* **1.** a man, especially one at a social gathering exclusively for men. *—v.* **2.** to go to a social function without a woman partner. *—adj.* **3.** for or of men only.

stag party *n.* a party exclusively for men.

stake *phr.* **pull up stakes,** to leave one's job, home, etc., and move away.

stakes *pl. n.* an assumed condition of competitiveness: *beauty stakes.* [from *stakes* a prize in a race or contest]

stall *n.* **1.** anything used as a pretext, pretence, or trick. *—v.* **2.** to act evasively or deceptively. **3.** *Sport.* to play below one's best in order to deceive for any reason. *—phr.* **4. stall for time,** to engage in a delaying tactic.

stamping ground *n.* the habitual place of resort; familiar territory.

stand *v.* **1.** to bear the expense of; pay for. *—phr.*

2. stand on one's dig, to claim respect for one's rights, dignity, etc.

3. stand out like dogs' balls/a sore thumb/a sore toe, to be prominent or conspicuous.

4. stand up, to fail to keep an appointment with, especially with a member of the opposite sex.

standover merchant *n.* one who bullies or intimidates; one who threatens violence to gain a desired result; hoodlum. Also, **standover man.**

star *phr.* **see stars,** to seem to see bright flashes of light, as after a heavy blow on the head.

stare *phr.* **stare one in the face, a.** to be inescapably obvious. **b.** to be impending and require immediate action.

starkers *adj.* **1.** stark-naked. **2.** absolutely mad; insane. [from *stark* absolutely, utterly]

starters *pl. n.* **1.** the first course of a meal. *—phr.* **2. for starters,** in the first place: *Well, for starters, he's a crook.*

starve *v.* **1.** to be hungry. *—phr.* **2. starve the lizards,** See **lizard.**

state *phr.* **the state of play** or **the art,** the current situation.

steady *n.* **1.** a regular boyfriend or girlfriend. *—phr.*

2. go steady, to go about regularly with the same boyfriend or girlfriend.

3. steady on, be calm! control youself!

steak and kidney *n.* Sydney. Also, **steak 'n.** [rhyming slang]

steal *n.* **1.** something acquired at very little cost or at a cost well below its true value. *—phr.* **2. steal (someone's) thunder,** to appropriate or use (another's) idea, plan, etc.

steam *n.* **1.** power or energy. **2.** cheap wine. *—adj.* **3.** antiquated; old-fashioned; belonging to the age of steam: *steam radio. —phr.* **4. let off steam,** to release repressed emotions, by behaving in an unrestrained manner.

steamed-up *adj.* excited or angry: *No need to get steamed-up about it.*

steamy *adj.* licentious.

Steelers, the *pl. n.* the Illawarra team in the N.S.W. Rugby Football League.

Steele Rudd *n.* a potato. [rhyming slang *Steele Rudd* spud; from the pen name of the Australian author A.H. Davis]

steep *adj.* **1.** unduly high, or exorbitant, as a price or amount. **2.** extreme or extravagant, as a statement.

steer *v.* **1.** to direct the course of. *—phr.*
2. bum steer, a misleading idea or suggested course of action.
3. steer clear of, to avoid.

step *phr.* **step on it,** to hasten; hurry.

step-ins *pl. n.* a woman's elasticised foundation garment, without fastenings; roll-ons; *N.Z.* easies.

stew *v.* **1.** to fret, worry, or fuss. *— n.* **2.** a state of uneasiness, agitation, or worry. *—phr.* **3. stew in one's own juice,** to suffer one's own misfortunes or the consequences of one's own actions without help.

stewed *adj.* intoxicated or drunk.

stick[1] *n* **1.** a person: *a decent stick; a dry old stick.* **2.** a surfboard. **3.** the penis. **4.** *U.S.* a marijuana cigarette. *—phr.*
5. give stick, a. to tease. **b.** to abuse or assault.
6. in a cleft stick, in a dilemma, awkward position, etc.
7. more than one can poke or **shake a stick at,** a lot of; many; much.
8. stick-in-the-mud, a cautious, unadventurous person.
9. the sticks, a. an area or district regarded as lacking in the amenities of urban life. **b.** the outback. **c.** *Football.* the goal posts.
10. up the stick, pregnant.
11. wrong end of the stick, a complete misunderstanding of facts, a situation, etc.

stick[2] *phr.*
1. stick around, to stay nearby; linger.
2. stick in one's throat, to be hard to accept.
3. stick it! an expression of contempt, dismissal, disgust, etc.
4. stick one's neck out, to expose oneself to blame, criticism, etc.; to take a risk.
5. stick one's nose in, to pry, interfere.
6. stick out, to be obvious, conspicuous, etc.: *to stick out like a sore thumb.*
7. stick out for, to continue to ask for; be persistent in demanding.
8. stick (something) out, to endure; put up with until the very end: *They were bored by the film but stuck it out for two hours.*
9. stick together, to remain friendly, loyal, etc., to one another.
10. stick up, to rob, especially at gunpoint.
11. stick up for, to speak or act in favour of; defend; support.
12. stick up to, to confront boldly; resist strongly.

13. (you can) stick that for a joke or **lark,** an expression indicating complete and often derisive rejection of a proposal, plan, etc.

stick book *n. Prison.* pornographic magazines or books.

sticker *n.* **1.** something that nonplusses or puzzles one. **2.** *N.Z.* a traffic-offence notice, usually stuck to the offending vehicle.

stickjaw *n.* any glutinous toffee, chewing gum, pudding, etc.

stick-picker *n.* a person employed to pick up sticks lying around on land which has been cleared or burnt off; emu-bobber.

stick-up *n.* a hold-up or robbery.

sticky[1] *adj.* **1.** difficult to deal with; awkward; troublesome. **2.** disagreeable; painful. *—phr.* **3. sticky on your boot,** *Football.* See **chewie.**

sticky[2] *n.* **1.** a look. *—phr.* **2. have a sticky,** to take a look. [from *stickybeak]*

stickybeak *n.* **1.** an inquisitive, prying person. *—v.* **2.** to pry or meddle.

sticky tape *n.* an adhesive tape, made of cellulose and usually transparent; durex. Also, **Scotch tape.**

sticky wicket *n.* See **wicket.**

stiff *adj.* **1.** drunk. *—n.* **2.** *Prison.* a letter sent illegally out of gaol. **3.** a dead body; corpse. **4.** a drunk. **5.** a racehorse that is certain to lose. **6.** Also, **stiffy, stiffie.** an erect penis. *—adv.* **7.** completely; extremely: *We were all scared stiff. —phr.*

8. big stiff, a foolish person.

9. stiff cheese/cheddar/luck. a. bad luck. **b.** an off-hand expression of sympathy. **c.** a rebuff to an appeal for sympathy.

10. stiff with, bristling with: *The area was stiff with cops.*

stiffen *phr.* **stiffen the lizards,** See **lizard.**

stiffy *n.* an erect penis. Also, **stiffie.**

sting *v.* **1.** to get money from, especially by begging, overcharging, or swindling. *—n.* **2.** *U.S.* a confidence trick. **3.** a drug, especially in a hypodermic injection, given to a racehorse. **4.** a strong drink.

stinger *n.* a stinging blow, remark, or the like.

stink *v.* **1.** to be very inferior in quality. **2.** to have a large quantity of something, especially money. *—n.* **3.** a commotion; fuss; scandal: *kick up a stink. —phr.* **4. play stink finger,** to engage in erotic play of the fingers with the female genitals.

stinker *n* **1.** a dishonourable, disgusting, or objectionable person. **2.** something difficult, as a task, problem, etc. **3.** a very hot and humid day.

stinking *adj.* **1.** disgusting; disgraceful. **2.** drunk. **3.** very rich. **4.** used to add emphasis to a statement: *stinking bastard.*

stinko *adj.* **1.** stinking. **2.** drunk.

stinkpot *n.* **1.** one who or that which stinks. **2.** an objectionable person.

stipe *n.* a stipendiary steward at a racecourse.

stir[1] *v.* **1.** to make trouble, tease, upset the equanimity of others. *—n.* **2.** a commotion. *—phr.*

3. **stir along,** to make frequent use of the gears when driving a heavy vehicle.
4. **stir the porridge** or **custard,** See **porridge.**
5. **stir the possum,** to instigate a debate on a controversial topic, especially in the public arena; create a disturbance.

stir[2] *n.* prison.

stir-crazy *adj.* crazy as a result of being institutionalised in gaol.

stirrer *n.* troublemaker, agitator, especially in class-room or work stations.

stitch *phr.*
1. **have not a stitch on,** to be naked.
2. **in stitches,** laughing unrestrainedly.

stodge *n.* **1.** heavy, indigestible, and unappetising food. **2.** uninteresting or difficult reading matter. —*v.* **3.** to stuff full with food, etc.

stoke *phr.* **stoke up,** to eat.

stoked *adj.* **1.** under the influence or as if under the influence of drugs or alcohol. **2.** enthusiastic; thrilled; delighted.

stomach *phr.* **one's stomach** or **belly thinks one's throat is cut,** to be very hungry.

stomp *v.* to stamp.

stone *n. Mining.* opal.

stone-cold *phr.* **stone-cold sober,** completely sober.

stoned *adj.* **1.** completely drunk or under the influence of drugs. —*phr.*
2. **stoned out of one's head/mind/brain,** to be completely irrational due to the influence of drugs or alcohol.
3. **stone the crows,** See **crows.**

stonk *n. Military.* heavy shelling; a severe bombardment.

stonkered *adj.* **1.** defeated; destroyed; overthrown. **2.** exhausted. **3.** confounded; discomfited. **4.** drunk. **5.** extremely lethargic or incapacitated, as after a large meal.

stony-broke *adj.* having no money whatever. Also, **stony, stone-broke.**

stooge *n.* **1.** an entertainer who feeds lines to a comedian and is often the object of his ridicule. **2.** one who acts on behalf of another, especially in an obsequious, corrupt, or secretive fashion. —*v.* **3.** to act as a stooge.

stool pigeon *n.* **1.** a person employed as a decoy or secret confederate, as by gamblers. **2.** *Originally U.S.* a police informer. Also, **stoolie.**

stop *phr.*
1. **pull out all (the) stops, a.** to speak with extreme emotion. **b.** to push oneself or a machine to the utmost.
2. **stop by,** to call somewhere briefly on the way to another destination.

stop-go *n.* **1.** a period of successive inflation and deflation. —*adj.* **2.** of or pertaining to such a period: *stop-go politics.*

stopout *n.* one who habitually stays out till late at night at parties, etc.

stopper *n.* any kind of sleeping pill.

storm *phr.*
1. **cook up a storm,** to engage in activities which will lead to a confrontation or quarrel.
2. **storm in a teacup,** a great deal of fuss arising out of a very unimportant matter.

story *n.* a lie; a fib.

storyteller *n.* one who tells untrue or fantastic stories, as if they were true.

stoush *n* **1.** a fight. **2.** an artillery bombardment. —*v.* **3.** to fight (someone or something). —*phr.* **4.** **the big stoush,** World War I.

stow *phr.* **stow it,** an impolite request to someone to be quiet.

strafe *v.* to punish.

straight *adj.* **1.** conforming to orthodox, conservative forms of behaviour, as heterosexuality, avoidance of illegal drugs, etc. **2.** reliable, as reports, information, etc. —*n.* **3.** one who conforms to orthodox forms of behaviour. **4.** a heterosexual. **5.** *Prison.* a sentence with no non-parole period. —*phr.*
6. **go straight,** to lead an honest life, especially after a prison sentence.
7. **play straight down the line,** to behave or act with directness and honesty.
8. **set** or **put (someone) straight,** to point out an error to (someone).

straight bat *phr.* **play with a straight bat,** to act honestly, straightforwardly.

straight face *n.* a deliberately serious expression, especially in an attempt to suppress laughter: *She managed to keep a straight face despite their antics.*

straight-out *adj.* **1.** frank; aboveboard. **3.** (of a bet) for a win only.

straightshooter *n.* a person who is honest and open in his dealings.

straight-up *adj.* **1.** honest; fair. **2.** of or pertaining to sexual intercourse in the more usual positions: *She likes only straight-up sex.*

straight wire *adj.* See **wire.**

strain *phr.* **strain the potatoes,** See **potatoes.**

Straitsman *n.* a person living on an island in Bass Strait.

strange *adv.* slightly unbalanced mentally: *She is a little strange.*

stranger *phr.* **a** or **the little stranger,** an unborn or new-born infant.

strap *phr.* **strapped for cash,** low on funds; broke.

straphanger *n.* a commuter. [in reference to a passenger in an overfull bus, train, or the like who has to stand holding on to a strap suspended from above]

strapper *n.* a tall, robust person.

strapping *adj.* **1.** tall, robust, and strongly built. **2.** very large of its kind; whopping. —*n.* **3.** a thrashing.

straw *phr.*
1. **a straw in the wind,** an indication of how things will turn out.
2. **grasp** or **clutch at straws,** to turn to desperate remedies or insubstantial expedients.
3. **the straw that broke the camel's back,** the final fact, circumstance, etc., which makes a situation unbearable. Also, **the last straw.**

straw bail *n.* bail papers signed by a corrupt policeman on worthless surety.

straw boss *n.* *Chiefly U.S.* a subordinate boss.

streak *n.* **1.** a run (of luck): *to have a streak of bad luck; a winning streak; a losing streak.* **2.** a tall, thin person. —*v.* **3.** to run stark naked through a crowd of people in a street, at a cricket match, etc., for dramatic effect. *—phr.*

4. be on a streak, *Mining.* to come across rock showing good colour opal.

5. long streak of pelican shit, See **pelican shit.**

6. streak of misery, a very tall, thin, unhappy person.

streaker *n.* one who runs naked through a crowded place.

street *phr.*

1. on the streets, a. earning one's living as a prostitute. **b.** destitute; homeless.

2. up one's street or **alley,** in the sphere that one knows or likes best.

3. win by a street, to win easily.

streetwalker *n.* one who walks the streets, especially a soliciting prostitute.

street-wise *adj.* skilled in living in an urban environment; knowing how to survive on the streets.

strength *phr.*

1. get with the strength, to side with the most powerful, influential person or group.

2. the strength of (something), the reliable information concerning (something): *What's the strength of his latest story?*

stretch *n.* **1.** a term of imprisonment. **2.** a form of address to a tall person.

stretcher case *n. Sport.* one who is injured or unconscious, as after a heavy tackle, etc.

strewth *interj.* See **struth.**

strides *pl. n.* trousers.

strife *phr.* **in strife,** in trouble.

strike *phr.*

1. strike a blow, to start or resume work.

2. strike-a-light, an exclamation of surprise, etc.

3. strike me lucky!, an exclamation of surprise, etc. [catchphrase of Australian comedian Roy Rene (Mo)]

4. strike me pink!, an exclamation of surprise, indignation, etc.

Strikers, the *pl. n.* the Leichhardt team in the National Soccer League of Australia.

strine *n.* **1.** Australian English, humorously and affectionately regarded. **2.** the form of it which appeared in the books of Alastair Morrison, pen-name 'Afferbeck Lauder', where it was written in scrambled form, as in *Gloria Soame* for *glorious home, muncer go* for *months ago, Emma Chisit,* for *how much is it?* etc.

string *n.* **1.** limitations on any proposal: *a proposal with no strings attached. —phr.*

2. keep on a string, to have someone under one's control, especially emotionally: *She kept him on a string and then refused to marry him.*

3. pull strings, to use social contacts and other means not directly connected with one's position, ability, etc., to gain some advantage, advancement, etc.

4. string along or **on,** to deceive in a progressive series of falsehoods; con.

5. string along with, cooperate with; agree with.

6. string out, to extend over a period of time; prolong.

string-pulling *n.* the act of bypassing normal channels, methods, etc., by using social contacts and other means not directly connected with one's position, ability, etc., to gain some advantage, advancement, etc.: *With a little string-pulling he might get us front-row seats.*

stringy-bark *adj.* **1.** rough, rustic or uncultured: *a stringy-bark settler; a stringy-bark carpenter.* —*n.* **2.** (formerly) a poor rural settler, often an ex-convict or native-born of convict parents.

stringy-bark cockatoo *n.* (formerly) a cockatoo farmer who farmed the poorest land on which only stringy-bark gums would grow.

strip *n.* a sporting uniform.

strong *phr.* **going strong,** continuing vigorously, in good health: *He is very old but still going strong.*

strongarm *adj.* **1.** having, using, or involving the use of muscular or physical force: *strongarm methods.* —*v.* **2.** *U.S.* to employ violent methods upon. **3.** to steal from by force.

stroppy *adj.* rebellious and difficult to control; awkward; complaining.

struck *phr.* **struck on,** in love or infatuated with.

struth *interj.* an exclamation expressing surprise or verification: *Did he say that? Struth!; Struth he did!* Also, **strewth, 'struth.**

stubbies *pl. n.* men's shorts with an elasticised waist. [Tradename]

stubby *n.* a small, squat beer bottle.

stubby cooler *n.* a polystyrene casing designed as insulation for a stubby. Also, **stubby holder.** *phr.*

1. get stuck into, a. to set about (a task) vigorously. **b.** to attack (someone) vigorously either physically or verbally. **c.** to eat hungrily.

2. stuck on, infatuated with.

stuck-up *adj.* conceited; haughty.

studbook *n.* a list of seniority in a public service department.

Students, the *pl. n.* **1.** the Sydney University team in the N.S.W. Rugby Football Union. **2.** the N.S.W. University team in the N.S.W. Rugby Football Union.

stuff *n.* **1.** property, as personal belongings, equipment, etc. **2.** actions, performances, talk, etc.: *to cut out the rough stuff.* **3.** one's own trade, profession, occupation, etc.: *to know one's stuff.* **4.** money. —*v.* **5.** (of males) to have sexual intercourse with. **6.** to cause to fail; render useless. —*phr.*

7. do one's stuff, to do what is expected of one; show what one can do.

8. not to give a stuff, to be unconcerned.

9. stuff it!, Also, **stuff this for a game of soldiers,** an exclamation indicating anger, frustration, etc.

10. stuff up, to blunder; fail.

11. that's the stuff, that is what is needed, right, proper, etc.

stuffed *v.* **1.** exhausted; ruined. —*phr.* **2. get stuffed,** to go away; leave someone alone.

stuffed shirt *n.* a pompous, pretentious person.

stuffing *phr.* **knock** or **beat the stuffing out of,** to destroy the self-confidence of or defeat utterly.

stuff-up *n.* a failure which has arisen from a foolish or thoughtless error.

stumblebum *n.* an incompetent, bungling, foolish person.

stumer *n.* **1.** a worthless cheque. —*phr.* **2. come a stumer,** to suffer a reversal of fortune, especially financial.

stumered *adj.* bankrupt; without any money.

stump *n.* **1.** (*usually pl.*) a leg: *to stir one's stumps.* **2.** union dues. —*v.* **3.** *U.S.* to stub, as one's toe. **4.** *Chiefly U.S.* to make speeches in an election campaign. —*phr.*
5. beyond or **back of the black stump,** in the far outback; in country areas beyond the reach of civilised comforts and facilities.
6. draw stumps, to finish. [from Cricket sense, to cease to play]
7. get up on the stump, to address a public meeting.
8. stump up, to pay up or hand over money required.

stumper *n. Cricket.* a wicket-keeper.

stump office *n.* an office into which union dues are paid.

stump orator *n. U.S.* one who travels around making political speeches; rabblerouser. [*stump* to make speeches in an election campaign + *orator*]

stung *adj.* **1.** drunk. **2.** tricked; cheated.

stunned *phr.* **like a stunned mullet, a.** in complete bewilderment or astonishment. **b.** in a state of inertia.

stunner *n.* a person or thing of striking excellence, beauty, attractiveness, etc.

stunning *adj.* of striking excellence, beauty, etc.

sub *n.* **1.** subeditor. **2.** submarine. —*adj.* **3.** substandard. —*v.* **4.** to act as a substitute for another. [shortened form]

subbie *n.* a subcontractor, especially a sub-contracting truck operator or building tradesman. Also, **subby.**

subteen *n.* a young person approaching adolescence.

suck *phr.*
1. fair suck of the sav/ saveloy/sausage/sauce bottle, an appeal for fairness or reason; fair go!
2. suck in, to cheat; swindle; deceive; defraud.
3. suck it and see, a humorous invitation to put something to the test.
4. suck off, to cause orgasm by oral stimulation of the genitalia.
5. suck up to, to flatter; toady; fawn upon.

sucker *n.* **1.** a person easily deceived or imposed upon; dupe. **2.** a lollipop, especially an all-day sucker.

sucker punch *n. Boxing.* an unorthodox punch or move which succeeds only because of its element of surprise.

suds *n.* **1.** the head of a glass of beer, etc. **2.** beer.

sugar *n.* a term of endearment.

sugar daddy *n.* a rich middle-aged or old man who lavishes money and gifts on a young woman or boy in return for sexual favours and companionship.

sullivan *n.* (used especially by Australians of non-British background) an Australian of British background. [from the

Sullivans, a television show which began in November 1976, in which the hero is a stereotype of the average Australian male]

summat *adj., pron., n.* somewhat or something.

sumpbuster *n.* a road having a particularly rough surface.

sun *phr.* **think the sun shines out of one's arse,** to have a very high opinion of oneself; to have a superiority complex.

sunbeam *n.* a plate, utensil, etc. which is not used at a meal, and does not need to be washed.

Sunday *phr.*
1. a month of Sundays, an extremely long time.
2. face as long as a month of Sundays, See **face.**

Sunday driver *n.* an unusually slow and cautious driver.

sundowner *n.* **1.** a swagman who arrives at a homestead at nightfall, too late for work, but obtains shelter for the night. **2.** an alcoholic drink taken in the evening, traditionally at sundown.

sunnies *pl. n.* sunglasses.

super *n.* **1.** superannuation. **2.** high-octane petrol. **3.** a superintendent. **4.** a supernumerary. **5.** superphosphate. **6.** a supervisor.—*v.* **7.** to treat (land) with superphosphate. [shortened form]

supercool *adj.* **1.** extremely sophisticated, fashionable, smart, etc. **2.** very calm; controlled.

super-duper *adj.* extremely fine, great, pleasing, etc.

superiority complex *n.* an exaggerated estimation of one's own worth.

sure *adv.* **1.** surely, undoubtedly, or certainly. **2.** *U.S.* inevitably or without fail. —*phr.*
3. a sure thing, a certainty; something assured beyond any doubt.
4. sure thing, *Chiefly U.S.* assuredly, certainly.

sure cop *n.* See **cop.**

sure enough *adv.* as expected; in actual fact: *He was expected to win, and sure enough he did.*

sure-fire *adj.* certain to succeed; assured: *He is a sure-fire winner for tomorrow's race.*

surface *v.* **1.** to wake up. **2.** to emerge at the end of a period of activity elsewhere. **3.** to appear in public, as arriving at one's job.

surfie *n.* **1.** a devotee of surfing, especially of surfboard riding. **2.** a person with tanned skin and sun-bleached hair, as from surfboard riding.

survive *v.* to remain unaffected or nearly so: *She doesn't love me, but I'll survive.*

suss *adj.* **1.** unreliable, needing confirmation: *Her story sounded very suss to me.* —*phr.* **2. suss out,** to attempt to determine the possibilities of a situation, especially one involving a particular challenge or presenting probable difficulties: *Before the minister proposed his bill, he sussed out the likely reaction of the opposition.* [shortened form *suspect*]

susso *n.* **1.** (formerly) form of payment given by the government to an unemployed person; dole. **2.** an unemployed person. —*phr.* **3. on the susso,** on the dole. Also, **suss.** [shortened form of *Sustenance Payment*]

swab *n.* a contemptible or useless person.

swag *n.* **1.** a bundle or roll carried across the shoulders or otherwise, and containing the personal belongings of a traveller through the bush, a miner, etc.; shiralee; bluey. **2.** any similar bundle of belongings. **3.** plundered property; booty. **4.** an unspecified but large number or quantity: *a swag of people.* [British dialect]

swagman *n.* **1.** a man who travels about the country on foot, living on his earnings from occasional jobs, or gifts of money or food. **2.** one who carries a swag. Also, **swaggie.**

swallow *v.* to accept without question or suspicion: *He'll never swallow that.*

swamp *v.* (of a swagman) to get a lift with, or have one's gear carried by, a bullock team, mail truck, etc., in return for giving minor assistance to the driver during the journey.

swank *n.* **1.** dashing smartness, as in bearing, appearance, etc.; style. —*adj.* **2.** pretentiously stylish.

swanky *adj.* **1.** conceited; boastful. **2.** expensive; smart; luxurious.

Swans, the *pl. n.* **1.** the Swan District team in the West Australian Football League. **2.** the Sydney team in the Victorian Football League (formerly South Melbourne). Also, **Sydney Swans.**

swat *v.* to hit with a smart or violent blow. Also, *U.S.,* **swot.**

swear *phr.*

1. swear by, to rely on; have confidence in.

2. swear off, to promise to give up something, especially intoxicating drink.

3. swear on a stack of bibles or **bag of boomerangs,** to state or assert positively; maintain as true.

sweat *v.* **1.** to exert oneself strenuously; work hard. **2.** to feel distress, as from anxiety, impatience, vexation, etc. **3.** to deprive (a person) of money, etc. as by exaction. **4.** to subject a person to severe questioning in order to extract information. —*n.* **5.** a state of perturbation, anxiety, or impatience. **6.** hard work. —*phr.*

7. no sweat!, it's no problem!

8. sweat blood, to be under a strain; be anxious; worry.

9. sweat it out, to hold out; endure until the end.

10. sweat on, to await anxiously.

sweep *n.* a disreputable person; scoundrel.

sweet *adj.* **1.** satisfactory as arranged: *She's sweet.* —*phr.* **2. sweet on,** in love with; fond of.

sweet cop *n.* See **cop.**

sweeten *v.* **1.** to bribe. **2.** *Poker.* to increase (a pot) by adding stakes before opening.

sweetener *n.* a bribe.

Sweet Fanny Adams *n.* very little; next to nothing. Also, **sweet F.A.** [euphemism for *sweet fuck-all;* Fanny Adams was a young British girl who was murdered and cut up into small pieces]

sweet fuck-all *n.* nothing: *He's done sweet fuck-all about it.*

sweetheart deal *n.* an agreement between employers and employees, which contains benefits well above those of the normal award.

sweetie *n.* **1.** a sweetheart (often used as a term of endearment). **2.** a sweet; confection.

sweet Jesus *interj.* an exclamation of surprise, vexation, indignation, etc. Also, **sweet Christ.**

sweet-talk *v.* **1.** to flatter excessively. **2.** to win over (someone) by ingratiating talk; con.

swell *n.* **1.** a fashionably dressed person. **2.** a person of high social standing. *—adj.* **3.** *U.S.* (of things) stylish; elegant; grand: *a swell hotel.* **4.** (of persons) fashionably dressed; of high standing, especially socially. **5.** *U.S.* first-rate; excellent: *swell food.*

swelled head *n.* an excessively high opinion of oneself; conceit.

swiftie *n.* **1.** an unfair act; a deceitful practice; a confidence trick. *—phr.* **2. put over** or **pull a swiftie,** to hoodwink; deceive. Also, **swifty.**

swig *n.* **1.** a large or deep drink, especially of alcoholic liquor, taken in one swallow; draught. *—v.* **2.** to drink heartily or greedily.

swill *n.* **1.** the public bar of a hotel: *in the swill.* *—phr.* **2. six o'clock swill,** *N.S.W.* (formerly) a hectic session of heavy accelerated consumption of alcohol in a hotel occasioned by the six p.m. closing of hotel bars (1916-1955).

swim *phr.* **in the swim,** actively engaged in current affairs, social activities, etc.

swing[1] *v.* **1.** to suffer death by hanging. **2.** (of the members of a group) to agree to exchange sexual partners on a casual basis. *—phr.* **3. in full swing,** fully active; operating at maximum speed or with maximum efficiency.

swing[2] *v.* **1.** to be characterised by a lively, modern, or knowledgeable attitude to life. **2.** (of a place) to have a lively atmosphere.

swinger *n.* **1.** an active, lively, or modern person. **2. a.** a person who exchanges sexual partners frequently. **b.** a member of a group of people who agree to exchange sexual partners on a casual basis. **3.** a motorcycle side-car rider.

swipe *v.* **1.** to steal. **2.** to borrow without the owner's knowledge. **3.** to criticise (someone); make a stinging remark about (someone).

swish *n.* **1.** *U.S.* a male homosexual. *—adj.* **2.** Also, **swishy.** smart; stylish; glamorous. *—phr.* **3. put (a bit of) swish into,** to cause to pay attention or take heed.

switch *n.* **1.** a switchboard. *—phr.*

2. switch off, to become oblivious of; ignore: *I switch off when she starts talking politics.*

3. switch on, a. to cause (a person) to be interested and enthused: *Bach switched him on.* **b.** (of a person) to become interested and enthused.

switched-on *adj.* with heightened awareness.

switcheroo *n.* a change; a turnabout.

switch hitter *n.* a bisexual person.

swiz *n.* **1.** a disappointment. **2.** a fraud; swindle.

swizzle *n.* **1.** a swindle. *—v.* **2.** to swindle.

sword *n.* **1.** (*pl.*) *Shearing.* hand shears. *—phr.*

2. cross swords, (of men) to urinate at the same time, in the same toilet or urinal.
3. give (something) the sword, to discard (something) as broken or no longer useful.
4. have had the sword, to be finished or ruined.

swy *n.* **1.** Also, **swy-game.** two-up. **2.** *Prison.* a prison sentence of two year's duration. **3.** (formerly) a two-shilling coin. [German *zwei* two]

Sydney *phr.* **Sydney or the bush!,** all or nothing, as in making a do or die attempt, gambling against the odds, etc.

Sydney Harbour *n.* a barber. [rhyming slang]

sync *phr.* **in sync,** in agreement. [shortened form of *synchronise*]

synchro *adj., n.* synchromesh. [shortened form]

syrup *n.* excessive or cloying sweetness or sentimentality.

system *n.* society at large or an organisation within it: *to buck the system.*

ta *interj.* thankyou.

tab[1] *phr.*
 1. keep tabs on, to keep account of or a check on: *keep tabs on your expenses.*
 2. pick up the tab, to pay the bill.

tab[2] *n.* tabulator.

tab[3] *n.* a tablet.

TAB *n.* a government-run betting shop. Also, **T.A.B.** [*T(otalisator) A(gency) B(oard)*]

tabbed *phr.* **have someone tabbed,** to identify: *I had him tabbed as a footballer.*

tabby *n.* a girl.

table *phr.*
 1. drink (someone) under the table, to drink in a somewhat competitive manner, the loser finishing up blind drunk.
 2. under the table, a. drunk to the extent of being incapable. **b.** given as a bribe.

tabo *n.* a girl.

tabs *phr.* **get tabs on,** to comprehend: *He couldn't get tabs on the way she had changed.*

tack[1] *phr.* **on the wrong tack,** following a false line of reasoning; under a wrong impression.

tack[2] *n.* food; fare: *hard tack.*

tacky *adj.* **1.** shabby; dowdy. **2.** in bad taste: *a tacky remark.*

taddie *n.* a tadpole.

Taffy *n.* a Welshman. [Welsh form of *Davy,* shortened form of *David,* proper name and name of patron saint of Wales]

tag *v.* to follow closely. [from *tag,* to attach a label, mark to]

tail *n.* **1.** (*pl.*) the reverse of a coin. **2.** the buttocks. **3.** a person who follows another, especially one who is employed to do so in order to hinder his escape or observe his movements. **4. a.** the vagina. **b.** a woman considered as a sex object: *a nice bit of tail.* —*v.* **5.** to follow close behind. **6.** to follow in order to hinder escape or to observe: *to tail a suspect.*

—*phr.* **7. seen more tails than Hoffmann,** *Horseracing.* said of a horse which is continually unsuccessful; a horse which usually finishes at the back of the field. [a pun on *'Tales of Hoffmann'* the title of an opera based on the works of German short-story writer Ernst Hoffmann]

8. tail out, to guide timber as it comes off the saw.

9. turn tail, a. to turn the back, as in aversion or fright. **b.** to run away; flee.

10. with one's tail between one's legs, in a state of utter defeat or humiliation; abjectly.

tailor-made *n.* a cigarette made by machine, i.e., not hand rolled. Also, **tailor.**

tail spin *n.* a sudden collapse into a state of utter confusion; flat spin.

take *n.* **1.** a cheat; a swindle. **2.** a profit, as from a short term business venture, a day's gambling, etc.: *The take was $100.* —*phr.*

3. on the take, receiving bribes, as a dishonest policeman, politician, etc.

4. take (someone) apart, a. to berate or abuse (someone). **b.** to physically assault (someone).

5. take care of, a. to control (someone's) behaviour, actions, etc., as by actual or threatened violence, blackmail, etc. **b.** to kill or destroy.

6. take down, a. to take advantage of; cheat; swindle. **b.** to lessen in power, strength, pride, arrogance, etc.; rebuke; humiliate: *I'll take him down a peg or two.*

7. take for a ride, a. to deceive; con. **b.** to kidnap. **c.** to murder.

8. take in, to deceive, trick or cheat.

9. take it easy, a. to relax. **b.** a request (to someone) to calm down, be less agitated, angry, etc.

10. take it out of, to exhaust; sap (one's) strength or energy.

11. take it out on, to vent wrath, anger or the like.

12. take off, a. to become popular: *The show really took off in Melbourne.* **b.** to escalate: *Prices took off.* **c.** to become excited. **d.** to imitate or mimic. **e.** to reach a level of excellence; success, flair, etc.: *The play took off in the last act.* **f.** *U.S.* to rob.

13. take on, a. to start a quarrel or fight with. **b.** to show great excitement, grief, or other emotion. **c.** to win popularity. **d.** to stand up to in a position of conflict, especially political.

14. take on board, to give consideration to: *I'll take that idea on board.*

15. take out, a. to vent: *to take out one's rage on the dog.* **b.** to destroy, eliminate, render harmless: *to take out a military post by bombing.* **c.** to win: *The Windies took out the series.*

16. take the bit between one's teeth, to throw off control; rush headlong.

17. take the bull by the horns, to act directly and promptly, particularly in a difficult situation.

18. take the wind out of (someone's) sails, See **wind.**

19. take to, a. to attack: *He took to his sister with a cricket bat.*

20. take/pick/gather up the threads, See **threads.**

takeaway *n.* a hot or cold meal purchased for consumption elsewhere.

take-down *n.* a fraudulent transaction.

take-off *n.* an imitating or mimicking; caricature.

taker *n.* **1.** (of a seedling plant etc.) one which strikes roots or begins to grow. **2.** an interested potential buyer.

talent *n.* at a party, dance, etc., women or men viewed as possible sexual partners.

tales *phr.* **tell tales (out of school),** See **tell.**

talk *v.* **1.** to reveal information: *We have ways to make you talk.* —*phr.*

2. talk about! used to add emphasis to a statement: *Talk about laugh.*

3. talk big, to speak boastfully.

4. talk down to, to speak condescendingly to.

5. talk off the top of one's head, a. to speak without prior preparation. **b.** to speak nonsense.

6. talk over (someone's) head, to discuss matters of which (someone) has no knowledge or understanding.

7. talk shop, See **shop.**

8. talk the leg off an iron pot; talk (someone's) ears off; talk until the cows come home, to talk at great length.

9. talk through the back of one's neck or **head,** to talk nonsense.

talkie *n.* a film having a soundtrack.

talking head *n.* a person interviewed or speaking at length on television.

talking shop *n.* parliament.

talking-to *n.* a scolding.

talk show *n.* an informal interview program on television or radio. Also, **chat show.**

tall *adj.* **1.** high, great, or large in amount: *a tall price.* **2.** extravagant; difficult to believe: *a tall story.* **3.** difficult to accomplish or fulfil: *a tall order.*

Tallarook *phr.* **things are crook in Tallarook** or **Muswellbrook,** the situation is not good.

tall poppy *n.* See **poppy.**

tallyman *n.* a man who lives with a woman outside marriage.

talon *n.* a finger or fingernail, especially when regarded as grasping or attacking.

tan *v.* to beat or thrash: *to tan one's hide.* [See **tanning**]

tangle *n.* **1.** a conflict, quarrel, or disagreement. —*phr.* **2. tangle with,** to conflict, quarrel, or argue with.

tank *n.* **1.** *Prison.* a safe. —*v.* **2.** to move like a tank: *a footballer tanking down the wing.* **3.** to deliberately lose (a match or a contest). —*phr.*

4. get tanked (up) to get drunk.

5. on the tank, on a drinking spree.

6. tank up, a. to fill the tank of a motor vehicle with fuel. **b.** to drink heavily.

tanked *adj.* intoxicated, especially with beer.

tankman *n.* *Prison.* a person who specialises in stealing from safes.

tanner *n.* (formerly) sixpence.

tanning *n.* a thrashing. [from *tanning,* the process of converting hides or skins into leather]

tap[1] *phr.* **off tap,** *Prison.* convicted.

tap[2] *v.* **1.** to extract money from, especially in a crafty manner. *—phr.* **2. on tap, a.** ready to be drawn off and served, as drink, especially beer, in a cask. **b.** furnished with a tap or cock, as a barrel containing drink, especially beer. **c.** ready for immediate use.

tap dancer *n.* cancer. [rhyming slang]

tape *phr.*
1. have (someone) taped, to understand thoroughly, especially a person's weakness or guile.
2. have (something) taped, to be in complete control of or be easily able to do (something).

tar[1] *phr.* **tarred with the same brush,** having similar or the same faults.

tar[2] *n.* a sailor. [said to be short for *tarpaulin*]

tarbrush *phr.* **a touch of the tarbrush,** (*offensive*) Negro or other coloured ancestry or appearance.

tarnation *interj.* **1.** damnation. *—adj.* **2.** damned: *I can't get the tarnation car to start.*

tarp *n.* a tarpaulin. [shortened form]

tart *n.* **1.** a girl or woman, now especially of low character. **2.** a prostitute. *—phr.* **3. tart up,** to adorn; make attractive, especially with cheap ornaments and cosmetics.

tartplate *n.* the pillion seat on a motor bike or motor scooter.

Tarzan *n.* a person of superior strength or agility. [name of the hero of the series of jungle stories by Edgar Rice Burroughs, 1845-1950, U.S. writer]

Tas *n.* Tasmania. Also, **Tassie, Tassy.**

Tasmania *phr.* **map of Tasmania,** the female pubic area.

tassel *n.* the penis.

taster *n.* one who samples and reports on manuscripts for a publisher.

Tasway *n.* Tasmania. [backformation of *Taswegian,* by analogy with *Norwegian*]

Taswegian *n.* a Tasmanian. [*Tas(manian)* + *wegian* (by analogy with *Norwegian, Glaswegian,* etc.)]

tata *interj.* good-bye

tater *n.* a potato. Also, **tatie.**

tatts *pl. n.* false teeth. Also, **tats, tatters.**

taws *phr.* **go back to taws** or **start from taws,** to go back to the beginning.

tax lurk *n.* a scheme or trick by which one avoids paying tax. Also, **tax dodge.**

tea *n.* **1.** marijuana. *—phr.* **2. cup of tea,** a task, topic, person, or object, etc., well suited to one's experience, taste, or liking: *That show wasn't my cup of tea.*

tea-and-sugar *phr.*
1. tea-and-sugar bandit, a small time criminal.
2. tea-and-sugar bushranger, a station owner who steals from his neighbours in order to set up his own herd.

teacup *phr.* **a storm in a teacup,** a great fuss about nothing very much.

tea-leaf *n.* **1.** a thief. *—v.* **2.** to steal. [rhyming slang]

team cream *n.* an occasion on which a number of males have sexual intercourse with one female; gang bang.

tear *v.* **1.** to move or go with violence or great haste: *to tear along the expressway.* —*n.* **2.** a spree: *on the tear.* —*phr.*
3. tear into, to attack violently, either physically or verbally.
4. tear off, a. to perform or do, especially rapidly or carelessly. **b.** to leave hurriedly.
5. tear strips off, to reprove severely.

tear-arse *n.* **1.** a tearaway. **2.** golden syrup or treacle.

tearaway *n.* **1.** an impetuous or unruly person. —*adj.* **2.** uncontrolled; impetuous. **3.** (of a sporting win, etc.) won by a long distance or a high score: *They had a tearaway victory.*

tear-jerker *n.* an excessively sentimental novel, film, or the like.

TEAS *phr.* **on TEAS,** in receipt of a Tertiary Education Assistance Scheme award.

teaser *n.* See **prickteaser.**

tec *n.* a detective. Also, **'tec.** [shortened form]

tech *n.* a technical college or school. [shortened form]

technical *phr.* **get technical,** to propound or apply a strict interpretation of the rules.

technicolour yawn *n.* **1.** the act of vomiting. **2.** vomit.

Teddy Bear *n.* a flashily dressed person; a lair. [rhyming slang]

teddy boy *n. British.* (in the mid 1950s) a boy in his teens or early twenties who dressed in a fashion resembling that of the Edwardian era and identified himself with others affecting a similar style of dress. Also, **ted.**

teeny-weeny *adj.* (*especially in children's speech*) very small. Also, **teensy-weensy.**

teeth *phr.*
1. be fed (up) to the (back) teeth with, to be heartily sick of; have had more than enough of.
2. cut one's teeth on, to gain experience on. See also **eyeteeth.**
3. get one's teeth into, to start to cope effectively with (a problem, etc.).
4. have the bit (in) between one's teeth, to tackle a task, problem, etc., in a determined and energetic fashion.
5. scarce or **rare as hen's teeth,** See **hen's teeth.**

teeth-to-tail ratio *n. Originally Military.* the proportion of active personnel to administration.

telegraph *v.* to give prior indication of (one's moves). [from the phrase *telegraph one's punches,* to give prior indication of one's plans or intentions, especially to an opponent]

tell *phr.*
1. tell off, to scold; rebuke severely.
2. tell tales (out of school), Also, **tell on.** to report the misdemeanours, true or fictitious, of one's friends, peers, relatives, etc.

telly *n.* television. Also, **tellie.**

temp *n.* a temporary member of an office staff, especially a secretary. [shortened form]

ten *adj.* **1.** excellent. —*n.* **2.** something or somebody worthy of admiration, high praise, etc. [as a rating on a scale from 1 to 10]

ten-four *interj.* an exclamation signifying agreement, acceptance, etc., especially used by C.B. radio operators. Also, **10-4.**

tenner *n.* **1.** a ten-dollar ($10) note. **2.** (formerly) a ten-pound (£10) note

ten-per-center *n.* **1.** an actor's agent. **2.** a person who sponges off other people's abilities. [from the ten per cent commission taken by an agent from the fee paid for work performed by the actor, writer, etc.]

tent *phr.* **born in a tent** or **barn,** See **barn.**

tents *phr.* **camp as a row of tents,** See **camp.**

terrible *adj.* **1.** very bad: *a terrible performance.* **2.** very great: *a terrible liar.*

terrif *adj.* wonderful; terrific. [shortened form]

terrific *adj.* **1.** extraordinarily great, intense, etc.: *terrific speed.* **2.** very good: *terrific food.*

Territory confetti *n.* ring pulls from beer cans.

Territory, the *n.* **1.** the Northern Territory. **2.** (formerly) Papua New Guinea.

terror *n.* a person or thing that is a particular nuisance: *That boy is a little terror.*

tester *n. Convict Obsolete.* a flogging of twenty-five lashes. See **bob, bull, canary.**

tetchy *adj.* irritable; touchy. Also, **techy.**

tether *phr.* **the end of one's tether** or **rope,** See **rope.**

thank *phr.*
1. have oneself to thank. to be oneself responsible or at fault.
2. have (someone) to thank for, to rightly place blame or responsibility for (something) on (someone).
3. thanks a million or **a bunch,** sarcastic thanks for something not wanted or disadvantageous to the speaker.

thatch *n.* the hair covering the head. [from *thatch,* the straw, rushes, leaves, etc., used to cover roofs, haystacks, etc.]

there *adj.* **1.** used to add emphasis: *that there man.* *—phr.* **2. all there, a.** of sound mind. **b.** shrewd; quick-witted.

thick *adj.* **1.** close in friendship, intimate. **2.** disagreeably excessive: *His demands are a bit thick.* **3.** slow of mental apprehension; stupid; dull; slow-witted: *His mind is very thick.* *—n.* **4.** a stupid, dull-witted person. *—phr.*
5. a thick ear, a swollen ear, as a result of being punched, slapped, etc.: *Shut up or you'll get a thick ear.*
6. lay it on thick, to be extravagant in flattery, praise, or the like.
7. thick as a brick; thick as two short planks or **poms; thick as a log of wood,** slow-witted.
8. thick with, having great numbers of; abounding with: *The place was thick with cops.*

thick-skulled *adj.* stupid; doltish.

thick-witted *adj.* stupid; dull.

thing *n.* **1.** an unaccountable attitude or feeling about something, as of fear or aversion: *I have a thing about spiders.* *—phr.*
2. do one's thing, to act in a characteristic manner; to do what is most satisfying to oneself.
3. do the right thing, to behave properly or responsibly.

4. **do the right thing by** or **do the handsome thing by,** to treat generously.

5. **know a thing or two,** to be shrewd.

6. **make a good thing out of,** to obtain an advantage from.

7. **make a (big) thing of,** to turn into a major issue: *OK, so I made a mistake, but there's no need to make a thing of it.*

8. **not get a thing out of, a.** to fail to elicit something desired, as information, from. **b.** to fail to enjoy, appreciate, etc.: *I went to a performance of a play in Czech, but didn't get a thing out of it.*

9. **old thing,** a familiar form of address.

10. **on a good thing,** (in betting on horses, dogs, etc.) backing a likely winner at favourable odds.

11. **one of those things,** an event or situation which is unavoidable or which is no longer remediable.

thingie *adj.* overly sensitive; anxious; tense: *He gets all thingie with the children on a long car trip.*

things *pl. n.* **1.** implements, utensils, or other articles for service: *Can you help me with the breakfast things.* **2.** personal possessions or belongings, often such as one carries along on a journey.

thingummyjig *n.* an indefinite name for a thing or person which a speaker cannot or does not designate more precisely. Also, **thingummybob, thingummy, thingumabob.**

think *n.* **1.** an act or process of thinking: *Go away and have a good think.* *—phr.* **2. not think twice about (something), a.** to consider only briefly, before taking action. **b.** to forget; shrug off (something).

thinking cap *phr.* **put on one's thinking cap,** to reflect upon or consider a matter; cogitate.

third-class *adj.* extremely shoddy and inferior.

third degree *n.* the use of bullying or torture by the police (or others) in some countries in order to extort information or a confession: *to give one the third degree.*

third half *n. Sport.* a social gathering after a match.

third man *n.* the referee in a boxing or wrestling match.

thirsty *adj.* causing thirst: *thirsty work.*

thou *n.* **1.** a thousand (dollars, kilometres, etc.). **2.** one thousandth of (an inch, etc.).

threads *phr.* **gather** or **pick up the threads, a.** to pull thoughts and ideas together. **b.** to get on with one's life after some serious personal disruption.

three *phr.* **three bangers short of a barbie,** See **barbie.**

three-day night *n.* a night full of excitement, danger, etc.

three-dog night *n.* a very cold night. [from the practice of bushmen of sleeping with their dogs; the colder the night, the more dogs needed]

three-on-the-tree *n.* a column shift for a motor vehicle with three forward gears (opposed to *four-on-the-floor*).

threepenny bits *phr.* **give (someone) the threepenny bits,** to arouse dislike, anger, disgust, etc., in (someone). [rhyming slang, *threepenny bits* the shits]

thrill *n.* a passion: *He had a thrill on Jane.*

throat *phr.*

1. cut one's (own) throat, to pursue a course of action which is injurious or ruinous to oneself.

2. get (something) by the throat, to master (something).

3. jump down (someone's) throat, to deliver a strong verbal attack on; berate; scold.

4. one's stomach thinks one's throat is cut, See **stomach.**

5. ram or **thrust (something) down (someone's) throat,** to force (something) on (someone's) attention.

6. stick in one's throat, a. to be difficult to express or utter. **b.** to be difficult to accept in one's mind.

throne *phr.* **the throne,** the toilet.

throw *v.* **1.** to permit an opponent to win (a race, contest, or the like) deliberately, as for a bribe. **2.** to astonish; disconcert; confuse. —*phr.*

3. throw away, to fail to use; miss (an opportunity, chance, etc.).

4. throw in, to add as an extra, especially in a bargain.

5. throw in one's hand, to concede defeat; surrender. [from card game]

6. throw in the towel or **sponge,** to give in; accept defeat [from boxing]

7. throw it in, a. to accept defeat. **b.** to cease an activity.

8. throw off at, to criticise or belittle.

9. throw or **chuck a seven,** See **seven.**

10. throw or **chuck a wobbly,** See **wobbly.**

11. throw oneself at (someone), to attempt to excite (someone's) interest in order to win their love.

12. throw oneself into, to work enthusiastically at.

13. throw out, to cause to make a mistake.

14. throw over, to end a relationship with.

15. throw up, to vomit.

thud *phr.* **come a thud,** to be disappointed in an expectation.

thumb *n.* **1.** *U.S.* a marijuana cigarette. —*v.* **2.** to hitch-hike: *to thumb one's way around Queensland.* —*phr.*

3. thumb in bum and mind in neutral, vague; absent-minded.

4. thumb one's nose, to put one's thumb to one's nose and extend the fingers in a gesture of defiance or contempt.

thumbnail *phr.* **(written with) a thumbnail dipped in tar,** rough, crude or untidy penmanship. [from the poem 'Clancy of the Overflow' by A.B. (Banjo) Patterson]

thumbs down *n.* rejection, disapproval, especially of a proposal. [from the gesture by which the Roman crowd authorised death, especially in the Colosseum circuses]

thumbs up *interj.* **1.** an exclamation indicating encouragement. —*n.* **2.** a gesture made by clenching the fingers and holding the thumb vertical, symbolising success. **3.** a similar gesture but made with a vigorous upward thrust of the hand, symbolising contempt.

thump *v.* **1.** to punch; thrash severely. **2.** to steal.

thumping *adj.* very great; remarkably or unusually large; exceptional.

thunderbox *n.* a toilet, especially one with a sanitary can.

thundering *adj.* extraordinary; very great.

thunder-mug *n.* a chamber-pot.

tick[1] *n.* **1.** a moment or instant: *Hang on a tick. —phr.*
2. on the tick, punctually.
3. tick off, to rebuke; scold.
4. tick over, to be inactive, often in preparation for action.
5. what makes one tick, what motivates one's behaviour.

tick[2] *n.* **1.** a score or account. **2.** credit or trust: *to buy on tick.*

ticker *n.* **1.** a watch. **2.** the heart.

ticket *n.* **1.** a certificate. **2.** discharge from the armed forces: *to get one's ticket.* **3.** blotting paper impregnated with LSD. **4.** the correct, right, or proper thing: *That's the ticket! —phr.*
5. have tickets on oneself, to be conceited.
6. put in one's ticket, a. to die. **b.** to be dismissed from one's job.

ticket, pakapoo *n.* See **pakapoo ticket.**

ticket snapper *n.* a ticket inspector on the railways. Also, **snapper.**

tickety-boo *adj. British.* fine; splendid: *Everything is tickety-boo.*

tickle *phr.*
1. tickled pink or **to death,** Also, **tickled.** greatly pleased or amused.
2. tickle the peter, to rob the till.
3. tickle the ivories, to play the piano.

tickler *n.* a difficult or puzzling situation, problem, etc.

tiddler *n.* a small child, especially one who is undersized.

tiddly *adj.* slightly drunk; tipsy.

tidy *adj.* considerable: *a tidy sum.*

tie *v.* **1.** to unite in marriage. *—phr.*
2. tie down, to hinder; confine; restrict; curtail.
3. tie one on, to get drunk.
4. tie up, a. to hinder. **b.** to bring to a stop or pause. **c.** to invest or place (money) in such a way as to make it unavailable for other purposes. **d.** to occupy or engage completely.
5. tied up, unable to accept engagements, etc.; occupied.

tiger[1] *n.* **1.** a shearer. *—v.* **2.** (formerly) to work very hard. *—phr.* **3. a tiger for punishment** or **work,** someone who works hard, especially beyond what is expected.

tiger[2] *n.* a swim. [rhyming slang *Tiger Tim* swim]

tiger country *n.* **1.** rough, thickly wooded bush. **2.** remote uncultivated country.

Tigers, the *pl. n.* **1.** the Balmain team in the N.S.W. Rugby Football League. **2.** the Claremont team in the West Australian Football League. **3.** the Eastern Districts team in the Queensland Rugby Football Union. **4.** the Eastern Suburbs team in the Queensland Rugby Football League. **5.** the Glenelg team in the South Australian National Football League. **6.** the Hobart team in the Tasmanian Australian National Football League. **7.** the Nightcliff team in the Northern Territory Football League. **8.** the

Richmond team in the Victorian Football League. **9.** the Werribee team in the Victorian Football Association.

tiggy *n.* **1.** a children's game involving chasing and catching. *—phr.* **2. tiggy (tiggy) touch wood,** *Australian Rules.* a game characterised by many free kicks awarded because of minor infringements.

tight *adj.* **1.** close; nearly even: *a tight race.* **2.** stingy or parsimonious. **3.** drunk; tipsy. *—phr.* **4. tight as a bull's arse in flytime; tight as a fish's arse; so tight one squeaks,** to be extremely mean with money; tight-arsed.

tight-arsed *adj.* **1.** mean; parsimonious. **2.** haughty.

tightwad *n.* a close-fisted or stingy person.

tile *n.* **1.** a tablet of LSD. *—phr.* **2. on the tiles,** having a wild, riotous, or debauched night's entertainment.

tilly *n. Qld.* a utility truck. Also, **til.**

timber *phr.* **big timber,** *Australian Rules.* the goalposts.

Timbuktu *n.* any faraway place. [from *Timbuktu,* town in central Mali, near the river Niger, Africa]

time *n.* **1.** a term of imprisonment. *—phr.*

2. do time, to serve a prison sentence.

3. give (someone) a hard time, to irritate, nag or make life difficult for (someone).

4. kill time, to occupy oneself in some manner so as to make the time pass quickly.

5. take time out, to spare the time, to make the effort (to do something).

6. time of one's life, a very enjoyable experience.

tin *phr.* **kick the tin,** See **kick.**

tin arse *n.* a lucky person. Also, **tin bum, tin back.**

tingle *n.* a telephone call.

tin hare *n.* **1.** a mechanical hare which greyhounds pursue in a race. **2.** a delivery truck carrying beer. **3.** the driver of such a truck.

tinhorn *n.* **1.** *U.S.* a pretentious or boastful person, especially a gambler, who claims power, influence, resources, etc., which he does not possess. *—adj.* **2.** insignificant; petty; cheap.

tinker's cuss *n.* something worthless or trivial: *His opinion is not worth a tinker's cuss.* Also, **tinker's damn.**

tin-kettle *v. N.Z.* to welcome (a newly-wed couple) by noisy banging of tin-cans, etc.

tin lizzie *n.* any cheap, old, or decrepit motor vehicle.

tinned dog *n.* canned meat.

tinnie *n.* a can of beer. Also, **tinny.**

tinny *adj.* lucky.

tin-pot *adj.* inferior; petty; worthless.

tinsel town *n.* **1.** Hollywood. *—adj.* **2.** artificial; ephemeral.

tintack *n.* (formerly) sixpence. [rhyming slang *tintack* zack]

tintacks *phr.* **get down to tintacks** or **brass tacks,** to deal with essentials.

tinted *adj.* (of a person) not, or not wholly, white; coloured.

tip *v.* **1.** *Prison.* to be transferred suddenly and unexpectedly from one gaol to another. *—phr.* **2. tip off** or **tip the wink, a.** to give (someone) private or secret information; inform. **b.** to warn of impending trouble, danger, etc.
3. tip the bucket, See **bucket.**

tip-off *n.* a hint or warning: *They got a tip-off about the raid.*

tipster *n.* one who makes a business of furnishing tips, as for use in betting, speculation, etc.

tiptop *adj.* of the highest quality or excellence: *in tiptop condition.*

tip-up *n. Prison.* a detailed search of a cell by prison officers looking for drugs, etc.

tired *phr.* **tired and emotional,** a euphemistic description of a person who is drunk.

'Tiser *n.* The Adelaide Advertiser newspaper.

tit *n.* **1.** a female breast. **2.** a push-button, as a light switch, etc.

titfer *n.* a hat. [rhyming slang *tit for tat* hat]

titivate *v.* **2.** to make smart or spruce. **2.** to make oneself smart or spruce. Also, **tittivate, tidivate.** [from *tidy,* modelled on *cultivate*]

titty *n.* a female breast.

tizz *n.* a state of somewhat hysterical confusion and anxiety, often expressed in frantic but ineffectual activity: *Don't get in a tizz.* Also, **tizzy.**

tizzy *n.* **1.** a tizz. *—adj.* **2.** gaudy; vulgar; tinselly.

TLC *n.* sympathetic attention. [*T(ender) L(oving) C(are)*]

T note *n.* a Treasury note.

toady[1] *n.* **1.** an obsequious sycophant; a fawning flatterer. *—v.* **2.** to be a toady. [shortened form of Obsolete *toadeater,* a mountebank's assistant who pretended to eat toads from which his master had supposedly removed the poison; hence a servile hanger-on]

toady[2] *n.* a toado, one of a species of poisonous, self-inflating fishes. Also, **toad fish.**

to-and-fro *n.* **1.** moustache; mo. *—v.* **2.** go. [rhyming slang]

to-and-from *n.* an Englishman. [rhyming slang *to and from,* pom]

toast-rack *n.* (formerly) a type of tramcar used in Sydney.

toby *n. Shearing.* a coloured mark (red, blue or yellow), placed upon badly shorn sheep; raddle.

toby jug *n. U.S.* a long, slender, cheap cigar.

tod *phr.* **on one's tod,** alone. [rhyming slang *Tod Sloan,* alone, from a British jockey of that name]

to-do *n.* bustle; fuss.

toe *phr.*
1. all done up like a sore toe, See **sore.**
2. hit the toe, a. Also, **toe it.** to go. **b.** *Prison.* to attempt to escape.
3. stick out like a sore toe, See **sore.**
4. toe the line, See **line.**
5. tread on (someone's) toes, to offend, especially by ignoring (another's) area of responsibility.
6. turn up one's toes, to die.

toe jam *n.* the dirt which collects between the toes.

toe-ragger *n.* **1.** (formerly) a swagman; a down-and-out. **2.**

Prison. a prisoner serving a short sentence.

toey *adj.* **1.** anxious; apprehensive. **2.** randy. **3.** keen, ready to go. **4.** (of a horse) having an excitable temperament; fast.

toff *n.* a rich, upper-class, usually well-dressed person; a gentleman. [short for *toffee-nosed*]

toffee-nosed *adj. Chiefly British.* snobbish; pretentious; upper-class.

tog *v.* to clothe; dress.

together *adj.* **1.** capable and calm: *She was a very together person.* *—phr.* **2. get it together,** See **get.**

toggery *n. British.* garments; clothes; togs.

togs *pl. n.* clothes: *football togs, swimming togs.*

toke *n.* a puff of a cigarette or joint.

tomato *n.* a cricket ball.

tomcat *phr.* **wouldn't say boo to a tomcat,** (of a person) to be inoffensive; timid; quiet.

Tom Collins *n.* gossip. [the name of a mythical character reputed to be the source of all idle rumours, later adopted as a pseudonym by Joseph Furphy, Australian writer]

Tom, Dick and Harry *n.* common people generally; anybody at all: *They invited every Tom, Dick and Harry.*

tomfoolery *n.* jewellery. [rhyming slang]

Tommy gun *n.* a Thompson machine gun.

tommyrot *n.* nonsense.

tomtits *pl. n.* **1.** diarrhoea. **2.** anger; exasperation. [rhyming slang *tomtits* the shits]

ton[1] *n.* **1.** a heavy weight: *That book weighs a ton.* **2.** (*pl.*) very many; a good deal: *tons of things to see.*

ton[2] *n.* **1.** one hundred. **2.** (formerly) a speed of one hundred miles an hour, especially on a motorcycle.

toney *adj.* of high class or pretending to it. Also, **tony.**

tonk *n.* **1.** a penis. **2.** a passive homosexual.

tonky *adj. N.Z.* toney

ton-up *adj. Chiefly British.* **1.** capable of travelling at a speed of a hundred miles an hour or more. **2.** of or pertaining to a person who derives pleasure or prestige from excessive speed.

too *adv.* **1.** indeed (used for emphasis): *I did so too!* *—phr.* **2. too right,** an emphatic expression of agreement.

too-hard basket *n.* an imaginary basket in which papers coming into an office are placed if the recipient finds them difficult and wishes to delay making a decision.

took *v.* See **take.**

tool *n.* **1.** the penis. *—v.* **2.** to drive or ride a vehicle.

toot *n.* a toilet.

tooth *phr.* **long in the tooth,** old.

tooth fairy *n.* a mythical or improbable agency: *Who do you think did it, the tooth fairy?*

toothpick *n.* a sculling boat designed for racing.

tootle *v.* **1.** to go or walk. **2.** to drive. *—phr.* **3. tootle off,** to depart.

too-too *adj.* excessive; absurd: *That dress is just too-too.*

tootsy *n.* **1.** a foot. **2.** a lesbian. *—phr.* **3. play tootsy, a.** (of two people) to touch feet secretly under a table as part of amorous play. **b.** (of a man) to have an affair. Also, **tootsie.**

top[1] *adj.* **1.** the best; excellent: *a top bloke; top fun. —phr.*
2. blow one's top, See **blow.**
3. from the top, from the beginning.
4. on top, successful; victorious; dominant.
5. the Top, See **top end.**
6. top off, a. to inform on; tell on. **b.** to finish; cap: *He topped off the sports day with a win in the marathon.*

top[2] *n.* an amplifier.

top dog *n.* person in the highest position; leader; boss.

top drawer *n.* the highest level, especially of social class.

top end *n.* the top end of the Northern Territory of Australia. Also, **Top End, the Top.**

top-ender *n.* a person living in the northern part of the Northern Territory of Australia. Also, **Top-Ender.**

top-hole *adj. Chiefly British.* first-rate.

topnotch *adj.* first-rate: *a topnotch job.*

topnotcher *n.* a person or thing of unsurpassed excellence.

top-off *n.* one who informs on another, usually apparently in jest or by accident.

topper *n.* **1.** a top hat. **2.** anything excellent.

top shelf *adj.* the best, as alcoholic spirits.

topsy *phr.* **grow like topsy,** See **grow.**

torch *v.* to set on fire.

torn *phr.* **that's torn it,** everything is ruined.

torpedo *n.* a sleeping pill.

tosh *n.* nonsense.

toss *v.* **1.** to outwit. **2.** to defeat. **3.** to go with a fling of the body: *to toss out of a room. —phr.*
4. argue the toss, to go on arguing after a dispute has been settled.
5. to not give a toss, to be unconcerned; not care.
6. toss it in, to give up.
7. toss off, a. (of a male) to ejaculate sperm; have an orgasm. **b.** to masturbate. **c.** to produce casually: *to toss off a poem.*

tossle *n.* the penis.

tosspot *n.* one who drinks to excess.

toss-up *n.* an even chance: *The election is a toss-up.*

tot *n.* **1.** a total. **2.** the act of adding. *—phr.* **3. tot up,** to add.

totem pole *phr.* **low on the totem pole,** the least important person or people. [from *totem pole,* of the American Indians, a pole carved and painted with figures]

t'othersider *n. W.A.* a person living on the other side of the Nullarbor Plain.

totter *n.* a rag-and-bone man; a scavenger.

touch *v.* **1.** to apply to for money, or succeed in getting money from; to beg. *—n.* **2.** the act of applying to a person for money, as a gift or loan. **3.** an obtaining of money thus. **4.** the money obtained. **5.** a person from whom such money can be obtained easily. *—phr.* **6. keep** or **stay in touch,** to maintain an association or friendship.

touch-and-go *adj.* precarious; risky: *a highly touch-and-go situation.* Also, **touch and go.**

tough *adv.* **1.** aggressively; threateningly: *to act tough.* —*phr.* **2. tough it out,** to persevere to the end against difficulties, especially while under possibly justified criticism.

toughie *n.* a callous person.

tough shit *interj.* **1.** bad luck; hard luck. **2.** a rebuff to an appeal for sympathy. Also, **tough titty.**

towel *phr.*
1. throw in the towel, See **throw.**
2. towel up, to attack or thrash, either physically or verbally.

towelling *n.* a thrashing. Also, *U.S.* **toweling.**

towie *n* **1.** a tow truck. **2.** a tow truck driver.

town *phr.*
1. go to town, a. to do something enthusiastically; splash out. **b.** to overindulge or lose one's self-restraint. **c.** to celebrate.
2. go to town (on), to berate; tell off.
3. paint the town red, to indulge in a concentrated set of rather wild social pleasures.
4. talk of the town, the subject of general gossip or rumour.

town bike *n.* (usually in a small town) **a.** a prostitute. **b.** a woman who will have sexual intercourse with any man that asks her.

townie *n.* **1.** an inhabitant of a university town. **2.** one who comes from a town and is ignorant of country ways. Also, **townee.**

trac *n. Prison.* an intractable prisoner.

trace *phr.* **kick over the traces,** to reject discipline; to act in an independent manner.

track *n.* **1.** *Prison.* a prison warder who will carry contraband messages or goods out of or into a prison for a prisoner. **2.** (*pl.*) scars or marks on the arms or legs caused by habitual use of a hypodermic needle. —*phr.*
3. in one's tracks, just where one is standing: *He was stopped in his tracks.*
4. make tracks, to leave or depart.
5. make tracks for, to head towards.
6. off the beaten track, See **beaten track.**
7. on the (wallaby) track, itinerant; on the move, as a swagman, etc.
8. (on) the wrong side of the tracks, *U.S.* (in) a low social position; (in) a low-class or poor neighbourhood.
9. run off the track, *Horseracing, etc.* to run wide at a turn.
10. the right (wrong) track, the right (wrong) idea, plan, interpretation, etc.
11. track (square) with, to keep company; cohabit.

tracs *pl. n. Prison.* solitary confinement cells for intractable prisoners.

trad *n.* **1.** traditional jazz. —*adj.* **2.** traditional; old-fashioned; conventional. [shortened form]

tragedy queen *n.* a person, especially a woman, given to flaunting despair, depression, etc., over misfortune, especially a failed relationship.

trail *v.* **1.** to follow along behind (another or others), as in a race. *—phr.* **2. trail a coat,** to provoke a heated reaction by persistent antagonistic remarks, in order to bring suspected latent hostility into the open.

trail foot *n.* a surfboard rider's back foot.

traipse *v.* **1.** to walk so as to be, or having become, tired; trudge. **2.** to walk (about) aimlessly; gad about. Also, **trapes.**

trammie *n.* a tram driver or conductor.

tramp *n.* a socially unacceptable woman.

trannie *n.* **1.** a transistor radio. **2.** a transformer. [shortened form]

trap *n.* **1.** the mouth: *Shut your trap!* *—phr.* **2. a trap for young** or **amateur players,** a danger or risk to the inexperienced.

traps *pl. n.* **1.** personal belongings; luggage. *—phr.* **2. round the traps,** in and about familiar places, as pubs, clubs, etc., especially as sources of information, etc.

trash *phr.* **white trash, a.** *U.S.* the poor white inhabitants of the southern U.S. **b.** the poor white inhabitants of a region or district where some of the inhabitants are coloured people.

trashy *n.* a garbage man.

travel *v.* to move with speed.

trawl *v.* to use a public address system attached to a moving vehicle to address people, as during a political election campaign.

tray *n.* See **trey.**

treacle *n.* cloying sentimentality as of music or behaviour. [from *treacle* a dark syrup obtained in refining sugar]

treadle *n.* a bicycle. [from *treadle* the lever worked by foot.

treat *n.* **1.** anything that affords particular pleasure or enjoyment. *—phr.*

2. a (fair) treat, excessively.

3. look a treat, to have a good appearance.

4. stand treat, to bear the expense of an entertainment.

treatment *phr.* **the treatment,** punishment; severe handling; thorough criticism: *The unions are getting the teatment from the media.*

tree *v.* **1.** to put into a difficult position. [from the hunting sense, to drive an animal up a tree] *—phr.* **3. out of one's tree,** demented; crazy; mad.

tremendous *adj.* **1.** extraordinarily great in size, amount, degree, etc. **2.** extraordinary; unusual; remarkable.

trendy *adj.* **1.** forming part of or influenced by fashionable trends; ultrafashionable. *—n.* **2.** one who embraces an ultrafashionable lifestyle. **3.** one who adopts a set of avant-garde social or political viewpoints.

trey *n.* (formerly) a threepenny piece; threepence. Also, **trey-bit.**

tribe *n.* a family.

trick *n.* **1.** a prostitute's customer. **2.** a person with whom one has had a casual encounter involving sexual intercourse; pick-up; one-night stand. *—phr.*

3. do the trick, to achieve the desired result.

4. **never miss a trick,** See **miss.**

5. **not be able to make a trick,** to have no success at all.

trick cyclist *n.* a psychiatrist.

Tricolours, the *pl. n.* the Eastern Suburbs team in the N.S.W. Rugby Football Union.

trigger-happy *adj.* reckless, irresponsible, or foolhardy, especially in matters which could lead to violence or war.

trillion *n.* a very large amount or number.

trim *v. U.S.* to defeat.

trimmings *pl. n.* agreeable accompaniments or additions.

trip *n.* **1.** a quantity of LSD prepared in some form for sale. **2.** a period under the influence of a hallucinatory drug. *—v.* **3. a.** to take LSD. **b.** to hallucinate under the influence of LSD or other drugs. **c.** to have an exhilarating experience similar to hallucination.

tripe *n.* **1.** anything poor or worthless, especially written work; nonsense; rubbish. **2.** exaggerated or untruthful statements, talk, etc. *—phr.* **3. beat the tripe out of, a.** to beat up. **b.** to defeat utterly.

tripehound *n.* a dog.

tripper *n.* one who goes on a pleasure trip or excursion: *day tripper.*

trippy *adj.* exhilarating, as of a hallucinogenic trip.

trog *n.* anyone thought to be primitive, barbaric, unintelligent, or insensitive. Also, **troglodyte.** [from *troglodyte* a caveman]

trog suit *n.* an outfit suitable for caving expeditions.

trolley *phr.*

1. **off one's trolley,** crazy; mad; insane.

2. **slip one's trolley,** to become mentally unbalanced or deranged.

trooper *phr.* **swear like a trooper,** to swear vigorously.

tropical *adj. Prison.* very risky.

troppo *adj.* mentally disturbed. [shortened form of *tropical* from mental illness resulting from long military service in the tropics.]

trot *phr.*

1. **a good (bad) trot,** a run of good (bad) luck.

2. **on the trot, a.** in a state of continuous activity. **b.** one after another, in quick succession: *He won three races on the trot.*

3. **trot out, a.** to bring forward for, or as for, inspection. **b.** to give voice to in a trite or boring way.

trots *pl. n.* **1.** diarrhoea. **2.** (in harness racing) a race-meeting for trotters.

trouble and strife *n.* a wife. [rhyming slang]

trousers *phr.* **wear the trousers,** to have control, as of the dominant partner in a marriage.

trout *n.* a woman of unattractive appearance: *an old trout.*

truck *v.* **1.** to walk confidently, jauntily, with somewhat exaggerated movement of the shoulders and arms. *—phr.* **2. (it) fell off the back of a truck,** See **fall.**

truckie *n.* a truck driver.

true *phr.* **true blue,** See **blue.**

trump *n. Prison.* a senior officer in a prison, or section of a prison.

trump card *n.* a decisive factor; important advantage. [from the card game]

trumpet *phr.* **blow one's own trumpet,** See **blow.**

trundle *v.* to walk in a leisurely fashion.

trundler *n. Cricket.* a spin bowler.

trunky *adj.* drunk. [shortened form of rhyming slang *elephant's trunk,* drunk]

truth drug *n.* sodium pentothal when used as a mild anaesthetic to render a person incapable of answering questions untruthfully.

try *phr.* **try (it) on,** to attempt to hoodwink or test the patience of, especially impudently.

try-on *n.* an attempt to hoodwink or test the patience of someone.

tryout *n.* a trial, practice, or test to ascertain fitness for some purpose.

tub *n.* **1.** a bath in a tub. **2.** *Prison.* a sanitary bucket in a prison cell. **3.** a slow, clumsy ship or boat. **4.** *Rowing.* a heavy boat used for training novices. *—v.* **5.** to wash or bathe in a tub. **6.** *Rowing.* to coach (oarsmen) in a tub. *—phr.* **7. go off the tub,** *Prison.* to commit suicide by hanging.

tube *n.* **1.** a can of beer: *Let's sink a few tubes.* **2.** a television set. **3.** *Shearing.* the flexible casing which leads from the main driving gear to the shearer's handpiece. *—phr.* **4. give the tube away,** *Shearing.* to give up shearing; retire.

tub-thumper *n.* a very vocal enthusiast.

tuck *n.* **1.** *Chiefly British.* food, especially sweet delicacies such as cakes, pastries, jam, etc. *—phr.*
2. tuck away, a. to imprison. **b.** to store or save.
3. tuck into or **away,** to eat or drink heartily or greedily.

tucker[1] *n.* food.

tucker[2] *phr.* **tucker out,** to weary; tire; exhaust.

tuckerbag *n.* a bag used for carrying food, as by a swagman.

tuck-in *n.* a hearty meal.

tug *v.* **1.** (of a male) to masturbate. *—phr.* **2. tug someone's head,** *Prison.* to distract someone's attention.

tumble *v.* **1.** to become suddenly alive to some fact, circumstance, or the like. *—phr.*
2. a tumble in the hay, sexual intercourse outdoors.
3. tumble to (someone), to understand the motives of (someone) suddenly.

tummy *n.* stomach. Also, **tum.**

tune *phr.*
1. call the tune, to be in a position to give orders, dictate policy, etc.; command; control.
2. change one's tune; sing another /or different tune, to change one's mind; reverse previously held views, attitudes, etc.
3. to the tune of, to the amount of.

tuppeny *adj.* See **twopenny.**

turd *n.* **1.** a piece of excrement. **2.** (*offensive*) an unpleasant person.

turd strangler *n.* a plumber.

turf *phr.* **turf out,** to throw out; eject.

turkey *n.* **1.** *U.S.* an unsuccessful theatrical production; flop. *—phr.* **2. head over turkey,** head over heels.

3. talk turkey, to talk seriously; talk business.

turn *n.* **1.** a nervous shock, as from fright or astonishment. —*v.* **2.** to change religious denomination, especially from Protestant to Roman Catholic: *We've not seen her since she turned.* —*phr.*

3. turn in, a. to go to bed. **b.** to hand over to; deliver; surrender.

4. turn it up!, stop it! shut up!

5. turn off, a. to arouse antipathy or revulsion in: *His teaching turns me off.* **b.** to lose interest in or sympathy with; develop a dislike for: *I've turned off gardening.*

6. turn on, a. to show or display suddenly: *to turn on the charm.* **b.** to excite or interest (a person): *That jazz really turns me on!* **c.** to experience heightened awareness under the influence of a drug (usually illegal), as marijuana or LSD. **d.** to take such a drug. **e.** to arouse sexually.

7. turn out, a. to get out of bed. **b.** to assemble, or cause to assemble; muster; parade; gather.

8. turn up, a. *British.* to make (a person) feel sick; nauseate. **b.** *Prison.* to give evidence that results in an acquittal for (someone).

9. turn up one's toes, See **toe.**

turned-up *v.* *Prison.* to be acquitted of a police charge.

turn-in *n.* a fight.

turn-on *n.* that which or one who excites interest, enthusiasm, etc., expecially sexual.

turnout *n.* a party, show, entertainment, etc.

turn-up *n.* **1.** a fight, row, or disturbance. —*phr.* **2. turn-up for the books,** a surprise; an unexpected reversal of fortune.

turps *n.* **1.** turpentine. —*phr.* **2. on the turps,** drinking intoxicating liquor excessively.

turpsy *adj.* drunk.

turtle *phr.* **turn turtle,** to retreat; become cowardly.

tute *n.* a tutorial. [shortened form]

tux *n.* a tuxedo. [shortened form]

twang *n.* opium.

twat *n.* **1.** the vagina. **2.** a woman considered as a sexual object. **3.** sexual intercourse. **4.** a despicable or unpleasant person. Also, **twot.**

tweaker *n.* a finger-spinner in cricket.

twee *adj.* affected; precious; excessively dainty; coy.

twerp *n.* an insignificant or stupid person. Also, **twirp.**

twicer *n.* a crook; double-crosser.

twig *v.* **1.** to look at; observe. **2.** to catch sight of; perceive. **3.** to understand. —*phr.* **4. snap one's twig, a.** lose one's temper. **b.** to become mentally unstable; deranged; demented.

twilight zone *n.* old age.

twinkle *v.* **1.** to urinate. —*n.* **2.** the act of urinating.

twin-set *phr.* **twin-set and pearls,** (of certain, usually young, middle-class women) typified by conservative dress, outlook, etc.

twist *n.* **1.** *Prison.* a life sentence. —*phr.*

2. round the twist, insane.

3. twist (someone's) arm, to persuade: *You've twisted my arm, I'll do it.*

twit *n.* **1.** a fool; twerp. **2.** head or mind: *Are you out of your twit?*

twitch *phr.* **the twitches,** a state of nerves causing muscular spasms.

two *phr.*

1. put two and two together, to draw a conclusion from certain circumstances.

2. two men and a dog, very few people.

Two Blues, the *pl. n.* **1.** the Parramatta team in the N.S.W. Rugby Football Union. **2.** the Prahran team in the Victorian Football Association.

two bob *n.* **1.** (formerly) a sum of money of the value of two shillings. **2.** (formerly) a silver coin of this value. *—phr.*

3. have two bob each way, to support contradictory causes at the same time, often in self-protection.

4. not the full two bob, weak of intellect.

5. not worth two bob, insignificant.

6. the full two bob, that which is genuine and of full value.

7. two bob's or **two cent's worth,** opinion, say or advice: *He got in his two bob's worth before the end of the meeting.* Also, **two-bob.**

two-bob *adj.* **1.** of poor quality; useless; unreliable: *goes like a two-bob watch. —phr.*

2. mad or **silly as a two-bob watch,** See **silly.**

3. two-bob boss, a minor official who delights in exerting his authority in an overbearing manner.

4. two-bob lair, a person whose clothes are flashy but cheap.

5. two-bob millionaire, a person temporarily flush with money.

two-fisted *adj.* *U.S.* strong and vigorous.

two-ie *v.* *Qld.* to gang up on.

twopenny *phr.*

1. not to matter a twopenny damn or **dump,** not to be worth any consideration.

2. twopenny upright, sexual intercourse while standing. Also, **tuppeny.**

two-pot screamer *n.* someone who gets uproariously drunk after consuming very little alcohol. Also, **two-schooner screamer.**

two shakes *n.* a very short time; a minute. Also, **two shakes of a lamb's tail.**

twot *n.* See **twat.**

two-time *v.* to deceive or doublecross, especially a friend or lover, by having a similar relationship with another.

two-up *n.* **1.** a gambling game in which two coins are spun in the air and bets are laid on whether they fall heads or tails; swy. *—adv.* **2.** (in motor-cycling) travelling with a pillion passenger.

two-up school *n.* an organised game of two-up: *Thommo's two-up school.*

two-wheeler *n.* a woman or girl; sheila. [rhyming slang]

U *adj.* appropriate to or characteristic of the upper class. See **non-U.**

ugh *interj.* an exclamation expressing disgust, aversion, horror, or the like.

ugly *phr.* **as ugly as a hatful** or **bagful of arseholes; as ugly as a plateful of mortal sins; as ugly as homemade sin,** very ugly (an insulting or abusive remark).

U-ie *n.* **1.** a U-turn. *—phr.* **2. chuck a U-ie,** to do a U-turn.

ultimate *phr.* **the ultimate,** the most successful, pleasing, handsome, etc.

um *phr.* **um and ah, a.** to be indecisive. **b.** to prevaricate.

umpie *n.* an umpire. Also, **umpy, ump.**

umpteen *n.* **1.** an indefinite, especially a very large or immeasurable, number. *—phr.* **2. (go) umpteen to the dozen,** (do something) at a great and often frenzied pace.

uncle *n.* **1.** a pawnbroker. *—phr.*

2. to talk like a Dutch uncle, talk severely (to someone).

3. uncle from or **in Fiji,** an imaginary rich relative, etc., said to be financing some venture, and used as a lure for unwary potential investors.

Uncle Tom *n.* (*usually offensive*) a Negro who is openly servile to whites.

Uncle Willy *adj.* silly. [rhyming slang]

unco *adj.* **1.** awkward; clumsy. *—n.* **2.** a clumsy person. [short for *uncoordinated*]

uncool *adj.* displaying an unsophisticated level of emotion, as delight, anger, etc.; gauche.

uncrowned *phr.* **uncrowned king,** one who is regarded as the unofficial leader of his own particular circle: *He was the uncrowned king of the wharfies.*

under *phr.*

1. down under, in Australia or New Zealand.

2. go under, a. to sink in or as in water. **b.** to fail, especially of a business. **c.** to be found guilty by a jury.

underdaks *pl. n.* underpants.

underground *adj.* **1.** *Originally U.S.* opposed to conventional social standards: *underground press.* —*n.* **2.** the world of those opposed to accepted social standards; counter culture. [?related to the *underground,* a secret organisation fighting the established government or, especially during World War II, the fascist occupation forces in Europe]

underground mutton *n.* rabbit.

undersell *phr.* **undersell oneself,** to lack confidence in one's own worth; underrate oneself.

under-the-counter *adj.* **1.** pertaining to goods kept hidden for sale in some improper way, as on the black market. —*adv.* **2.** sold illegally, as black-market goods. Also, **under the counter.** [from the conventional hiding place of illegal stores in shops, as opposed to *on display*]

under-the-table *adj.* **1.** drunk. **2.** pertaining to payments made secretly or in some improper way.

undies *pl. n.* underwear.

undone *phr.* **come undone,** to be detected; make a mistake that leads to arrest and conviction.

undress *v.* to shear: *to undress a sheep.*

unearthly *adj.* unreasonable; absurd: *to get up at an unearthly hour.*

unemployment office *n.* the employment office.

ungodly *adj.* dreadful; outrageous: *They telephoned at the ungodly hour of 4 a.m.* Also, **unholy.**

unhealthy *adj.* dangerous.

unloaded *adj.* knocked over heavily, as in football.

unmentionables *pl. n.* underwear.

unreal *adj.* **a.** unbelievably awful. **b.** unbelievably wonderful.

unstuck *phr.* **come unstuck, a.** to fail; suffer defeat or disaster, often as a result of questionable practice or being too clever. **b.** to lose mental stability; feel disturbed or insecure.

untogether *adj.* not appropriately oriented: *Cynthia was so untogether she didn't like boys.*

unwashed *phr.* **the great unwashed,** (*offensive*) **a.** the masses; the rabble. **b.** the English.

up *prep.* **1.** engaged in sexual intercourse with: *He was up her like a lizard up a log.* **2.** angry with: *She was up him for being late.* —*v.* **3.** to be under the influence of a hallucinatory drug: *up on a trip.* **4.** to make larger; step up: *to up output.* **5.** *Gambling.* to raise; go better than (a preceding wager). —*phr.*

6. all up, at an end; at the point of defeat or failure.

7. be up oneself, to have an unjustifiably high opinion of oneself; be self-deluding.

8. get up (someone), to have sexual intercourse with (someone).

9. get up (someone's) nose, See **nose.**

10. on the up and up, a. *Chiefly U.S.* tending upwards; improving; having increasing success. **b.** honest, frank, or credible.
11. up a gumtree, a. confused; incorrect. **b.** in a quandary. **c.** left helpless.
12. up against, faced with: *They are up against enormous problems.*
13. up against it, in difficulties; in severe straits.
14. up and, to get or start up: *She upped and got herself a new job.*
15. up each other or **up one another,** behaving in a sycophantic or toadying fashion to each other.
16. up for, liable to pay: *You'll be up for $100 if you break that.*
17. up shit creek, See **shit.**
18. up the duff, pregnant.
19. up the river, in gaol.
20. up the spout, See **spout.**
21. up there Cazaly, See **Cazaly.**
22. up to, a. engaged in; doing: *What are you up to?* **b.** capable of: *He is not up to the job.*
23. up to mud/putty/shit, Also, **upter.** broken down; of poor quality; worthless.
24. up you (for the rent)!, Also, **up yours; up your arse!** an exclamation indicating insolent or disgusted dismissal.
25. up your nose (with a rubber hose), a generalised insult. (The response is **twice as far with a chocolate bar.**)
26. who's up who (and who's paying the rent), a. an enquiry as to the personal alliances in a political or business group, etc. **b.** an enquiry as to the alliances and sexual relationships within a particular group of people.

up-beat *adj.* optimistic; cheerful.

up-country *adj.* unsophisticated.

up-front *adj.* **1.** straightforward; open; honest. *—adv.* **2.** in advance: *to get money up-front for a project.*

upgrade *phr.* **on the upgrade,** improving; up-and-coming.

up-market *adj.* superior in style or production; pretentious.

upper *n.* **1.** a stimulant, as amphetamine, etc. **2.** a pleasant or exhilarating experience. *—phr.* **3. be on one's uppers,** to be reduced to poverty or want (i.e. having the soles out of one's shoes).

uppity *adj.* affecting superiority; presumptuous; self-assertive. Also, **uppish.**

uprights *pl. n. Football.* the goalposts.

upstairs *adv.* **1.** into the air. *—n.* **2.** *Prison.* a higher court. *—phr.* **3. kick upstairs,** See **kick.**

uptight *adj.* **1.** tense, nervous, or irritable. **2.** angry. **3.** conforming to established conventions.

upya *interj.* up you (an offensive exclamation).

urge *phr.* **give someone an urge,** to let someone in ahead of one in a queue.

urger *n.* **1.** a racecourse tipster. **2.** one who takes advantage of others. **3.** one who incites others; an agitator.

use *phr.* **use up,** to exhaust; tire out.

user *n.* a drug addict.

ute *n.* a utility truck or utility van. [shortened form]

vac *n.* vacation. [shortened form]

vag *n.* **1.** a vagrant. —*v.* **2.** to arrest on a vagrancy charge. [shortened form]

Vales, the *pl. n.* the Springvale team in the Victorian Football Association.

vamoose *v. Chiefly U.S.* to make off; decamp; depart quickly. [Spanish *vamos,* let us go]

vamp *n.* **1.** a woman who uses her charms to seduce and exploit men. —*v.* **2.** to act as a vamp. [short for *vampire*]

Varsity, the *pl. n.* the University team in the Queensland Rugby Football Union. Also, **Red Heavies.**

Vatican roulette *n.* the rhythm method of contraception.

Vee Dub *n.* a Volkswagen. [from the initials *V.W.*]

vegemite *n.* **1.** a child, especially one who is good or well-behaved: *You're a clever little vegemite, aren't you.* **2.** a humorous term for people in general: *Look at the happy little vegemites working away in there.* [from the tradename and advertising jingle of *Vegemite,* a yeast extract used as a spread]

vegetable *n.* **1.** a dull or uninspiring person. **2.** a person who, due to physical injury or mental deficiency, is entirely dependent on the agencies of others for subsistence.

vegies *pl. n.* vegetables.

velvet *n.* **1.** a very agreeable or desirable position or situation. **2.** money gained through gambling or speculation. **3.** clear gain or profit. **4.** See **black velvet.**

verbal *n. Prison.* **1.** a verbal confession, usually made to the police and recorded by them, and sometimes alleged to be fabricated. —*v.* **2.** (of the police) to insert into a prisoner's statement admissions which he did not make, and present it to a court as evidence.

verbal diarrhoea *n.* nonsense talked at length.

vet[1] *n.* **1.** a veterinary surgeon. —*v.* **2.** to examine (a person): *The applicants were well vetted.* **3.** to examine (a product, proposal, or the like) with a view to acceptance, rejection, or correction.

vet[2] *n. U.S.* a veteran.

V.G. *n.* a valuation authorised by the Valuer-General.

vibes *pl. n.* the quality, mood or atmosphere of a place or person, thought of as producing vibrations to which one unconsciously responds: *The vibes of that town were all wrong.*

video *n. U.S.* television.

video land *n.* the market audience for video-cassette films.

Vikings, the *pl. n.* the Gold Coast team in the Queensland Rugby Football League.

villain *n. Originally British.* a criminal.

vim *n.* force; energy; vigour in action.

vino *n.* wine. [Italian]

V.I.P. *n.* very important person. [initials]

visiting card *n.* any article left behind and subsequently recognised as belonging to or associated with whoever left it: *Susan's dog left a visiting card.*

visiting fireman *n.* a visiting member of high status in an organisation, such as a visiting professor in academic circles, a visiting director of a multinational company, etc.

visitor *n.* menstruation; period.

vital statistics *pl. n.* the measurements of a woman's figure, as at the bust, waist, and hips. [term popularised in the 1920s in Australia following the Berlei Anthropometrical Survey conducted in conjunction with the University of Sydney to determine basic female figure types]

vocab *n.* vocabulary: *He has the vocab of a six-year old.*

vomit-making *adj.* unattractive or unpleasant, as an idea, colour, etc.: *I think the whole concept is pretty vomit-making.*

vote *phr.* **vote with one's feet, a.** to express one's disapproval by leaving. **b.** a public exhibition of sympathy, opposition, etc., as a mass meeting, demonstration or march.

wack *n.* an erratic, irrational, or unconventional person. Also, **wacker.**

wacky *adj. Chiefly U.S.* erratic, irrational, or unconventional; crazy. Also, **whacky.**

waddy *n.* a heavy stick or club. [Aboriginal]

wade *phr.* **wade in** or **into, a.** to begin energetically. **b.** to attack strongly.

waffle *v.* **1.** to speak or write vaguely, pointlessly, and at considerable length. **2.** to talk or write nonsense. *—n.* **3.** nonsense; twaddle. Also, **woffle.**

wag *phr.* **wag it,** to deliberately stay away from school, work, etc., without permission.

wagga *n.* a blanket made from hessian bags or similar material. Also, **Wagga blanket, Wagga rug.** [from *Wagga Wagga,* a town in N.S.W.]

wagon *n.* **1.** a tea-trolley. **2.** a station wagon. *—phr.*
3. climb or **jump on the band wagon,** See **band wagon.**
4. on the (water) wagon, abstaining from alcoholic drink.

waipiro *n. N.Z.* strong alcoholic drink. [Maori *wai* water + *piro* stinking]

wait *v.* to defer or postpone in expectation of the arrival of someone: *to wait dinner for the guests.*

wake *phr.*
1. enough or **fit to wake the dead,** See **dead.**
2. wake up to oneself, to adopt a more sensible and responsible attitude.
3. wake up to (someone), to become aware of the motives, true nature, etc., of (someone).

wake-up *n.* **1.** one who is fully aware: *He's a real wake-up. —phr.*
2. take a wake-up to, to understand; perceive meaning or purpose of.

wakey-wakey *n.* **1.** the time to rise from bed. —*interj.* **2.** wake up!

Waler *n. Obsolete.* a breed of sturdy horses originating in New South Wales and used as cavalry mounts in the Boer War and World War I and also exported for use by the Indian Army. Also, **waler.** [named after the colony of New South *Wales*]

walk *phr.*
1. take a walk, to disappear.
2. walk all over, a. to behave in a domineering and aggressive fashion towards. **b.** to defeat.
3. walk away with, to win easily.
4. walk off, to abandon a farm, etc., because of severe economic difficulties: *The original owner walked off in 1896 because of the drought.*
5. walk off with, a. to remove without permission; steal. **b.** to win, as in a competition. **c.** to outdo one's competitors; win easily.
6. walk out, to leave in protest; leave angrily. [from *walk out,* to go on strike]
7. walk over, to win easily. [See **walkover]**
8. walk the board, *Surfing.* to walk along the board while riding a wave, usually as a means of controlling the board's performance.
9. walk the streets, a. to wander about the streets, especially as a result of being homeless. **b.** to be a prostitute, especially one who solicits on the streets.

walkabout *phr.* **go walkabout, a.** to wander around the country in a nomadic manner. **b.** to be misplaced or lost. [from *walkabout,* a period of wandering, often as undertaken by Aborigines]

walking papers *n.* dismissal from one's place of work; the sack: *He was given his walking papers.* Also, **walking ticket.**

walkover *n.* **1.** *Racing.* a going over the course at a walk or otherwise by a contestant who is the only starter. **2.** an unopposed or easy victory.

walk-up start *n.* **1.** something easily taken or achieved. **2.** a person easily deceived: *That customer is a walk-up start.*

wall *phr.*
1. go to the wall, a. to give way or suffer defeat in a conflict or competition. **b.** to fail in business, or become bankrupt.
2. send or **drive (someone) up the wall,** send into a state of exasperation, confusion, etc.: *Those kids drive me up the wall.*
3. up against the wall, in a very difficult predicament.

Wallabies, the *pl. n.* the Australian representative Rugby Union Football team. [name adopted in 1908]

wallaby *phr.* **on the wallaby (track),** on the move as a swagman.

wallaby day *n. Qld.* a day when country people go to town, often Saturday.

wallah *n.* a person employed at or concerned with a particular thing (used especially in combination with another word): *laundry wallah; cleaning wallah.* Also, **walla.** [Hindustani *wala*]

wallflower *n.* a person, especially a woman, who looks on at a dance, especially from failure to obtain a partner.

wallop *v.* **1.** to beat soundly; thrash. **2.** to strike with a vigorous blow. **3.** to defeat thoroughly, as in a game. —*n.* **4.** a vigorous blow. **5.** a forceful impression or impact. **6. a.** beer. **b.** any alcoholic drink.

walloper *n.* **1.** one who or that which wallops. **2.** a policeman.

walloping *n.* **1.** a sound beating or thrashing. **2.** a thorough defeat. —*adj.* **3.** of large size; whopping.

wally *n.* **1.** *British.* a fool; a stupid person. —*adj.* **2.** angry; annoyed. [contraction of *Walter,* man's name]

waltz[1] *v.* **1.** to gain with great ease: *He waltzed off with the first prize.* **2.** to move nimbly or quickly. [from the dance step]

waltz[2] *phr.* **waltz Matilda,** to wander about as a tramp with a swag. [?German *walzen,* to move in a circular fashion, as of apprentices travelling from master to master + German *Mathilde,* female travelling companion, bed roll, from the girl's name; ?taken to goldfields by German speakers from South Australia]

wampum *n. U.S.* money. [short for *wompanpeag,* string of shell beads, from Algonquian Indian. The beads were used by North American Indians as money]

wanda *n.* a white man. [Aboriginal: white ghost]

wangle *v.* **1.** to bring about, accomplish, or obtain by contrivance, scheming, or often, indirect or insidious methods. **2.** to fake; falsify; manipulate. **4.** to manipulate or continue something for dishonest purposes. —*n.* **5.** an act or instance of wangling. [blend of *wag* and *dangle*]

wank *v.* **1.** to masturbate. —*n.* **2.** an act or instance of masturbation. **3.** a hobby: *flying is his wank.* **4.** behaviour which is self-indulgent or egotistical. —*phr.* **5. wank oneself,** to maintain an illusion; deceive oneself.

wanker *n.* **1.** a deluded fool. **2.** a self-indulgent or egotistical person.

want *phr.* **want out of,** to wish to withdraw from (a difficult situation, obligation, etc.).

war *phr.* **in the wars,** involved in a series of misfortunes or minor injuries.

warb *n.* **1.** a dirty or unkempt person. **2.** a derelict person. [backformation from *warby*]

warby *adj.* **1.** unkempt; decrepit. **2.** of dubious worth. **3.** unwell; squeamish. [British dialect *warbie,* maggot]

war horse *n.* a veteran soldier, politician, etc., especially an aggressive one.

warm *adj.* **1.** relatively close to something sought, as in a game. **2.** uncomfortable or unpleasant. —*phr.* **3. warm a seat,** to occupy a position, usually in a temporary capacity, and without actively discharging its responsibilities.

war paint *n.* **1.** make-up; cosmetics. **2.** full dress; finery.

warpath *phr.* **on the warpath,** in a state of wrath; angry; indignant.

warrigal *n.* anything wild and untamed. [Aboriginal, applied to a dingo]

Warriors, the *pl. n.* **1.** the Heidelberg United team in the National Soccer League. **2.** the Waratah team in the Northern Territory Football League. **3.** the

Woodville team in the South Australian National Football League.

warts *pl. n.* **1.** defects, of body or character. *—phr.* **2. warts and all,** including all defects.

warwicks *pl. n.* the arms. [rhyming slang, *Warwick Farm,* (a racecourse in Sydney), arm]

wash *v.* **1.** to stand being put to the proof; bear investigation: *That excuse might wash with some people.* *—phr.*

2. come out in the wash, to be revealed eventually; become known.

3. wash one's dirty linen, See **linen.**

4. wash one's hands of, to dismiss from consideration; reject.

washed-out *adj.* **1.** utterly fatigued; exhausted. **2.** tired-looking; pale; wan.

washed-up *adj.* **1.** having failed completely; finished; ruined. **2.** exhausted.

washer *n.* a face-washer made of towelling.

wash-out *n.* **1.** a failure or fiasco. **2.** a helpless or inefficient person.

wash-up *n.* the outcome or aftermath.

WASP *n.* **1.** *Originally U.S.* a member of the establishment conceived as being white, Anglo-Saxon and Protestant. *—adj.* **2.** of or pertaining to this establishment. [*W(hite) A(nglo) S(axon) P(rotestant)*]

waste *v.* *Chiefly U.S.* to murder.

wasted *adj.* lethargic or exhausted, as a result of taking drugs.

watch *phr.* **watch it!,** be warned.

water *phr.*

1. above water, out of embarrassment or trouble, especially of a financial nature.

2. go to water, to lose courage; abandon one's resolve.

3. in deep or **hot water, a.** in trouble; in a difficult situation. **b.** touching on an area of consideration which is contentious.

4. like water, abundantly; freely: *to spend money like water.*

5. like water off a duck's back, See **duck.**

6. throw cold water on, to discourage.

7. water the horse, to urinate.

water-bag *n.* a teetotaller.

watergate *n.* **1.** a scandal involving charges of corruption against a political leader. **2.** a downfall of a political leader, caused by a scandal. *—v.* **3.** to conceal; cover up. [from *Watergate,* hotel in Washington, the bugging of which led to a political scandal and the resignation of U.S. President, Richard Nixon]

watering hole *n.* a hotel. Also, **watering-place.**

water-rat *n.* *U.S.* a vagrant or thief who frequents a waterfront.

watertable *n.* a ditch, especially one at the side of the road: *I fell off my bike and rolled into the watertable.*

water wagon *phr.* **on the water wagon,** See **wagon.**

waterworks *n.* **1.** tears, or the source of tears. **2.** the bladder or its functioning. *—phr.* **3. turn on the waterworks,** to cry loudly and profusely, often for the sake of gaining sympathy or getting one's own way.

wax *n.* a fit of anger. [origin uncertain; ?from phrase *wax angry*]

waxhead *n.* a surfboard rider.

way *n.* **1.** condition, as to health, prosperity, etc.: *to be in a bad way.* *—phr.*
2. have a way with, a. to have a skill in dealing with: *She has a way with children.* **b.** to have a charming or persuasive manner.
3. no way, not at all; never.
4. on the way out, a. becoming obsolete; ready for rest or retirement. **b.** losing popularity.
5. out of the way, a. murdered: *to put a person out of the way.* **b.** out of the frequented way; off the beaten track.

wayback *n.* **1.** the outback. *—phr.* **2. the waybacks,** *N.Z.* remote rural districts.

way-out *adj.* **1.** advanced in technique, style, etc. **2.** unusual; odd; eccentric.

WC *n.* a toilet. Also, **wc** [*W(ater) C(loset)*]

weaker sex *n.* (*offensive*) the female sex.

weakie *n.* a weak or cowardly person. Also, **weaky.**

weak-kneed *adj.* yielding readily to opposition, intimidation, etc.

wear *v.* to accept, tolerate, or be convinced by: *He told me a lie but I wouldn't wear it.*

weaselword *v.* to talk one's way out of trouble, usually by talking nonsense in such a way that it appears pertinent.

weather *phr.*
1. keep one's weather eye open, to be on one's guard; keep a sharp lookout. [from *weather eye,* ability to forecast the weather]
2. make heavy weather of, to have a lot of difficulty coping with (something).
3. under the weather. a. indisposed; ill; ailing. **b.** drunk.

weave *phr.* **get weaving,** to make a start, especially hurriedly, enthusiastically, etc.

weed *n.* **1.** a cigar or cigarette. **2.** a marijuana cigarette. *—phr.* **3. the weed, a.** tobacco. **b.** marijuana.

weekend revolutionary *n.* one who advocates revolution while enjoying the life-style of the established society under attack.

weeny *adj.* **1.** very small; tiny. *—n.* **2.** Also **weenie.** See **gruff nut.**

weepie *n.* a sentimental film or play.

wee-wee *n.* (*in children's speech*) urine. Also, **wee-wees.**

weigh *phr.*
1. weigh into, a. to attack, physically or verbally. **b.** to begin to eat with hearty appetite.
2. weigh in with, to offer or contribute (an opinion, etc.).

weight *phr.*
1. pull one's weight, to do one's fair share of work.
2. put (a person's) weights up, *N.Z.* to inform on (someone).
3. throw one's weight around or **about, a.** to behave in an aggressive or selfish fashion. **b.** to use one's influence, personality, etc., to gain one's own ends without regard for others.

weird *adj.* startlingly or extraordinarily singular, odd, or queer: *a weird get-up.*

weirdo *n.* one who behaves in a strange, abnormal, or eccentric way. Also, **weirdie, weirdy.**

well-endowed *adj.* **1.** (of a woman) bosomy. **2.** (of a man) with big genitals.

well-fixed *adj.* well-off; affluent.

well-heeled *adj.* wealthy; prosperous.

well-hung *adj.* (of a male) with big genitals.

wellies *pl. n.* wellington boots.

well-in *adj.* **1.** having influential friends; well-connected. **2.** affluent; well-off.

well-oiled *adj.* drunk.

welsh *v.* to cheat by evading payment, especially of a gambling debt. Also, **welch, welsh on.**

welter *phr.* **make a welter of it,** to indulge in to excess.

werris *n.* an act of passing water; urination. [rhyming slang *Werris Creek,* leak]

west *phr.*
1. the West, Western Australia.
2. go west, a. to die. **b.** to disappear; be lost.

westie *n.* someone from the western suburbs of Sydney.

Westralia *n.* Western Australia.

wet *adj.* **1.** weak; feeble; spiritless. —*n.* **2.** Also, **Wet.** one who advocates government intervention in the economic affairs of a country. —*phr.*
3. silly as a wet hen, See **silly.**
4. the wet, Also, **the Wet.** the rainy season in central and northern Australia, from December to March.
5. wet behind the ears, naive, lacking maturity, experience, or the like.
6. wet enough to bog a duck, very wet, as in a prolonged period of rain.
7. wet one's whistle, See **whistle.**
8. wet the baby's head, to celebrate a baby's birth, as by having a drink.

wet blanket *n.* **1.** a person or thing that dampens ardour or has a discouraging or depressing effect. —*v.* **2.** to dampen the ardour of.

wet-nurse *v.* to cosset or pamper.

wet weekend *n* **1.** something (usually boring or unpleasant) which appears to last for a very long time: *The dinner party dragged on like a wet weekend.* —*phr.* **2. a face as long as a wet weekend,** See **face.**

whack *v.* **1.** to strike with a smart, resounding blow or blows. —*n.* **2.** a smart, resounding blow: *a whack with his hand.* **3.** a trial or attempt: *to take a whack at a job.* **4.** a portion or share. —*phr.*
5. whack down, to put down quickly: *whack it down here.*
6. whack in, put in; insert.
7. whack off, to masturbate.
8. whack up, to divide up; share.

whacked *adj.* exhausted; defeated: *I am whacked from all that work.*

whacking *adj.* large.

whacko! *interj.* an expression denoting pleasure, delight, etc. Also, **wacko, whacko-the-diddle-oh, whacko-the-did.**

whale[1] *n.* **1.** something extraordinarily big, great, or fine of its kind: *a whale of a lot, a whale of a time.* **2.** a Murray cod. —*v.* **3.** to fish for Murray cod. **4.** to live as an itinerant. See **Murrumbidgee whaler.**

whale[2] *v.* **1.** to whip, thrash, or beat soundly. —*phr.* **2. whale into, a.** to beat up, bash. **b.** to attack verbally, berate. **c.** to throw oneself into something energetically.

whaler *n.* a swagman or itinerant. See **Murrumbidgee whaler.**

wham *phr.* **wham bam thankyou ma'am, a.** the act of sexual intercourse, especially when quick and unemotional. **b.** anything done quickly and without fuss.

whammy *n.* **1.** a forceful influence, blow, spell, etc.: *The witch's whammy immobilised him.* *—phr.* **2. put the whammy on (someone** or **something),** to render useless, motionless or powerless.

wharfie *n.* wharf labourer. Also, **wharfy.**

what *phr.*

1. so what? an exclamation of contempt, dismissal, or the like.

2. what for, severe treatment, punishment, or violence: *He hit me, so I gave him what for.*

3. what it takes, the necessary ability, personality, or the like: *He may look stupid, but he's got what it takes to hold the job down.*

4. what's what, the true position.

what-d'ye-call-it *n.* a name used in place of one temporarily forgotten: *Please pass me the what-d'ye-call-it.* Also, **what-d'ye-m'-call-it, whatsit.**

whatever *pron.* anything or anyone: *Bring friends, family, whatever.*

whatnot *n.* anything; no matter what; what you please: *a chronicler of whatnots.*

wheat-cocky *n.* a wheat farmer on a small scale of operation. [See *cocky*]

wheel *n.* **1.** a person of considerable importance or influence: *a big wheel.* *—phr.*

2. be on (someone's) wheel, to stand over, annoy or harass (someone).

3. silly as a wheel, very silly.

4. wheel and deal, to act as a wheeler-dealer.

wheeler-dealer *n.* **1.** *Chiefly U.S.* one in a position of power who controls and directs the actions of others. **2.** one who actively pursues his own advancement by moving constantly from one profitable business transaction to another.

wheelie *n.* **1.** a violent, usually noisy skidding of the driving wheels of a motor car while accelerating as around a corner or from a standing start: *to do a wheelie.* **2.** a manoeuvre performed on a bicycle, motorbike, etc., in which the rider maintains his balance while lifting the front wheel off the ground.

wheelman *n.* the driver of a getaway car.

wheels *pl. n.* a motor vehicle.

wheeze *n. British.* a trick, dodge, or idea, especially a cunning or artful one.

whelp *n.* (*offensive*) a youth.

whew *interj.* a whistling exclamation or sound expressing astonishment, dismay, etc.

whiffy *adj.* smelly.

whim-wham *n.* any odd or fanciful object or thing; something showy or useless.

whingeing Pom *n.* (*offensive*) an Englishman thought to be always criticising and complaining about life in Australia.

whip *n.* **1.** a jockey. *—v.* **2.** *Chiefly U.S.* to beat, outdo, or defeat, as in a contest. *—phr.*

3. **crack the whip,** to urge to greater effort.

4. **fair crack of the whip!,** an appeal for fairness.

5. **whip around,** to make a collection of money.

6. **whip** or **flog the cat, a.** to make a fuss or commotion. **b.** to reproach oneself.

7. **whip off,** to take; steal.

8. **whip up, a.** to create quickly: *I whipped up a meal when I heard he was coming.* **b.** to arouse to fury, intense excitement, etc.: *His speech soon whipped up the crowd.*

9. **whips of,** an abundance: *whips of room; whips of money.*

whippy *n.* **1.** a wallet. **2.** (in games) the finishing post; base. *—phr.* **3. all in, the whippy's taken,** an expression used to signal the end of a round in certain games of hide and seek.

whip-round *n.* an impromptu collection of money. Also, **whip-around.**

whipstitch *n. U.S.* an instant.

whirl *phr.* **give it a whirl,** to make an attempt.

whirlybird *n. U.S.* a helicopter.

whirly-whirly *n.* a spiralling wind, often collecting dust, refuse, etc.; a willy-willy.

whisker *n.* **1.** a very small quantity or distance: *He won the race by a whisker.* *—phr.* **2. have whiskers on it,** to be old-fashioned or useless: *That idea has whiskers on it.*

whisky drinker *n.* a dark, thickset cicada, with a bright red nose; cherry nose.

whistle *phr.*

1. **blow the whistle on,** See **blow.**

2. **wet one's whistle,** to satisfy one's thirst, usually with an alcoholic drink.

3. **whistle for,** to ask or wish for (something) in vain.

4. **whistle in** or **against the wind,** to protest in vain.

white *adj.* **1.** *British.* honourable; trustworthy. *—phr.* **2. bleed white,** to deprive or be deprived of resources.

white angel *n.* a drink made by mixing methylated spirits and white shoe polish.

white ant *phr.* **to have white ants,** to be silly, crazy, eccentric.

white-ant *v.* to subvert or undermine from within (an organisation or enterprise). [from the destructive wood-eating insect]

white bread *n.* a sanitary napkin.

white-haired *adj.* favourite; darling: *a white-haired boy.* Also, **white-headed.**

white lady *n.* methylated spirits, as a drink; sometimes mixed with another liquid.

white leghorn *n.* a woman who plays lawn bowls. [a humorous reference to the supposed similarity in appearance to the *white leghorn,* a pure white domestic fowl]

white lightning *n. U.S.* cheap, inferior whisky.

white trash *n.* **1.** (*offensive*) poor white people collectively, especially in the southern U.S. **2.** one such person.

whitewash *n.* **1.** (in various games) a defeat in which the loser fails to score. *— v.* **2.** (in various games) to subject to a whitewash.

whitey *n.* (*offensive*) a white man.

whiz *n.* a person who shows outstanding ability in a particular field or who is notable in some way; expert: *a computer whiz.* [?abbreviation of *wizard*]

whiz-kid *n.* a person, especially young, who achieves spectacular success in a given enterprise.

who *phr.* **who's up who and who's paying the rent,** See **up.**

whodunit *n.* a novel, play, etc., dealing with a murder or murders and the detection of the criminal.

whole *phr.* **out of whole cloth,** *U.S.* without foundation in fact: *A story made out of whole cloth.*

whole hog *n.* **1.** entireness; completeness. *—phr.* **2. go the whole hog,** to involve oneself to the fullest extent; do something thoroughly or completely.

whoop *phr.*

1. not worth a whoop, not worth a thing; utterly valueless.

2. whoop it or **things up, a.** to raise an outcry or disturbance. **b.** to have a party or celebration. [from *whoop* a loud cry or shout]

whoopee *n.* **1.** uproarious festivity. *—phr.* **2. make whoopee,** to engage in uproarious merry-making. [extended variant of *whoop*]

whoops *interj.* an exclamation of mild surprise, dismay, etc. Also, **whoops-a-daisy, woops.**

whop *v.* **1.** to dash, pitch or throw with force. **2.** to strike forcibly. **3.** to defeat soundly, as in a contest. **4.** to strike out or move suddenly. **5.** to plump suddenly down; flop. *—n.* **6.** a forcible blow or impact. **7.** the sound made by it. **8.** a bump; a heavy fall.

whopper *n.* **1.** something uncommonly large of its kind. **2.** a big lie.

whopping *adj.* very large of its kind; huge.

whores bath *n.* a quick wash, usually only the armpits and crotch.

whoreson *n.* (*offensive*) a person.

whosits *n.* a name used in place of one temporarily forgotten.

WHS *n.* (used by women of men) Wandering Hands Society. [initials]

wick *n.* **1.** a penis. *—phr.*

2. dip one's wick, (of a man) to have sexual intercourse.

3. get on one's wick, to irritate.

wicked *adj.* **1.** distressingly severe, as cold, pain, wounds, etc. **2.** ill-natured, savage, or vicious: *a wicked horse.* **3.** extremely trying, unpleasant, or troublesome.

Wickers, the *pl. n.* the Berwick team in the Victorian Football Association.

wicket *phr.*

1. sticky wicket, a delicate, difficult or disadvantageous situation or set of circumstances: *He's on a sticky wicket now that his father has disinherited him.*

2. a good wicket, an advantageous situation or set of circumstances. [from the cricketing sense, a wicket affected by rain (sticky) is unpredictable and difficult to play]

Wicks, the *pl. n.* the Randwick team in the New South Wales Rugby Football Union.

widey *n.* a wide motor car tyre.

widgie *n.* (in the 1950s) a female equivalent of a bodgie, characterised by American influence on behaviour, style of dress, etc.

widow-maker *n.* **1.** *Timber Industry.* a dead branch on a tree which is likely to snap off and kill the feller beneath. **2.** *Australian Rules.* a very high kick which, as it descends, puts the player taking it in danger from both the force of the ball and from the actions of the converging members of the opposing team.

wife *phr.* **the wife's best friend,** the penis.

wife-beater *n.* a long thin loaf of bread. Also, **husband-beater.**

wife starver *n.* a maintenance defaulter.

wig *v. British.* to scold, reprimand or reprove severely.

wigging *n. British.* a scolding or reproof.

wigwam *n.* **1.** *U.S.* a structure, especially of large size, used for political conventions, etc. *—phr.* **2. a wigwam** or **whim wham for a goose's bridle,** See **goose.**

Wilcannia shower *n.* a dust storm [from *Wilcannia,* a town in western N.S.W.]

wild *adj.* **1.** intensely eager or enthusiastic. **2.** very annoyed. *—phr.*
3. go wild, to respond with extreme emotions, usually pleasure: *The crowd went wild when the home team won.*
4. run wild, to behave in an unrestrained or uncontrolled manner: *He allows his children to run wild.*
5. wild and woolly, a. rough; untidy; unkempt. **b.** uncivilised; unrestrained.

wildcatter *n. U.S.* one who prospects for oil or ores; a prospector.

wilderness *phr.* **in the wilderness, a.** in a state or place of isolation; away from the centre of things. **b.** out of political office.

wild goose *n.* **1.** a foolish person. *—phr.* **2. a wild goose chase,** a foolish or futile endeavour; a waste of time.

wild oats *phr.* **sow one's wild oats,** to have a dissolute life, especially to be promiscuous in youth.

willie *n.* (*especially in children's speech*) a penis.

willies *pl. n.* feelings of uneasiness or fear: *That creaking door is giving me the willies.*

willow *n.* a cricket bat. [from *willow,* a wood traditionally used to make cricket bats]

willy[1] *n.* a wallet.

willy[2] *n.* **1.** a sudden outburst of emotion, as enthusiasm, annoyance, etc. *—phr.*
2. have or **take a willy,** to be very upset or alarmed.
3. throw a willy, to become heated or excited. [?short for willy-willy, a spiralling wind]

willy-willy *n.* **1.** a spiralling wind, often collecting dust, etc. **2.** a cyclonic storm. [Aboriginal]

Wimmera shower *n.* a dust storm. [from *Wimmera,* a district in northern Victoria]

wimp *n. Originally U.S.* a meek, unaggressive, weak, spineless person, especially a male. [?from the character *Wimpy* in the 'Popeye' comic strip]

wind[1] *n.* **1.** the solar plexus, where a blow may cause shortness of breath. *—phr.*
2. between wind and water, in a vulnerable or precarious position.
3. close to the wind, a. taking a calculated risk. **b.** transgressing or nearly transgressing conventions of taste, propriety, or the like.
4. get wind of, to hear about (something which was meant to be secret).
5. get the wind up, to take fright.
6. in the teeth of the wind, against opposition.
7. in the wind, a. likely to happen; imminent. **b.** circulating as a rumour.
8. put the wind up, to frighten.
9. raise the wind, to obtain the necessary finances.
10. take the wind out of (someone's) sails, to frustrate, disconcert, or deprive of an advantage.

wind[2] *phr.*
1. wind down, to relax after a period of tension or activity.
2. wind up, a. to conclude action, speech, etc. **b.** to end: *wind up in the poorhouse.* **c.** to bring to a state of great tension; key up; excite.

windbag *n.* an empty, voluble, pretentious talker. [from *windbag,* the bag of a bagpipe]

Windies, the *pl. n.* the West Indian representative cricket team.

windmill *v.* **1.** *Aeronautics.* (of an aeroplane propeller or turbojet) to rotate freely under the influence of a passing airstream. *—n.* **2.** *U.S.* a talkative person.

windy *adj.* frightened; nervous.

windy woof *n.* (in cricket) a wild swing by a batsman, usually missing the ball.

winedot *n.* an alcohol addict. [pun on *Wyandotte,* an American breed of fowl]

wineskin *n.* one who drinks great or excessive quantities of wine.

wing *n.* **1.** an arm of a human being. *—v.* **2.** *Theatre.* to perform (a part, etc.) relying on prompters in the wings. *—phr.*
3. clip (someone's) wings, to restrict (someone's) independence or freedom of action.
4. in the wings, unobtrusively ready to take action when required; in reserve.
5. under one's wing, in or into one's care or protection.

wing-ding *n.* a wild party.

wink *n.* **1.** a bit: *I didn't sleep a wink.* *—phr.*
2. forty winks, a short sleep or nap.
3. tip (someone) the wink, to give (someone) information or a vital hint.

winker *n.* **1.** a trafficator on a vehicle. **2.** an eyelash or an eye.

winkle *phr.* **winkle out,** to prise or extract (something) out of, as a winkle from its shell with a pin. [short for *periwinkle*]

winklepicker *n.* a shoe with a long pointed toe.

winner *n.* something successful or highly valued: *This song is a real winner.*

winning streak *n.* a run of success, often attributed with having its own fortunate momentum: *He was on a winning streak.*

wino *n.* one addicted to drinking wine.

wipe *v.* **1.** to refuse to have anything to do with; dismiss; reject. *—phr.*
2. wipe out, to kill.
3. wipe the floor with, to defeat utterly; overcome completely.

wipe-out *n.* **1.** *Surfing.* a fall from a surfboard because of loss of balance. **2.** a failure; fiasco.

wire *n.* **1.** *Originally U.S.* a telegram. **2.** *Originally U.S.* the telegraphic system: *to send a message by wire.* **3.** *Horseracing.* the finishing post. *—v.* **4.** to send by telegraph, as a message. **5.** to send a telegraphic message to. *—phr.*
6. have or **get one's wires crossed,** to become confused; misunderstand.
7. pull wires, *Chiefly U.S.* to exert hidden influence; pull strings.
8. straight wire, honest; true; fair dinkum.

wire-tapper *n.* *U.S.* a swindler who professes to secure by tapping telephones or some similar means advance information for betting or the like.

wise *adj.* **1.** Also, **wise to,** in the know (about something implied); alerted; cognisant: *They tried to keep it secret, but he was wise; I'm wise to your tricks. —phr.*
2. get wise, a. to face facts or realities. **b.** to learn something.
3. put wise, a. to explain something (to someone, especially a naive person). **b.** to warn.
4. wise up, to become aware, informed, or alerted; face the realities.

wisecrack *n.* **1.** a smart, pungent, or facetious remark. *—v.* **2.** to make wisecracks.

wise guy *n.* *Chiefly U.S.* a cocksure or impertinent person of either sex.

with *adj.* **1.** comprehending of: *Are you with me? —phr.* **2. be** or **get with it, a.** to become aware of a situation. **b.** to concentrate. **c.** to be able to cope. **d.** to become fashionable or up-to-date.

with-it *adj.* trendy; sophisticated; up-to-date: *with-it gear.*

wits *phr.* **out of one's wits,** in or into a state of great fear or incoherence: *frightened out of his wits.*

wizard *adj.* *Chiefly British.* superb; marvellous.

wobbles *pl. n.* a disease of the spinal cord in horses and cattle, usually held to be a recessive hereditary trait, but also caused by eating the leaves of certain palm trees. Also, **wabble.**

wobbly *n* **1.** *N.Z.* a tantrum. *—phr.* **2. chuck** or **throw a wobbly, a.** to become angry; have a tantrum; make a scene. **b.** (of a machine, etc.) fail to function properly; break down.

Wobbly *n.* a member of the Industrial Workers of the World (trade union).

wog[1] *n.* **1.** (*offensive*) a native of North Africa or the Middle East, especially an Arab. **2.** a person of Mediterranean extraction, or of similar complexion and appearance. [?short for *golliwog*]

wog[2] *n.* **1.** a germ, especially a germ leading to a minor disease such as a cold or a stomach upset. **2.** such a cold, stomach upset, etc. **3.** an insect or small animal.

wogball *n.* soccer. [*wog* + (foot)*ball*; soccer being seen as especially popular with Australians of European descent]

wolf *n.* **1.** a man who is boldly flirtatious or amorous towards many women. —*v.* **2.** to eat ravenously.

Wolves, the *pl. n.* the Wollongong City team in the National Soccer League.

woman *phr.*

1. little woman, See **little.**

2. old woman, a. a man who is pedantic or tends to fuss, gossip, etc. **b.** one's wife.

wombat[1] *n.* **1.** someone who is slow-moving or slow-witted. —*phr.* **2. blind as a wombat,** very blind. [from the large, burrowing Australian marsupial]

wombat[2] *adj.* **1.** disabled. **2.** dead. —*n.* **3.** a corpse. [rhyming slang *wombat,* hors de combat, French: out of the battle]

wongi *n.* **1.** a talk, a chat. —*v.* **2.** to talk. [Aboriginal]

wonk *n.* **1.** (*offensive*) a white man. **2.** a male homosexual.

wonky *adj.* **1.** shaky; unsound. **2.** askew; awry. **3.** unwell; upset. [Old English *wonkel,* shaky, unsteady]

wood *phr.* **have the wood on (someone),** to be in possession of evidence or information which can be used to damage (someone).

wood-and-water joey *n.* a station hand, usually very young, performing menial tasks; an odd-job man; rouseabout. [from *joey* a young kangaroo]

wood duck *n.* one who is easily duped; a naive customer. Also, **woodie.**

woodener *n.* a knock-out blow.

woodenhead *n.* a blockhead; a dull or stupid person.

wooden-headed *adj.* thick-headed; dull; stupid.

wooden spoon *n.* **1.** the fictitious prize awarded to the individual or team coming last in a sporting competition. —*phr.* **2. win** or **get the wooden spoon,** to come last in a contest.

woodpeckers' day *n. Prison.* the day reserved for hearing the cases of people who are pleading guilty. [the majority of prisoners appearing in court on this day are 'nodding the head', that is, pleading guilty]

woods *phr.*

1. in the woods, confused; lost; in trouble.

2. neck of the woods, See **neck.**

Woodser, Jimmy *n.* See **Jimmy Woodser.**

Woodsmen, the *pl. n.* the Collingwood team in the Victorian Football League. Also, **Mighty Woodsmen, Woods.**

Woods, the *pl. n.* the Eastwood team in the New South Wales Rugby Football Union.

woodwork *phr.* **come out of the woodwork,** to appear after a long absence, as if having been in hiding: *The nutters are really coming out of the woodwork now.*

woofters *pl. n.* greyhound races.

wool *n.* **1.** the human hair, especially when short, thick, and curly. —*phr.*

2. all wool and a yard wide, (formerly) genuine; completely honest.

3. keep or **lose one's wool,** to keep (or lose) one's temper; not become (or become) angry.
4. pull the wool over one's eyes, to deceive or delude one.

woollies *pl. n.* woollen clothing.

Woolloomooloo Yank *n.* an Australian who modelled his behaviour on that of American soldiers on leave in Sydney during World War II. [from *Woolloomooloo,* an inner Sydney suburb]

woolly *adj.* **1.** *U.S.* like the rough atmosphere of the early West: *the wild and woolly West.* —*n.* **2.** an article of clothing made of wool.

woollynose *n.* a fettler.

woolshed hop *n.* a barn-dance.

woops *interj.* See **whoops.**

Woop Woop *n.* any remote or backward town or district. Also, **woop woop.**

woop woop pigeon *n.* a kookaburra.

woozy *adj.* muddled, or stupidly confused. **2.** out of sorts physically, as with dizziness, nausea, or the like. **3.** slightly or rather drunk. [?from *wooze,* variant of *ooze*]

wop *n.* **1.** (*usually offensive*) an Italian or any foreigner thought to be of Italian appearance. —*adj.* **2.** of or pertaining to any Latin country, its culture, or inhabitants. [?from Italian *guapo* dandy]

wop-wops *pl. n.* (*sometimes singular*) *N.Z.* any remote town or district: *I live out in the wop-wops.*

word *v.* **1.** Also, **word up.** to speak to, especially when informing beforehand: *He worded up the magistrate.* —*phr.*
2. eat one's words, See **eat.**
3. have words with, to remonstrate with; to argue.
4. my word!, a. an expression of agreement. **b.** an expression of surprise, mild annoyance, etc.
5. put in a (good) word for, to recommend; mention in a favourable way.
6. take the words out of (someone's) mouth, to say exactly what another was about to say.
7. the last word, the very latest, most modern, or most fashionable; the best, or most sophisticated: *This machine is the last word in automation.*

work *phr.*
1. have one's work cut out, to be pressed; have a difficult task.
2. make short work of, to dispose of or deal with quickly.
3. work a point, to take an unfair advantage.
4. work the oracle, to achieve (often secretly and with cunning) a desired end.

workhorse *n.* a person who works very hard.

working girl *n.* a prostitute.

works *phr.* **1. in the works, a.** in the process of production: *He has another book in the works.* **b.** official procedures, routines, formalities, etc.: *Your letter has gone astray in the works.* **2. the works, a.** everything there is; the whole lot. **b.** a violent assault.

world *n.* **1.** all that is important, agreeable, or necessary to one's happiness: *You're the world to me.* —*phr.*
2. dead to the world, a. unaware of one's surroundings; sleeping heavily. **b.** totally drunk. **c.** utterly tired; exhausted.

3. go around the world for fourpence/sixpence/a zack, to get drunk. [from the 'fourpenny dark', a cheap fortified wine which was served in miniature mugs decorated with this saying]

3. dead to the world, a. unaware of one's surroundings; sleeping heavily. **b.** totally drunk. **c.** utterly tired; exhausted.

4. on top of the world, elated; delighted; exultant.

5. out of this world, excellent; supremely or sublimely good.

6. set the world on fire, to be a great success.

world-beater *n.* a surpassingly good thing, person, etc.

worms *phr.* **a can of worms,** a difficult and complicated situation.

worriment *n.* **1.** trouble; harassing annoyance. **2.** worry, anxiety.

worry *phr.*

1. no worries!, an expression of confidence that everything will go well.

2. worry along or **through,** to progress by constant effort, in spite of difficulties.

worrywart *n.* one who constantly worries unnecessarily.

worse *phr.*

1. none the worse for, positively benefited by.

2. the worse for wear, drunk.

wouldn't *phr.* **wouldn't it!,** an exclamation indicating dismay, disapproval, disgust, etc.: *Wouldn't it (rot your socks)!*

wounded *phr.* **the walking wounded,** people who, though emotionally or physically distressed in some way, nevertheless continue with the normal conduct of their lives. [from *the walking wounded,* soldiers who, though injured, can walk from the battlefield]

wow *n.* **1.** something that proves an extraordinary success. *—interj.* **2.** an exclamation of surprise, wonder, pleasure, dismay, etc. *—v.* **3.** to win approval or admiration from.

wowser *n.* a prudish teetotaller; a killjoy. [?British dialect *wow,* to make a complaint; whine; popularly supposed to be an acronym of *W(e) O(nly) W(ant) S(ocial) E(vils) R(emedied),* a slogan invented by John Norton, Australian journalist and politician, 1862-1916]

w.p.b. *n.* a wastepaper basket.

w.p.b. file *n.* the wastepaper basket, viewed as a place for filing useless or unwanted material.

wrap *n* **1.** enthusiastic approval; a wrap-up. *—phr.*

2. wrap up, to conclude or settle: *to wrap up a financial transaction.*

3. under wraps, kept secret.

wrapped *phr.*

1. wrapped in, enthused about: *He's not really wrapped in the idea.*

2. wrapped up in, involved with or implicated in: *She's totally wrapped up in the new baby.*

wrap-up *n.* an enthusiastic approval or recommendation: *He gave the new product a good wrap-up.*

wriggle *phr.* **get a wriggle on,** to hurry.

wrinkle *n.* a problem needing to be solved before progress can be made: *Just a few wrinkles to iron out and then we can start.*

wrinklie *n.* anyone as old as or older than one's parents. Also, **wrinkly.**

write *phr.*
1. have (something) written all over one (it), to show as a clear characteristic: *Delight was written all over her face; his guilt was written all over him.*
2. write off, to consider as dead.
3. write oneself off, a. to get very drunk. **b.** to have a motor accident. **c.** to give a poor account of oneself.

write-off *n.* **1.** something irreparably damaged, as an aircraft, car, etc. **2.** a person who is incapacitated through drunkenness, injury, etc. **3.** an incompetent person; a no-hoper. [from *write-off,* something written off in account books]

wrong *phr.*
1. get on the wrong side of, to incur the hostility of.
2. get (someone) wrong, to misunderstand (someone).
3. get the wrong end of the stick, to misunderstand.
4. get up on the wrong side (of bed), to be bad-tempered.
5. wrong in the head, crazy; mad.

wrong 'un *n.* **1.** a dishonest person. **2.** *Cricket.* a delivery bowled by a wrist-spinner which looks as if it will break one way but in fact goes the other; googly.

X *n.* **1.** one's signature: *Put your X on the dotted line.* **2.** a kiss. **3.** any unknown factor, agency or thing. **4.** someone whose identity is unknown, or who wishes to keep his identity a secret: *Mr. X; Madame X.*

x out *v.* to erase; forget: *She should x out that man.*

X-rated *adj.* sexually explicit.

ya *pron.* you.

yabber *v.* **1.** to talk; converse. —*n.* **2.** talk; conversation. [Aboriginal]

yabby *n. Cricket.* a wicket-keeper. [a humorous reference to the supposed similarity in appearance of a gloved wicket-keeper to the *yabby,* an Australian freshwater crayfish]

yachtie *n.* a yachtsman or yachtswoman. Also, **yachty, yottie.**

yack-ai *interj.* an exclamation drawing attention or expressing enthusiasm. [?Aboriginal]

yah *interj.* an exclamation of impatience or derision.

yahoo *n.* **1.** a rough, coarse or uncouth person. —*interj.* **2.** an exclamation expressing enthusiasm or delight. —*phr.* **3. yahoo around,** to act in a rough, loutish manner. [from *Yahoo,* one of a race of brutes having the form of man and all his degrading passions, in 'Gulliver's Travels' (1726) by Jonathan Swift]

yair *adv.* yes. Also, **yeah.**

yak *n.* **1.** empty conversation. —*v.* **2.** to talk or chatter, especially pointlessly and continuously. Also, **yakety-yak.** [imitative]

yakka *n.* work. Also, **yacker, yakker.** [Aboriginal]

yamidgee *n.* an Aborigine.

yammer *v.* to whine or complain.

yank *v.* **1.** to pull or move with a sudden jerking motion; tug sharply. —*n.* **2.** a jerk or tug.

Yank *n., adj.* American. [short for *Yankee*]

Yankeeland *n.* the United States.

yank tank *n.* a large car of American manufacture.

yap *v.* **1.** to talk snappishly, noisily, or foolishly. —*n.* **2.** snappish, noisy, or foolish talk. **3.** the mouth.

yarn *n.* **1.** a story or tale of adventure, especially a long one about incredible events. **2.** a lie. —*v.* **3.** to tell stories; gossip. —*phr.* **4. spin a yarn, a.** to tell stories. **b.** to make up a story, especially as an excuse.

yarra *adj.* mad. Also, **yarrah.**

Yarra banker *n. Vic.* a man addressing passers-by from a soapbox on the banks of the Yarra; an agitator. [from the bank of the *Yarra* River, Melbourne, a popular spot for soapbox orations]

Yarrasider *n.* a Melburnian. [from the *Yarra* River which flows into Port Philip Bay at Melbourne]

yawn *n.* a boring event; a bore.

yawp *v.* **1.** to utter a loud, harsh cry or sound; bawl. **2.** to talk noisily and foolishly. —*n.* **3.** a noisy, foolish utterance. Also, **yaup.** [imitative]

yeah *adv.* yes. Also, **yair.**

yecch *interj.* an exclamation of disgust, aversion, horror, or the like.

yegg *n.* **1.** *U.S.* a burglar, especially a petty one. **2.** a thug. Also, **yeggman.** [origin obscure; ?variant of *yekk* beggar]

yellow *adj.* **1.** *U.S.* (*often offensive*) having the yellowish skin characteristic of mulattos or dark-skinned quadroons. **2.** cowardly; mean or contemptible. **3.** (of newspapers, etc.) sensational, especially morbidly or offensively sensational.

yellow-bellied *adj.* cowardly.

yellow-belly *n.* a coward.

yellow monday *n.* a cicada common to southern and eastern Australia. Also, **green grocer.**

yellow peril *n.* the putative danger of a Chinese invasion of Australia.

yellow press *n.* publications that exploit, distort or exaggerate news to create sensations and attract readers. [said to be an allusion to the *Yellow Kid*, a cartoon (1895) in the *New York World*]

yellow satin *n.* a Chinese woman considered as a sex object. See also, **black velvet.**

yen *n.* **1.** desire; longing. —*v.* **2.** to desire. [?alteration of *yearn*]

yeo *n.* ewe: *bare-bellied yeo.* [British dialect variant of *ewe*]

yep *interj.* yes.

yes-man *n.* one who always agrees with his superiors; an obedient or sycophantic follower.

Y-fronts *pl. n.* men's underpants with two overlapping layers of cloth at the bottom front which, with a vertical seam at the top front, form an inverted Y. [Trademark]

yickadee *interj. N.T.* an exclamation of greeting or farewell.

Yid *n.* (*offensive*) a Jew.

yike *n.* brawl, argument.

yikes *interj.* a mild exclamation of surprise or concern.

yippee *interj.* an exclamation used to express joy, pleasure, or the like.

yob *n.* a loutish, aggressive, or surly youth or man. Also, **yobbo.** [?*boy* spelt backwards]

yoggy *n.* a young criminal.

yonks *pl. n.* ages or a long time: *We haven't seen them for absolutely yonks.*

yonnie *n.* a stone, especially one for throwing.

yoo-hoo *interj.* a call used to attract someone's attention.

yottie *n.* See **yachtie.**

you *n.* something resembling or closely identified with the person addressed: *That dress simply isn't you.*

you-beaut *adj.* wonderful; amazing; excellent.

you-beaut country *n.* Australia.

you'lldo *n.* a bribe.

youngie *n.* a child.

yours truly *pron.* I, myself, or me.

youse *pron.* (*in non-standard use*) plural form of **you.**

yucky *adj.* disgusting; unpleasant; repulsive. Also, **yukky.**

yuk *interj.* (*especially in children's speech*) an expression of disgust.

yummy *adj.* (especially of food) very good. Also, **yum.**

yumpie *n.* *Chiefly U.S.* a trendy city person, conceived as being young and upwardly mobile. [Y(oung) U(pwardly) M(obile) P(rofessional)]

yum yum *interj.* delicious.

yuppie *n.* *Chiefly U.S.* a successful city person, conceived as being young and a member of a profession. [Y(oung) U(rban) P(rofessional)]

zack *n.* **1.** (formerly) a sixpence. **2.** a five cent piece. **3.** *Prison.* a prison sentence of six month's duration.

zambuck *n.* a St John ambulance officer. [from *Zambuck,* tradename of ointment in a black and white container, calling to mind the black and white uniform worn by St John ambulance personnel]

zap *v.* **1.** to destroy with a sudden burst of violence; annihilate. **2.** to move quickly. *—n.* **3.** vitality; force. *—phr.* **4. zap up,** to make livelier and more interesting.

zapped *adj.* tired to the point of exhaustion.

zappy *adj.* lively and interesting.

Zebras, the *pl. n.* the Sandringham team in the Victorian Football Association.

zed *n.* **1.** something or someone of little or no importance or value; a nonentity. *—phr.* **2. push up** or **stack zeds,** to go to sleep.

zero *phr.* **zero in (on),** to focus attention (on).

zero hour *n.* the time at which any contemplated move is to begin. [from *zero hour* the time set for the beginning of a military attack]

ziff *n.* a (short) beard.

zilch *n. Chiefly U.S.* nothing.

zillion *n.* an unimaginably large amount. [the letter *z,* the last in the alphabet + *(m)illion*]

zing *n.* vitality; enthusiasm: *She has lots of zing.*

zinger *n.* something exciting or excellent.

zip *n.* **1.** a sudden brief hissing sound, as of a bullet. *—v.* **2.** to move quickly: *zip away.*

zippy *adj.* lively; bright.

zit *n.* **1.** a pimple. **2.** (*pl.*) acne.

zizz *n.* a nap; sleep.

zombie *n.* (*offensive*) a person having no independent judgment, intelligence, etc. Also, **zombi.** [from *zombie* walking dead]

zonked *adj.* **1.** exhausted; faint with fatigue. **2.** drunk. **3.** under the influence of drugs. Also, **zonked-out.**

zoomie *n.* a fancy exhaust pipe on a motor car.

zot *v.* **1.** to knock, or kill: *Quickly, zot that fly.* —*interj.* **2.** an exclamation expressing suddenness: *When suddenly, zot, out jumped a red kangaroo.* —*.phr.* **3. zot off,** to depart quickly.

zulu time *n. Military. Aeronautics.* Greenwich Mean Time. [from the code used by radio operators, *Zulu,* standing for Z, to represent zero degrees, the longitude at Greenwich.

zzz *n.* a sleep. [from the convention used, especially by cartoonists, to represent sleep or the sound of snoring